Diseases and Disorders of Cattle

Diseases and Disorders of Cattle

Edited by Eleanor Clark

SYRAWOOD
PUBLISHING HOUSE
New York

Published by Syrawood Publishing House,
750 Third Avenue, 9th Floor,
New York, NY 10017, USA
www.syrawoodpublishinghouse.com

Diseases and Disorders of Cattle
Edited by Eleanor Clark

International Standard Book Number: 978-1-64740-075-0 (Hardback)

Cataloging-in-Publication Data

Diseases and disorders of cattle / edited by Eleanor Clark.
    p. cm.
Includes bibliographical references and index.
ISBN 978-1-64740-075-0
1. Cattle. 2. Cattle--Diseases. 3. Veterinary medicine. 3. Livestock. I. Clark, Eleanor.
SF961 .D57 2022
636.208 96--dc23

TABLE OF CONTENTS

Permissions

List of Contributors

Index

PREFACE

Buiatrics is the branch of veterinary science that deals with cattle and cattle diseases. Some of the common diseases which afflict cattle are bovine spongiform encephalopathy, blackleg, bluetongue, and foot rot. Diseases like digital dermatitis in cattle are caused by bacteria from the genus treponena and it majorly affects the dairy cattle. Other diseases like ringworm are caused by fungus trichophyton and verrucosum. The major causes of such diseases are unsanitary conditions like poor hygiene, inadequate hoof trimming and many more. Cattle health is characterized as a veterinarian issue as well as a public health issue since diseases like ringworm may be transferred to humans who are exposed to infected cattle. This book contains some path-breaking studies related to diseases and disorders in cattle. It also presents researches and studies performed by experts across the globe. Those in search of information to further their knowledge will be greatly assisted by this book.

This book is a comprehensive compilation of works of different researchers from varied parts of the world. It includes valuable experiences of the researchers with the sole objective of providing the readers (learners) with a proper knowledge of the concerned field. This book will be beneficial in evoking inspiration and enhancing the knowledge of the interested readers.

In the end, I would like to extend my heartiest thanks to the authors who worked with great determination on their chapters. I also appreciate the publisher's support in the course of the book. I would also like to deeply acknowledge my family who stood by me as a source of inspiration during the project.

Editor

A descriptive analysis of the growth of unrecorded interactions amongst cattle-raising premises in Scotland and their implications for disease spread

Jessica Enright[1*] and Rowland R Kao[2]

Abstract

Background: Individual animal-level reporting of cattle movements between agricultural holdings is in place in Scotland, and the resulting detailed movement data are used to inform epidemiological models and intervention. However, recent years have seen a rapid increase in the use of registered links that allow Scottish farmers to move cattle between linked holdings without reporting.

Results: By analyzing these registered trade links as a number of different networks, we find that the geographical reach of these registered links has increased over time, with many holdings linked indirectly to a large number of holdings, some potentially geographically distant. This increase was not linked to decreases in recorded movements at the holding level. When combining registered links with reported movements, we find that registered links increase the size of a possible outward chain of infection from a Scottish holding. The impact on the maximum size is considerably greater than the impact on the mean.

Conclusions: We outline the magnitude and geographic extent of that increase, and show that this growth both has the potential to substantially increase the size of epidemics driven by livestock movements, and undermines the extensive, invaluable recording within the cattle tracing system in Scotland and, by extension, the rest of Great Britain.

Keywords: Cattle movement network, Modeling, Scotland

Background

Explicit tracing of livestock movements is becoming increasingly common in many countries. This high level of detail offers opportunities to improve the efficiency and effectiveness of infectious disease surveillance and control, with the potential for substantial cost savings and, by reduction of disease burden, improvement in public and livestock health [1, 2]. Analysis of these detailed records of animal movements as a network is also becoming common, with the terminology and methods of network analysis making an impact on veterinary epidemiology [3, 4].

Since January 2001, it has been legally required to record movements of cattle between holdings in Great Britain. These movements are recorded by the British Cattle Movement Service (BCMS) to allow individual-level tracing of animals for public safety and disease control. This careful recording is consistent with European council directives [5], and is implemented in British and Scottish legislation. These recorded movements are also used as the basis of several epidemiological analyses (examples include [6–9]). When livestock movements are interpreted as a network, it is common to cast the holdings as nodes of the network and movements as arcs between those nodes. For full details of network terminology as used in veterinary epidemiology, we direct the reader to [3, 4].

* Correspondence: jae@cs.stir.ac.uk
[1]Computing Science and Mathematics, University of Stirling, Stirling FK9 4LA, UK

Because reporting animal movements that are frequent and repetitive can impose a significant administrative burden on farmers, several programs exist to allow regular or short-distance movements, particularly within a single business, to go unreported. One such program is Cattle Tracing System (CTS) Links, which are granted to account for movements between holdings either for the use of shared facilities or for additional land (commonly used for grazing). When a Link is registered, one holding is listed as a main holding, and the other as a secondary holding. Once a link is established, cattle may be moved from the main holding to the secondary holding and back without reporting the movement to BCMS. As with most livestock movements in Scotland, movement of an animal from a CTS Linked holding should trigger a "13-day standstill" on the destination holding – a period of 13 days during which animals (except for those in a specially exempted category) may not be moved away from that destination holding.

Previous work [10] has shown that CTS Links present in 2008 could pose a significant epidemiological risk in a foot-and-mouth disease outbreak, with particular potential to increase the geographic extent of an outbreak. The increased danger posed by CTS Links comes from both the possibility that animals are moved when they would not be if their movement had to be reported, and from the impediment to rapid animal tracing in the case of an outbreak. In addition, because the frequency of link use is not recorded, movements between linked holdings represent an unquantifiable risk, and so could undermine efforts to optimize risk-based surveillance and risk management.

Since 2008 the number and connectivity of CTS Links in Scotland has increased dramatically. Motivated by the potential for CTS Links to contribute to epidemiological risk and the increase in their number since the last significant study of them, we investigate the state of CTS Links in Scotland. Our objective is to characterize the current network of links in Scotland, and explore its growth over time by plotting the change in the geographic distribution of links, and in epidemiologically-relevant network measures.

Methods
Data sources and adaptations
Several agricultural datasets were sourced to analyse the current state of the CTS Links network in Scotland. We used the 2010 Agricultural Census to find geographical locations of holdings, and a 2014 extract of individual-level cattle movement records (CTS) from the British Cattle Movement Service to estimate the number of animals on each holding, and for the recorded movements themselves. Information on CTS Links consisting of the pairs of registration numbers of holdings in open Links

in December of 2009, 2010, 2012, 2013 and 2014 was provided by Scottish Government.

Some holdings that are in a CTS Link are not listed in the Agricultural Census because not every small holding is recorded in every year's Agricultural Census. Where possible, we have estimated geographic locations using the county and parish of holdings, by taking a mean easting and northing across holdings in the same county and parish, and perturbing it at random by up to 3 miles. These estimated locations make up less than 5 % of holdings mapped.

Availability of data and materials
Due to the commercially sensitive nature of the data used in this work, we cannot make the data publically available. The data on CTS Links and the Agricultural Census are held by Scottish Government, and the data on cattle movement records by the British Cattle Movement Service.

Network analysis
Because we use a number of standard network analysis methods, we briefly review some terms. We generate two types of networks: *directed networks* when using known animal movements with a known direction, and *undirected networks* when using only CTS Links, which have movement in both directions over the Link.

In an undirected network, the *degree* of a node (here a holding) is the number of network neighbours it has (here the number of holdings it is Linked to) A *component* of an undirected network is a set of nodes of the network that are joined up both directly and indirectly. More precisely, two nodes in a network are in the same component if there is some path between them in the network. This concept is epidemiologically useful because in a disease spreading on a network the size of the largest component is an upper bound on the number of holdings infected over that network.

When analyzing a directed network, which includes explicit movement direction we calculate, as described in [3], the size of an *infection chain* from a holding, which is the total number of other holdings in the network that could be infected by that holding, either directly or indirectly. We also use a similar notion: the size of an *infector chain* of a holding, which is the total number of other holdings in the network that could infect that holding, either directly or indirectly. The infector chain is, in some sense, a backwards version of the infection chain. These two measures are an indicator of the susceptibility of the network to disease [3].

All analyses were performed using Python code written by the authors, making use of **networkx** and **matplotlib** libraries.

Results and discussion

Networks investigated

Our analysis includes 41 different networks: five undirected networks, each composed only of the Links in one of our five study years, and 36 directed networks, three for each month of 2014: one composed of that month's reported cattle movements in Scotland, a second composed of that month's reported movements with simulated movements along the 2014 CTS Links added in, and a third derived from the network of reported movements by contracting all holdings directly or indirectly joined-up by Links in 2014 into a single super-holding. This extreme contraction is equivalent to assuming that there is constant movement along all links, and is the approach predominantly used by Orton et al. [10].

When adding in links to reported movements, we add a movement from the main holding to the secondary holding direction on each link on the 14th of the month, and one returning on the 21st. There are no data currently available on the regularity or timing of use of CTS Links, so this timing of inclusion should be considered only as a demonstration.

Characterizing the network in combination with recorded movement

When we consider the combination of all the CTS Links into a network, we see that not only are holdings linked to each other directly by their individual links, but that the links chain together to form larger components: thus a holding with only a small number of CTS Links may be indirectly linked to a large number of other holdings in a component. Because unrecorded animal movement and therefore pathogen movement is possible throughout a component, the size and geographic extent of these components is one of our primary interests.

We have used the Agricultural Census and each of five years of CTS Links information to plot the maps of links shown in Fig. 1. As seen in the top row of Fig. 1, the overall network has grown substantially in size and connectivity since 2009. Both long-range links that connect geographically distant holdings and short-range links that densely cover areas with high concentrations of cattle holdings have increased. Throughout the time period of study short-range links are more common than long-range links: in 2014, approximately 65 % of links were between holdings within five miles of each other.

In the bottom row of Fig. 1, we see that the geographic extent of the largest component has increased dramatically over time, with a large increase between 2010 and 2012. The number of holdings within 3 and 10 km of a holding in the largest component in each of our study years is shown in Table 1. The size of the network increased consistently from 2009 to 2013, but, surprisingly,

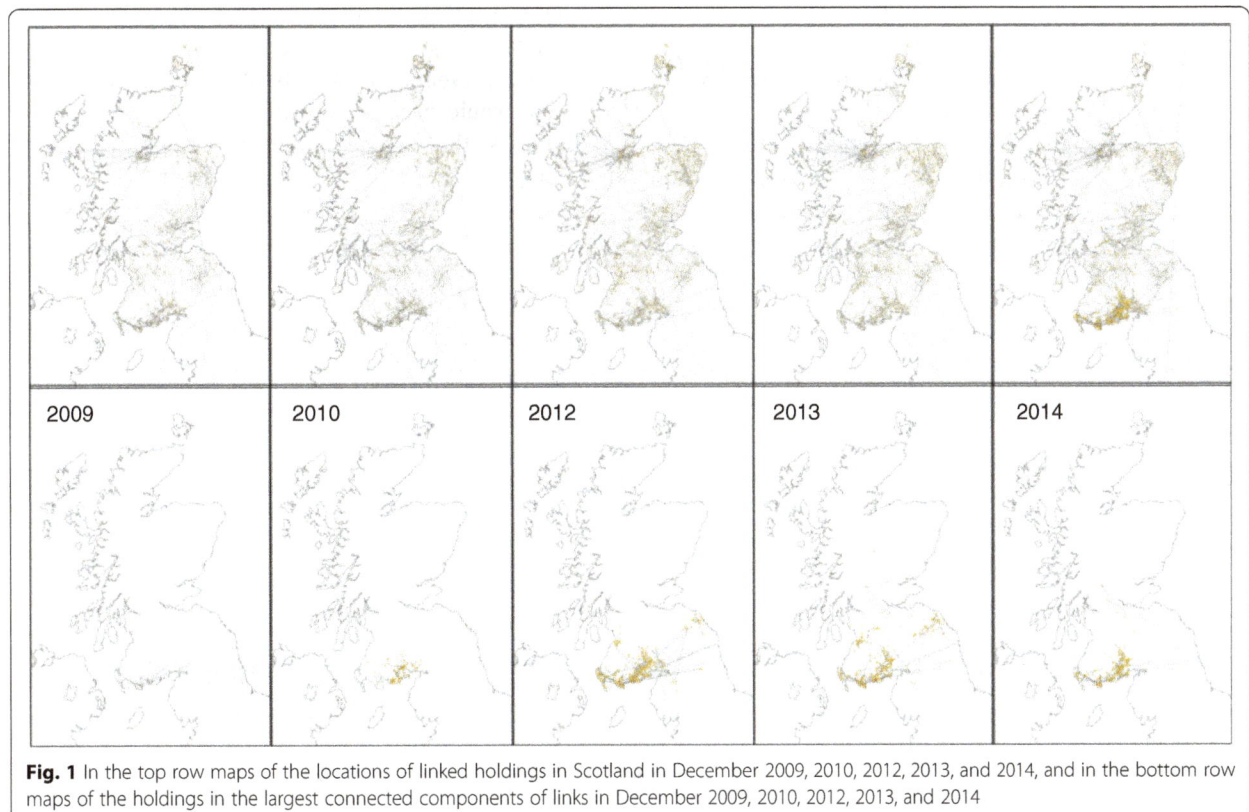

Fig. 1 In the top row maps of the locations of linked holdings in Scotland in December 2009, 2010, 2012, 2013, and 2014, and in the bottom row maps of the holdings in the largest connected components of links in December 2009, 2010, 2012, 2013, and 2014

Table 1 The number of cattle holdings within 3 and 10 km of a holding that is in the largest CTS Links component in each of 2009, 2010, 2012, 2013, and 2014

	Number of holdings within distance of a holding in the largest component	
Year	3 km	10 km
2009	1,265	3,437
2010	1,413	3,532
2012	3,380	6,691
2013	3,697	10,153
2014	2,356	5,561

decreased slightly from 2013 to 2014. The size of the largest component has shown an overall increasing trend, but has decreased slightly between 2012 and 2013, and again between 2013 and 2014.

In Fig. 2 we plot degrees and component sizes for the CTS Links network in each of our five study years, reporting the mean and maximum for each.

While the largest component has grown over time to 334 in 2013, and 245 in 2014 from less than 100 in 2009, the mean size of a component has not increased over the years.

Neither the mean nor the maximum number of holdings a holding is linked to (its *degree*) has changed substantially over our study period. Thus when looking at any single holding in the CTS Links network, the situation in 2009 looks much the same as in 2013. It is only when we consider the overall picture that we see a change.

We investigated changes in reported movement volume for holdings without CTS Links, and found

no relationship between the establishment of a link and any change in reported movements. It does not appear to be the case that a holding that establishes a CTS Link then reports fewer movements the following year.

We have calculated the size of infection and infector chains for holdings within each month of 2014, for the directed networks described above: the network of only reported movements without CTS Links, the network of reported movements with movements along CTS Links added in, and the network of reported movements with holdings in the same links component contracted to a superholding.

In Fig. 3 we see that sizes of infection and infector chains increase when CTS Links are added to recorded movements, with an even larger change in the maximum of these numbers than in the mean. The change in these numbers highlights the potential danger in CTS Links, both for disease spread and for legal traceability. Without CTS Links cattle holdings in Scotland could have received infection from or transmitted it to a mean of 7 holdings within the average month. With CTS Links added, this more than doubles to 16 holdings. When all holdings linked by a chain are aggregated into a super-holding, this number increases more dramatically to 68.

This increase impacts not only holdings that are members of a CTS Link, but also those that are not via animal movement chains that involve holdings that are in CTS Links. The mean size of an infection chain over all holdings represents an estimate of the number of holdings that would have to be investigated using tracing in an outbreak: this number would increase significantly if fenceline contacts were taken into account.

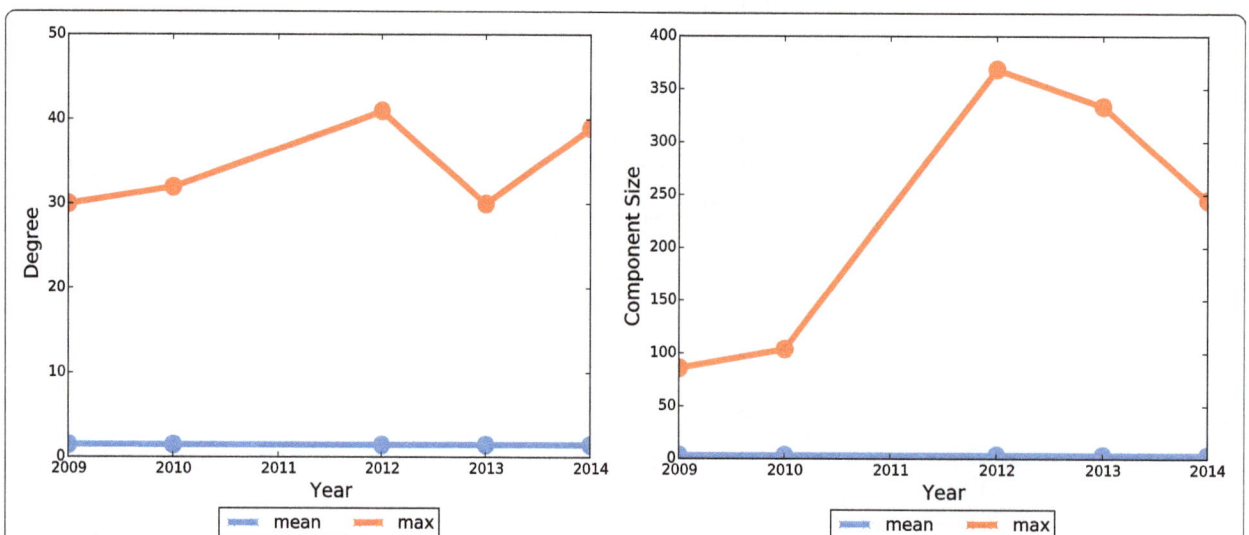

Fig. 2 Mean and maximum degree (left) and component (right) size change for Scottish holdings in the CTS Links networks over Decembers of 2009, 2010, 2012, 2013, and 2014

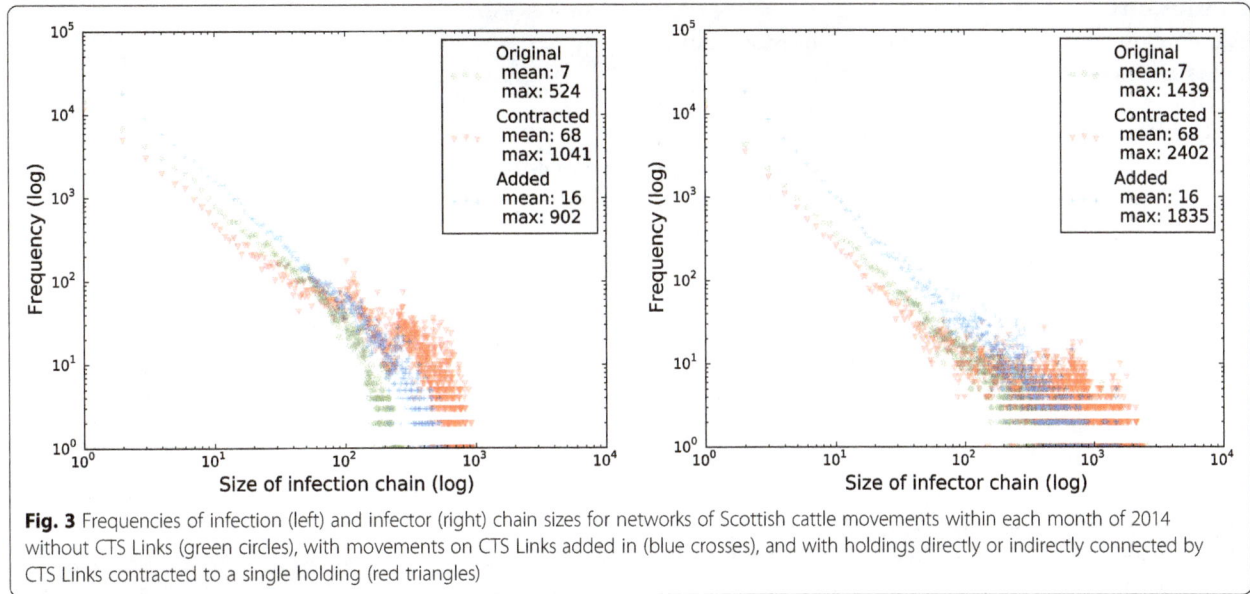

Fig. 3 Frequencies of infection (left) and infector (right) chain sizes for networks of Scottish cattle movements within each month of 2014 without CTS Links (green circles), with movements on CTS Links added in (blue crosses), and with holdings directly or indirectly connected by CTS Links contracted to a single holding (red triangles)

Focus on the largest component

As an example of the growth of the system, we focus our attention on the largest connected component in 2014, and examine its development over several years.

As we can see in Fig. 4, the largest component in 2014 formed over a number of years, not just by addition of single holdings joining a larger component, but also by several smaller connected components being joined together by the addition of new CTS Links. It is noticeable that most links seem to persist over several years, though a few

occur do not, and therefore appear only in cooler colours in the central aggregate network of Fig. 4. Most of these links are issued only for a single year, and so persistent links are renewed by the farmer every year. We also see that, while there are a small number of holdings in the largest connected component with many links, the majority of the holdings have only a small number of links, and removing the holdings with a large number of links would decrease the size of the component, but would still leave the majority of holdings in that component joined-up.

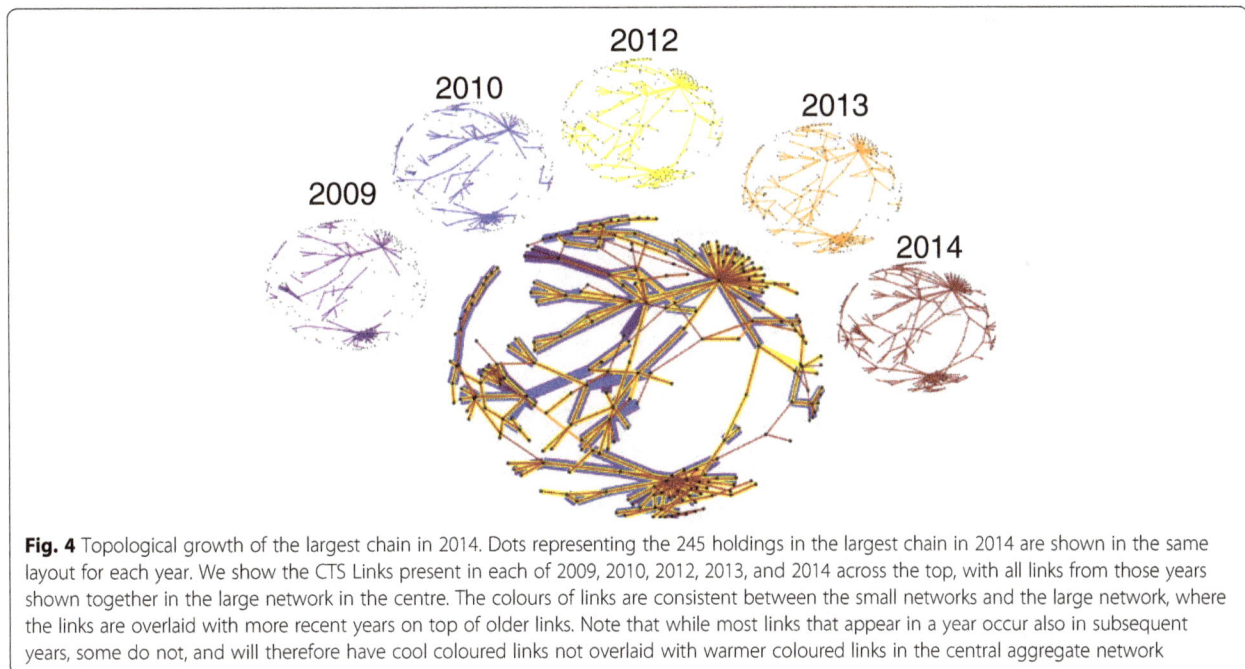

Fig. 4 Topological growth of the largest chain in 2014. Dots representing the 245 holdings in the largest chain in 2014 are shown in the same layout for each year. We show the CTS Links present in each of 2009, 2010, 2012, 2013, and 2014 across the top, with all links from those years shown together in the large network in the centre. The colours of links are consistent between the small networks and the large network, where the links are overlaid with more recent years on top of older links. Note that while most links that appear in a year occur also in subsequent years, some do not, and will therefore have cool coloured links not overlaid with warmer coloured links in the central aggregate network

Conclusions

The development of detailed records of livestock movements provides important opportunities to implement risk-based surveillance and testing, but is dependent on available data being sufficiently unbiased to make analyses of these data robustly predictive. Here we show that exemptions from recording in GB have the potential to compromise this robustness, and that their usage has been growing year-on-year from 2009 to 2013.

Despite the fact that the average number of holdings a linked holding is linked to has not changed over time, the large increase in the number of holdings involved in links has lead to an alarming growth in the overall CTS Links network in Scotland. The largest connected component in the current network reflects this effect: it was formed by a small number of links joining up relatively small components, and could be disassembled into smaller, more manageable components by the removal of only a few links. However, these link removals would have to be highly strategic: simply removing links to holdings with a large number of links would not be adequate.

An important consequence of the dramatic increase in link usage is the possibility of increased disease spread or more widespread tracing required in an outbreak is substantial: the mean number of holdings in a possible outward infection chain from a single holding within a month more than doubles when CTS Links are taken into account, increasing even for holdings that are not directly involved in a CTS Link. This network growth undermines the extensive, invaluable recording within the cattle tracing system in Scotland and, by extension, the rest of Great Britain. However, our investigations show that its impact could be mitigated by the removal of relatively few links.

Monitoring links on a holding-by-holding level would not have revealed the growth of this network, and so for disease control and robust traceability, this network should be monitored as a system. The overall picture is one of a system that has become more than the sum of its parts: while most links are in themselves not a large epidemiological risk, when combined into an overall network the potential for disease spread or traceability failure is significant.

Abbreviations
BCMS: British Cattle Movement Service; CTS: Cattle tracing system; GB: Great Britain.

Competing interests
The authors declare they have no competing interests.

Authors' contributions
RK conceived of the study, guided the analyses, and cooperatively drafted the manuscript. JE performed the analyses, and cooperatively drafted the manuscript. All authors read and approved the final manuscript.

Acknowledgments
The authors thank the anonymous referees for helpful suggestions, and gratefully acknowledge funding from the Scottish Government as part of EPIC: Scotland's Centre of Expertise on Animal Disease Outbreaks.

Author details
[1]Computing Science and Mathematics, University of Stirling, Stirling FK9 4LA, UK. [2]Institute of Biodiversity, Animal Health, and Comparative Medicine, University of Glasgow, Jarrett Building, Glasgow G61 1QH, UK.

References
1. Bessell PR, Orton R, O'Hare A, Mellor DJ, Logue D, Kao RR. Developing a framework for risk-based surveillance of tuberculosis in cattle: a case study of its application in Scotland. Epidemiol Infect. 2013;141(2):314–23.
2. Stärk KD, Regula G, Hernandez J, Knopf L, Fuchs K, Morris RS, et al. Concepts for risk-based surveillance in the field of veterinary medicine and veterinary public health: Review of current approaches. BMC Health Serv Res. 2006;6:20.
3. Dubé C, Ribble C, Kelton D, McNab B. A review of network analysis and terminology and its application to foot-and-mouth disease modelling and policy development. Transbound Emerg Dis. 2009;56(3):73–85.
4. Martinez-López B, Perez AM, Sánchez-Vizcaíno JM. Social network analysis. Review of general concepts and use in preventative veterinary medicine. Transbound Emerg Dis. 2009;56(4):109–20.
5. European Economic Community issued Council Directive 92/102/EEC. Bovine Animals (Records, Identification and Movement) Order. Scotland: The Cattle Identification Regulations; 2007. p. SSI 174.
6. Robinson SE, Everett MG, Christley RM. Recent network evolution increases the potential for large epidemics in the British cattle population. J R Soc Interface. 2007;4:669–74.
7. Green DM, Kizz IZ, Kao RR. Modelling the initial spread of foot-and-mouth disease through animal movements. Proc Royal Soc B. 2006;273(1602):2729–35.
8. Kao RR. The role of mathematical modelling in the control of the 2001 FMD epidemic in the UK. Trends Microbiol. 2002;10(6):279–86.
9. Brooks-Pollock E, Roberts GO, Keeling MJ. A dynamic model of bovine tuberculosis spread and control in Great Britain. Nature. 2014;511:228–31.
10. Orton RJ, Bessell PR, Birch CPD, O'Hare A, Kao RR. Risk of Foot-and-Mouth Disease Spread Due to Sole Occupancy Authorities and Linked Cattle Holdings. PLoS ONE. 2012; doi:10.1371/journal.pone.0035089

2

Clinical and laboratory findings in 503 cattle with traumatic reticuloperitonitis

Ueli Braun[1]* (iD), Sonja Warislohner[1], Paul Torgerson[2], Karl Nuss[1] and Christian Gerspach[1]

Abstract

Background: The study evaluated the results of clinical examination and haematological and serum biochemical analyses in 503 cattle with traumatic reticuloperitonitis (TRP).

Results: The most common clinical findings were abnormal demeanour and general condition (87%), decreased rumen motility (72%), poorly digested faeces (57%), decreased rumen fill (49%), fever (43%) and tachycardia (26%). In 58% of the cattle, at least one of three tests for reticular foreign bodies (pinching of the withers, pressure on the xiphoid and percussion of the abdominal wall) was positive, and in 42% all three tests were negative. The most common haematological findings were decreased haematocrit in 45% of cattle and leukocytosis in 42%. An increase in the concentration of fibrinogen in 69% of cattle and total protein in 64% were the main biochemical findings. The glutaraldehyde test time was decreased with coagulation occurring within 6 min in 75% of cattle.

Conclusions: In many cases, a diagnosis of TRP is not possible based on individual clinical or laboratory findings because even the most common abnormalities are not seen in all cattle with TRP.

Keywords: Cattle, Traumatic reticuloperitonitis, Foreign body tests, Haematological findings, Biochemical findings

Background

Traumatic reticuloperitonitis (TRP) remains one of the most important internal disorders of cattle in addition to abomasal displacement. One of the first large studies found that the incidence of TRP was as high as 80% [1], but more recent reports have shown that it is now approximately 2–12% [2–5]. Despite this decrease, the clinical implications remain the same. Traumatic reticuloperitonitis most commonly results from perforation of the reticulum by metal objects such as nails or wire that have been accidently incorporated into feed and ingested [6–9]. This may lead to localised peritonitis, sometimes involving neighbouring organs, and in severe cases it results in generalised peritonitis. Clinical signs of TRP have been described in a number of reference texts [6, 8, 9]. Acute disease usually results in distinct signs that include anorexia, decreased milk production, fever, ruminal atony and tympany, abdominal pain, arched back, abdominal guarding and tense abdomen [6, 8, 9]. The clinical signs in cattle with chronic disease on the other hand are often less

apparent. A short audible grunt is considered characteristic of acute TRP and may be a spontaneous response to reticular contractions or changes in posture such as lying down and getting up [6, 10, 11]. There may be additional signs in cattle with sequelae such as traumatic pericarditis [12], liver abscesses [13] or cranial functional stenosis [6, 9]. Cattle with acute localised peritonitis typically have neutrophilia with a regenerative left shift [9] and those with acute diffuse peritonitis have leukopenia with a degenerative left shift [9]. The most important biochemical findings are increased concentrations of total protein and fibrinogen. Tests for reticular foreign bodies may elicit a grunt, although other painful disorders of the thorax and abdomen may stimulate the same reaction [6, 10, 14]. Pinching of the withers, gradual application of pressure followed by sudden release of pressure on the area between the xiphoid and the umbilicus using a pole and percussion of the abdominal wall with a rubber hammer over the region of the reticulum are the most useful foreign body tests for TRP [15]. Others include the zone test developed by Kalchschmidt, leading the animal up and down a steep incline and ferroscopy to detect the presence of metal [15]. Pinching of the withers, abdominal percussion and the pole test have been the clinical tests used in more than 20,000 cattle over more

* Correspondence: ubraun@vetclinics.uzh.ch
[1]Department of Farm Animals, Vetsuisse-Faculty, University of Zurich, Winterthurerstrasse 260, CH-8057 Zurich, Switzerland
Full list of author information is available at the end of the article

than 30 years in our clinic. Other diagnostic tests include radiography and ultrasonography, which will be addressed in a separate paper. The clinical and laboratory findings in cattle with TRP have been thoroughly described in a number of standard texts. However, the majority of information is largely based on empirical evidence, and systematic evaluation of the clinical and laboratory findings in cattle with a definitive diagnosis of TRP has not yet been done. In particular, the frequency of positive responses to foreign body tests, which are considered an essential part of a diagnostic work-up in cattle with TRP, has not been determined. The goals of the present study were to describe the clinical and laboratory findings in 503 cattle with TRP, to establish the frequency of positive responses to foreign body tests and to determine which foreign body test elicited the most positive responses.

Methods
Animals
This was a retrospective study of 503 cattle that had a main diagnosis of TRP, which meant that the clinical signs were attributable to TRP and not another concomitant disease or disorder. The cattle were all greater than 1 year of age and had been admitted to the Veterinary Teaching Hospital, University of Zurich, from January 1, 2001 to December 31, 2014. The diagnosis of TRP was based on the results of ultrasonography, radiography, laparoruminotomy and/or postmortem examination. Cattle with TRP that had concomitant diseases causing anterior abdominal or caudal thoracic pain were excluded; this included 27 cows with broncho-pneumonia and 24 cows with abomasal ulcers. All cattle that had been part of previous reports were not included in the present study. Traumatic reticuloperitonitis was diagnosed based on radiographic evidence of a foreign body that penetrated or perforated the reticular wall or was seen outside of the reticulum in 225 cattle and on ultrasonographic changes of the reticular wall in 403 cattle. Foreign bodies that penetrated or perforated the reticular wall were removed during laparoruminotomy in 196 cattle, and in 10 others, a reticular abscess was drained transcutaneously under ultrasonographic guidance. In all 61 cattle that were euthanased because of a poor prognosis, TRP was confirmed during postmortem examination. In all cattle, the diagnosis of TRP was based on more than one criterion. The results of ultrasonography, radiography, surgical treatment and postmortem examination as well as the outcome of treatment were described in a dissertation [16]. There were 496 females and 7 males, which ranged in age from 1.0 to 14.9 years (median, 4.1 years) with 97% of the cattle being more than 2 years of age. Breeds included Swiss Braunvieh (208), Holstein-Friesian (155), Simmental (124), Jersey (3), Eringer (1), Hinterwälder (1) and crossbred

cattle (11). The length of illness ranged from 1 to 90 days (median, 4 days). The majority of cows ($n = 168$, 33%) had calved 0 to 8 weeks before becoming ill; this incidence was significantly higher than that of other reproductive stages ($P < 0.01$). There were no significant differences between the other reproductive stages. Of the 503 cattle, 58 had received no treatment before referral, 50 had been treated with an antibiotic, 88 had received a magnet and 209 had received an antibiotic and a magnet. A non-steroidal anti-inflammatory drug or metamizole was used in addition to other treatments in 183 cattle or exclusively in 11 cattle.

Clinical examination
The cattle underwent a thorough clinical examination [17]. The general health condition was evaluated by determining demeanour, appearance of hair coat and muzzle, skin elasticity, position of the eyes in relation to the sockets and skin surface temperature. Each animal was observed for signs of pain such as spontaneous grunting and bruxism. The general health condition was considered to be mildly to moderately abnormal when appetite and degree of alertness were decreased and severely abnormal when there was anorexia and apathy or constant bruxism or grunting. The rumen was assessed for degree of fill, number and intensity of contractions and layering of contents. Sensitivity in the reticular region was assessed by preventing the animal from breathing for a short period by placing a plastic rectal sleeve over the mouth and nose and listening for grunting during the following deep breath. This was followed by the foreign body tests, which included the pole test, pinching of the withers and percussion of the abdominal wall over the region of the reticulum with a rubber hammer. Each test was carried out four times, and the reaction of the animal was observed each time. A test was considered positive when it elicited a short grunt three out of four times. The response to a test was considered questionable when it elicited a grunt two out of four times and negative when the animal did not grunt or grunted only once. Swinging and percussion auscultation as well as a rectal examination were also carried out. Faeces were assessed for colour, consistency, amount, fibre particle length and abnormal contents.

Urinalysis
In 445 cattle, a urine sample was collected during spontaneous micturition, but in 33 cases catheterisation of the bladder was carried out. The colour and transparency of the urine were assessed macroscopically, and the specific gravity was determined using a refractometer (HRMT 18, A. Krüss Optronic GmbH, Hamburg, Germany). A urine test strip (Combur9®, Roche, Basel) was used to determine urine pH and the presence of

protein, erythrocytes, glucose, ketones, leukocytes, nitrite, urobilinogen and bilirubin.

Rumen fluid analysis

A sample of rumen fluid (200 to 300 ml) was collected using a Dirksen probe [15] and assessed for colour, odour, consistency and pH. In addition, a methylene blue reduction time and the concentration of chloride were determined. The concentration of chloride in rumen fluid was carried out using an MK-II-Chloride Analyser 9265 (Sherwood, Cambridge).

Haematological and serum biochemical analyses

The following blood samples were collected from all cattle: 5 ml of EDTA blood for haematological analysis, 10 ml of whole blood for serum biochemistry, 2 ml of whole blood mixed with 0.2 ml heparin for venous blood gas analysis and 5 ml of EDTA blood for the glutaraldehyde test. Haematological analysis included the determination of PCV, total leukocyte count and the concentrations of fibrinogen and total protein using an automated blood analyzer (CELL-Dyn 3500, Abbott Diagnostics Division, Baar). A differential leukocyte count was done in cattle with leukopenia (< 5000 leukocytes/µl blood) or leukocytosis (> 10,000 leukocytes/µl blood). The concentrations of serum urea nitrogen and bilirubin and the activities of the enzymes aspartate aminotransferase (ASAT), γ-glutamyltransferase (γ-GT) and glutamate dehydrogenase (GLDH) were determined at 37 °C using an automated analyser (Cobas-Integra-800-Analyser, Roche Diagnostics, Basel) and the manufacturer's reagents (Roche-Reagents) according to the International Federation of Clinical Chemistry and Laboratory Medicine (IFCC). Venous blood gas analysis was done using an automated analyser (RapidLab 248, (Siemens Schweiz AG, Zurich). A glutaraldehyde test (Glutaltest*, Graeub AG, Bern) was performed according to the manufacturer's instructions. Results were compared to reference intervals recently reported [18].

Statistical analysis

The program IBM SPSS Statistics 22.0 was used for analysis. Frequencies were determined for each clinical and laboratory variable. The Wilk-Shapiro test was used to test the data for normality. Means ± standard deviations were calculated for normal data (rectal temperature, urine-pH, urine specific gravity, lymphocyte count, fibrinogen concentration and venous blood pH) and medians for non-normal data (heart rate, respiratory rate, rumen pH, rumen chloride concentration, haematocrit, white blood cell, neutrophil count, total protein, urea nitrogen and bilirubin concentrations, ASAT, γ-GT and GLDH activities, glutaraldehyde test time and pCO_2, HCO_3^- and base excess of venous blood). Differences in

seasonal incidence of TRP and differences in occurrence at various reproductive stages were analysed using a one-way analysis of variance and the post hoc Bonferroni test. The 3-month periods of January to March, April to June, July to September and October to December, and the reproductive stages, which included the first 8 weeks postpartum, > 8 weeks postpartum and open, and 3 months, 4 to 6 months and 7 to 9.5 months of gestation were compared. A value of $P < 0.05$ was considered significant.

Results

Over the 14-year-study period, cattle with TRP constituted a yearly average of 7.1% (range, 5.1–9.1%) of the bovine patients treated for internal disorders.

Seasonality

There were significantly more cattle ($n = 232$, 46%) with TRP seen in the months of January to April than May to August ($n = 146$, 29%) or September to December ($n = 125$, 25%) ($P < 0.01$) (Fig. 1).

Clinical findings

The general demeanour was normal in 65 (13%) cattle, mildly to moderately abnormal in 425 (84%) and markedly abnormal in 13 (3%). The rectal temperature varied from 36.4 to 41.3 °C (39.0 ± 0.7 °C) (Table 1) and was mildly to severely increased (39.1–41.3 °C) in 217 (43%) cattle. The heart rate ranged from 40 to 162 bpm (median, 76 bpm), and bradycardia was found in 24 (5%) cattle and tachycardia in 130 (26%). The respiratory rate was 12 to 100 breaths per minute (median, 28 breaths per min), and 102 (21%) cattle had bradypnoea and 42 (8%) had tachypnoea. The type of respiration was costoabdominal in 86% of cattle, costal in 1% and abdominal in 13%.

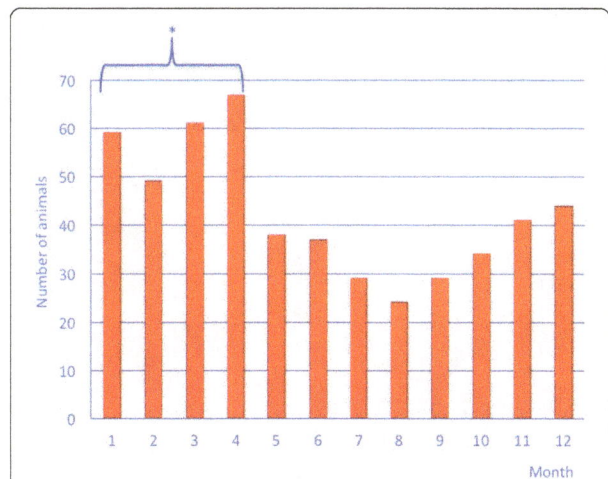

Fig. 1 Seasonal occurrence of traumatic reticuloperitonitis in cattle over a period of 14 years. * Difference to May to August and to September to December $P < 0.05$

Table 1 Rectal temperature and heart and respiratory rates in cattle with traumatic reticuloperitonitis

Variable	Finding	Range	Number of cattle	Percent
Rectal temperature (39.0 ± 0.7 °C) (n = 503)	Normal	38.0–39.0	263	52
	Decreased	36.4–37.9	23	5
	Mildly increased	39.1–39.5	146	29
	Moderately increased	39.6–40.0	45	9
	Severely increased	40.1–41.3	26	5
Heart rate (median = 76 bpm) (n = 502)	Normal	60–80	348	69
	Decreased	40–59	24	5
	Mildly increased	81–100	108	22
	Moderately increased	101–120	15	3
	Severely increased	121–162	7	1
Respiratory rate (median = 28 breaths per min) (n = 501)	Normal	21–40	357	71
	Decreased	12–20	102	21
	Increased	41–100	42	8

Rumen fill was decreased in 242 (49%) cattle, and in 70 (14%) the rumen appeared fuller than normal (Table 2). Rumen motility was decreased in 354 (72%) cattle, and the layering of rumen contents was abnormal or absent in 227 (45%). Ruminal tympany was present in 51 (10%) cattle in addition to other ruminal abnormalities.

Table 2 Results of rumen evaluation in cattle with traumatic reticuloperitonitis

Variable	Finding	Number of cattle	Percent
Rumen fill (n = 495)	Normal	183	37
	Reduced	242	49
	Fuller than normal	70	14
Rumen motility (n = 489)	Normal	135	28
	Reduced	325	66
	Absent	29	6
Ruminal contractions per 2 min. (n = 502)	Normal (2–3)	268	53
	None	146	29
	1	72	14
	4 to 6	16	3
Stratification (n = 499)	Normal	185	37
	Reduced	195	39
	Absent	32	6
	Tympanic	51	10
	Firm content	36	7

Signs of pain occurred spontaneously in 179 (36%) cattle and included arching of the back (n = 68), bruxism (n = 80) and grunting (n = 8). Twenty-three cattle had both arching of the back and bruxism. Walking on a lead rope elicited grunting in another five cattle. Bruxism was elicited in 43 (9%) cattle, and at least one grunt, rarely more, was heard in 20 (4%) after temporary interruption of breathing using a plastic rectal sleeve over the mouth and nose. Of the foreign body tests, a positive response was seen with the pole test in 210 (43%), pinching of the withers in 191 (39%) and pain percussion in 120 (24%); this meant that the tests elicited grunting a minimum of three of the four times the test was done (Table 3). A positive response was seen with one of the three tests in 57 (19%) cattle, with two of the tests in 52 (17%) and with all three tests in 66 (22%). Thus, a minimum of one of the tests was positive in 175 (58%) cattle, and all three tests were negative in 129 (42%).

Swinging and percussion auscultation on the left side of the animal was positive in 26 (5%) cattle. The same test on the right side of the animal was positive in 87 (17%) cattle.

Intestinal motility was decreased or absent in 252 (50%) cattle. Faecal consistency was watery to loose in 121 (30%), abnormally thick in 70 (14%) and partly liquid and partly firm in 16 (3%). The degree of comminution of faeces based on fibre particle length was abnormal in 286 (57%), and the amount of faeces produced was decreased in 177 (35%).

Rectal examination revealed distension of the rumen in 98 (19%) cattle. The rumen felt firm in 51 (10%) and tympanic in 25 (5%) and was L-shaped in 15 (3%). There was loss of negative pressure in 21 (4%) cattle and crepitus in 8.

Urinalysis
Urine pH was higher than normal in 203 (42%) cattle, and the urine specific gravity was lower than normal in

Table 3 Results of foreign body tests in cattle with traumatic reticuloperitonitis

Test	Finding	Number of cattle	Percent
Pole test (n = 488)	Positive	210	43
	Questionable	72	15
	Negative	206	42
Pinching of the withers (n = 495)	Positive	191	39
	Questionable	80	16
	Negative	224	45
Pain percussion (n = 494)	Positive	120	24
	Questionable	91	19
	Negative	283	57

Positive: at least three of four tests elicited a grunt
Questionable: two of four tests elicited a grunt
Negative: none or one of four tests elicited a grunt

212 (46%) (Table 4). Proteinuria was found in 57% and haematuria in 72% of cattle; however, in most cases these abnormalities were mild and likely due to contamination. A few cattle had glucosuria (11%) or ketonuria (9%).

Rumen fluid analysis

The pH of rumen fluid was greater than 7.0 in 312 (72%) cattle (Table 5). Based on the results of the methylene blue reduction test, rumen fluid was considered inactive in 236 (73%) cattle. The concentration of chloride was decreased in 88 (20%) cattle and increased in 113 (26%) compared with the reference interval.

Haematological and serum biochemical analyses

The most common abnormalities in blood analysis were a decrease in haematocrit in 224 (45%) cattle and leukocytosis in 209 (42%) (Table 6). Neutrophilia occured in 90% of animals with leukocytosis and lymphopenia was noted in 36%. The most common abnormalities in the biochemical profile were an increase in the concentration

Table 5 Results of rumen fluid analysis in cattle with traumatic reticuloperitonitis

Variable	Finding	Number of cattle	Percent
pH (n = 432) median = 8	Normal (5.5–7.0)	118	28
	Increased (7.1–10.0)	312	72
Methylene blue reduction (min.) (n = 323)	Hyperreactive (< 3 Min.)	19	6
	Moderately reactive (3–6 Min.)	68	21
	Inactive (> 6 Min.)	236	73
Chloride concentration (n = 435) median = 20 mmol/l	Normal (15–25)	234	54
	Decreased (6–14)	88	20
	Increased (26–99)	113	26

of fibrinogen in 345 (69%) and total protein in 319 (64%) cattle. The coagulation time in the glutaraldehyde test was less than 6 min in 371 (75%) cattle. Venous blood pH was lower than normal in 170 (36%) cattle and higher than normal in 88 (18%) (Table 7).

Discussion

In 1954, a study of slaughter cattle found that the incidence of TRP was 80% [1], but when administration of magnets was introduced in the 1960s, the incidence decreased sharply [19]. The results of the present study showed that the yearly incidence of TRP as 7.1%, which was in agreement with the results of other studies in which it ranged from 2 to 12% [2–5]. Traumatic reticuloperitonitis occurred more often in the months of December to April and there was a marked decrease in cases during the summer months, similar to the results of other studies [6, 20], although one study found no association with time of year [21]. There is a lower risk of ingestion of a foreign body during the summer when cattle are grazing than in the winter when they are fed prepared feed. Traumatic reticuloperitonitis is extremely rare in cattle that are kept on pasture year-round [9] because they are more likely to detect foreign bodies in grass than in hay [22]. Contamination of feed with metal foreign bodies is greater with preparation and storage of hay than in fresh forage, although wire is no longer used to tie hay bales. In 33% of the cattle, TRP occurred in the first 2 months postpartum, which was the reproduction stage with the highest number of cases and in agreement with the results of another study (31.3%) [23]. There were no differences among the stages of pregnancy, which was similar to the findings of one study [20] but contrasted the results of one other in which TRP was observed more often in the last trimester of pregnancy [6]. Differentiation of acute and chronic TRP was deliberately omitted for two reasons. Firstly, the definitions of acute and chronic TRP vary considerably depending on which author is cited, and secondly, reliable differentiation is not

Table 4 Results of urinalysis in cattle with traumatic reticuloperitonitis

Variable Mean ± sd	Finding	Number of cattle	Percent
Transparency (n = 474)	Transparent	463	98
	Opaque	11	2
pH (n = 478) 7.9 ± 1.15	Normal (7.0–8.0)	203	42
	Decreased (5.0–6.9)	73	16
	Increased (8.1–9.0)	203	42
Specific gravity (n = 461) 1020.6 ± 11.18 g/l	Normal (1020–1040)	229	50
	Decreased (1000–1019)	212	46
	Increased (1021–1060)	20	4
Protein concentration (n = 478)	Negative (< 30 mg/dl)	204	43
	+ (approx. 30 mg/dl)	251	53
	++ (approx. 100 mg/dl)	21	4
Erythrocytes (n = 478)	Negative	348	73
	+ (approx. 5–10)	61	13
	++ (approx. 25)	29	6
	+++ (approx. 50)	40	8
Glucose concentration (n = 478)	Negative (< 50 mg/dl)	425	89
	+ (approx. 50 mg/dl)	24	5
	++ (approx. 100 mg/dl)	18	4
	+++ (≥ 300 mg/dl)	12	2
Ketone bodies (n = 477)	Negative (< 10 mg/dl)	438	91
	+ (approx. 10 mg/dl)	19	4
	++ (approx. 50 mg/dl)	12	3
	+++ (≥ 150 mg/dl)	8	2

+ Mildly increased
++ Moderately increased
+++ Severely increased

Table 6 Haematological and blood biochemical findings in cattle with traumatic reticuloperitonitis

Variable (mean ± sd or median)	Finding	Number of cattle	Percent
Haematocrit (%) (n = 501) median = 30%	Normal (30–35)	218	43
	Decreased (18–29)	224	45
	Increased (36–60)	59	12
White blood cell count (/μl) (n = 501) median = 9400/μl	Normal (5000–10,000)	272	54
	Decreased (1500–4999)	20	4
	Increased (10,001–29,200)	209	42
Neutrophil count (/μl) (n = 211) median = 8568/μl	Normal (1230–3350)	17	8
	Decreased (380–1229)	4	2
	Increased (3351–26,280)	190	90
Lymphocyte count (/μl) (n = 222) 2576 ± 1016/μl	Normal (2190–5120)	137	62
	Decreased (150–2189)	81	36
	Increased (5121–7040)	4	2
Fibrinogen concentration (n = 499) (8.6 ± 3.1 g/l)	Normal (4–7)	131	26
	Decreased (2.0–3.9)	23	5
	Increased (7.1–17.0)	345	69
Total protein concentration (n = 501) (median = 84 g/l)	Normal (60–80)	177	35
	Decreased (45–59)	5	1
	Increased (81–122)	319	64
Urea concentration (n = 501) (median = 4.0 mmol/l)	Normal (2.4–6.5)	373	74
	Decreased (0.7–2.3)	58	12
	Increased (6.6–33.5)	70	14
Bilirubin concentration (n = 500) (Median = 4.6 μmol/l)	Normal (≤ 6.5)	359	72
	Increased (6.6–71.5)	141	28
ASAT activity (n = 501) (median = 69.0 U/l)	Normal (≤ 103)	425	85
	Increased (104–812)	76	15
γ-GT activity (n = 500) (median = 24.0 U/l)	Normal (≤ 30)	407	81
	Increased (31–154)	93	19
GLDH activity (n = 191) (median = 15.0 U/l)	Normal (≤ 25.0)	127	67
	Increased (25.1–522.0)	64	33
Glutaraldehyde test (n = 498) median = 3.5 min.	Normal (≥ 10 min.)	91	18
	6.1–9.9 min.	36	7
	3.1–6.0 min.	128	26
	≤ 3 min.	243	49

ASAT Aspartate aminotranferase, *γ-GT* γ-glutamyltransferase, *GLDH* glutamate dehydrogenase

Table 7 Venous blood gas analysis in cattle with traumatic reticuloperitonitis

Variable (median; mean ± sd)	Finding	Number of cattle	Percent
pH (n = 474) 7.42 ± 0.05	Normal (7.41–7.45)	216	46
	Decreased (7.20–7.40)	170	36
	Increased (7.46–7.58)	88	18
pCO_2 (n = 472) (median = 43.3 mmHg)	Normal (35–45)	254	54
	Decreased (24.1–34.9)	45	9
	Increased (45.1–75.6)	173	37
HCO_3^- (n = 474) (median = 26.5 mmol/l)	Normal (20.0–30.0)	354	75
	Decreased (10.0–19.9)	25	5
	Increased (30.1–58.9)	95	20
Base excess (n = 464) (median = 3.1 mmol/l)	Normal (−2 bis + 2)	127	27
	Decreased (−9.1 bis −2.1)	65	14
	Increased (2.1 bis 28.8)	272	59

temperature between 39.1 and 39.5 °C, which is considered typical of TRP. The majority of cattle (52%) had a normal rectal temperature, which was likely attributable to having been ill for several days and treatment before admission to the clinic. In a clinical study of 1446 cattle with acute TRP, the mean rectal temperature was 39.5 °C [20], and in another study of cattle with TRP that had been ill for less than 24 h, it was 39.3 °C [23]. A recently published reference text states that the rectal temperature is 39.5 to 40.0 °C in cattle with acute TRP and within the reference interval or mildly increased in cattle with chronic TRP [9]. A persistent mild increase in rectal temperature is characteristic of chronic inflammation [9].

The mean heart rate was 76 bpm, which was in the upper range of the reference interval and in agreement with the values reported in another study [23] and in a reference text, which states that the mean heart rate in cattle with acute localised TRP is 80 bpm [9]. The mean heart rate in a clinical study of 1446 cattle with acute TRP was slightly higher at 82.4 bpm [20]. A heart rate of more than 90 bpm together with a rectal temperature of more than 40 °C indicates severe complications [9] such as generalised peritonitis or concurrent traumatic pericarditis.

The mean respiratory rate was 28 breaths per minute, which was in the upper range of the reference interval. Constable et al. [9] reported a mean respiratory rate of approximately 30 breaths per minute in cows with TRP. Cattle with TRP do not inhale as deeply as normal in an effort to mitigate pain elicited by movement of the fully-expanded lungs and the diaphragm. An increase in respiratory rate is a compensatory mechanism, which sometimes is misdiagnosed as bronchopneumonia. In cattle with tachypnoea that do not respond to antibiotic therapy, a differential diagnosis should include TRP and other diseases.

possible based on history and clinical examination alone. In our experience, cows with chronic TRP are often misdiagnosed as having acute TRP. Reliable differentiation of cattle with acute and chronic TRP can only be determined with a post-mortem examination, which was only carried out in 61 of 503 cattle.

The mean rectal temperature at the time of admission was 39.0 ± 0.7 °C, and only 29% of the cattle had a

Rumen motility is often decreased or absent in cattle with TRP [6, 9]. This was true in the majority (72%) of cattle in the present study and is explained by inhibition of the gastric centre in the medulla oblongata via the vagal nerve because of pain associated with a foreign body [24]. However, decreased rumen motility is a non-specific finding seen in many other diseases of the gastrointestinal tract as well as in systemic disorders. Ruminal tympany, due to decreased eructation because of pain, may be seen in cattle with TRP [6, 8, 9] but occurred in only 51 (10%) of our cases and was recognised as mild bulging of the left paralumbar fossa.

Cattle with TRP often have increased fibre particle length in the faeces [6, 11, 25] because of dysfunction of the sorting mechanisms between the reticulum and omasum, which leads to the movement of incompletely digested feed into the omasum. An increase in fibre particle length occurred in the faeces of 60% of the cattle with TRP in the present study. Fibre particle length is an important indicator of disease of the reticulum, but an increase in length can also be due to dental disease or an increase in the speed of passage of ingesta through the gastrointestinal tract, such as occurs in diarrhoea [17].

Abdominal pain is a cardinal sign of TRP [6, 9–11, 14] and may manifest as arching of the back, grunting or bruxism, which may occur spontaneously or be elicited via foreign body tests. Arching of the back is a sign of parietal pain, and spontaneous grunting is a response to pain caused by reticular contractions. Bruxism is a sign of pain associated with many diseases and is uncommon in cattle with TRP [6]. Signs of pain were seen in 179 (36%) of the cattle in the present study and the most common were bruxism ($n = 80$) and arching of the back ($n = 68$). Spontaneous grunting ($n = 8$) was uncommon; grunting may be difficult to hear, and thus auscultation of the larynx [6] or trachea [26] or placing the palm over the larynx to palpate vibrations [27] is recommended.

The pole test, pinching of the withers and pain percussion were considered by several authors to be the most important part of the clinical examination in cattle suspected of having TRP [6, 28–30]. However, there are few studies that have investigated the response to foreign body testing and the presence of a foreign body in the reticulum [23, 31]. In one study, grunting was elicited by pinching of the withers in 41% and deep palpation with a fist caudal to the sternum in 45% of cows [23], and in another, 16 (61%) of 26 cows had at least one positive foreign body test [30]. In the present study, the pole test was most often positive (43%) followed by pinching of the withers (39%) and pain percussion of the reticular region (24%); at least one test was positive in 58% of cattle, whereas all tests for foreign bodies were negative in 42% of cases. These are sobering results but one must remember that to elicit grunting in cattle with chronic TRP, considerable strength may be required when conducting foreign body tests [6] and in cows with chronic localised peritonitis, the grunt test may be positive, negative or equivocal [9]. Positive swinging and percussion auscultation on the left side was the result of concurrent left displacement of the abomasum in 13 cattle and ruminal atony in another 13. On the right side, positive swinging and percussion auscultation was attributable to intestinal atony or diarrhoea in 82, right displacement of the abomasum in three and caecal dilation in two. The loss of negative pressure in 21 cattle and crepitus in 8 were considered to be signs of peritonitis.

The leukogram and the plasma protein and fibrinogen concentrations are an aid in the diagnosis of TRP in cattle [8]. Acute cases are typically characterised by neutrophilia with a left shift and hyperfibrinogenaemia; however, the leukocyte count may vary in inflammatory disease from severely decreased to severely increased [32]. Leukocyte numbers vary with species and reflect the balance between production and release from the bone marrow and consumption [32]. In contrast to dogs, which have a rapid regenerative capacity and a relatively high bone marrow reserve of neutrophils, cattle have a slow regenerative capacity and a relatively low reserve. Thus dogs with chronic infection usually have persistent neutrophilia, whereas cattle may have normal neutrophil numbers and a normal differential cell count, or even neutropenia with a left shift because of a slow regenerative response. The majority of cattle (54%) in the present study had normal leukocyte counts, which supports the findings of other studies [23, 33–35] in which neutrophilia was not a consistent feature of TRP. Forty-two percent of cattle had leukocytosis with greater than 10,000 leukocytes/μl blood and 90% of these had neutrophilia. Acute localised peritonitis is commonly accompanied by neutrophilia [36, 37], often with a left shift [13, 38]. Leukopenia was rare and only occurred in 4% of cattle in the present study. Leukopenia may occur after the first 1 or 2 days of severe acute inflammation because of migration of circulating neutrophils to the site of inflammation combined with reduced bone marrow response [39]. Stress-related endogenous corticosteroids also can suppress neutrophil numbers. Considering that more than half of a large sample of cows with TRP did not have neutrophilia makes determination of leukocyte numbers a poor diagnostic indicator. However, it is important to note that 272 cattle (54%) had a total leukocyte count within the reference interval (5000–10,000 leukocytes/μl blood) but did not have a differential leukocyte count done to minimise costs. It is not improbable that a left shift or lymphopenia may have been present in some of these animals. Erythropenia is a common sequel of chronic inflammation in cattle [8, 37] and occurred in 4% of the cows in the present study. Chronic illness is the most common cause of mild and clinically insignificant anaemia and the multifactorial pathogenesis of this disorder has been described in detail [40]. Unlike leukocyte numbers, hyperfibrinogenaemia and hyperproteinaemia are good indicators of TRP in cattle and have been linked to this disorder in several studies [41–43]; in fact, in the present study the most important findings of blood analysis were increased fibrinogen concentrations in 69% and increased protein concentrations in 64% of cattle with TRP. Fibrinogen is an acute-phase protein [44] and may be increased as early as 2 to 3 days after the onset of illness [8]. Fibrinogen is often increased in the absence of changes in leukocyte numbers and therefore is the better indicator for inflammation. A recent study on the diagnostic value of different acute-phase proteins concluded that the diagnostic accuracy of fibrinogen was significantly lower than that of serum amyloid A and haptoglobin [43], but other authors favour mainly haptoglobin and fibrinogen as biomarkers for TRP [45, 46] and consider serum amyloid A to be non-specific [46]. High plasma protein concentrations primarily reflect high globulin concentrations and are typical of chronic TRP [41, 47]. However, increases in fibrinogen and plasma protein concentrations are not specific for TRP and the concentrations may be normal or decreased in cows with TRP as seen in this study. The glutaraldehyde clotting test is a point-of-care diagnostic test of considerable practical value used to detect increased gamma globulin and/or fibrinogen concentrations based on polymerisation of fibrinogen and gamma globulins with aldehyde resulting in clot formation of the test sample. The clotting time decreases proportionally with increases in fibrinogen and gamma globulin concentrations. There is a significant correlation between onset of coagulation of the test sample and fibrinogen and gamma globulin concentrations in the blood of cattle. Coagulation time was shorter than 6 min in 75% of tested cattle, and in 49% occurred within 3 min; the respective positive predictive values of these test results for the diagnosis of an inflammatory process are 87.9 and 97.8% [48]. In the latter study, 62.9% of cattle with severe inflammatory changes had clotting times of less than 3 min. However, this test is not specific for TRP and results of blood analysis including the glutaraldehyde test must be interpreted in view of the clinical picture. Reduced clotting time in the glutaraldehyde test, an arched back, fever, reduced rumen motility and positive foreign body tests allow a tentative diagnosis of TRP, but the absence of any of these findings does not rule out TRP. For this reason, the diagnosis should be confirmed by ultrasonographic and radiographic examination of the reticulum. A limitation of the study was that 3.6% of the 503 cattle had concomitant left or right displacement of the abomasum or caecal dilation, which may have affected laboratory findings. Such a low percentage likely had minimal effect on the average and mean values, but is certain to have affected the range of variation. The concomitant disorders would have led to an increase in the maximum values of haematocrit, concentration of urea nitrogen in serum, blood gas variables and concentration of rumen chloride.

Conclusions

Traumatic reticuloperitonitis cannot be diagnosed on the basis of individual clinical or laboratory criteria because even the most common abnormalities are not seen in all cattle with TRP. Furthermore, many of these findings are not specific for TRP and may occur in other disorders of the abdomen and thorax. A tentative diagnosis may be possible based on all the clinical and laboratory findings.

Acknowledgements
The authors thank the technicians of the Medical Laboratory for the haematological and biochemical analyses and the agricultural assistants for their help with the clinical examinations.

Funding
Not applicable since it was a retrospective analysis of medical records.

Authors' contributions
UB initiated, planned and supervised the study and prepared the manuscript, SW analysed the medical histories of the cows as part of her dissertation, PT was responsible for the statistics, KN and CG made substantial contributions to acquisition and interpretation of data and were involved in drafting and revising the manuscript. All authors read and approved the final manuscript.

Consent for publication
All owners signed a consent form allowing us to use the animals and all the associated medical data for scientific analysis and publication.

Competing interests
The authors declare that they have no competing interests.

Author details
[1]Department of Farm Animals, Vetsuisse-Faculty, University of Zurich, Winterthurerstrasse 260, CH-8057 Zurich, Switzerland. [2]Section of Epidemiology, Vetsuisse-Faculty, University of Zurich, Winterthurerstrasse 260, CH-8057 Zurich, Switzerland.

References
1. Maddy KT. Incidence of perforation of the bovine reticulum. J Am Vet Med Ass. 1954;124:113–5.
2. Poulsen JS. Prevention of traumatic indigestion in cattle. Vet Rec. 1976;98:149–51.
3. Neumann N. Untersuchungen über das Vorkommen von Netzmagen-Fremdkörpern bei Schlachtrindern in Bayern, Dr Med Vet Thesis. Munich: University of Munich; 1979.
4. Starke A, Rehage J. Diagnose und Therapie bei abszedierender Reticuloperitonitis traumatica. Tierärztl Prax. 2000;28(G):93–5.
5. Cramers T, Mikkelsen KB, Andersen P, Enevoldsen E, Jensen HE. New types of foreign bodies and the effect of magnets in traumatic reticulitis in cows. Vet Rec. 2005;157:287–9.

6. Dirksen G. Krankheiten von Haube und Pansen beim ruminanten Rind. In: Dirksen G, Gründer HD, Stöber M, editors. Innere Medizin und Chirurgie des Rindes. Berlin: Parey Buchverlag; 2002. p. 396–455.

7. Braun U, Milicevic A, Forster E, Irmer M, Reichle S, Previtali M, Gautschi A, Steininger K, Thoma R, Zeller S, Lazzarini A, Manzoni C, Ohlerth S. An unusual cause of traumatic reticulitis/reticuloperitonitis in a herd of Swiss dairy cows nearby an airport. Schweiz Arch Tierheilk. 2009;151:127–31.

8. Francoz D, Guard CL. Traumatic reticuloperitonitis (hardware disease, traumatic reticulitis). In: Smith BP, editor. Large Animal Internal Medicine. St. Louis: Elsevier Mosby; 2015. p. 805–7.

9. Constable PD, Hinchcliff KW, Done SH, Grünberg W. Diseases of the alimentary tract – ruminant. In: Veterinary medicine. A textbook of the diseases of cattle, horses, sheep, pigs and goats. St. Louis: Elsevier; 2017. p. 436–621.

10. Henniger RW, Mullowney PC. Anterior abdominal pain in cattle. Compend Contin Educ Pract Vet. 1984;6:453–63.

11. Garry F. Evaluating motility disorders of the bovine forestomach. Vet Med. 1990;85:634–42.

12. Braun U. Traumatic pericarditis in cattle: clinical, radiographic and ultrasonographic findings. Vet J. 2009;182:176–86.

13. Braun U, Pusterla N, Wild K. Ultrasonographic findings in 11 cows with a hepatic abscess. Vet Rec. 1995;137:284–90.

14. Ward JL, Ducharme NG. Traumatic reticuloperitonitis in dairy cows. J Am Vet Med Assoc. 1994;204:874–7.

15. Dirksen G, Gründer HD, Stöber M. Verdauungsapparat. In: Dirksen G, Gründer HD, Stöber M, editors. Die klinische Untersuchung des Rindes. Berlin, Hamburg: Paul Parey; 1990. p. 288–400.

16. Warislohner S. Reticuloperitonitis traumatica beim Rind – eine Analyse von 503 Krankengeschichten, Dr Med Vet Thesis. Zurich: University of Zurich; 2017.

17. Rosenberger G. Clinical examination of cattle. Berlin, Hamburg: Paul Parey; 1990.

18. Braun U, Beckmann C, Gerspach C, Hässig M, Muggli E, Knubben-Schweizer G, Nuss K. Clinical findings and treatment in cattle with caecal dilatation. BMC Vet Res. 2012;8:75.

19. Albright JL, Briggs JL, Jessup RV. Long-term effects of magnets and management in the control of traumatic gastritis (hardware disease) in large commercial dairy herds. J Dairy Sci. 1962;45:547–9.

20. Leuenberger W, Martig J, Schneider E. Untersuchungen zur Diagnose und Therapie der Reticulitis traumatica des Rindes. 1. Teil. Schweiz Arch Tierheilk. 1978;120:213–29.

21. Gröhn YT, Bruss ML. Effect of diseases, production, and season on traumatic reticuloperitonitis and ruminal acidosis in cattle. J Dairy Sci. 1990;73:2355–63.

22. Schneider E. Zur Reticulitis traumatica: Was passiert mit dem krankmachenden Fremdkörper. Schweiz Arch Tierheilk. 1963;105:500–6.

23. Hjerpe CA. Studies on acute bovine traumatic reticuloperitonitis. II. Signs of traumatic reticuloperitonitis. J Am Vet Med Assoc. 1961;139:230–2.

24. Constable PD, Hoffsis GF, Rings DM. The reticulorumen: normal and abnormal motor function. Part I. Primary contraction cycle. Compend Contin Educ Pract Vet. 1990;12:1008–14.

25. Herzog K, Kaske M, Bischoff C, Kehler W, Hoeltershinken M, Starke A, Stöber M, Rehage J. Post surgical development of inflammatory adhesions and reticular function in cows suffering from traumatic reticuloperitonitis. Dtsch Tierärztl Wschr. 2004;111:57–62.

26. Radostits OM. Detection and localization of abdominal pain. In: Radostis OM, Mayhew IG, Houston DM, editors. Veterinary clinical examination and diagnosis. London: WB Saunders; 2000. p. 444–5.

27. Stöber M. Beitrag zur Diagnose der Reticuloperitonitis traumatica des Rindes: die Betastung der Luftröhre als einfaches Hilfsmittel zur Feststellung des schmerzhaften Stöhnens bei den Fremdkörperproben. Dtsch Tierärztl Wschr. 1961;68:497–8.

28. Götze R. Die Fremdkörperoperation beim Rinde praxisreif durch extraperitoneale Pansennaht. Dtsch Tierärztl Wschr. 1934;42:353–7. 374–9.

29. Andres J. Zur modernen Diagnose, Prophylaxe und Therapie der Reticulitis traumatica des Rindes. Schweiz Arch Tierheilk. 1952;94:287–95.

30. Williams F. Einige diagnostische Hilfen zur Feststellung der traumatischen Retikuloperitonitis und Perikarditis beim Rind. Dtsch Tierärztl Wschr. 1974;81:558.

31. Braun U, Götz M, Marmier O. Ultrasonographic findings in cows with traumatic reticuloperitonitis. Vet Rec. 1993;133:416–22.

32. Weiser G. Interpretation of leukocyte responses in disease. In: Thrall MA, Weiser G, Allison RW, Campbell TW, editors. Veterinary hematology and clinical chemistry. Ames: Wiley-Blackwell; 2012. p. 127–39.

33. Brown JM, Kingrey BW, Rosenquist BD. The hematology of chronic bovine reticuloperitonitis. Am J Vet Res. 1959;20:255–64.

34. Smith DF, Becht JL, Whitlock RH. Traumatic reticuloperitonitis. In: Anderson NV, editor. Veterinary gastroenterology. Philadelphia/London: Lea and Febiger; 1992. p. 715–9.

35. Ramprabhu R, Dhanapalan P, Prathaban S. Comparative efficacy of diagnostic tests in the diagnosis of traumatic reticuloperitonitis and allied syndromes in cattle. Isr J Vet Med. 2003;58:68–72.

36. Tharwat M, Ahmed AF, El-Tookhy OS. Chronic peritonitis in buffaloes and cattle: clinical, hematological, ultrasonographic findings and treatment. J Anim Vet Adv. 2012;11:2775–81.

37. Reddy LSSVP, Reddy BS, Naik BR, Prasad CS. Haematological and clinical alterations with traumatic reticuloperitonitis in cattle. Int J Vet Sci. 2014;3:203–5.

38. Fecteau G. Management of peritonitis in cattle. Vet Clin North Am Food Anim Pract. 2005;21:155–71.

39. Tornquist SJ, Rigas S. Interpretation of ruminant leukocyte responses. In: Weiss DJ, Wardrop KJ, editors. Schalm's veterinary hematology. Ames: Wiley-Blackwell; 2010. p. 307–13.

40. Thrall MA. Nonregenerative anemia. In: Thrall MA, Weiser G, Allison RW, Campbell TW, editors. Veterinary hematology and clinical chemistry. Ames: Wiley-Blackwell; 2012. p. 81–6.

41. Dubensky RA, White ME. The sensitivity, specificity and predictive value of total plasma protein in the diagnosis of traumatic reticuloperitonitis. Can J Comp Med. 1983;47:241–4.

42. Jafarzadeh SR, Nowrouzian I, Khaki Z, Ghamsari SM, Adibhashemi F. The sensitivities and specificities of total plasma protein and plasma fibrinogen for the diagnosis of traumatic reticuloperitonitis in cattle. Prev Vet Med. 2004;65:1–7.

43. Nazifi S, Ansari-Lari M, Asadi-Fardqi J, Rezaei M. The use of receiver operating characteristic (ROC) analysis to assess the diagnostic value of serum amyloid a, haptoglobin and fibrinogen in traumatic reticuloperitonitis in cattle. Vet J. 2009;182:315–9.

44. Allison R. Laboratory evaluation of plasma and serum proteins. In: Thrall MA, Weiser G, Allison RW, Campbell TW, editors. Veterinary hematology and clinical chemistry. Ames: Wiley-Blackwell; 2012. p. 460–75.

45. Hirvonen J, Pyörälä S. Acute-phase response in dairy cows with surgically-treated abdominal disorders. Vet J. 1998;155:53–61.

46. Kirbas A, Ozkanlar Y, Aktas MS. Acute phase biomarkers for inflammatory response in dairy cows with traumatic reticuloperitonitis. Isr J Vet Med. 2015;70:23–9.

47. Gokce HI, Gokce G, Cihan M. Alterations in coagulation profiles and biochemical and haematological parameters in cattle with traumatic reticuloperitonitis. Vet Res Commun. 2007;31:529–37.

48. Doll K, Schillinger D, Klee W. Der Glutaraldehyd-Test beim Rind – seine Brauchbarkeit für Prognose und Diagnose innerer Entzündungen. Zbl Vetmed A. 1985;32:581–93.

3

Validation of blood vitamin A concentrations in cattle: comparison of a new cow-side test (iCheck™ FLUORO) with high-performance liquid chromatography (HPLC)

Jens Raila[1]*[iD], Chiho Kawashima[2], Helga Sauerwein[3], Nadine Hülsmann[4], Christoph Knorr[4], Akio Myamoto[2] and Florian J. Schweigert[1]

Abstract

Background: Plasma concentration of retinol is an accepted indicator to assess the vitamin A (retinol) status in cattle. However, the determination of vitamin A requires a time consuming multi-step procedure, which needs specific equipment to perform extraction, centrifugation or saponification prior to high-performance liquid chromatography (HPLC).

Methods: The concentrations of retinol in whole blood ($n = 10$), plasma ($n = 132$) and serum ($n = 61$) were measured by a new rapid cow-side test (iCheck™ FLUORO) and compared with those by HPLC in two independent laboratories in Germany (DE) and Japan (JP).

Results: Retinol concentrations in plasma ranged from 0.033 to 0.532 mg/L, and in serum from 0.043 to 0.360 mg/L (HPLC method). No significant differences in retinol levels were observed between the new rapid cow-side test and HPLC performed in different laboratories (HPLC vs. iCheck™ FLUORO: 0.320 ± 0.047 mg/L vs. 0.333 ± 0.044 mg/L, and 0.240 ± 0.096 mg/L vs. 0.241 ± 0.069 mg/L, lab DE and lab JP, respectively). A similar comparability was observed when whole blood was used (HPLC vs. iCheck™ FLUORO: 0.353 ± 0.084 mg/L vs. 0.341 ± 0.064 mg/L). Results showed a good agreement between both methods based on correlation coefficients of $r^2 = 0.87$ ($P < 0.001$) and Bland-Altman blots revealed no significant bias for all comparison.

Conclusions: With the new rapid cow-side test (iCheck™ FLUORO) retinol concentrations in cattle can be reliably assessed within a few minutes and directly in the barn using even whole blood without the necessity of prior centrifugation. The ease of the application of the new rapid cow-side test and its portability can improve the diagnostic of vitamin A status and will help to control vitamin A supplementation in specific vitamin A feeding regimes such as used to optimize health status in calves or meat marbling in Japanese Black cattle.

Keywords: Cattle, Vitamin A, Biomarker, Blood, Method comparison, Cow-side assay

* Correspondence: jens.raila@uni-potsdam.de
[1]Institute of Nutritional Science, University of Potsdam, Arthur-Scheunert-Allee 114-116, 14558 Nuthetal, Germany
Full list of author information is available at the end of the article

Background

Vitamin A (retinol) is an essential micronutrient not only to ascertain vision but also to modulate growth and development [1–4]. Cattle as well as other herbivorous animals are unable to synthesize retinol de novo and they have to obtain their vitamin A primarily from dietary β-carotene, which is converted into retinol within the enterocytes [5]. Independently from provitamin A activity, β-carotene can be absorbed intact in the gut and is considered to have positive effects on reproductive performance and immune response in cattle [6]. All green forages are rich in β-carotene and thus provide a high vitamin A value mainly during pasture conditions [7]. Low concentrations of vitamin A in blood resulting from dietary β-carotene restriction or lacking in vitamin A supplementation and are related to an increased risk of disease [8]. Neonate calves are particularly prone to develop symptoms of vitamin A deficiency, because they have low blood vitamin A concentrations and insufficient hepatic vitamin A stores at birth [9]. Low hepatic vitamin A storage is associated with stillborn and postnatal calf losses as a consequence of an insufficient vitamin A intake of the mother cows especially during non-grazing conditions [10]. In addition, cases of dermatopathy associated with hypovitaminosis A were recently described in juvenile Angus calves, which could be successfully treated with parenteral vitamin A supplementation [11]. Thus, the measurement of blood retinol concentrations in mother cows as well as in neonate and juvenile calves is a prerequisite for evaluation of their health status.

Low vitamin A plasma concentrations are generally found in Japanese Black cattle and are associated with an increased intramuscular or marbling adipose deposition without influencing other adipose depots [12–15]. Thus, to optimize the desired intramuscular fat marbling that increases the carcass value, Japanese Black fattening cattle are provided with a high concentrate diet deficient in vitamin A or β-carotene. It was suggested that optimal marbling scores are obtained when animals are kept very close to health deteriorating low vitamin A plasma concentrations in a range of 100–150 µg/L (40–50 IU/dl) over a period of usually 12 month during the fattening period [16]. However, in Japanese Black cattle low vitamin A plasma concentrations have been reported to be associated with an impaired immune function [17] and increased morbidity and mortality [18]. Therefore, blood vitamin A levels in Japanese Black cattle needs to be carefully monitored and nutritionally controlled. Additional reduced vitamin A plasma concentrations are also observed as a consequence of the acute phase reaction e.g. during inflammation or around parturition [19–21]. Thus, due to the restrict feeding management, the induced vitamin A deficiency in Japanese Black cattle can cause a serious health problem if not monitored and nutritionally controlled properly.

Under practical management conditions, vitamin A deficiency is difficult to diagnose and marginal deficiency may exist without obvious deficiency symptoms. Measurement of retinol in blood plasma is an appropriate marker to assess vitamin A status but it is currently a time consuming and cost-intensive multi-step procedure. Specific equipment is needed to perform extraction and centrifugation or saponification. Vitamin A is finally measured by colorimetric methods [22] or high performance liquid chromatography (HPLC) [23]. All preanalytic steps and the final analyses have to be performed in qualified analytical laboratories by highly trained personal. Alternatively suggested methods such as thin layer chromatography are similar complicated and not very reliable [24]. So far, however, no method has been described which is also able to measure vitamin A directly at cow-side in the barn or even determining vitamin A directly from whole blood prior to centrifugation.

This study presents data for a new, fast, easy to perform and laboratory-independent test to measure retinol in whole blood, plasma or serum. Results are compared with those obtained by HPLC as the reference method at two independent laboratories. The new assay is based on a separation and extraction principal published for the determination of β-carotene in blood plasma [25] as well as human and cow milk [26, 27].

Methods

The comparison was performed at two independent laboratory sites at the University of Potsdam, Germany (lab DE) and at the Obihiro University, Japan (lab JP).

Animals and sampling
Lab DE

A total of 132 blood samples were collected from dairy cows ($n = 40$) and bulls ($n = 92$) into EDTA-evacuated tubes (Saarstedt, Nümbrecht, Germany) and immediately separated by centrifugation ($1500 \times g$; 10 min; 4 °C). The plasma was frozen at −80 °C and analyzed for their vitamin A concentrations with both methods (HPLC and iCheck™ FLUORO) within three months. In addition, 10 blood samples were taken from dairy cattle and collected into EDTA-evacuated tubes (Saarstedt) and analyzed freshly for their vitamin A content within 24 h after acquisition both in whole blood and in plasma after removal of the erythrocytes by centrifugation ($1500 \times g$; 10 min; 4 °C). All samples were obtained from institutional farms (Research Station Frankenforst, Faculty of Agriculture, University of Bonn, Koenigswinter, Germany and Verein Ostfriesischer Stammviehzuechter, Leer, Germany).

Lab JP

A total of 61 blood samples from Holstein dairy cows ($n = 29$) and Japanese Black cattle ($n = 32$) were obtained by caudal venipuncture using non-heparinized and silicone-coated 9-mL tubes (Venoject, Autosep, Gel and Clot. Act., VP-AS109K; Terumo Corporation, Japan). Samples were collected from the institutional farm of the Obihiro University of Agriculture and Veterinary Medicine, Japan and Akiyama Farm Animal Clinic, Japan. To obtain serum, blood samples were coagulated for 15 min at 38 °C in an incubator. All tubes were centrifuged at $2000 \times g$ for 20 min at 4 °C, and serum samples were kept at −30 °C until analysis within three months.

Vitamin A determination by HPLC

HPLC lab DE

A modified gradient reverse-phase HPLC system (Shimadzu, Duisburg, Germany) was used [28]. Briefly, vitamin A as retinol or retinyl esters was extracted from plasma and thereafter separated on a reverse-phase column (ReproSil 70 C18 column, 200 × 3.0 mm; inside diameter 5 μm; Dr. Maisch GmbH, Ammerbuch, Germany). The solvent system consisted of solvent A with methanol and solvent B with ethyl acetate at a flow rate of 0.5 mL/min (pump LC 20-AD, Shimadzu). Retinol was identified based on retention time in comparison to an external standard (Sigma, Munich, Germany) by use of a photodiode array detector (PDA SPDM-20A, Shimadzu). Vitamin A was quantified by measuring the absorption at 325 nm.

HPLC lab JP

The concentrations of retinol in serum were determined by HPLC, as described previously [29]. Extraction efficiency was 96%. Intra- and inter-assay CVs averaged 2.1% and 3.3%, respectively.

Vitamin A determination by a portable fluorometer

The novel cow-side test for vitamin A consists of an extraction unit that contains all necessary chemicals for extraction and separation and a portable fluorometer (iCheck™ FLUORO, BioAnalyt GmbH, Teltow, Germany). The combination of these two components enables to extract vitamin A from whole blood without prior separation of plasma in a single step at cow-side. The analysis of whole blood, blood plasma or serum with the new and innovative assay system was done as recommended by the manufacturer. A total volume of 500 μl of whole blood, plasma or serum was injected with an included syringe into the extraction and measuring vials. Thereafter it was shaken intensively for 10 s and then let settle for 5 min until complete phase separation. Finally, the vials were inserted into the fluorometer and measured. The quantification of vitamin A in whole blood, plasma or serum is based on the specific autofluorescent characteristics of excitation and emission wavelengths. Since that autofluorescence is based on the retinol moiety, any form of vitamin A is included in the quantification. Results were recorded as μg/L whole blood, plasma or serum.

Statistical analyses

The data were analysed using SPSS version 23.0 software (SPSS, Munich, Germany). The results obtained by HPLC and the new fluorometric method were judged for method acceptability as suggested for clinical laboratories [30]. Associations between the results were determined with Pearson correlation method. Mean results obtained by HPLC and fluorometer were compared with a paired t test. The difference of results obtained from HPLC and fluorometer were analyzed by one sample t test. A Bland-Altman bias plot [31] was done to determine the analytical accuracy of the two analytical methods. The analytical detection limit was defined as proposed in the concept of 'functional sensitivity'. This is defined as the lowest concentration of an assay that can be measured with an inter-assay CV of 20% [32]. Values of $P < 0.05$ were assigned as statistically significant.

Results and discussion

The concentrations of retinol in plasma or serum as determined by HPLC and the new method ranged from 0.033 mg/L to 0.532 mg/L for plasma in Lab DE and from 0.043 mg/L to 0.360 mg/L for serum in Lab JP (Additional file 1: Table S1). No significant differences were observed between both methods at two different laboratories for the average values obtained (HPLC vs. new method: 0.320 ± 0.047 mg/L vs. 0.333 ± 0.044. mg/L; mean ± SD in Lab DE and 0.240 ± 0.096 mg/L vs. 0.241 ± 0.069 mg/L; mean ± SD in Lab JP).

The analytical advantage of the new innovative test method is to measure retinol from whole blood directly. This avoids the time consuming and limiting preanalytical step of centrifugation to obtain plasma or serum. In consequence, the new assay can be directly used in the barn. When the subset of 10 whole blood samples were assayed for retinol by the new test method and compared to plasma levels assessed by HPLC in Lab DE, a similar good agreement was found as for the use of plasma (HPLC plasma: 0.353 ± 0.084 mg/L vs. new test plasma: 0.340 ± 0.063 mg/L vs. new test whole blood 0.341 ± 0.064 mg/L; mean ± SD; see Additional file 2: Table S2). There were significantly correlations between concentrations of retinol measured in whole blood by the new method and retinol measured in plasma by HPLC ($r^2 = 0.84$; $P < 0.001$) as well as retinol measured in plasma by the new method ($r^2 = 0.87$; $P < 0.001$). The

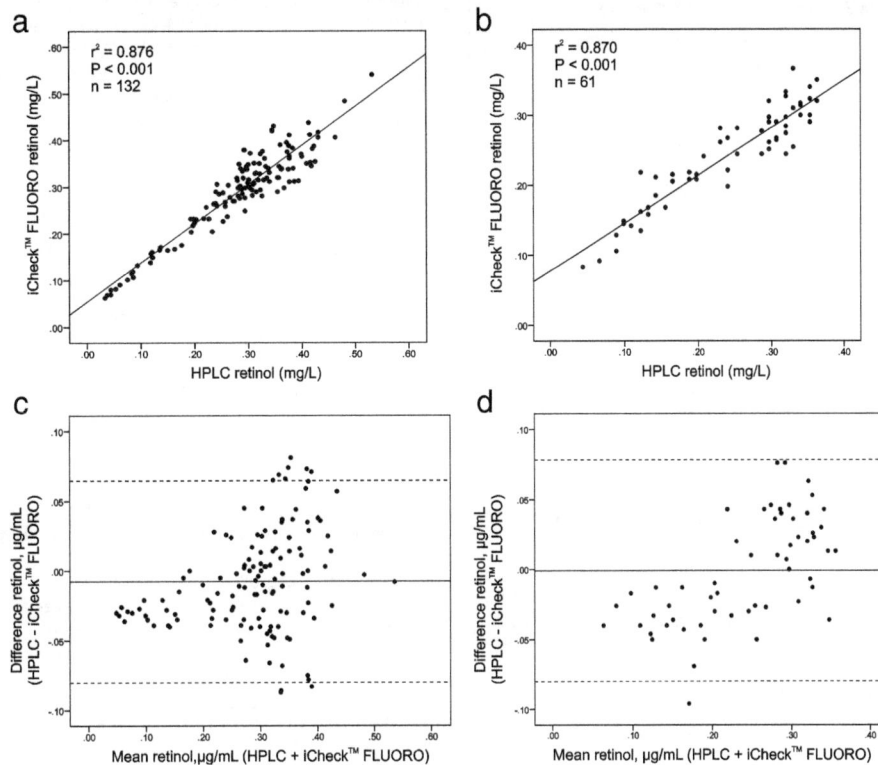

Fig. 1 Correlation analyses demonstrate close association between plasma retinol as measured by HPLC and iCheck™ fluorometer in lab Germany (**a**) and lab Japan (**b**). Bland-Altman plot showing the mean difference (solid line) and the 95% confidence interval (dotted lines) of the retinol values in blood plasma obtained by HPLC and iCheck™ fluorometer in lab Germany (**c**) and lab Japan (**d**)

average packed cell volume of the 10 whole blood samples in this study was 31.7 ± 2.7% (mean ± SD).

The comparison between plasma or serum samples assayed with the two different methods in Lab DE (plasma) and Lab JP (serum) indicated a strong correlation between methods (Fig. 1a, b). Results show that the novel cow-side test for vitamin A correlated very well with HPLC analysis (r^2 = 0.876 and 0.870, Lab DE and Lab JP, respectively, both $P < 0.001$). The calculated differences in retinol concentrations obtained by HPLC and fluorometer were not significantly different (P-values >0.05 for both Lab DE and Lab JP) indicating an acceptable level of agreement between the two assay methods. Based on the Bland-Altman plot (Fig. 1c, d) no systematic error and a good level of agreement occurred between the two methods and only 4% of the differences in measured values felt outside the 95% acceptability limits.

Based on multiple determinations of selected samples the intra-assay and inter-assay precision were calculated for each method (Lab DE). In average, the coefficient of variance (CV) was in an order of magnitude generally accepted for average analysis of this kind. The CV was slightly higher in the HPLC method compared to the new method (5.3% vs. 2.3%, respectively).

Conclusions

The comparison of the new cow-side test (iCheck™ FLUORO) to measure vitamin A in cattle whole blood, blood plasma or serum showed a very good agreement, reliability and accuracy with the cumbersome, time-intensive and expensive HPLC analytical technique. The assay allows the determination of vitamin A directly in whole blood within 5 min even at cow-side at the barn without a further sample preparation. Because all chemicals are sealed in the vial, the staff is not directly exposed to potentially hazardous organic chemicals. Additionally, the limited volume of organic solvent necessary is positive with regard to environmentally and ecologically critical waste. The novel test realizes important aspects of a cow-side assay: it is sensitive, specific and rapid. Finally, the test delivers clear end-point results which can help to optimize nutritional interventions in cattle.

Additional files

Additional file 1: Table S1. Comparison of retinol concentrations in plasma ($n = 132$, obtained from Lab Germany) and serum ($n = 61$, obtained from Lab Japan) measured by HPLC and iCheck™ FLUORO.

Additional file 2: Table S2. Concentrations of retinol in plasma (*n* = 10) measured by HPLC and iCheck™ FLUORO and whole blood measured by iCheck™ FLUORO.

Abbreviations
EDTA: Ethylenediaminetetraacetic acid; HPLC: High-performance liquid chromatography; iCheck™ FLUORO: Portable fluorometer for quantitative vitamin A determination; Lab DE: Laboratory located in Germany; Lab JP: Laboratory located in Japan

Acknowledgments
The authors are grateful to the technicians of the Department of Physiology and Pathophysiology, University of Potsdam for their excellent technical assistance, constant helpfulness, and overall kindness. The authors thank Dr. K. Akiyama (Akiyama Farm Animal Clinic, Japan) for supporting our research. Blood samples were made available by the Research Station Frankenforst, Faculty of Agriculture, University of Bonn, Koenigswinter, Germany. We thank the staff and Dr. Johanna F.L. Heinz, Institute of Animal Science, University of Bonn for their support. The authors thank VOST (Verein Ostfriesischer Stammviehzuechter) for providing blood samples of bulls.

Funding
The authors received no financial support for authorship except from the free supply of consumables for the new test by BioAnalyt GmbH. This study was supported by a Grant-in-Aid for Scientific Research (CK) and the Global COE program from the Japan Society for the Promotion of Science, and the Foerderverein Biotechnologieforschung e.V. (FBF), Bonn, Germany. We acknowledge the support of the Deutsche Forschungsgemeinschaft and Open Access Publishing Fund of University of Potsdam.

Authors' contributions
JR carried out HPLC analysis, analyzed the data, performed statistical analysis and participated in drafting the manuscript. HS, CK, NH, XK participated in study design and sample acquisition and contributed to the drafting of the manuscript. AM, FJS participated in its design and coordination and contributed to draft the manuscript. All authors read and approved the manuscript.

Authors' information
Not applicable.

Competing interests
Florian J. Schweigert is shareholder of BioAnalyt. Other authors declare no potential conflict of interest with respect to the authorship and publication of this article.

Consent for publication
Not applicable.

Ethics approval and consent to participate
Lab DE: The study was done according to the German Animal Welfare Law (released on 05/18/2006, last changes on 12/03/2015) and the animal care and use protocol was approved by the Animal Welfare Committee (Landesamt für Natur, Umwelt und Verbraucherschutz Nordrhein-Westfalen, Germany, AZ 8.87–51.05.20.10.258).
Lab JP: All experimental procedures complied with the Guidelines for the care and use of agricultural animals of Obihiro University, and were approved by the Committee on the Ethics of Animal Experiments (approval numbers: #20–13; #20–128 and #21–14). All samples from Holstein cows were obtained from cattle kept at an institutional farm. Therefore, owner consent to participate the study was not required. Blood samples from Japanese Black cattle were supplied from the owner of Akiyama farm, who is a veterinarian.

Author details
[1]Institute of Nutritional Science, University of Potsdam, Arthur-Scheunert-Allee 114-116, 14558 Nuthetal, Germany. [2]Obihiro University of Agriculture and Veterinary Medicine, Obihiro, Hokkaido 080-8555, Japan. [3]Institute for Animal Science, Physiology and Hygiene, University of Bonn, Katzenburgweg 7-9, 53115 Bonn, Germany. [4]Department of Animal Sciences, Biotechnology and Reproduction of Livestock, Georg-August-University Goettingen, Burckhardtweg 2, 37077 Goettingen, Germany.

References
1. Ikeda S, Kitagawa M, Imai H, Yamada M. The roles of vitamin A for cytoplasmic maturation of bovine oocytes. J Reprod Dev. 2005;51:23–35.
2. Clagett-Dame M, Knutson D. Vitamin A in reproduction and development. Nutrients. 2011;3:385–428.
3. Blomhoff R, Blomhoff HK. Overview of retinoid metabolism and function. J Neurobiol. 2006;66:606–30.
4. Gomez E, Caamano JN, Rodriguez A, De Frutos C, Facal N, Diez C. Bovine early embryonic development and vitamin A. Reprod Domest Anim. 2006; 41(Suppl 2):63–71.
5. von Lintig J. Provitamin A metabolism and functions in mammalian biology. Am J Clin Nutr. 2012;96:1234S–44S.
6. Kawashima C, Matsui M, Shimizu T, Kida K, Miyamoto A. Nutritional factors that regulate ovulation of the dominant follicle during the first follicular wave postpartum in high-producing dairy cows. J Reprod Dev. 2012;58:10–6.
7. Elgersma A, Soegaard K, Jensen SK. Fatty acids, alpha-tocopherol, beta-carotene, and lutein contents in forage legumes, forbs, and a grass-clover mixture. J Agric Food Chem. 2013;61:11913–20.
8. Weiss WP. Requirements of fat-soluble vitamins for dairy cows: a review. J Dairy Sci. 1998;81:2493–501.
9. Gallina AM, Helmboldt CF, Frier HI, Nielsen SW, Eaton HD. Bone growth in the hypovitaminotic A calf. J Nutr. 1970;100:129–41.
10. Waldner CL, Blakley B. Evaluating micronutrient concentrations in liver samples from abortions, stillbirths, and neonatal and postnatal losses in beef calves. J Vet Diagn Invest. 2014;26:376–89.
11. Baldwin TJ, Rood KA, Kelly EJ, Hall JO. Dermatopathy in juvenile Angus cattle due to vitamin A deficiency. J Vet Diagn Invest. 2012;24:763–6.
12. Adachi K, Katsura N, Nomura Y, Arikawa A, Hidaka M, Onimaru T. Serum vitamin A and vitamin E in Japanese black fattening cattle in Miyazaki prefecture as determined by automatic column-switching high performance liquid chromatography. J Vet Med Sci. 1996;58:461–4.
13. Kato Y, Ito M, Hirooka H. Genetic parameters of serum vitamin A and total cholesterol concentrations and the genetic relationships with carcass traits in an F1 cross between Japanese Black sires and Holstein dams. J Anim Sci. 2011;89:951–8.
14. Oka A, Maruo Y, Miki T, Yamasaki T, Saito T. Influence of vitamin A on the quality of beef from the Tajima strain of Japanese Black cattle. Meat Sci. 1998;48:159–67.
15. Pickworth CL, Loerch SC, Fluharty FL. Effects of timing and duration of dietary vitamin A reduction on carcass quality of finishing beef cattle. J Anim Sci. 2012;90:2677–91.
16. Adachi K, Kawano H, Tsuno K, Nomura Y, Yamamoto N, Arikawa A, Tsuji A, Adachi M, Onimaru T, Ohwada K. Relationship between serum biochemical values and marbling scores in Japanese Black steers. J Vet Med Sci. 1999;61:961–4.
17. Yano H, Ohtsuka H, Miyazawa M, Abiko S, Ando T, Watanabe D, Matsuda K, Kawamura S, Arai T, Morris S. Relationship between immune function and serum vitamin A in Japanese black beef cattle. J Vet Med Sci. 2009;71:199–202.
18. Adachi K, Kawano H, Tsuno K, Nomura Y, Katsura N, Arikawa A, Tsuji A, Onimaru T. Values of the serum components in Japanese black beef steers at farms with high productivity and low frequencies of disease and death in Miyazaki Prefecture. J Vet Med Sci. 1997;59:873–7.
19. Trevisi E, Amadori M, Bakudila AM, Bertoni G. Metabolic changes in dairy cows induced by oral, low-dose interferon-alpha treatment. J Anim Sci. 2009;87:3020–9.
20. Rezamand P, Hoagland TA, Moyes KM, Silbart LK, Andrew SM. Energy status, lipid-soluble vitamins, and acute phase proteins in periparturient Holstein and Jersey dairy cows with or without subclinical mastitis. J Dairy Sci. 2007;90:5097–107.
21. Kawashima C, Nagashima S, Sawada K, Schweigert FJ, Miyamoto A, Kida K. Effect of beta-carotene supply during close-up dry period on the

onset of first postpartum luteal activity in dairy cows. Reprod Domest Anim. 2010;45:282–7.

22. Katsoulos PD, Roubies N, Panousis N, Karatzanos P, Karatzias H. Long-term fluctuations and effect of age on serum concentrations of certain fat-soluble vitamins in dairy cows. Vet Clin Pathol. 2005;34:362–7.

23. Schweigert FJ, Zucker H. Concentrations of vitamin A, beta-carotene and vitamin E in individual bovine follicles of different quality. J Reprod Fertil. 1988;82:575–9.

24. Suzuki J, Katoh N. A simple and cheap methods for measuring serum vitamin A in cattle using only a spectrophotometer. Nihon Juigaku Zasshi. 1990;52:1281–3.

25. Raila J, Enjalbert F, Mothes R, Hurtienne A, Schweigert FJ. Validation of a new point-of-care assay for determination of beta-carotene concentration in bovine whole blood and plasma. Vet Clin Pathol. 2012;41:119–22.

26. Laillou A, Renaud C, Berger J, Moench-Pfanner R, Fontan L, Avallone S. Assessment of a portable device to quantify vitamin A in fortified foods (flour, sugar, and milk) for quality control. Food Nutr Bull. 2014;35:449–57.

27. Engle-Stone R, Haskell MJ, La Frano MR, Ndjebayi AO, Nankap M, Brown KH. Comparison of breast milk vitamin A concentration measured in fresh milk by a rapid field assay (the iCheck FLUORO) with standard measurement of stored milk by HPLC. Eur J Clin Nutr. 2014;68:938–40.

28. Schweigert FJ, Steinhagen B, Raila J, Siemann A, Peet D, Buscher U. Concentrations of carotenoids, retinol and alpha-tocopherol in plasma and follicular fluid of women undergoing IVF. Hum Reprod. 2003;18:1259–64.

29. De Ruyter MG, De Leenheer AP. Determination of serum retinol (vitamin A) by high-speed liquid chromatography. Clin Chem. 1976;22:1593–5.

30. Jensen AL, Kjelgaard-Hansen M. Method comparison in the clinical laboratory. Vet Clin Pathol. 2006;35:276–86.

31. Bland JM, Altman DG. Statistical methods for assessing agreement between two methods of clinical measurement. Lancet. 1986;1:307–10.

32. Spencer CA, Takeuchi M, Kazarosyan M, MacKenzie F, Beckett GJ, Wilkinson E. Interlaboratory/intermethod differences in functional sensitivity of immunometric assays of thyrotropin (TSH) and impact on reliability of measurement of subnormal concentrations of TSH. Clin Chem. 1995;41:367–74.

4

Relative abundance of *Mycobacterium bovis* molecular types in cattle: a simulation study of potential epidemiological drivers

Hannah Trewby[1]* ⓘ, David M. Wright[2], Robin A. Skuce[3,4], Carl McCormick[3], Thomas R. Mallon[3], Eleanor L. Presho[3], Rowland R. Kao[1], Daniel T. Haydon[1] and Roman Biek[1]

Abstract

Background: The patterns of relative species abundance are commonly studied in ecology and epidemiology to provide insights into underlying dynamical processes. Molecular types (MVLA-types) of *Mycobacterium bovis,* the causal agent of bovine tuberculosis, are now routinely recorded in culture-confirmed bovine tuberculosis cases in Northern Ireland. In this study, we use ecological approaches and simulation modelling to investigate the distribution of relative abundances of MVLA-types and its potential drivers. We explore four biologically plausible hypotheses regarding the processes driving molecular type relative abundances: sampling and speciation; structuring of the pathogen population; historical changes in population size; and transmission heterogeneity (superspreading).

Results: Northern Irish herd-level MVLA-type surveillance shows a right-skewed distribution of MVLA-types, with a small number of types present at very high frequencies and the majority of types very rare. We demonstrate that this skew is too extreme to be accounted for by simple neutral ecological processes. Simulation results indicate that the process of MVLA-type speciation and the manner in which the MVLA-typing loci were chosen in Northern Ireland cannot account for the observed skew. Similarly, we find that pathogen population structure, assuming for example a reservoir of infection in a separate host, would drive the relative abundance distribution in the opposite direction to that observed, generating more even abundances of molecular types. However, we find that historical increases in bovine tuberculosis prevalence and/or transmission heterogeneity (superspreading) are both capable of generating the skewed MVLA-type distribution, consistent with findings of previous work examining the distribution of molecular types in human tuberculosis.

Conclusion: Although the distribution of MVLA-type abundances does not fit classical neutral predictions, our simulations show that increases in pathogen population size and/or superspreading are consistent with the pattern observed, even in the absence of selective pressures acting on the system.

Keywords: Bovine tuberculosis, Multiple-locus variable number tandem repeats, Neutral theory of biodiversity, Species abundance distributions, Population genetics

* Correspondence: hannah.trewby@glasgow.ac.uk
[1]Boyd Orr Centre for Population and Ecosystem Health, Institute of Biodiversity and Animal Health, University of Glasgow, Glasgow, UK
Full list of author information is available at the end of the article

Background

Bovine tuberculosis (bTB) is one of the most important diseases facing the livestock industry in Britain and Ireland. Control is complicated by infection in a wildlife reservoir, the Eurasian badger [1–3], and there has been much debate regarding the relative roles of cattle and badgers in driving pathogen population dynamics. Analysis of the relative abundance distributions (RADs) of pathogen molecular types could provide insight into the mechanisms underlying bTB spread. These techniques originated in the ecological literature to explain patterns of species diversity and are presently under-used in epidemiology, despite the wealth of data on the relative abundances of pathogen molecular types collected for diseases of both human and animal importance.

Analysis of abundance distributions should be applied with caution however, as RADs in isolation do not necessarily provide sufficient information to identify or distinguish between the specific ecological or epidemiological events that shape them [4–6]. In fact, simple neutral processes, such as those outlined in Hubbell's neutral theory of biodiversity and biogeography (hereafter referred to as NTB [7]), have shown a surprising ability to predict the RADs recorded in a wide variety of ecological datasets [8].

NTB is a theory of "ecological drift" and, in ecology, is traditionally set against the non-neutral hypotheses of niche theory and selective pressures. It is based around the idea of per-capita equivalence: all individuals, irrespective of species, have an equal chance of birth, death, reproduction, and immigration and/or speciation (with the analytical solutions put forward by Hubbell also relying on an assumption of constant population size) [7]. Further developments have extended the theory to consider the explicit spatial structure of the population through spatially limited dispersal of individuals [7, 8]. Although it is widely accepted that real ecosystems are not truly neutral and species are not identical, the success of NTB demonstrates that simple processes are often sufficient to drive patterns of diversity observed in natural systems.

RADs have also been considered in some epidemiological contexts, for example to examine the relative abundance of different molecular types of tuberculosis-causing mycobacteria in human host populations. As with ecological datasets, the observed distributions tend to be right-skewed, with a small number of molecular types present at high frequency and the majority of strains being rare. Luciani et al. showed that the RADs seen in several human tuberculosis datasets did not fit the predictions generated by NTB, and demonstrated that an increasing bacterial population size could account for the distribution of at least one of these datasets [9]. Ypma et al. also examined the RAD of different molecular types of human tuberculosis, this time in the Netherlands. They argued that an increasing bacterial population was unlikely in this system, but demonstrated that variation in the numbers of offspring per infection (superspreading) could explain the skewed distribution observed [10]. In a rare example from veterinary epidemiology, Smith et al. suggested that "clonal expansion" of certain bacterial lineages, for example due to natural selection or the expansion of lineages into new host species or geographical areas, accounted for the RAD of different molecular types of *Mycobacterium bovis*, the causal agent of bTB, recorded in cattle in Great Britain (GB) [11].

In this study we extend these approaches, exploring four biologically plausible hypotheses to examine the processes driving molecular type relative abundances of *M. bovis* in the Northern Ireland (NI) cattle population. BTB is endemic in cattle in Northern Ireland [12], and intensive molecular typing of *M. bovis* cultures from culture-confirmed bTB breakdowns has been carried out since 2003 [13]. Similar to the molecular types of *M. bovis* in GB [11], molecular types in NI have shown a marked right-skewed RAD [14], however the possible epidemiological processes underlying this skew have not been examined. In common with *M. bovis* in other parts of the UK and Ireland [15–18], infections in NI also give the visual impression of clustering geographically by molecular type [13].

Molecular typing of *M. bovis* in NI is conducted through a combination of spoligotyping, based on the presence of multiple spacer oligonucleotides within the Direct Repeat region of the genome [19], and Multiple-Locus Variable Number Tandem Repeat Analysis (MVLA)-typing [20, 21], which measures the number of repeats at various MVLA loci present in the mycobacterial genome. MVLA-typing gives a higher level of discrimination than spoligotyping, although both methods are prone to a degree of homoplasy, i.e. the occurrence of the same MVLA-type in unrelated lineages [22, 23]. In NI, the specific MVLA-loci used in MVLA-typing have been chosen to optimise the discriminatory power of the technique in this population [14], and MVLA-types are used in conjunction with spoligotyping. Throughout this paper the molecular type of an isolate, as determined by the combination of spoligotype and MVLA-type results, is referred to as the "MVLA-type".

M. bovis as a species shows very little genetic diversity, especially in Britain and Ireland [15]. Studies in NI have identified no significant differences between the molecular types of *M. bovis* in either the size of herd outbreaks or in the response to the bTB skin test, although there are indications that the number and anatomical distribution of bTB lesions may differ by molecular type [24, 25]. In a low diversity population such as that in NI, selective pressures are less likely to play a prominent role in shaping the diversity of the *M. bovis* population. However, historical increases in bTB incidence over recent decades [26] may have affected the relative abundances of the different molecular types. Furthermore, superspreading at the herd level has been indicated to play a role in the spread of the disease in Great Britain [27], and this may also impact the RAD observed here.

In this study we investigate the processes that potentially underlie the RAD of molecular types of *M. bovis* isolated from cattle in NI. We first evaluate the null hypothesis that the observed RAD of molecular types in NI can be accounted for by neutral processes alone, as put forward in NTB, by comparing the observed distribution of molecular types to that predicted under NTB. We go on to use simulation models to examine four alternative hypotheses regarding mechanisms involved in shaping the distribution of MVLA-types in NI: 1) the process of MVLA-type speciation and/or the manner in which MVLA-loci were chosen; 2) the existence of an unsampled pool of infection in a separate but linked population (for example a wildlife reservoir); 3) the recent expansion of the *M. bovis* population [9]; and 4) superspreading, i.e. variability in the number of onward infections transmitted from an infected herd [10].

Methods

All analyses and simulations were conducted in R v3.3.1 [28].

Data on molecular types of *M. bovis* in NI

In NI, all cattle over the age of 6 weeks are routinely tested for bTB on an annual basis, and bTB surveillance is additionally conducted routinely at slaughter. A "breakdown" of bTB in a cattle herd is defined as the period of time beginning with the first detection of bTB in an animal from the herd, and ending when the entire herd has passed two consecutive bTB tests, the first completed at least 60 days after removal/isolation of the infected animal and the second at least 120 days after

removal/isolation of the infected animal (if the breakdown was not confirmed, this may be reduced to a single bTB test completed at least 60 days after removal/isolation of the infected animal), plus completion of cleansing and disinfection of the premises as specified by notice. Since 2003, all NI bTB breakdowns from which *M. bovis* was successfully cultured have had at least one isolate typed using both spoligotyping and MVLA-typing using standard protocols [14]. In NI, the specific loci used for MVLA-typing are MV2163B/QUB11B, MV4052/QUB26A, MV2461/ETRB, MV2165/ETRA, MV2163/QUB11A and MV323/QUB3232, although since 2006 an extra MVLA locus, MV1895/QUB1895, has been added to the NI panel in order to split one of the common types into two geographically distinct types [13]. To ensure consistency across the sampled timeframe, in this study we consider MVLA-types as defined using the original seven-locus MVLA panel plus spoligotype.

Details of herd breakdowns, restricted to those breakdowns beginning in the years 2003–2010 inclusive, were made available from the APHIS database [29]. Each molecular typing record was assigned to a breakdown based on date and herd identifier, and the dataset was downsampled to include only the first typing record for each breakdown, as multiple isolates per breakdown were only routinely typed after 2008. A total of 10,049 out of 17,484 breakdowns (57.5%) for this period were linked to at least one molecular typing record (Table 1), and 18,699 out of 20,322 molecular typing records (92%) were successfully assigned to a breakdown (2656 typed breakdowns had more than one molecular typing result recorded). Sixteen unique spoligotypes (Additional file 1) and 183 unique MVLA-types (Additional file 2) were recorded over the study period.

Null hypothesis - neutral ecological drift

The R package untb [30] was used to test whether the observed RAD of *M. bovis* molecular types conformed to predictions generated by NTB. This package applies

Table 1 Numbers of *M. bovis* spoligotypes, MVLA-types, and typed breakdowns recorded by year in NI

Year	No. unique spoligotypes	No. unique MVLA-types	No. typed breakdowns	Total no. of breakdowns
2003	12	57	1347	2902
2004	13	61	1621	2954
2005	12	61	1432	2556
2006	9	72	1285	2033
2007	10	72	1235	1862
2008	11	72	1115	1789
2009	10	72	1080	1771
2010	9	72	934	1617
Total (unique) 2003–2010	16	185	10,049	17,484

the theory put forward by Hubbell [7] to generate stochastic realisations of the predicted RAD for a set group of species. This is achieved by first calculating the fundamental biodiversity number θ, which is estimated from the number of species present in the community S using statistical inference based on Ewens' sampling formula [30, 31]. Theta is a composite parameter discussed extensively by Hubbell [7], and is related to the total number of individuals present in the community (J_M) and the speciation rate (v) by $\theta = J_M \frac{v}{1-v}$ [8]. When θ is combined with J_M, it allows stochastic generation of predicted RADs expected for the community under neutral theory (in the absence of dispersal limitation), using the algorithm specified by Hubbell ([7], p289). Predicted RADs are calculated without direct reference to the relative abundances observed in the community under study, therefore comparison between these predictions and the observed RADs provide a means to test whether the observed abundances deviate from the distribution expected under NTB [8].

For the purposes of this analysis, different molecular types of M. bovis were treated as different species, and the number of herd breakdowns of each molecular type in NI was taken to represent the number of individuals per species. Molecular types were defined based on either spoligotyping alone (spoligotypes), or combined spoligotyping and MVLA-typing (MVLA-types). Using the untb package, the fundamental biodiversity number θ was first estimated from the observed RAD, and taking this value of θ and the total number of typed herd breakdowns (n = 10,049), 1000 stochastic realisations of the RADs predicted under NTB were generated. The 95th percentile interval of these distributions was then plotted to compare the observed RAD with NTB predictions for the community.

The above steps were carried out for the RADs of: i) all recorded M. bovis spoligotypes across the entire study period (2003–2010); ii) all recorded MVLA-types across the entire study period; and iii) MVLA-types subdivided by year of observation (the latter to identify whether there were differences between RADs over the course of the study period and to what extent this subdivision affected the fit to neutral predictions).

Alternative hypotheses - basic simulation model structure

To test alternative hypotheses regarding the processes involved in shaping the distribution of MVLA-types in NI, four stochastic simulation models were constructed. Specific details of each model are given in the sections below, but we describe the basic model structure here as it is comparable across all subsequent models. Within the model, a single 'individual' represents a single (MVLA-typed) herd breakdown of bTB, and species

equates to MVLA-type (defined by the combination of spoligotype and MVLA-type). A speciation event represents a mutation in MVLA-type, and 'birth' and 'death' events are the start and end of herd breakdowns respectively, with each new breakdown infected from a "parent" individual. The population size in the model is equivalent to the number of (MVLA-typed) bTB breakdowns on herds at any one time, and the birth rate is equivalent to the herd-level incidence rate.

The total population size at time t is given by: $J_{M,t} = \sum_{i=1}^{S_t} n_{i,t}$, where S_t is the total number of species present at time t, $n_{i,t}$ is the number of individuals in the ith species at time t, and $i = 1...S_t$. Simulations were started with individuals in the population at one of two extremes: either all individuals started as the same species, with $S_0 = 1$ and $n_{i,0} = J_{M,0}$; or every individual started as a different species, with $S_0 = J_{M,0}$ and $n_{i,0} = 1$. Comparing results from both starting conditions allowed us to check the model for convergence.

An overview of the steps involved in one timestep of the basic simulation model is given in Fig. 1. First, a vector, n_t^d, representing the number of individuals in each species that "die" in the current timestep, is identified (box 1, Fig. 1) and removed from the population (box 2, Fig. 1). The number of individuals to be removed is given by the death rate d multiplied by the total population size $J_{M,t}$ (box 1). A vector representing the number of new births in each species at this timestep, n_t^b, is then identified, with the total number of individuals to be born given by the birth rate b multiplied by total population size $J_{M,t}$. The number of offspring per species in the basic model is proportional to the frequency of that species in the population (box 3). Each of the newly "born" individuals then has the option to convert to a new species with fixed speciation probability v (boxes 4 and 5). If this occurs, a new species containing one individual is added to the population (box 6); otherwise the original species is updated to add another individual (box 7). The process is then repeated with the updated population vectors as input. To ensure convergence, for each model the simulations were run until the two starting population conditions had converged on the same species abundance distribution (allowing for stochastic variation).

The simulation model described above generates the same outcome as the algorithm implemented in the untb package [7, 30] (see previous section). It additionally provided a framework that can be altered to reproduce the mechanisms outlined in Hypotheses 1–4 (MVLA-type speciation/selection, unsampled reservoir population, increasing prevalence, and superspreading), and to assess to what extent they are able to replicate the manner in which the observed distribution of NI MVLA-

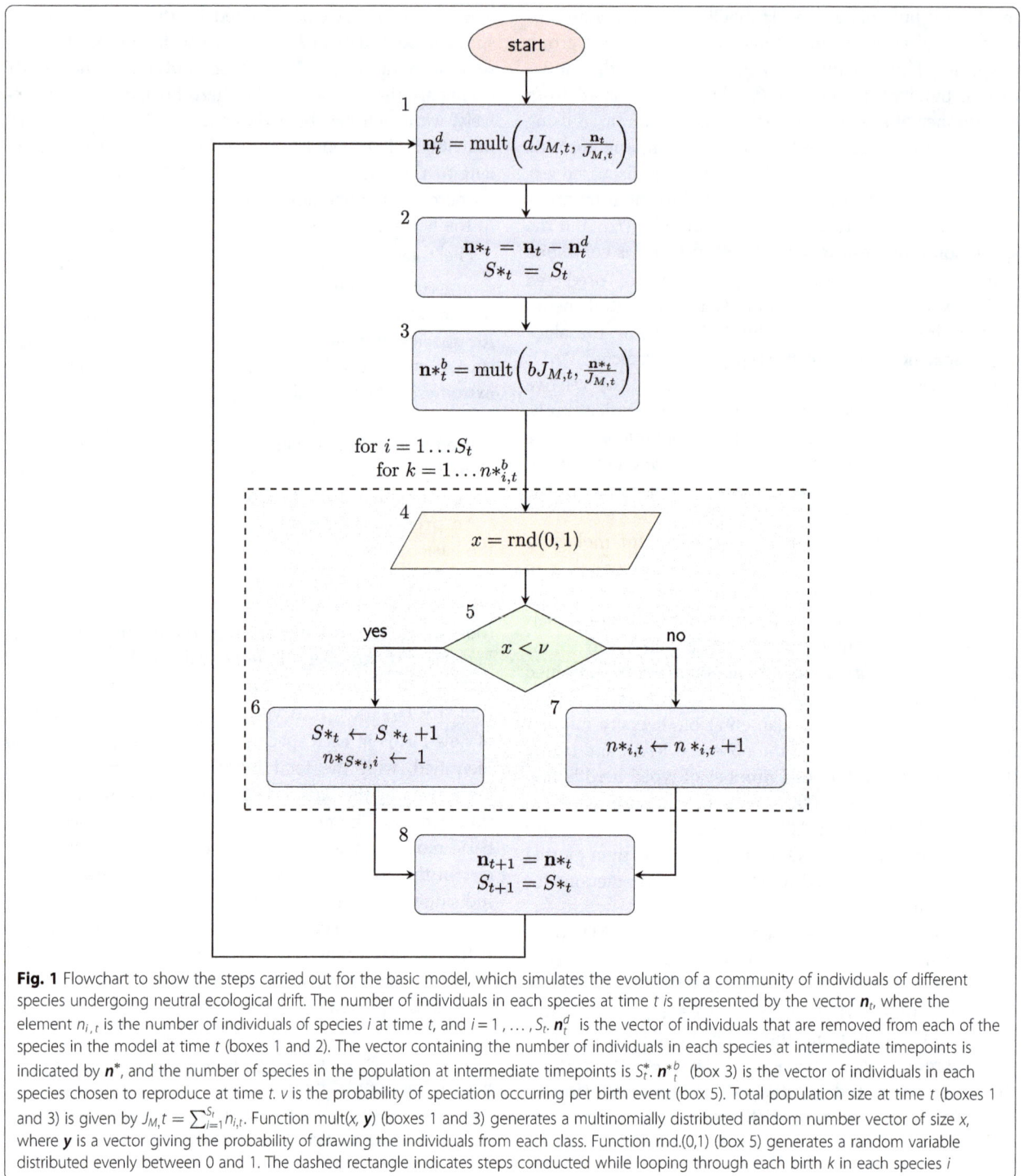

Fig. 1 Flowchart to show the steps carried out for the basic model, which simulates the evolution of a community of individuals of different species undergoing neutral ecological drift. The number of individuals in each species at time t is represented by the vector $\boldsymbol{n}_t$, where the element $n_{i,t}$ is the number of individuals of species i at time t, and $i = 1, \ldots, S_t$. $\boldsymbol{n}_t^d$ is the vector of individuals that are removed from each of the species in the model at time t (boxes 1 and 2). The vector containing the number of individuals in each species at intermediate timepoints is indicated by $\boldsymbol{n}^*$, and the number of species in the population at intermediate timepoints is S_t^*. $\boldsymbol{n}_t^{*b}$ (box 3) is the vector of individuals in each species chosen to reproduce at time t. v is the probability of speciation occurring per birth event (box 5). Total population size at time t (boxes 1 and 3) is given by $J_{M},t = \sum_{i=1}^{S_t} n_{i,t}$. Function mult($x$, $\boldsymbol{y}$) (boxes 1 and 3) generates a multinomially distributed random number vector of size x, where $\boldsymbol{y}$ is a vector giving the probability of drawing the individuals from each class. Function rnd.(0,1) (box 5) generates a random variable distributed evenly between 0 and 1. The dashed rectangle indicates steps conducted while looping through each birth k in each species i

types diverged from NTB predictions. The individual models themselves are described in the following section.

On completion of a simulation, the value of the biodiversity number θ was estimated for each model run using the untb package, and in order to summarise the runs for each model and to assess their divergence from NTB these values of θ were then averaged across all runs

for each specific model. Predictions of the RAD expected under NTB were generated in the untb package as above using this average value of θ, and the 95th percentile interval of the NTB predictions was then plotted and compared to the 95th percentile interval of the simulation results.

The parameters for the baseline model were chosen to approximate the situation in NI, with each timestep approximately representing 1 week. Using data on the numbers of typed breakdowns over the study timeframe the number of deaths, $dJ_{M,t}$, and the number of births, $bJ_{M,t}$, per timestep were both set to 24 individuals (equivalent to the mean herd-level incidence of MVLA-typed breakdowns per week; Additional file 3 grey line); the population size $J_{M,t}$ was set to 890 individuals (equivalent to the mean herd-level breakdown prevalence across NI; Additional file 3 black line); and the speciation probability v as set as 0.015 (equivalent to the mean proportion of new MVLA-typed breakdowns attributable to novel MVLA-types; Additional file 4). Although the values of d, b, $J_{M,t}$ and v were not formally fitted to the data, these parameters form an intrinsic part of NTB. Varying these should not alter the qualitative agreement between the simulation results and the NTB predictions, except where b is not equal to d: this latter scenario is explored in the Hypothesis 3 (increasing population size) simulations described below.

The model for Hypothesis 1 (MVLA-type speciation and/or choice of loci) assumes a stepwise mutation process, but none of the other models described below discriminate between new species occurring through mutation or through immigration from outside the simulated community. For all models with the exception of Hypothesis 4 (superspreading), the number of new births in each species is directly proportional to the frequency of the species in the population, after accounting for stochastic variation (box 4 Fig. 1).

Stochastic simulations representing each of the four hypotheses were run 500 times (250 runs starting with all individuals of the same species, and 250 runs starting with all individuals as different species), and the results were summarised to give output equivalent to the incidence of (MVLA-typed) breakdowns over the course of a year (timesteps corresponding to 52 weeks).

Hypothesis 1 - MVLA-type speciation and/or choice of MVLA-loci

Mutation in MVLA-type occurs through variation in the number of repeats present at the MVLA loci. This process can give rise to homoplasies, with unrelated lineages converging on the same MVLA-type [23], and this could act to augment the more common MVLA-types. To test whether homoplasy in MVLA-types and/or the method by which MVLA-typing loci were chosen in NI would affect the RAD of different MVLA-types, the basic model above was altered so that species was denoted by a string of 12 integers. The 12 integers represent the number of tandem repeats present at the 12 candidate MVLA-loci from which the current NI typing loci were originally selected [14]. In simulations started with all individuals in the same species, all 12 "loci" were set to 5 for all individuals, while in simulations started

with all individuals in different species, each of the 12 loci for each individual was chosen as a randomly sampled integer between 1 and 10. When a speciation event occurred in the model (box 6 Fig. 1), this generated a stepwise mutation in MVLA-type, with one "locus" chosen at random from the string of 12, and the number of repeats at this locus then increased or decreased by one. Note that in this simulation we do not address the scenario that different MVLA-loci may have differential rates of gain and/or loss of repeats [32].

Since we assume that each locus has an equivalent chance of gaining/losing a repeat and so generating a new species, having more MVLA loci will proportionally increase the overall rate of speciation. Therefore, as we consider 12 loci here (as opposed to the seven loci that determined the rate of occurrence of new species in Additional file 4 and the basic model), for these simulations, the speciation rate v was proportionally increased to 0.0257 per birth event. All other parameters were as described above.

On completion of each simulation, to mimic the manner in which the MVLA-typing loci were chosen in NI [14], the seven most diverse "loci" from each model run were chosen using the Hunter-Gaston index, a measure of the probability that two individuals sampled at random are of the same type when assuming sampling without replacement [33]. This is a commonly used measure of diversity in microbiology, and was used in the selection of MVLA-typing loci in NI [14]. Following identification of the most diverse loci, the species designation of each individual in the model output was then re-named based solely on these seven loci and comparison to NTB predictions was carried out as above.

Hypothesis 2 - Unsampled reservoir of infection

Hypothesis 2 posited that an unsampled reservoir of M. bovis infection might account for the observed RAD of MVLA-types in NI. To test this, we extended the baseline model to consider two separate populations of M. bovis, linked through immigration of individuals between them (Fig. 2, boxes b and c).

We consider this hypothesis in terms of an unsampled badger reservoir for illustrative purposes, however the same processes could also easily be represented by another type of unsampled population. The MVLA-type abundances reported here were sampled solely from the cattle population. Assuming bi-directional transmission between cattle and badgers (Additional file 5), in infected badgers, novel MVLA-types may emerge through speciation from types circulating in badgers or through import of strains from the cattle population, whereas in cattle novel MVLA-types could occur through speciation from circulating cattle strains, import from the badger population, and also additionally through import of

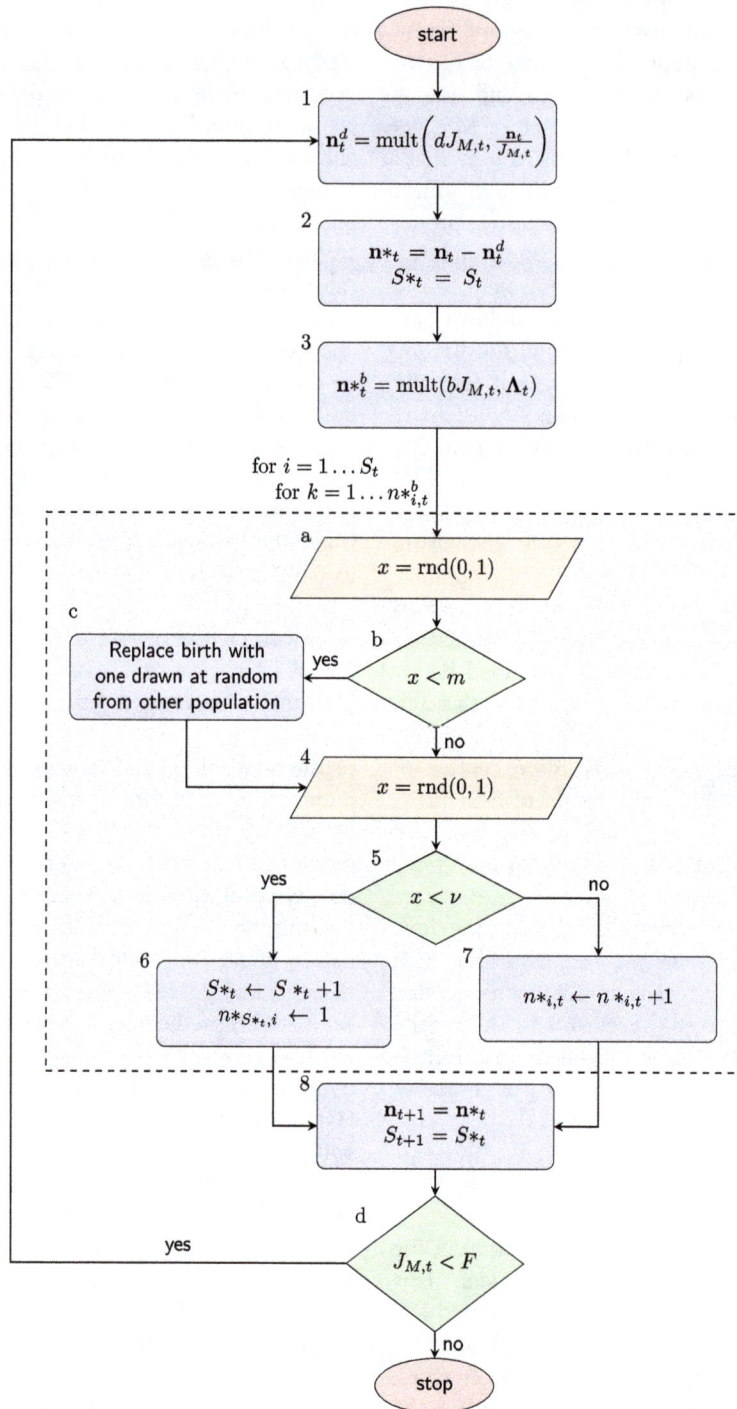

Fig. 2 Flowchart based on Fig. 1 (basic model), showing the modifications made to simulated Hypotheses 2–4. Boxes 1–8 are similar to Fig. 2. For Hypothesis 2 (unsampled reservoir), parameter m (box b) represents the likelihood of an individual migrating into the system from the separately modelled population (box c). For hypothesis 3 (increasing population size), parameter F (box d) represents the final target population size for the model. For Hypothesis 4 (superspreading), Λ_t (box 4) gives the vector of the probability of reproduction for each species, which is the sum of the reproduction probabilities λ_j for each individual j in each species l, divided by the sum of reproduction probabilities for all individuals in the system

MVLA-types circulating in cattle in other countries. Therefore the rate of occurrence of new MVLA-types may differ between cattle and badgers, and this difference in the occurrence of novel MVLA-types could affect the RAD observed.

In this model we make no distinction between novel species occurring as a result of speciation or immigration, both being combined within the speciation rate parameter v. To mimic the effects of differing rates of the occurrence of novel MVLA-types between simulated badger and cattle infections, different values of speciation rates v were investigated for the target population (representing infections in cattle) and the linked population (representing infection in badgers), and values of migration m between the two populations were also varied. Migration rates were tested for values of 0.0208 (representing approximately one migration per timestep) and 0.0833 (approximately two migrations per timestep). Increasing the rate of migration between the populations brings the simulation closer to an homogeneously mixing population, and therefore would be expected to result in RADs more similar to that predicted under NTB. Table 2 shows the combinations of these parameters examined here. Except for these parameters the two simulated populations were identical.

Hypothesis 3 – Recent expansion of the *M. bovis* population

To investigate the effects of historical bTB increases on the RAD of MVLA-types, the basic model was altered to approximate the increase in bTB prevalence that has affected NI in the decades prior to 2002 [26].

Based on data from [12], peak bTB levels in NI were estimated at approximately 3200 herd breakdowns in 2002. To keep the results of this model proportional to our other simulations we took 57.5% of this value (the proportion of NI herd breakdowns linked to at least one molecular typing record in the 2003–2010 data), giving 1840 breakdowns as the final (maximum) population size for the increasing population size model. Data

Table 2 Combinations of parameter values used in simulations for the Hypothesis 2 model (unobserved reservoir of infection)

	v_t	v_l	$m_{l,t}$	$m_{t,l}$
A	0.01	0.01	0.0208	0.0208
B	0.02	0.01	0.0208	0.0208
C	0.1	0.01	0.0208	0.0208
D	0.01	0.01	0.0208	0.0833
E	0.02	0.01	0.0208	0.0833
F	0.02	0.01	0.0833	0.0208

Speciation rates for the target population (v_t) and for the linked population (v_l) were varied as were migration rates from the target population to the linked population ($m_{l,t}$) and from the linked population to target population ($m_{t,l}$). Six combinations of these values were tested, labelled A-F

available from http://www.bovinetb.info/docs/number-of-bovine-tb-reactors-slaughtered-in-northern-ireland-from-1961-to-nov-2011.pdf (originally made available by the Department of Agriculture and Rural Development NI) provide historic summaries of numbers of cattle testing positive to bTB (Additional file 6). These figures do not include bTB infections identified at slaughter, however we took them to broadly represent historical trends, which would indicate that the increase in bTB in NI started in 1986, at which point bTB levels were approximately 9.3% of their maximum.

The basic constant-size simulation model described above was first run at the starting population size $J_{M,0} = 171$ for 100,000 timesteps to allow equilibration of the simulated RAD. A model simulating increasing population size was then run starting from this population, until the target end population size (1840, box d F, Fig. 2) was reached. To generate an exponentially increasing population size approximately equivalent to the historical situation in NI, the model was run with death rate $d = 0.02697$ (box 1 Fig. 2) and birth rate $b = 0.02883$ (box 3, Fig. 2). Varying the end population size (F) would alter the rate of population increase over the simulation: a higher value of F would lead to a higher rate of population increase, which would be expected to exacerbate any divergences between the simulation results and those predicted under NTB prediction.

Hypothesis 4 - Superspreading

To assess the effect of superspreading on the RAD of NI MVLA-types, the basic model was extended to incorporate transmission heterogeneity through the introduction of systematic variation in the number of offspring per individual breakdown. Superspreading in this context is taken to represent variability in the number of herds infected by each infected herd, however we do not distinguish between the various possible causes of this herd-level transmission heterogeneity.

To simulate this transmission heterogeneity, at the point at which new births were identified in the model (box 3 Fig. 2), each new birth was also assigned an individual reproduction number, λ, which was then retained throughout the lifespan of that individual. This individual reproduction number was drawn from a gamma distribution with a mean of 1 (equal to the effective reproduction rate in a population with a constant prevalence) and with shape parameter k, after [34]. Simulations were run for k values of 0.1, 0.5, 1, and 10, to assess the effect of different degrees n transmission heterogeneity on model outputs. As before, a set number of individuals died ($dN_t = 24$) and were born ($bN_t = 24$) at each timestep, however, in this model the chance of an individual being chosen to reproduce at each timestep was directly proportional to its pre-assigned reproduction probability λ, and therefore $\Lambda_{i,t}$, the number of individuals of species i reproducing at each timestep t, is determined by the sum of the reproduction numbers λ for each individual j in species i divided by the sum of the reproduction probabilities for all individuals in the population:

$$\Lambda_{i,t} = \frac{\sum_{j=1}^{N_{i,t}} \lambda_{i,j}}{\sum_{i=1}^{n_t} \sum_{j=1}^{N_{i,t}} \lambda_{i,j}}$$

The number of offspring generated by a single individual over the course of its lifespan is related to the reproduction number of the individual as well as the lifespan of the individual (as death rate d is constant, the individuals' lifespan will follow a geometric distribution).

Results and discussion

Null hypothesis - neutral ecological drift

The null model of neutral ecological drift described by NTB [7, 8] was unable to explain the observed relative abundances of *M. bovis* MVLA-types recorded in NI, whether considering the distribution across the whole study period (Fig. 3a), or when divided into separate years (Fig. 3b). Although previous work has shown that individual MVLA-types in NI have expanded or contracted significantly over time [13], we found no obvious differences in the MVLA-type RADs observed over different years of the study period (Fig. 3b). In the observed MVLA-type distributions, the common types appear more dominant than would be expected under NTB, with a tail of MVLA-types that are more rare than expected. The RAD of *M. bovis* spoligotypes in NI does not fall outside the 95th percentile interval of the NTB predictions (Fig. 3c), however, with lower numbers of spoligotypes than MVLA-types the predicted neutral distributions show greater variation and therefore the power to detect deviations from neutral expectations is reduced for the spoligotype RAD.

Hypothesis 1 - MVLA-type speciation and/or selection of MVLA-loci

The process through which MVLA-types are identified and new types arise is based on variation in the number of repeats present at different MVLA loci. This gives rise to the possibility of unrelated lineages converging on the same MVLA-type [23], which could act to augment the more common MVLA-types. Additionally, the MVLA-loci used for MVLA-typing in NI were originally chosen from a panel of 12 candidate loci to optimally discriminate the *M. bovis* population in NI [14]. It was unclear whether either or both of these processes would affect the RAD of MVLA-types and their fit to neutral predictions. The results of the simulations replicating these processes show minimal differences from the distribution predicted under NTB (Fig. 4), and therefore it

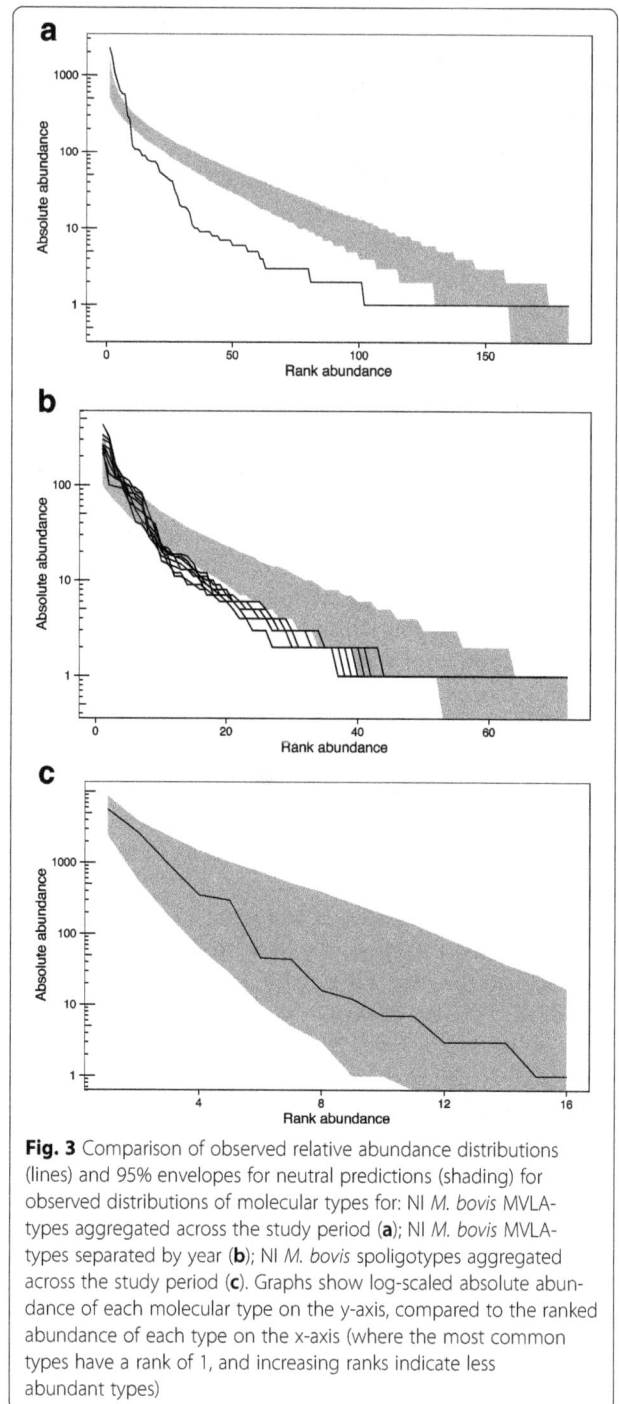

Fig. 3 Comparison of observed relative abundance distributions (lines) and 95% envelopes for neutral predictions (shading) for observed distributions of molecular types for: NI *M. bovis* MVLA-types aggregated across the study period (**a**); NI *M. bovis* MVLA-types separated by year (**b**); NI *M. bovis* spoligotypes aggregated across the study period (**c**). Graphs show log-scaled absolute abundance of each molecular type on the y-axis, compared to the ranked abundance of each type on the x-axis (where the most common types have a rank of 1, and increasing ranks indicate less abundant types)

appears that the process of MVLA-type speciation and the manner in which the panel of MVLA-typing loci were chosen in NI cannot account for the shape of the observed RAD seen in NI MVLA-types.

Hypothesis 2 - Unsampled reservoir of infection

All molecular typing results described in this study originated from cattle infections in NI, where only limited information is available on *M. bovis* types infecting hosts

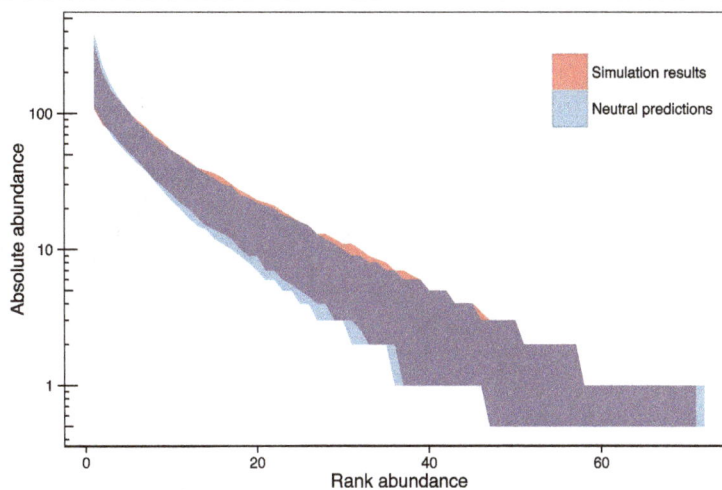

Fig. 4 Comparison between Hypothesis 1 simulations (MVLA-type speciation and choice) and neutral predictions. The 95% envelope for the relative abundance distributions generated by 500 simulations is shown in red, and the 95% envelope for predictions under neutral theory in blue. Log-scaled absolute abundance is shown on the y-axis and ranked abundance on the x-axis (species ranked in order of decreasing abundance)

other than cattle, such as badgers. The results of the model simulating an unsampled reservoir suggested that this scenario in itself could not account for the observed distribution of *M. bovis* MVLA-types in NI, a result which held true even when the rate of occurrence of novel MVLA-types was allowed to differ between the two host populations (Fig. 5). In fact, this simple metacommunity structure actually generated RADs that were slightly more even than that predicted by NTB, for all combinations of migration and speciation probabilities explored here.

Similar findings have been reported for spatial structuring of a community via spatially limited dispersal of individuals [4], which likewise generates more even distributions of species across the metacommunity than predicted by NTB in the absence of dispersal limitation. We therefore suggest that a reservoir of infection or other forms of metacommunity structure, including the spatial structuring of the *M. bovis* population, would (in the absence of other factors) be expected to draw the RAD in the opposite direction to that seen in the NI MVLA-type data, acting to even out the skewed distribution observed.

The absence of a more even distribution of MVLA-types in the observed data does not rule out metapopulation structure (either as a reservoir of infection in a different species, and/or through other forms of population structuring) in the epidemiology of bTB in NI. In fact, many lines of evidence point towards a reservoir of bTB infection in the badger population [35], and spatial structuring of the *M. bovis* population is evident in the geographical clustering of molecular types recorded in NI [13]. Rather, the observed pattern of the MVLA-type RADs indicates that, if some form of population structure is involved in the spread of *M. bovis*, other

processes must also be acting to overcome its levelling influence and push the RADs towards the skewed distributions present in these data (Fig. 3).

Hypothesis 3 - recent expansion of the *M. bovis* population

The available data on numbers of cattle testing positive for bTB in NI suggest the prevalence of bTB increased from the late 1980s to 2002 (Additional file 6, also [26]). One of the key assumptions made by NTB is a constant size population [7], and Luciani et al. demonstrated that violating this assumption through an increasing population size (i.e. increasing prevalence) could generate the patterns of relative abundances recorded in an outbreak of human TB in California [9].

The results of the Hypothesis 3 simulations (Fig. 6) agree with this, demonstrating that an increase in population size, approximating historical increases in bTB prevalence in NI, is capable of generating a RAD that deviates from NTB predictions in a similar manner to that of observed distribution of NI MVLA-types. However, since 2002 levels of bTB have been declining in NI (Additional file 6, [26]). Model simulations mimicking a declining population (without the prior increase) indicated that a decreasing prevalence would have the opposite effect, generating more even RADs than expected under NTB (results not shown). Thus the more recent decrease in bTB prevalence is likely to even out to some extent the skew in RAD generated by earlier increases.

Hypothesis 4 – Superspreading

Variation in the number of secondary infections generated by infected individuals, or superspreading, is manifest in the epidemiology of a range of diseases [34, 36].

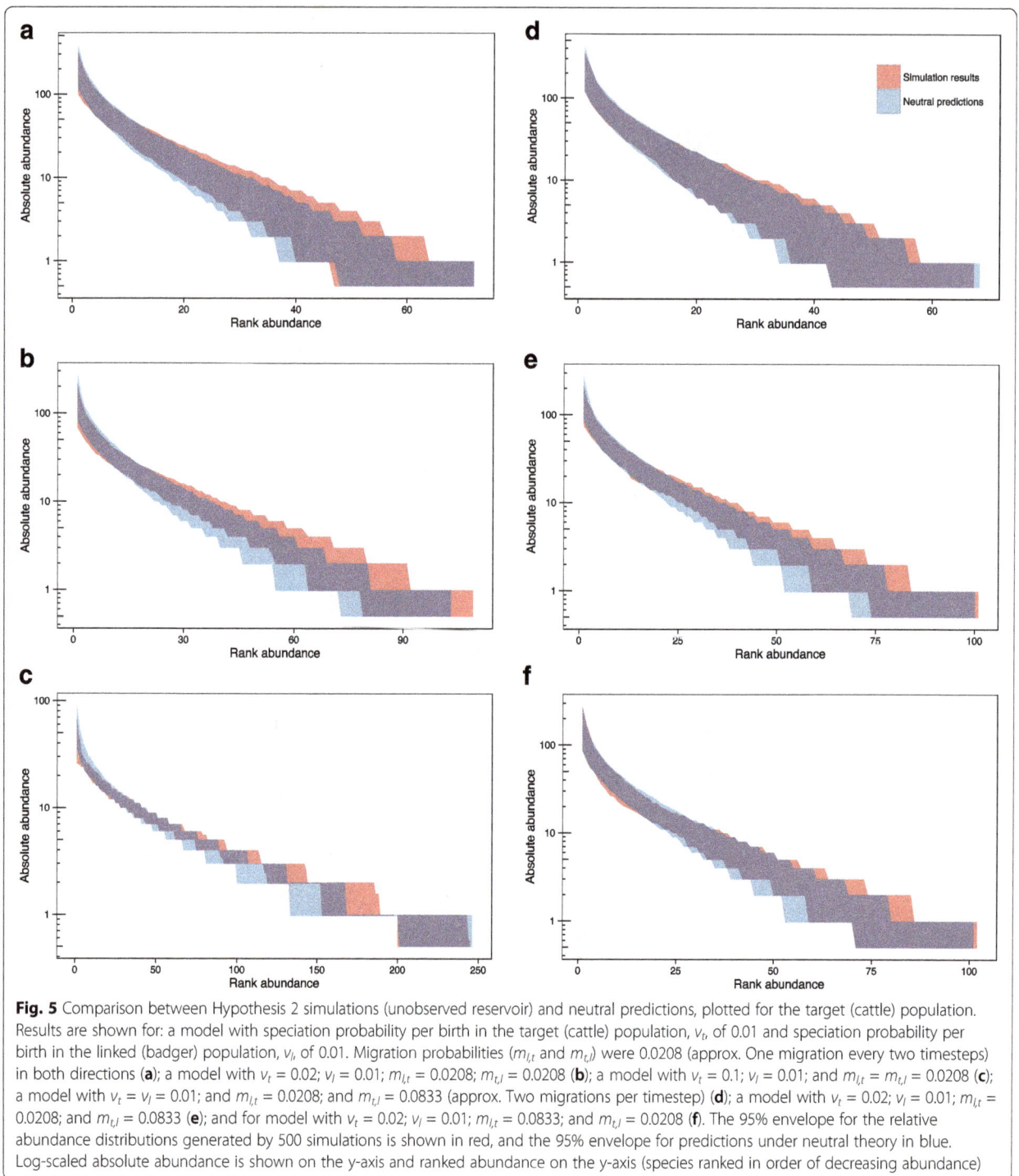

Fig. 5 Comparison between Hypothesis 2 simulations (unobserved reservoir) and neutral predictions, plotted for the target (cattle) population. Results are shown for: a model with speciation probability per birth in the target (cattle) population, v_t, of 0.01 and speciation probability per birth in the linked (badger) population, v_l, of 0.01. Migration probabilities ($m_{l,t}$ and $m_{t,l}$) were 0.0208 (approx. One migration every two timesteps) in both directions (**a**); a model with $v_t = 0.02$; $v_l = 0.01$; $m_{l,t} = 0.0208$; $m_{t,l} = 0.0208$ (**b**); a model with $v_t = 0.1$; $v_l = 0.01$; and $m_{l,t} = m_{t,l} = 0.0208$ (**c**); a model with $v_t = v_l = 0.01$; and $m_{l,t} = 0.0208$; and $m_{t,l} = 0.0833$ (approx. Two migrations per timestep) (**d**); a model with $v_t = 0.02$; $v_l = 0.01$; $m_{l,t} = 0.0208$; and $m_{t,l} = 0.0833$ (**e**); and for model with $v_t = 0.02$; $v_l = 0.01$; $m_{l,t} = 0.0833$; and $m_{t,l} = 0.0208$ (**f**). The 95% envelope for the relative abundance distributions generated by 500 simulations is shown in red, and the 95% envelope for predictions under neutral theory in blue. Log-scaled absolute abundance is shown on the y-axis and ranked abundance on the y-axis (species ranked in order of decreasing abundance)

Superspreading has also been implicated in the spread of bTB in Britain, with a minority of infected cattle herds thought to be responsible for a large proportion of the onward transmission of the disease [27]. Ypma et al. also demonstrate that superspreading could account for skewed abundances of molecular types of human TB in the Netherlands [10].

The results of superspreading simulations where individuals' chances of reproduction were distributed with mean of 1 and k values of 1, 0.5, and 0.1 are shown in Fig. 7a–c, respectively, with the distribution of the number of offspring per individual over the course of the simulations given in inset graphs. These results indicate that superspreading at these levels could also generate

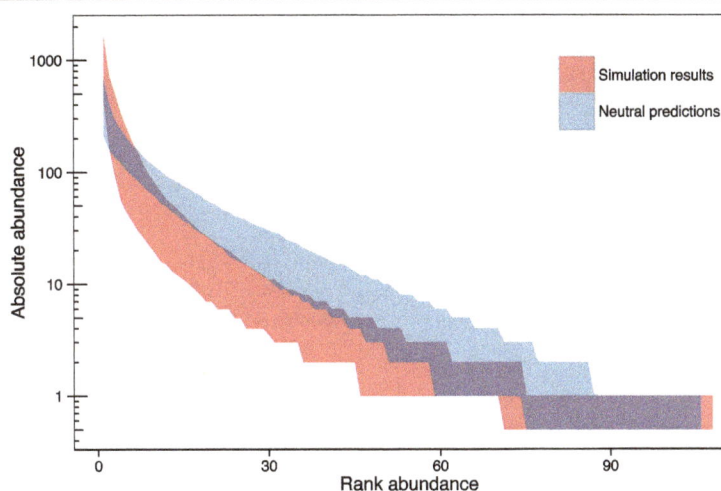

Fig. 6 Comparison between Hypothesis 3 simulations (increasing population size) and neutral predictions. The 95% envelope for the relative abundance distributions generated by 500 simulations is shown in red, and the 95% envelope for predictions under neutral theory in blue. Log-scaled absolute abundance is shown on the y-axis and ranked abundance on the x-axis (species ranked in order of decreasing abundance)

divergences from NTB similar to that seen in the RAD of MVLA-types in NI (Fig. 3a and b).

Conclusions

In this study we show that a simple model of neutral ecological drift cannot account for the observed distribution of NI MVLA-types, and that the observed data showed a more skewed distribution than expected under NTB. Simulation models were then used to identify which, if any, of four specific biologically-plausible hypotheses could reproduce the manner in which the observed RAD diverged from NTB predictions. We note that the hypotheses examined in this study are not mutually exclusive, and were specifically chosen due to the likelihood that they are all involved in the epidemiology and evolution of *M. bovis* in NI.

We conclude that MVLA-type homoplasy and the manner in which MVLA-typing loci were selected in NI, taken in isolation, have little impact on the expected abundances of different MVLA-types. By contrast, structuring of the community, for example due to a reservoir population or alternatively through processes such as spatially limited dispersal of individuals [7], is likely to generate a more even RAD than expected under NTB. The finding of a very uneven distribution of MVLA-types despite the probable involvement of the latter scenarios in bTB epidemiology in NI indicates that processes other than these must be involved and responsible for generating the skewed RAD recorded in these data. The results presented here for Hypotheses 3 and 4 indicate that some combination of historical increases in the prevalence of bTB in NI and/or variation in the number of onward transmission events generated by each infected herd (i.e. superspreading) may have

contributed to the unevenness evident in the distribution of *M. bovis* MVLA-types in NI. Studies of the RAD of human TB also concluded that an increasing population size [9] and superspreading [10] may be responsible for the skewed distribution of molecular types observed in some human TB datasets.

None of the various hypotheses tested here break the basic tenet of neutral drift, as in each simulation different molecular types were considered equal and interchangeable, although they did deviate from the original assumptions of Hubbell's NTB in other ways. Another potential mechanism through which the observed skew in distribution of MVLA-types could be generated is through selective pressures acting on different molecular types, with more fit types present at higher than expected abundance and less fit types rarer than expected. This has been explored extensively in the ecological literature as the main alternative hypothesis to neutral theory [8], and was also put forward by Smith et al. [15] as a possible explanation for the distribution of molecular types of *M. bovis* in GB. Although not modelled here, this remains an additional explanation for the observed pattern present in the MVLA-type data. However, in a low diversity population such as that of *M. bovis* in NI, selective pressures are not expected to play a strong role in driving the patterns of diversity observed, and our results demonstrate that other processes as outlined above can generate similar skewed distributions without needing to invoke selective pressures.

Data on the abundances of different molecular types are relatively easy to obtain for many pathogens of veterinary and human importance. We demonstrate how the rich body of work in ecology relating to relative abundances of different species in an ecosystem can be used

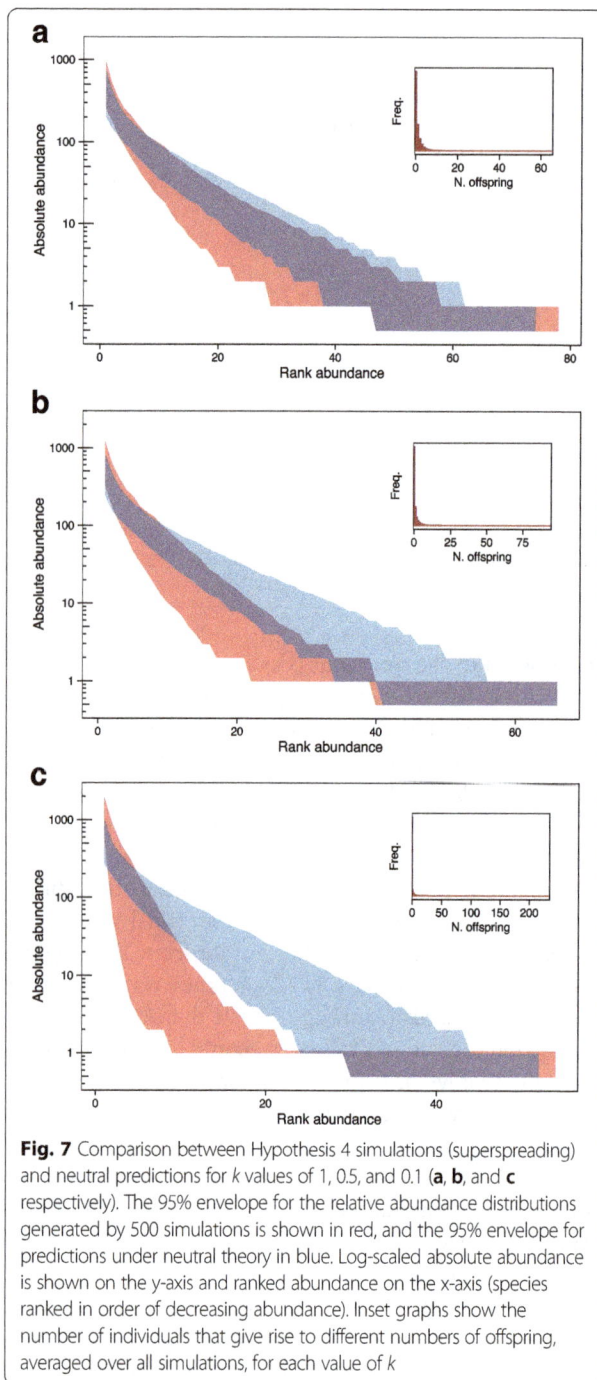

Fig. 7 Comparison between Hypothesis 4 simulations (superspreading) and neutral predictions for *k* values of 1, 0.5, and 0.1 (**a**, **b**, and **c** respectively). The 95% envelope for the relative abundance distributions generated by 500 simulations is shown in red, and the 95% envelope for predictions under neutral theory in blue. Log-scaled absolute abundance is shown on the y-axis and ranked abundance on the x-axis (species ranked in order of decreasing abundance). Inset graphs show the number of individuals that give rise to different numbers of offspring, averaged over all simulations, for each value of *k*

to test hypotheses regarding the driving forces underlying the distribution of relative abundances of different pathogen molecular types. Although RADs in isolation do not definitively disentangle the relative roles of the processes under consideration, they do enable the identification of candidate mechanisms likely to be involved in driving the observed patterns, as we show here. Combining these approaches with additional data on, for example, spatial and genetic relationships between samples may allow more detailed discrimination between candidate processes, however

other methodologies may be more appropriate to fully understand the relative roles of the different processes involved in the system.

Additional files

Additional file 1: Table of number of breakdowns recorded for each NI spoligotype, 2003–2010

Additional file 2: Table of number of breakdowns recorded for each NI MVLA-type, 2003–2010

Additional file 3: Graph showing prevalence (black) and monthly incidence (grey) of MVLA-typed herd breakdowns in NI over the study period.

Additional file 4: Graph showing number of breakdowns attributed to novel MVLA-types per month in NI over the study period

Additional file 5: Summary of the linked cattle-badger population and the source of new MVLA-types in each.

Additional file 6: Graph of historical incidence of cattle testing positive for bTB in NI.

Acknowledgements

We thank the Department of Agriculture and Rural Development (DARD) for access to Animal and Public Health Information System (APHIS) data; the support provided by DARD Veterinary Service, staff at WD Meats and VSD-AFBI specialist histology/pathology and bacteriology staff involved in bTB case confirmation.

Funding

HT was funded by a BBSRC DTG studentship and Novartis Animal Health Case studentship. Molecular typing surveillance of confirmed bTB cases in NI was supported initially by DARD R&D Project DARD0407 and subsequently under the Assigned Work Programme.

Authors' contributions

RB, DH, RRK, and HT designed the study; RS, CM, TRM, and ELP generated the data; HT and DW analysed the data; RB, DH, RRK, and HT wrote the manuscript; and all authors read and approved the final manuscript.

Consent for publication

Not applicable.

Competing interests

The authors declare they have no competing interests.

Author details

[1]Boyd Orr Centre for Population and Ecosystem Health, Institute of Biodiversity and Animal Health, University of Glasgow, Glasgow, UK. [2]School of Medicine, Dentistry, and Biomedical Sciences, Queen's University Belfast, Belfast, UK. [3]Veterinary Sciences Division, Agri-Food and Biosciences Institute, Stormont, Belfast, UK. [4]School of Biological Sciences, Queen's University Belfast, Belfast, UK.

References

1.   Donnelly CA, Nouvellet P. The contribution of badgers to confirmed tuberculosis in cattle in high-incidence areas in England. PLoS Curr Outbreaks. 2013;1:1–15.

2.   Donnelly CA, Woodroffe R, Cox DR, Bourne FJ, Cheeseman CL, Clifton-Hadley RS, Wei G, Gettinby G, Gilks P, Jenkins H, Johnston WT, le Fevre AM,

McInerney JP, Morrison WI. Positive and negative effects of widespread badger culling on tuberculosis in cattle. Nature. 2006;439:843–6.

3. Griffin JM, Williams DH, Kelly GE, Clegg TA, O'Boyle I, Collins JD, More SJ. The impact of badger removal on the control of tuberculosis in cattle herds in Ireland. Prev Vet Med. 2005;67:237–66.

4. McGill BJ, Etienne RS, Gray JS, Alonso D, Anderson MJ, Benecha HK, Dornelas M, Enquist BJ, Green JL, He F, Hurlbert AH, Magurran AE, Marquet PA, Maurer BA, Ostling A, Soykan CU, Ugland KI, White EP. Species abundance distributions: moving beyond single prediction theories to integration within an ecological framework. Ecol Lett. 2007;10:995–1015.

5. Magurran AE, Magurran AE. Species abundance distributions: pattern or process? Funct Ecol. 2005;19:177–81.

6. Rosindell J, Hubbell SP, He F, Harmon LJ, Etienne RS. The case for ecological neutral theory. Trends Ecol Evol. 2012;27:203–8.

7. Hubbell SP. The unified neutral theory of biodiversity and biogeography. New Jersey: Princeton University Press; 2001.

8. Rosindell J, Hubbell SP, Etienne RS. The unified neutral theory of biodiversity and biogeography at age ten. Trends Ecol Evol. 2011;26:340–8.

9. Luciani F, Francis AR, Tanaka MM, Faso B, Republic C. Interpreting genotype cluster sizes of mycobacterium tuberculosis isolates typed with IS6110 and spoligotyping. Infect Genet Evol. 2008;8:182–90.

10. Ypma RF, Altes HK, van Soolingen D, Wallinga J, van Ballegooijen WM. A sign of superspreading in tuberculosis: highly skewed distribution of genotypic cluster sizes. Epidemiology. 2013;24:395–400.

11. Smith NH, Dale J, Inwald J, Palmer S, Gordon SV, Hewinson RG, Smith JM. The population structure of Mycobacterium Bovis in great Britain: clonal expansion. Proc Natl Acad Sci. 2003;100:15271–5.

12. Abernethy DA, Upton P, Higgins IM, McGrath G, Goodchild AV, Rolfe SJ, Broughan JM, Downs SH, Clifton-Hadley R, Menzies FD, de la Rua-Domenech R, Blissitt MJ, Duignan A, More SJ. Bovine tuberculosis trends in the UK and the Republic of Ireland, 1995-2010. Vet Rec. 2013;172:312.

13. Skuce RA, Mallon TR, McCormick CM, McBride SH, Clarke G, Thompson A, Couzens C, Gordon AW, McDowell SWJ. Mycobacterium Bovis genotypes in Northern Ireland: herd-level surveillance (2003 to 2008). Vet Rec. 2010;167:684–9.

14. Skuce RA, McDowell SW, Mallon TR, Luke B, Breadon EL, Lagan PL, McCormick CM, McBride SH, Pollock JM. Discrimination of isolates of Mycobacterium Bovis in Northern Ireland on the basis of variable numbers of tandem repeats (VNTRs). Vet Rec. 2005;157:501–4.

15. Smith NH, Gordon SV, de la Rua-Domenech R, Clifton-Hadley RS, Hewinson RG. Bottlenecks and broomsticks: the molecular evolution of Mycobacterium Bovis. Nat Rev Microbiol. 2006;4:670–81.

16. Woodroffe R, Donnelly CA, Johnston WT, Bourne FJ, Cheeseman CL, Clifton-Hadley RS, Cox DR, Gettinby G, Hewinson RG, le Fevre AM, McInerney JP, Morrison WI. Spatial association of Mycobacterium Bovis infection in cattle and badgers Meles Meles. J Appl Ecol. 2005;42:852–62.

17. Goodchild AV, Watkins GH, Sayers AR, Jones JR, Clifton-Hadley RS. Geographical association between the genotype of bovine tuberculosis in found dead badgers and in cattle herds. Vet Rec. 2012;170:259.

18. Olea-Popelka FJ, Flynn O, Costello E, McGrath G, Collins JD, O'Keeffe J, Kelton DF, Berke O, Martin SW. Spatial relationship between Mycobacterium Bovis strains in cattle and badgers in four areas in Ireland. Prev Vet Med. 2005;71:57–70.

19. Kamerbeek J, Schouls L, Kolk A, van Agterveld M, van Soolingen D, Kuijper S, Bunschoten A, Molhuizen H, Shaw R, Goyal M, van Embden J. Simultaneous detection and strain differentiation of mycobacterium tuberculosis for diagnosis and epidemiology. J Clin Microbiol. 1997;35:907–14.

20. Supply P, Lesjean S, Savine E, Kremer K, van Soolingen D, Locht C. Automated high-throughput genotyping for study of global epidemiology of mycobacterium tuberculosis based on mycobacterial interspersed repetitive units. J Clin Microbiol. 2001;39:3563–71.

21. Frothingham R, Meeker-O'Connell WA. Genetic diversity in the mycobacterium tuberculosis complex based on variable numbers of tandem DNA repeats. Microbiology. 1998;144:1189–96.

22. Trewby H, Wright D, Breadon EL, Lycett SJ, Mallon TR, McCormick C, Johnson P, Orton RJ, Allen AR, Galbraith J, Herzyk P, Skuce RA, Biek R, Kao RR. Use of bacterial whole-genome sequencing to investigate local persistence and spread in bovine tuberculosis. Epidemics. 2016;14:26–35.

23. Reyes JF, Chan CHS, Tanaka MM. Impact of homoplasy on variable numbers of tandem repeats and spoligotypes in mycobacterium tuberculosis. Infect Genet Evol. 2012;12:811–8.

24. Wright DM, Allen AR, Mallon TR, McDowell SWJ, Bishop SC, Glass EJ, Bermingham ML, Woolliams JA, Skuce RA. Field-isolated genotypes of Mycobacterium Bovis vary in virulence and influence case pathology but do not affect outbreak size. PLoS One. 2013;8:e74503.

25. Wright DM, Allen AR, Mallon TR, McDowell SWJ, Bishop SC, Glass EJ, Bermingham ML, Woolliams JA, Skuce RA. Detectability of bovine TB using the tuberculin skin test does not vary significantly according to pathogen genotype within Northern Ireland. Infect Genet Evol. 2013;19:15–22.

26. Abernethy DA, Denny GO, Menzies FD, McGuckian P, Honhold N, Roberts AR. The Northern Ireland programme for the control and eradication of Mycobacterium Bovis. Vet Microbiol. 2006;112:231–7.

27. Brooks-Pollock E, Roberts GO, Keeling MJ. A dynamic model of bovine tuberculosis spread and control in great Britain. Nature. 2014;511:228–31.

28. R Core Team, R Develoment Core Team: R: A Language and Environment for Statistical Computing. 2014:http://www.r-project.org/.

29. Houston R. A computerised database system for bovine traceability. Rev Sci Tech l'Office Int des Epizoot. 2001;20:652–61.

30. Hankin RKS. Introducing untb, an R package for simulating ecological drift under the unified neutral theory of biodiversity. J Stat Softw. 2007;22:1–15.

31. Ewens WJ. The sampling theory of selectively neutral alleles. Theor Popul Biol. 1972;3:87–112.

32. Reyes JF, Tanaka MM. Mutation rates of spoligotypes and variable numbers of tandem repeat loci in mycobacterium tuberculosis. Infect Genet Evol. 2010;10:1046–51.

33. Hunter PR, Gaston MA. Numerical index of the discriminatory ability of typing systems: an application of Simpson's index of diversity. J Clin Microbiol. 1988;26:2465–6.

34. Lloyd-Smith JO, Schreiber SJ, Kopp PE, Getz WM. Superspreading and the effect of individual variation on disease emergence. Nature. 2005;438:355–9.

35. Godfray HCJ, Donnelly CA, Kao RR, Macdonald DW, Mcdonald RA, Petrokofsky G, Wood JLN, Woodroffe R, Young DB, McLean AR, McDonald, Robbie A, Petrokofsky G, JLN W, Woodroffe R, Young DB, AR ML, Mcdonald RA, Petrokofsky G, JLN W, Woodroffe R, Young DB, AR ML, Fann B. A restatement of the natural science evidence base relevant to the control of bovine tuberculosis in great Britain. Proc R Soc B. 2013;280:20131634.

36. Woolhouse MEJ, Dye C, Etard JF, Smith T, Charlwood JD, Garnett GP, Hagan P, Hii JLK, Ndhlovu PD, Quinnell RJ, Watts CH, Chandiwana SK, Anderson RM. Heterogeneities in the transmission of infectious agents: implications for the design of control programs. Proc Natl Acad Sci. 1997;94:338–42.

5

Significance of clinical observations and biochemical alterations in buffalo calves with dietary abomasal impaction

Maged R. El-Ashker[1]*[ID], Mohamed F. Salama[2], Mohamed E. El-Boshy[3,4] and Eman A. Abo El-Fadle[5]

Abstract

Background: The present study aimed to throw light on the clinical characteristics of abomasal impaction in buffalo calves and its associated biochemical alterations. For this reason, a total of 20 male buffalo calves *(Bubalus bubalis)* with abomasal impaction were studied. The investigated calves were at 6 to 12 months of age and were belonged to three private farms in Dakahlia Governorate besides sporadic cases admitted to the Veterinary Teaching Hospital, Faculty of Veterinary Medicine, Mansoura University, Egypt. Ten apparently healthy buffalo calves were also included as controls. According to the clinical outcome, the diseased calves were categorized into survivors ($n = 11$) and non-survivors ($n = 9$). Blood samples were collected from all animals to estimate blood gases besides a panel of selected biochemical parameters. The definitive diagnosis of dietary abomasal impaction was achieved by either left flank exploratory laparotomy or by necropsy.

Results: Both survivors and non-survivors demonstrated common clinical findings including distension of ventro-lateral aspect of the right abdomen, and varying degrees of dehydration. The great majority of survivors (81%) and 100% of non-survivors were anorexic and had rumen stasis as well as hard texture upon ballottement of the left flank. Approximately 45% of non-survivors had frothy salivation, expiratory grunting and were being tender when strong percussion was applied on the right flank. Diseased calves had metabolic alkalosis, while plasma potassium and chloride were significantly lower in non-survivors than those of survivors ($P < 0.05$). Serum malondialdehyde, superoxide dismutase and uric acid were significantly higher in diseased buffalo than controls and in non-survivors than survivors ($P < 0.05$). Serum total protein, albumin, creatinine, urea, aspartate aminotransferase, gamma-glutamyl transferase, and total bilirubin levels were also higher in non-survivors than those of survivors ($P < 0.05$).

Conclusion: Buffalo calves with dietary abomasal impaction were associated with marked clinical and biochemical alterations that could be helpful for an accurate diagnosis of the disease.

Keywords: Abomasum, Alkalosis, Buffalo, Impaction, Oxidative stress

Background

Abomasal impaction is a rarely reported clinical condition in adult ruminants that occurs due to accumulation of excess solid materials in the abomasum with subsequent enlargement of the organ [1, 2]. Distention of abomasum secondary to a luminal foreign body is more appropriately termed "luminal obstruction" rather than "abomasal impaction" because of its focal nature [3]. It is mainly observed in cows fed on low quality roughage with decreased water consumption [4, 5]. The disease may be observed as a primary disorder whenever straw or similar poor quality fibrous feed constitutes a large portion of the winter diet [6–8]. Other reports have indicated that the disease can be caused by non-food foreign substances, such as sand and gravel [9–11].

Under field conditions particularly in some developing countries, owners feed their feedlot animals a ration composed mostly of poor quality roughage mixed with some grains to reduce the feeding costs and to satisfy beef grading standards [2]. A combination of low digestibility and excessive intake of these roughages could lead

* Correspondence: maged.elashker1978@gmail.com; magid_rez@mans.edu.eg
[1]Department of Internal Medicine and Infectious Diseases, Faculty of Veterinary Medicine, Mansoura University, Mansoura 35516, Egypt
Full list of author information is available at the end of the article

to their accumulation in the forestomach and abomasum with a resulting rumeno-abomasal impaction [8]. Other factors that contribute to dietary impactions include high energy demands of growing heifers and cold weather [12]. The disease can also occur as a secondary condition in association with vagal indigestion and traumatic reticuloperitonitis [13, 14]. In the past, vagus nerve damage has been believed to be the key player in the development of abomasal impaction [15, 16]; however, mechanical fixation of the reticulum to the ventral abdominal floor in animals with traumatic reticuloperitonitis interferes with the normal sieving action of the reticulum with subsequent accumulation of fibers in the abomasum leading to its distention [14, 17, 18].

Up to now, there has been limited information about the clinical settings of abomasal impaction in water buffalo and its associated biochemical abnormalities. In line with these considerations, the present study aimed to characterize the clinical findings of abomasal impaction in buffalo calves and to explore the associated biochemical alterations. Here, we hypothesize that abomasal impaction will be associated with characteristic clinical and biochemical alterations that might have diagnostic potential.

Methods
Animals and study design
The present study was carried out on a total of 20 male buffalo calves *(Bubalus bubalis)*, belonging to three private farms in Dakahlia Governorate, Egypt ($n = 18$) as well as sporadic cases admitted to the Veterinary Teaching Hospital, Faculty of Veterinary Medicine, Mansoura University, Mansoura, Egypt ($n = 2$) during 2013 and 2014. For comparison, ten apparently healthy buffalo calves from the same population, aged between six to 12 months, were randomly selected and served as a control group.

The main presenting complaints were anorexia for several days, cud-dropping, scanty feces and moderate ventro-lateral distension of the right side of the abdomen. Medical history included feeding on poor quality and coarse roughage consisted of chopped rice straw in alternation with corn silage ad libitum. To obtain a uniform population for this study, data analyses were restricted to buffalo aged between 6 and 12 months, and those who did not receive medication prior to clinical examination.

Dietary abomasal impaction was tentatively diagnosed on the basis of nutritional history, clinical findings, and the results of laboratory investigations, while confirmation was achieved by left flank exploratory laparotomy ($n = 2$) or by necropsy ($n = 7$). The left flank laparo-rumenotomy was performed in standing position under linear infiltrating anesthesia according to the previously described method

[19]. According to the clinical outcome, the diseased calves were categorized into survivors ($n = 11$) and non-survivors ($n = 9$).

Clinical examinations
Thorough physical examinations of diseased as well as clinically healthy calves were carried out according to the standard methods described by Radostits et al. [20]. In brief, a general and close up observations were conducted to evaluate the body condition, demeanor, color of visible mucous membrane, circumference of the abdomen, eating behavior, besides the pattern of urination and defecation. Vital signs including rectal temperature, heart rate, and respiratory rate were essentially recorded. The rectal temperature was measured by using a commercially available mercury thermometer; while the heart rate and respiratory rate were measure by a respective auscultation of the heart and lung. Deep palpation, strong percussion, as well as ballottement were performed on both sides of the abdomen. Tests for reticular foreign bodies were also implemented.

Sampling and measurements
Arterial blood samples
Two mL of blood were collected from all calves via auricular artery into a heparinized syringe and used for measuring blood gases, such as partial pressure of carbon dioxide tension (pCO_2), partial pressure of oxygen tension (pO_2), blood pH, and calculating acid base parameters including bicarbonate (HCO_3^-) and base excess (BE) using blood gas analyzer (AVL 995-Hb, AVL List GmbH Medizintechnik, AVL Medical Instruments UK Ltd) adjusted to the rectal temperature of the investigated calves. The air bubbles were expelled and the syringes were kept at 4 °C until analysis, which was performed within 30 min. Plasma electrolytes such as sodium (Na^+), potassium (K^+), and chloride (Cl^-) levels were measured by using AVL 984-S Electrolyte Analyzer (AVL List GmbH Medizintechnik, AVL Medical Instruments UK Ltd). The anion gap (AG) was then calculated according to the eq. $AG = (Na^+ + K^+) - (Cl^- + HCO_3^-)$, as previously adopted [21]. Blood lactate and glucose levels were also measured spectrophotometrically by using commercial kits supplied by Spinreact (Spinreact, Barcelona, Spain).

Venous blood samples
Blood samples were collected from all the examined calves through jugular vein puncture into tubes either containing sodium ethylenediaminetetraacetic acid (EDTA) or plain tubes. The blood samples collected in the EDTA tubes were used for hematological evaluation of packed cell volume (PCV %), total and differential leucocytic counts; while those collected in plain tubes were left to coagulate to separate serum. Only clear non hemolyzed

serum samples were collected after centrifugation at 3000 rpm, and kept frozen in aliquots at –20 C^0 until required for biochemical analyses. Serum samples were used for estimation of calcium, phosphorus, malondialdehyde (MDA), superoxide dismutase (SOD) activity, uric acid (UA), reduced glutathione (GSH), vitamin C (Vit. C), nitric oxide (NO), aspartate amino transferase (AST), gamma-glutamyl transferase (GGT), sorbitol dehydrogenase (SDH), total bilirubin, total protein, albumin, urea, and creatinine concentrations.

Biochemical analyses of the aforementioned parameters were carried out spectrophotometrically by using commercial kits according to the manufacturer's instructions. Calcium and phosphorus were measured by using kits supplied by Genzyme (Genzyme Diagnostics Co, USA). For MDA, SOD, UA, GSH, Vit. C, and NO measurements, the kits were supplied by Bio Diagnostic (Bio Diagnostic, Cairo, Egypt). For AST, GGT, SDH, and total bilirubin measurements, the kits were supplied by Randox (Randox Laboratories Ltd., UK). For estimation of total protein, albumin, urea and creatinine concentrations, kits were supplied by Spinreact (Spinreact, Barcelona, Spain).

Medical management

Besides offering fresh food and water, the following medications were given to affected animals: oral liquid paraffin (El Gomhorea Co. for Chemicals and Pharmaceuticals, Egypt) at a dose of 10 ml/ kg daily for 3 days; a daily I/V infusion of a balanced electrolyte solution (Misr Co. for Chemicals and Pharmaceuticals, Egypt) at a dose of 100 ml/kg for 3 days; the half of that fluid was given in the first 4 h and the remaining fluids were given 3 times daily. To help reestablish the normal rumen flora, a transfaunation was adopted to sick animals. The rumen transfaunate was obtained from a healthy ruminant during abattoir slaughter and was transferred, as soon as possible post-collection, to the recipient animal and was given either at surgery ($n = 2$) or via oroesophageal intubation ($n = 18$), at approximately 8 L as an oral dose. Follow-up information was collected through owner contact, and via referring veterinarians regarding the animal's health status post treatment.

Statistical analysis

Data were statistically analyzed using Statistical Package for Social Sciences (SPSS) version 17.0 (USA). Data were tested for normality by using Shapiro–Wilk test, and P-values were non-significant; therefore, parametric analysis of variance (general linear model) was used for ANOVA test. The differences between groups were compared by using one-way ANOVA with *post-hoc* Bonferroni multiple comparison test. Least square means ±

standard error for each variable were calculated by following formula:

$$Y_{ij} = \mu + X_{i1} + X_{i2} + X_{i3}... + Eij$$

Where μ is the overall mean of the population and X_{i1}, X_{i2}, X_{i3} are values of the explanatory (predictor, independent) variables for individual. Y_{ij} is numerical response (outcome, dependent) variable for individual. E_{ij} is the random error in the observation of the predictors. P values at ≤0.05 were considered statistically significant.

Results

An overview of the detailed clinical settings as well as results of laboratory findings is summarized in Tables 1, 2, 3, 4, 5 and 6. Out of the 20 studied calves, 11 survived, while nine animals perished. Both survivors and non-survivors demonstrated common clinical findings including distension of ventro-lateral aspect of the right abdomen, and varying degrees of dehydration. The great majority of survivors (81%) and 100% of non-survivors were anorexic and had rumen stasis as well as hard mass upon ballottement of the left flank, but rumen fluid splashing sound was evident in approximately 18% among survivors. Scanty feces with hard contents were also evident in 55% of survivors and in 11% among non-survivors. On the other hand, defecation was absent in 45% of survivors and in 89% among non-survivors. Approximately 45% of non-survivors had frothy salivation, expiratory grunting and were being tender when strong percussion was applied on the right flank. Cud-dropping was observed in 33% of non-survivors and in 9% of survivors. The heart rate, respiratory rate and rectal temperature were within the normal reference range in both groups (Table 1).

PCV % was higher in both survivors and non-survivors than controls by 5% and 10%, respectively and was higher in non-survivors compared to survivors. On the other hand, no significant differences were observed in total and differential leucocytic counts between diseased and control buffalo (Table 2). Metabolic alkalosis was detected in diseased buffaloes from both groups as indicated by a significant increase ($P < 0.05$) in blood pH, pCO_2, HCO_3^-, and more than 50-fold increase in BE (Table 3). Plasma K^+ and Cl^-, blood glucose were significantly lower in non-survivors than those of survivors ($P < 0.05$). However, anion gap and blood L-lactate were significantly higher in non-survivors than survivors ($P < 0.05$). Phosphorus and Na^+ levels did not change among different groups (Table 4). Serum MDA, SOD, and UA were significantly higher in diseased buffalo than controls and in non-survivors than survivors ($P < 0.05$). On the other hand, levels of calcium, GSH, Vit. C, and

Table 1 Clinical findings and outcome of abomasal impaction in buffalo calves ($n = 20$)

Variables	Survivors ($n = 11$)	Non- Survivors ($n = 9$)
Heart rate (Beat/min)	65.72 ± 5.98	66.44 ± 8.24
Respiratory rate (Cycle/min)	14.27 ± 2.14	14.77 ± 1.48
Rectal temperature °C	38.31 ± 0.34	38.42 ± 0.35
Appetite	Inappetance (2/11); Anorexia (9/11)	Anorexia (9/9)
Distension of the abdomen	Present (11/11)	Present (9/9)
Defecation	Dry, scanty feces (6/11); Absent (5/11)	Dry and scanty feces (1/9); Absent (8/9)
Dropping cuds	Present (1/11)	Present (3/9)
Frothy salivation	Present (3/11)	Present (4/9)
Expiratory grunting	Absent (11/11)	Present (4/9)
Auscultation of the rumen	Hypomotile (2/11); Stasis (9/11)	Stasis (9/9)
Deep palpation and strong percussion of right flank region	Tender (1/11)	Tender (4/9)
Ballottement of left flank region	Fluid splashing sound (2/11); Firm mass (9/11)	Firm mass (9/9)
Ballottement of the lower portion of the right abdominal flank	Firm mass (11/11)	Firm mass (9/9)

NO were significantly lower in the diseased animals than controls ($P < 0.05$) (Table 5). Serum total protein, albumin, creatinine, urea, AST, GGT, and total bilirubin levels were also higher in the diseased animals, and were higher in non-survivors than survivors ($P < 0.05$) (Table 6).

Discussion

Here, we aimed at studying the clinical and biochemical alterations associated with abomasal impaction in buffalo calves. To date, there has been scarce literature about abomasal impaction in buffalo. In one report, omasal impaction in water buffalo was described [5], and in another one, sand abomasal impaction was reported in two cows [11].

The buffalo calves examined in the present study exhibited common clinical features that are in agreement with previous reports in cows with omasal and abomasal impaction [10, 11, 22]. Interestingly, the non-survivors

displayed an expiratory grunting and painful palpation of the right flank region besides dropping of the cuds and regurgitation through the nostrils and the buccal cavity. These signs could possibly be helpful for early prediction of animals with a poor prognosis. In a previous report, it has been stated that cows displaying clinical signs of chronic weight loss, rumen tympany and decreased fecal output are recommended to be evaluated for abomasal impaction [11]. However, in a retrospective study, it has been reported that decreased food intake was considered the most prevailed clinical sign observed in cows with abomasal impaction [2].

In the present study, the increase in PCV% in the diseased calves could be due to dehydration and subsequent concentration of the blood. Similar findings were observed in cows with abomasal displacement [23, 24], and in cows with abomasal impaction [2]. However, in a previous study conducted on cattle and buffalo with omasal impaction, the authors have shown that the

Table 2 Least Square Means ± Standard Error of some hematologic variables in buffalo calves with abomasal impaction compared with clinically healthy controls

Variables	Control ($n = 10$)	Survivors ($n = 11$)	Non-Survivors ($n = 9$)
PCV (%)	35.10 ± 0.58^a	40.09 ± 0.67^b	45.89 ± 0.68^c
TLC (10^3cell/µL)	7.93 ± 0.18	7.49 ± 0.29	7.68 ± 0.37
Neutrophils (10^3cell/µL)	3.49 ± 0.12	3.31 ± 0.14	3.36 ± 0.14
Band Cells (10^3cell/µL)	0.06 ± 0.03	0.02 ± 0.02	0.03 ± 0.02
Lymphocytes (10^3cell/µL)	4.18 ± 0.17	3.91 ± 0.36	3.98 ± 0.47
Monocytes (10^3cell/µL)	0.18 ± 0.07	0.045 ± 0.02	0.24 ± 0.1
Eosinophils (10^3cell/µL)	0.02 ± 0.02	0.21 ± 0.1	0.04 ± 0.03

a, b, c: Variables with different superscript in the same row are significantly different at $P < 0.05$. PCV%packed cell volume; TLC total leucocyte count

Table 3 Least Square Mean values ± Standard Error of blood gases as well as various acid-base parameters in buffalo calves with abomasal impaction compared with clinically healthy controls

Variables	Control ($n = 10$)	Survivors ($n = 11$)	Non-Survivors ($n = 9$)
pH (mmHg)	7.36 ± 0.001^a	7.48 ± 0.006^b	7.56 ± 0.14^c
PCO$_2$ (mmHg)	36.20 ± 0.42^a	47.27 ± 0.76^b	51.00 ± 0.69^c
PO$_2$ (mmol/L)	96.60 ± .748^c	85.18 ± .872^b	80.78 ± .925^a
HCO$_3^-$(mmol/L)	24.40 ± 0.34^a	35.73 ± 0.82^b	40.44 ± 0.71^c
BE (mmol/L)	0.20 ± 0.19^a	10.36 ± 0.73^b	13.89 ± 0.81^c
Anion Gap (mmol/L)	16.51 ± 0.49^a	21.21 ± 0.69^b	28.36 ± 1.83^c

a, b, c: Variables with different superscript in the same row are significantly different at $P < 0.05$. PCO$_2$ partial pressure of carbon dioxide tension, PO$_2$ partial pressure of oxygen tension, BE base excess

Table 4 Least Square Means ± Standard Error of serum electrolytes and some metabolic variables in buffalo calves with abomasal impaction compared with clinically healthy controls

Variables	Control (n = 10)	Survivors (n = 11)	Non-Survivors (n = 9)
Sodium (mmol/L)	135.70 ± 0.45	135.55 ± 0.73	136.78 ± 0.465
Potassium (mmol/L)	4.52 ± 0.17^c	3.00 ± 0.14^b	2.21 ± 0.01^a
Chloride (mmol/L)	99.40 ± 0.96^c	81.73 ± 0.91^b	70.33 ± 1.8^a
Lactate (µg/ml)	219.40 ± 2.1^a	330.45 ± 16.1^b	690.00 ± 65.6^c
Glucose (mmol/L)	3.35 ± 0.04^c	3.00 ± 0.05^b	2.61 ± 0.11^a
Calcium (mmol/L)	2.42 ± 0.14^b	1.93 ± 0.17^a	1.84 ± 0.15^a
Phosphorus (mmol/L)	1.27 ± 0.033	1.25 ± 0.030	1.26 ± 0.04

$^{a, b, c}$: Variables with different superscript in the same row are significantly different at $P < 0.05$

Table 6 Least square means ± Standard Error of some liver and kidney function tests in buffalo calves with abomasal impaction compared with clinically healthy controls

Variables	Control (n = 10)	Survivors (n = 11)	Non-Survivors (n = 9)
AST (U/L)	58.05 ± 0.6^a	68.45 ± 1.23^b	80.98 ± 4.9^c
GGT (U/L)	50.90 ± 0.53^a	60.18 ± 1.24^b	69.63 ± 3.4^c
SDH (U/L)	12.07 ± 0.4^a	27.09 ± 1.41^b	32.00 ± 4.3^b
Total bilirubin (µmol/L)	7.27 ± 0.03^a	7.47 ± 0.024^b	7.78 ± 0.1^c
Total protein (g/L)	68.55 ± 0.51^a	78.90 ± 0.81^b	84.80 ± 1.69^c
Albumin (g/L)	39.24 ± 0.30^a	45.64 ± 0.75^b	51.13 ± 2.92^c
Urea (mmol/L)	12.21 ± 0.24^a	28.27 ± 1.24^b	36.22 ± 3.55^c
Creatinine (µmol/L)	61.44 ± 0.67^a	76.18 ± 1.6^b	97.62 ± 6.24^c

$^{a, b, c}$: Variables with different superscript within the same row are significantly different at $P < 0.05$. *AST* aspartate amino transferase, *GGT* gamma-glutamyl transferase, *SDH* Sorbitol dehydrogenase

majority of the diseased animals demonstrated neutrophilic leukocytosis and lymphopenia and attributed such hematologic alterations to the potential inflammatory complications of impacted feed material [25]. Our results clearly demonstrated that the diseased buffalo suffered from metabolic alkalosis, as evidenced by increased blood pH together with increased blood HCO_3^-, and BE. Moreover, apparent hypokalemia, hypochloremia, hyperlactic acidemia, and elevated AG were also observed, and were significantly higher in non-survivors than survivors. In fact, abomasal impaction is considered one of the common causes of alkalosis. It is mainly caused by the continuous secretion of HCl and potassium into the abomasum and inability to evacuate its contents. The observed increase in pCO_2 could be compensatory due to hypoventilation secondary to alkalosis. Moreover, such an increase in pCO_2 and a decrease in pO_2 could lead to anaerobic glucose oxidation and increased lactate levels and AG as observed in our study. Nearly similar findings were previously reported in cows with abomasal impaction [2], and in cattle and buffalo with omasal impaction [25] except that hypochloremia and hypokalemia were not observed in the former study; while, hypochloremia

Table 5 Least Square Means ± Standard error of selected pro-oxidant/antioxidant variables in buffalo calves with abomasal impaction compared with clinically healthy controls

Variables	Control (n = 10)	Survivors (n = 11)	Non-Survivors (n = 9)
MDA (µmol/L)	1.99 ± 0.041^a	2.71 ± 0.17^b	3.53 ± 0.115^c
NO (µmol/L)	4.861 ± 0.078^b	4.37 ± 0.054^a	3.14 ± 0.2^a
SOD (u/ml)	196.12 ± 1.16 ^{a}	220.82 ± 1.51 ^{b}	249.4 ± 7.8^c
Vit. C (mg/dl)	3.22 ± 0.01^b	2.74 ± 0.035^a	2.59 ± 0.05^a
GSH (mg/dl)	2.57 ± 0.014^b	2.40 ± 0.01^a	2.15 ± 0.04^a
Uric acid (mg/dl)	1.4 ± 0.01 ^{a}	1.71 ± 0.14 ^{b}	2.29 ± 0.08 ^{c}

$^{a, b, c}$: Variables with different superscript within the same row are significantly different at $P < 0.05$. *MDA* malondialdehyde, *NO* nitric oxide, *SOD* superoxide dismutase, *Vit. C* Vitamin C, *GSH* reduced glutathione

with normal potassium levels were evident in the later report. The hypochloremia observed in the diseased buffalo could be either due to abomasal atony that impairs the flow of HCl into the duodenum or anorexia of the affected animals [8]. The observed hypokalemia could be attributed to the shift of potassium from extracellular to the intracellular space secondary to alkalosis [26]. It has been stated that when abomasal impaction is developed, its contents become drier than normal, with subsequent decline in the rate of emptying. This could lead to hypocalcemia, hypokalemia, and hypochloremia [2].

In the present study, the observed biochemical alterations were in part similar to those reported previously [2, 21, 24]. However, hyperglycemia rather than hypoglycemia was observed [2, 24]. It has been suggested that hypocalcemia could be associated with the development of abomasal impaction as it decreases abomasal motility [27].

Our results demonstrated that abomasal impaction is associated with increased MDA level which is indicative of oxidative stress. The alteration of MDA concentrations was associated with increase SOD activity and decrease in GSH and Vit. C levels. It is well known that free radicals are normally formed from metabolic processes inside the body. Reactive oxygen species (ROS) are derived from oxygen and superoxide anion during mitochondrial electron transport [28]. Accumulation of ROS leads to cellular damage of DNA, proteins, and lipids. There are several antioxidants that scavenge free radicals. Among the enzymatic antioxidants, SOD plays an important role in dismutation of superoxide anions. In addition, there are also non-enzymatic antioxidants, such as GSH and Vit. C [29]. GSH is oxidized with glutathione peroxidase to reduce H_2O_2 into two water molecules that requires selenium as a co-factor [30, 31]. Vitamin C is one of the water-soluble vitamins that acts as a direct free radical scavenger [32].

In the present study, the elevated total protein and albumin levels in the diseased buffalo calves could be due to dehydration and increased blood volume. In contrast, other researchers have stated that cows with abomasal impaction showed normal serum total protein concentrations with slightly elevated serum fibrinogen compared with controls [2]. Similarly, in a recent study, it was found that both cattle and buffalo with omasal impaction had normal serum total protein levels with hypoalbuminemia and hyperfibrinogenemia compared with clinically healthy controls [25]. The authors attributed those findings to the chronic starvation or failure of the liver to synthesize adequate amounts of proteins.

The observed elevation in serum creatinine and urea levels could be attributed to a decrease in renal blood flow as a part of compensatory mechanisms to maintain circulation in hypovolemia associated with dehydration, leading to azotemia. Moreover, the significant increase in the levels of liver enzymes in the diseased buffaloes is indicative of liver damage. Such hepatic damage could be due to anorexia and constipation that lead to absorption of toxic substances from the rumen and GIT. The hepatic failure with decreased lactate uptake along with hypoperfusion due to dehydration could also contribute to the observed increase in the blood lactate level [25, 33].

Conclusion

Buffalo calves with dietary abomasal impaction were associated with marked clinical and biochemical alterations that could aid in an accurate diagnosis of the disease.

Abbreviations

AG: Anion gap; AST: Aspartate amino transferase; BE: Base excess; Cl $^-$: Chloride; GGT: Gamma-glutamyl transferase; GSH: Reduced glutathione; HCO$_3^-$: Bicarbonate; K$^+$: Potassium; MDA: Malondialdehyde; Na$^+$: Sodium; NO: Nitric oxide; pCO$_2$: Partial pressure of carbon dioxide tension; PCV: Packed cell volume; pO$_2$: Partial pressure of oxygen tension; SDH: Sorbitol dehydrogenase; SOD: Superoxide dismutase activity; UD: Uric acid; Vit. C: Vitamin C

Acknowledgements
The authors would like to thank staff members of Surgery Department, Faculty of Veterinary Medicine, Mansoura University for providing surgical assistance. We also would like to thank Dr. Sara Balesaria for editorial assistance.

Consent to participate
Animals were enrolled with owner's consent and were given a document containing information about the disease definition, its economic impact as well as the potential clinical consequences.

Funding
Not applicable

Authors' contributions
MRE designed and coordinated the study. MRE also responsible for clinical examinations and samples collection and prepared the manuscript. MRE and MFS wrote the manuscript. EAA performed statistical analysis. All authors contributed equally in data collection, analyses, and interpretation besides reviewing of the manuscript. All authors approved the final version of the manuscript for publication.

Competing interests
The authors declare that they have no competing interests.

Author details
[1]Department of Internal Medicine and Infectious Diseases, Faculty of Veterinary Medicine, Mansoura University, Mansoura 35516, Egypt. [2]Department of Biochemistry, Faculty of Veterinary Medicine, Mansoura University, Mansoura 35516, Egypt. [3]Department of Clinical Pathology, Faculty of Veterinary Medicine, Mansoura University, Mansoura 35516, Egypt. [4]Department of Laboratory Medicine, Faculty of Applied Medical Science, Umm Al-Qura University, Makkah 21955, Kingdom of Saudi Arabia. [5]Department of Animal Husbandry and Development of Animal Wealth, Faculty of Veterinary Medicine, Mansoura University, Mansoura 35516, Egypt.

References
1. Ashcroft RA. Abomasal impaction of cattle in Saskatchewan. Can Vet J. 1983; 24:375.
2. Wittek T, Constable PD, Morin DE. Abomasal impaction in Holstein-Friesian cows: 80 cases (1980–2003). J Am Vet Med Assoc. 2005;227:287–91.
3. Fox FH. Abomasal disorders. J Am Vet Med Assoc. 1965;147:383–8.
4. Frasser CM. The Merck veterinary manual, 9th edn ed. Rahway: Merck and Co Inc.; 1986.
5. Turkar S, Uppal SK. Blood biochemical and ruminal liquor profile in buffaloes (Bubalus Bubalis) showing omasal impaction. Vet Res Commun. 2007;31:967–75.
6. Baker GJ, Lewis MR. A case of abomasal impaction and its surgical correction. Vet Rec. 1964;76:416–8.
7. Merritt AM, Boucher WB. Surgical treatment of abomasal impaction in the cow. J Am Vet Med Assoc. 1967;150:1115.
8. Radostits OM, Gay C, Hinchcliff K, Constable P. Veterinary medicine - a textbook of the diseases of cattle, horses, sheep, pigs and goats. 10th ed. Edinburgh: Elsevier Saunders; 2007.
9. Hunter R. Sand impaction in a herd of beef cattle. J Am Vet Med Assoc. 1975;166:1179.
10. Simkins KM, Nagele MJ. Omasal and abomasal impaction in beef suckler cows. Vet Rec. 1997;141:466–8.
11. Simsek A, Sekin S, Icen H, Kochan A, Celik OY, Yaman T. Abomasal impaction due to sand accumulation in two cows. Large Anim Rev. 2015;21:125–7.
12. Mitchell KJ. Dietary abomasal impaction in a herd of dairy replacement heifers. J Am Vet Med Assoc. 1991;198:1408–9.
13. Pope DC. Abomasal impaction of adult cattle. Vet Rec. 1961;73:1175–7.
14. Rehage J, Kaske M, Stockhofe-Zurwieden N, Yalcin E. Evaluation of the pathogenesis of vagus indigestion in cows with traumatic reticuloperitonitis. J Am Vet Med Assoc. 1995;207:1607–11.
15. Neal PA, Edwards GB. Vagus indigestion in cattle - a review of 52 cases. Vet Rec. 1968;82:396.
16. Breukink HJ, Kuiper R. Digestive disorders following obstruction of flow of ingesta through the abomasum and small intestine. Bov Pract. 1980;15:139–43.
17. Hutchins DR, Blood DC, Hyne R. Residual defects in stomach motility after traumatic reticulo-peritonitis of cattle. Pyloric obstruction, diaphragmatic hernia and indigestion due to reticular adhesions. Aust Vet J. 1957;33:77–82.
18. Kaske M, Midasch A. Effects of experimentally-impaired reticular contractions on digesta passage in sheep. Br J Nutr. 1997;78:97–110.
19. Ducharme NG, Fubini SL. Surgery of the forestomach compartments. In: Fubini SL, Ducharmi NG, editors. Farm animal surgery. USA: Elsevier Science, Saunders; 2004. p. 184–94.
20. Radostits OM, Mayhew IG. Houston DM veterinary clinical examination and diagnosis. London: WB Saunders; 2000.
21. Feldman BF, Rosenberg DP. Clinical use of anion and osmolal gaps in veterinary medicine. J Am Vet Med Assoc. 1981;178:396–8.
22. Braun U, Rauch S, Schade B, Sydler T. Clinical findings in a cow with severe abomasal sand impaction. Tierärztliche Praxis Großtiere. 2008;36:241–4.
23. Jubb TF, Malmo J, Davis GM. Vawser AS left-side displacement of the abomasum in dairy cows at pasture. Aust Vet J. 1991;68:140–2.

24. Rohn M, Tenhagen BA, Hofmann W. Survival of dairy cows after surgery to correct abomasal displacement: 2. Association of clinical and laboratory parameters with survival in cows with left abomasal displacement. J Vet Med A. 2004;51:300–5.

25. Hussain SA, Uppal SK, Randhawa C, Sood NK, Mahajan SK. Clinical characteristics, hematology, and biochemical analytes of primary omasal impaction in bovines. Turk J Vet Anim Sci. 2013;37:329–36.

26. Kaneko JJ, Harvey JW, Michael LB. Clinical biochemistry of domestic animals. 5th ed. New York: Academic Press; 1997.

27. Madison JB, Troutt HF. Effects of hypocalcaemia on abomasal motility. Res Vet Sci. 1988;44:264–6.

28. Valko M, Leibfritz D, Moncol J, Cronin MTD, Mazur M, Telser J. Free radicals and antioxidants in normal physiological functions and human disease. Int J Biochem Cell B. 2007;39:44–84.

29. Halliwell B, Gutteridge JMC. Free radical in biology and medicine. New York: Oxford Univ Press; 2007.

30. Smith KL, Hogan JS, Weiss WP. Dietary vitamin E and selenium affect mastitis and milk quality. J Anim Sci. 1997;75:1659–65.

31. Wichtel JJ. A review of selenium deficiency in grazing ruminants part 1: new roles for selenium in ruminant metabolism. N Z Vet J. 1998;46:47–52.

32. Blokhina O, Virolainen E, Fagerstedt KV. Antioxidants, oxidative damage and oxygen deprivation stress: a review. Ann Bot. 2003;91:179–94.

33. Vary TC, Siegel JH, Rivkind A. Clinical and therapeutic significance of metabolic patterns of lactic acidosis. Perspect Crit Care. 1988;1:85–132.

6

A frameshift mutation in MOCOS is associated with familial renal syndrome (xanthinuria) in Tyrolean Grey cattle

Leonardo Murgiano[1], Vidhya Jagannathan[1], Christian Piffer[2], Inmaculada Diez-Prieto[3], Marilena Bolcato[4], Arcangelo Gentile[4] and Cord Drögemüller[1*] (iD)

Abstract

Background: Renal syndromes are occasionally reported in domestic animals. Two identical twin Tyrolean Grey calves exhibited weight loss, skeletal abnormalities and delayed development associated with kidney abnormalities and formation of uroliths. These signs resembled inherited renal tubular dysplasia found in Japanese Black cattle which is associated with mutations in the *claudin 16* gene. Despite demonstrating striking phenotypic similarities, no obvious presence of pathogenic variants of this candidate gene were found. Therefore further analysis was required to decipher the genetic etiology of the condition.

Results: The family history of the cases suggested the possibility of an autosomal recessive inheritance. Homozygosity mapping combined with sequencing of the whole genome of one case detected two associated non-synonymous private coding variants: A homozygous missense variant in the uncharacterized *KIAA2026* gene (g.39038055C > G; c. 926C > G), located in a 15 Mb sized region of homozygosity on BTA 8; and a homozygous 1 bp deletion in the *molybdenum cofactor sulfurase (MOCOS)* gene (g.21222030delC; c.1881delG and c.1782delG), located in an 11 Mb region of homozygosity on BTA 24. Pathogenic variants in *MOCOS* have previously been associated with inherited metabolic syndromes and xanthinuria in different species including Japanese Black cattle. Genotyping of two additional clinically suspicious cases confirmed the association with the *MOCOS* variant, as both animals had a homozygous mutant genotype and did not show the variant *KIAA2026* allele. The identified genomic deletion is predicted to be highly disruptive, creating a frameshift and premature termination of translation, resulting in severely truncated MOCOS proteins that lack two functionally essential domains. The variant *MOCOS* allele was absent from cattle of other breeds and approximately 4% carriers were detected among more than 1200 genotyped Tyrolean Grey cattle. Biochemical urolith analysis of one case revealed the presence of approximately 95% xanthine.

Conclusions: The identified *MOCOS* loss of function variant is highly likely to cause the renal syndrome in the affected animals. The results suggest that the phenotypic features of the renal syndrome were related to an early onset form of xanthinuria, which is highly likely to lead to the progressive defects. The identification of the candidate causative mutation thus enables selection against this pathogenic variant in Tyrolean Grey cattle.

Keywords: Bovine, Congenital disease, Hereditary, Kidney, Rearing success

* Correspondence: cord.droegemueller@vetsuisse.unibe.ch
[1]Institute of Genetics, Vetsuisse Faculty, University of Bern, Bern, Switzerland
Full list of author information is available at the end of the article

Background

Sporadic cases of inherited renal syndromes have been reported in domestic animals including cattle. In Danish Holsteins and Danish Red dairy cattle, an autosomal recessive form of renal lipofuscinosis has been described, accompanied by renal dysfunction and reduced longevity (OMIA 001407–9913) [1]. A single case of renal amyloidosis has been described in Iranian cattle (OMIA 000040–9913) [2] and various forms of renal dysplasia have been reported in other cattle (OMIA 001135–9913) [3–9]. In Japanese Black cattle, renal dysplasia occurred due to two independent autosomal recessive mutations of the *claudin 16* gene (*CLDN16*), but no phenotypic differences between both types were reported [10–12]. Members of the claudin gene family play important roles in the formation of tight junctions in the kidney [13]. In Japanese Black cattle, inherited xanthinuria is associated with growth retardation and death at approximately 6 months of age (OMIA 001819–9913) [14]. This autosomal recessive inherited disease has been associated with a 3 bp deletion in the coding region of the *molybdenum cofactor sulfurase* (*MOCOS*) gene. In a single genetically uncharacterized Galician Blond beef calf, signs of listlessness and weight loss and renal failure with bilateral nephrolithiasis, composed of 100% xanthine were reported [15].

In a previous report [16] the authors described kidney abnormalities in two eight months old female Alpine Grey (better known as Tyrolean Grey) twin cattle presenting a number of signs: growth retardation, overgrowth of hooves, gradual loss of weight and impaired skeletal development, despite a normal or only slightly decreased appetite and intense vitamin integration.

Clinical biochemistry data e.g. associated with variable blood phosphate concentrations indicated a renal failure. Pathological examination of the animals showed hypotrophic, firm and pale kidneys, with a roughened and granular surface. Subsequent histology showed interstitial infiltrates of immature mesenchymal tissue and disseminated mineralization. Mild dilation of the pelvis with free yellow calculi and disseminated medullar mineralization were also observed. Small stones (0.3 to 0.5 cm in diameter) were also found in the ureters and in the bladder. The observed renal syndrome in Tyrolean Grey cattle resembled inherited renal tubular dysplasia in Japanese Black cattle. Despite phenotypic similarities, no mutation in *CLDN16* was identified [16]. Therefore, the aim of this study was to use whole genome sequencing to unravel the genetic etiology of this condition.

Methods

Animals and SNP genotyping

Initially, blood samples were collected from two affected twin calves and their parents. Genotyping of these four animals was performed using the BovineHD BeadChip (Illumina), including 777,962 evenly distributed SNPs, at GeneSeek. PLINK software [17] was used to search for extended intervals of homozygosity with shared alleles as described previously [18]. Additionally, archived DNA samples from 1201 normal Tyrolean Grey cattle were used for genotyping the *MOCOS* variant. During the course of the review process of this paper, 3 additional blood samples were taken from two clinically suspicious cases and their dam. For the mutation analysis and comparison of sequencing data, 106 normal cattle from 20 genetically diverse *Bos taurus* breeds were used: Angler (*n* = 5), Angus (*n* = 3), Brown Swiss (*n* = 8), Charolais (*n* = 1), Chianina (*n* = 1), Cika (*n* = 1), Danish Red Dairy (*n* = 3), Eringer (*n* = 2), Galloway (*n* = 2), Hereford (*n* = 1), Hinterwalder (*n* = 3), Holstein (*n* = 47), Limousin (*n* = 4), Pezzata Rossa Italiana (*n* = 1), Piemontese (*n* = 1), Romagnola (*n* = 5), Scottish Highland (*n* = 2), Simmental (*n* = 12), Tyrolean Grey (*n* = 1), and Vorderwalder (*n* = 3).

Whole genome re-sequencing

A fragment library with a 260 bp insert size was prepared and one lane of Illumina HiSeq2500 paired-end reads collected (2 × 100 bp); the fastq files were created using Casava 1.8 (illumina). Sequence reads were then mapped to the reference cow genome assembly UMD3.1 as previously described [19], at average sequence coverage of 19-fold. The SAM file generation, conversion to BAM, duplicate detection, variant calling (data for each sample was obtained in variant call format, version 4.0), and variant effect prediction were carried out as described previously [20]. The genome data has been made freely available under accession PRJEB11962 at the European Nucleotide Archive [21].

Genotyping

The *MOCOS* variant was genotyped by Sanger sequencing of a 287 bp PCR product using a forward primer (5-CATCATTTCACTTCCTTTTGGA-3) and a reverse primer (5-TAGGTGATCAGGTGGCCTCT-3) flanking the *MOCOS* variant. PCR products were amplified with AmpliTaqGold360Mastermix (Life Technologies) and the products directly sequenced using the PCR primers on an ABI 3730 capillary sequencer (Life Technologies). Sequence data were analyzed using Sequencher 5.1 (GeneCodes). In addition, fragment size analyses were performed for the genotyping of the *MOCOS* variant on the ABI 3730 and analyzed with the GeneMapper 4.0 software (Life Technologies).

Urolith analysis

The uroliths were washed with deionized distilled water and dried in an oven at 45 °C for 48 h. After visual inspection of the external surface by stereomicroscope, appearance, color and size were recorded, stones weighed and one of the largest stones cut (to check for layers that were analyzed independently). The test samples were ground in an agate pestle and mortar to obtain a homogenous powder. A small amount of the powder (approximately 2 mg) was mixed with 200 mg of potassium bromide and this mixture pressed in a ten-tone press to form a thin pill/pellet. Fourier transform infrared (FT-IR) spectroscopy of this pellet was performed in a spectrophotometer (FT-IR 2000, Perkin Elmer, United Kingdom) and the obtained spectrum compared with a specific library for uroliths (IR Kidney Stones 1668 Spectra, Nikodom, Czech Republic).

Results and discussion

Pedigree analysis and homozygosity mapping suggests a recessive inheritance

The pedigree of the established Tyrolean Grey cattle family was consistent with a possible monogenic autosomal recessive inheritance (Fig. 1). Both parents were healthy and could be traced back to a single common male ancestor born ~40 years ago. The genotypes of more than 770,000 evenly spaced SNPs showed that the affected twins were monozygotic twins with an identity by state

(IBS) of 100%. Since we assumed a homozygous recessive condition, the affected calves were expected to be identical by descent (IBD) and homozygous for the causative mutation and flanking chromosomal segments. The search for extended regions of homozygosity with simultaneous allele sharing was carried out, and the results compared with the homozygous region between the cases and the parents. Thereby, 29 genome regions greater than 1 Mb were identified where both affected animals were homozygous, but the parents were not (Fig. 2, Additional file 1).

Although it was assumed that recessive inheritance was considered most likely, a second possibility that the renal syndrome was due to a spontaneous dominant acting mutation could not be excluded. In fact, since after genotyping the twins were identified as identical twins, the causative mutation could have occurred *de novo* in the parental germinal cells. This alternative hypothesis was acknowledged in further analyses.

Whole genome sequencing reveals a deleterious *MOCOS* mutation

Due to the severe effect of the mutation, we suggested that most likely a non-synonymous loss of function mutation

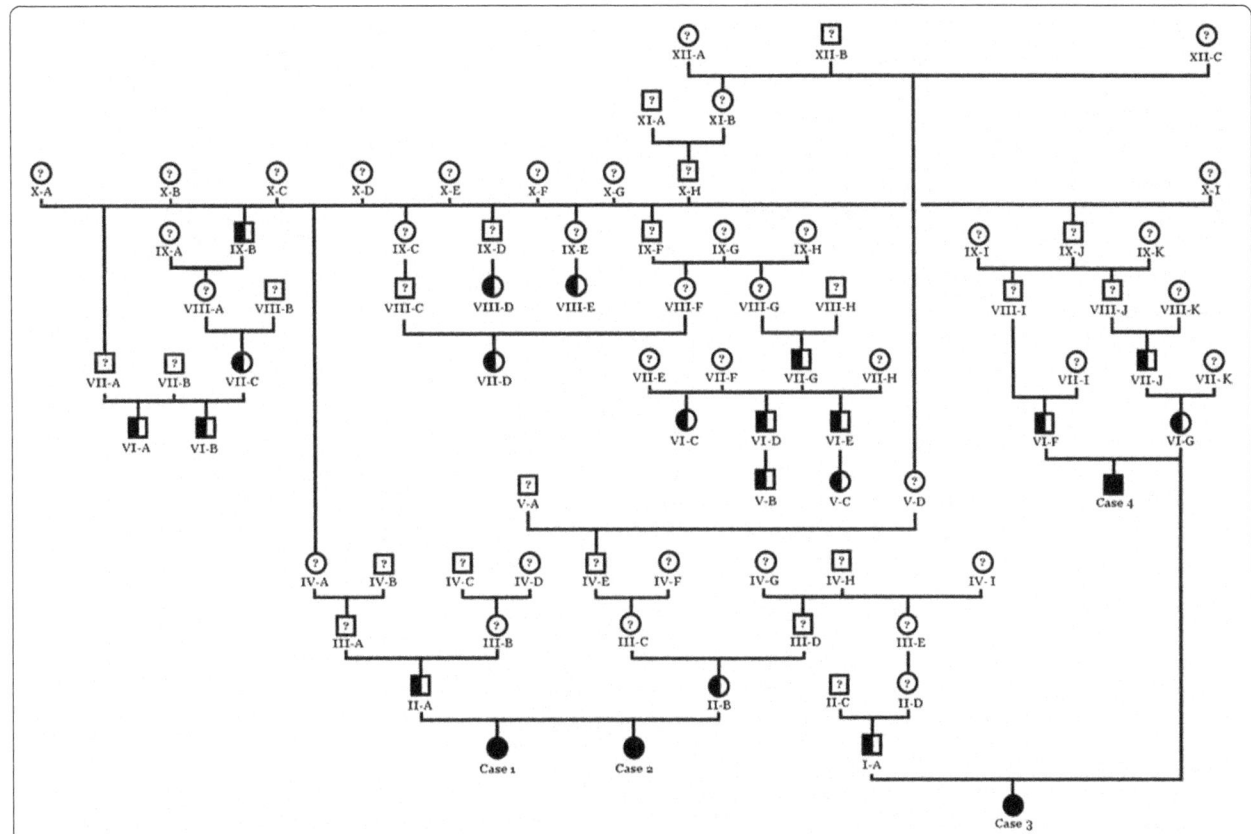

Fig. 1 Family tree of the affected Tyrolean Grey cattle with renal syndrome (xanthinuria). Males are represented by *squares*, females by *circles*. Affected animals are shown with fully *black symbols*. *Half-filled symbols* represent healthy obligate heterozygous carriers and *open symbols with a question mark* represent relatives with an unknown genotype. Cases 1 and 2 are the initially studied twin cases, whereas case 3 and 4 are the additional cases reported later

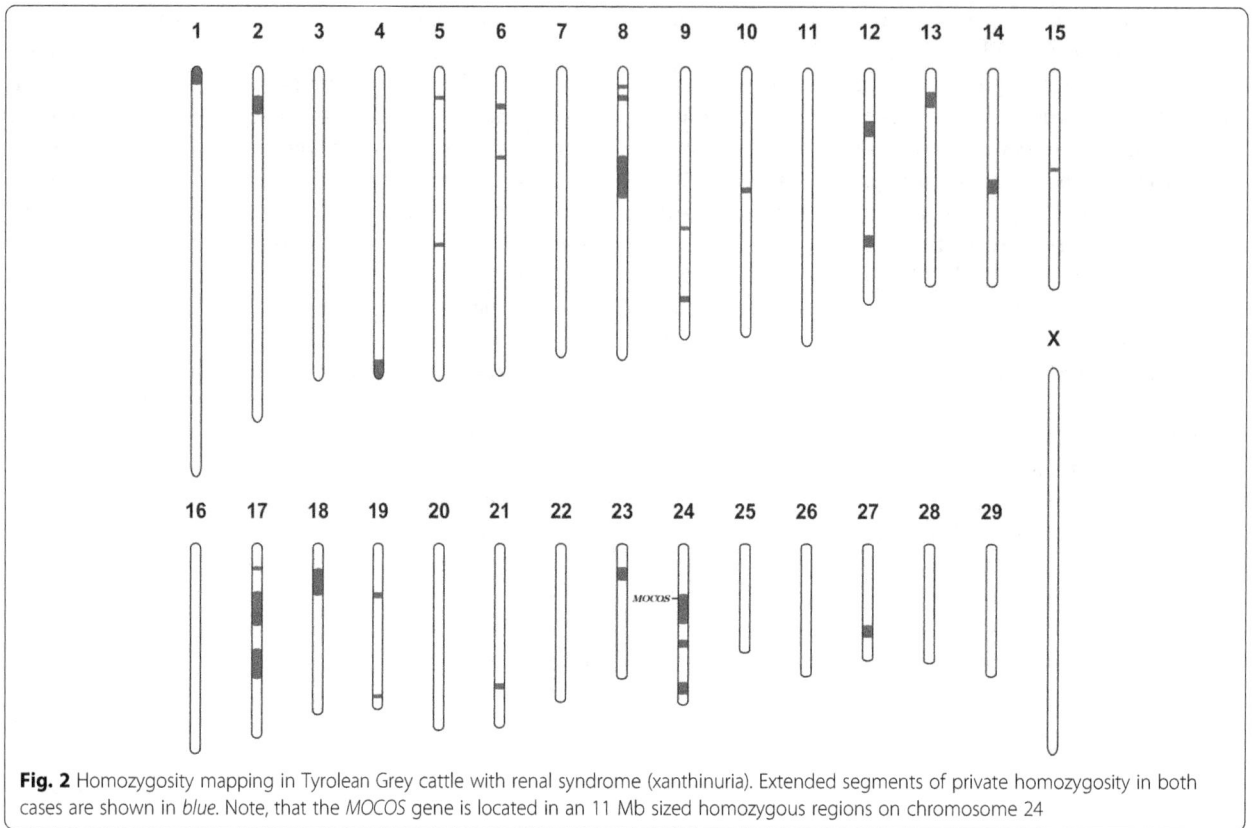

Fig. 2 Homozygosity mapping in Tyrolean Grey cattle with renal syndrome (xanthinuria). Extended segments of private homozygosity in both cases are shown in *blue*. Note, that the *MOCOS* gene is located in an 11 Mb sized homozygous regions on chromosome 24

affecting the coding sequence of a gene would be responsible for the renal syndrome. The whole genome of case 1 was sequenced, and initially the variant calls were filtered according to the most likely scenario of recessive inheritance. A total of 36,940 homozygous SNPs and short indel variants lying in the coding regions of annotated genes were called across the entire genome. A triple-step filtering to search for private variants was applied: (*I*) Only homozygous variants falling into the homozygous IBD regions presented above were considered; (*II*) A comparison was then made between the remaining variants and variant data of 106 cow genomes of various different cattle breeds that were sequenced in our laboratory in the course of other ongoing studies; (*III*) Lastly, the remaining variants were compared with the run4 variant database of the 1000 bull genome project [22] including 1119 additional genomes of different breeds. Finally, the number of private homozygous variants was dropped to 435. These included two nonsynonymous variants (Additional file 2): A missense variant in exon 8 of the *KIAA2026* gene (c.926C > G; p.S309C), located in a 15 Mb sized region of homozygosity on BTA 8 (g.39038055C > G); and a 1 bp deletion in exon 9 of the *molybdenum cofactor sulfurase* (*MOCOS*) gene, located in an 11 Mb region of homozygosity on BTA 24 (g.21222030delC). The function of KIIAA2026 is not well characterized, however recent studies identified KIIAA2026 as interacting with ubiquitin-like UBTD1 in a yeast-2-

hybrid screen [23], and a possible role in cancer [24]. On the other hand, the MOCOS enzyme is known to affect kidney metabolism and kidney development and a *MOCOS* mutation was previously shown to be associated with a renal condition in cattle, specifically xanthinuria, characterized by developmental problems, renal failure and the presence of kidney stones [14]. Therefore *MOCOS* represents a very good candidate gene for the condition, and the detected *MOCOS* variant is much more likely to be responsible for the observed phenotype.

The *MOCOS* variant was confirmed by Sanger sequencing (Fig. 3). The 1 bp deletion in the bovine *MOCOS* gene affects both annotated transcripts [25]: ENSBTAT00000048768 (*MOCOS-201*; c.1881delG) and ENSBTAT00000065375 (*MOCOS-202*; c.1782delG). For both *MOCOS* transcripts, the deletion is predicted to cause a frameshift introducing premature stop codons (p.Ser628Valfs9* and p.Ser595Valfs9*).

The possibility that the condition was caused by a recent *de novo* mutation could not be excluded, therefore the private heterozygous variants present in the genome of the sequenced case were also analyzed. The number of heterozygous variants identified in the coding regions of the affected case were 124,393; these were then filtered against the in house 106 control genomes. Of the remaining 445 variants, 328 were subsequently excluded because of their presence in the 1000 bull genome

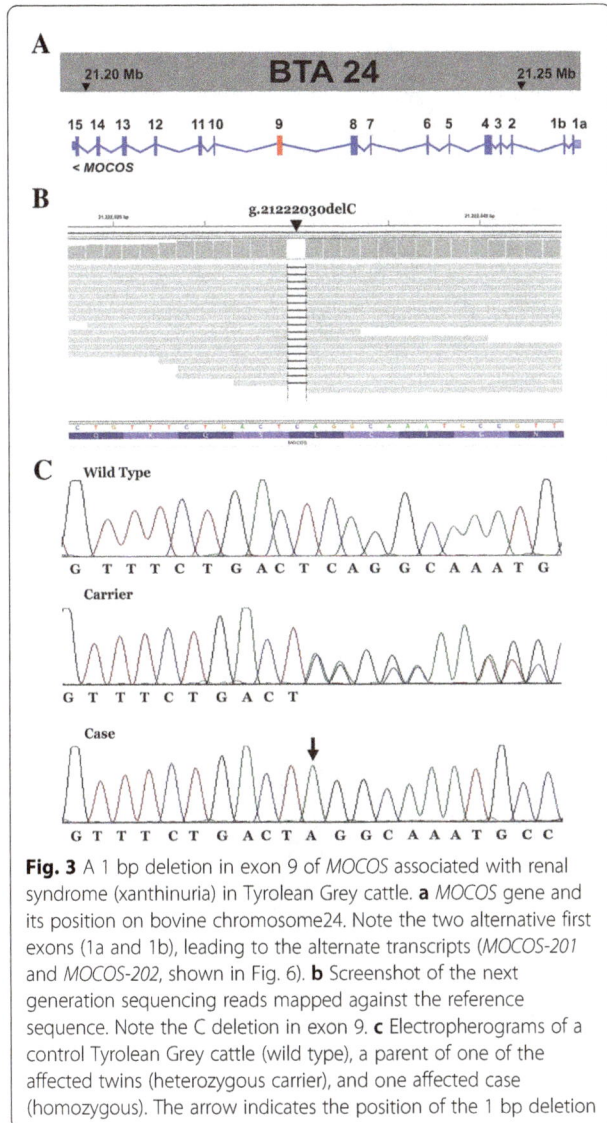

Fig. 3 A 1 bp deletion in exon 9 of *MOCOS* associated with renal syndrome (xanthinuria) in Tyrolean Grey cattle. **a** *MOCOS* gene and its position on bovine chromosome24. Note the two alternative first exons (1a and 1b), leading to the alternate transcripts (*MOCOS-201* and *MOCOS-202*, shown in Fig. 6). **b** Screenshot of the next generation sequencing reads mapped against the reference sequence. Note the C deletion in exon 9. **c** Electropherograms of a control Tyrolean Grey cattle (wild type), a parent of one of the affected twins (heterozygous carrier), and one affected case (homozygous). The arrow indicates the position of the 1 bp deletion

variant database. Finally, a total of 59 private heterozygous coding variants remained. These did not contain any non-synonymous variants and no other variants in genes which might represent suitable candidates for the renal phenotype (Additional file 3).

Thus, the final conclusion was that among all the possible variants, the recessive inherited deleterious *MOCOS* variant represented the most probable causative mutation for the renal syndrome. Therefore, the observed renal syndrome most likely represents a xanthinuria which causes the severe (early-onset) metabolical signs and the occurrence of kidney stones. Genotyping this variant in an extended cohort of Tyrolean Grey cattle confirmed that the *MOCOS* deletion was associated perfectly with the condition: only the two twins carried two copies of the mutation, both parents were heterozygous carriers and none of the 1201 controls showed the homozygous mutant genotype (Table 1). A total of 50 genotyped controls (~4%) were heterozygous carriers. In each case of 13 carriers with available pedigree records a relationship to the previously identified male ancestor was noticed (Fig. 1).

Additional cases confirmed *MOCOS*-association and kidney stone analysis supporting xanthinuria diagnosis

After the mapping and identification of the *MOCOS* variant, two novel cases were brought to the authors' attention. Pedigree analysis demonstrated that these maternal half siblings were related to the affected twins and also showed genealogical connections to the previously identified male ancestor (Fig. 1). The animals displayed similar clinical signs as described for the affected twins [16], whereas the overgrowth of hooves was noticed only in case 3 (Fig. 4). The six-month-old heifer (case 3) was euthanized at 6 months of age and cross sections of urinary bladder and kidney indicated the presence of uroliths (Fig. 5a+b). The 21-month-old male (case 4) is still alive. Interestingly this latter animal did never show problem of hooves overgrowth, but only retarded growth and discontinuous disturbed appetite. Both affected animals were genotyped as homozygous carriers for the *MOCOS* deletion, the dam carried a single copy of the mutant allele and the two sires were known carriers, already genotyped before. In addition, we genotyped both additional cases and their dam for the missense variant in the *KIAA2026* gene. Thereby we showed that all three animals were homozygous for the wild type allele and could exclude this variant as a candidate mutation.

As the kidney stones of case 3 were freshly available, they were analyzed in order to detect their composition. The uroliths contained a combination of grit and some small oval-shaped stones (the largest measuring $8 \times 5 \times 3$ mm). They were yellowish-brown and smooth with a glossy surface and no layers were observed (Fig. 5c+d). The quantitative mineral analysis by infrared spectroscopy using the FT-IR technique, which shows a high sensitivity and allows an accurate identification of stone composition, demonstrated a greater than 95% xanthine content (Fig. 5e).

Table 1 Association of the 1 bp deletion in *MOCOS* with the renal syndrome (xanthinuria) in Tyrolean Grey cattle

	MOCOS c.1881delG /c.1782delG		
	G/G	G/del	del/del
Affected animals			4
Obligate carriers		3	
Tyrolean Grey population controls	1151	50	
Controls of other breeds[a]	1225		
Total	2376	53	4

[a]106 in-house controls and 1119 genomes of the 1000 bull genome project

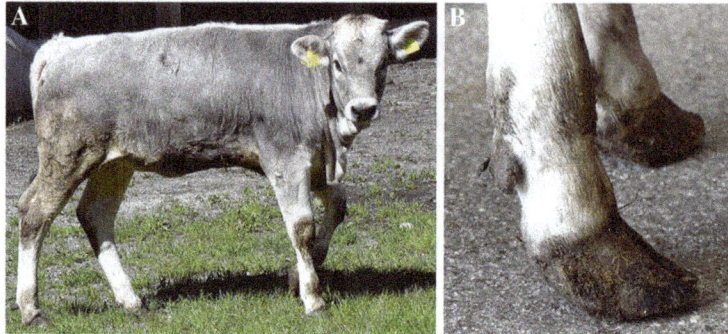

Fig. 4 Six-month-old Tyrolean Grey heifer with renal syndrome (xanthinuria). **a** The affected animal showed retarded growth and discontinuous disturbed appetite. **b** Note the overgrown hooves. The owner reported an exaggerated growth and therefore cut the hooves

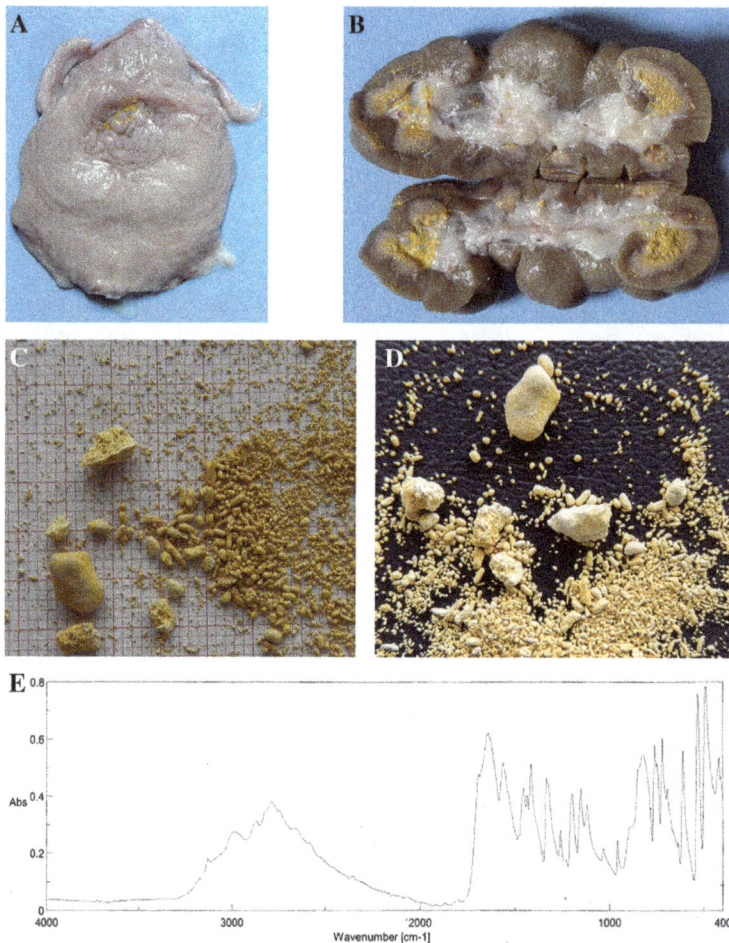

Fig. 5 Xanthinuria in a six-month-old Tyrolean Grey heifer. Note the presence of stones visible as diffused mineralization in the granular sandy *yellow* mass in the urinary bladder (**a**) and kidney (**b**) of case 3 showing uroliths (**c+d**). Quantitative mineral analysis by FT-IR spectroscopy of a kidney stone demonstrated a greater than 95% xanthine content (**e**)

In conclusion, the results obtained from the two additional cases confirmed the suspected xanthinuria and the disease association with the 1 bp deletion in the *MOCOS* gene.

Role of *MOCOS* in renal conditions

Molybdenum (Mo) is present in a large number of metalloenzymes and is ubiquitous in many phyla: bacteria, archaea, fungi, algae, plants and animals [26]. Mo-containing enzymes are essential for life, holding a key position in the metabolism of the individual organism [27]. To achieve biological activity and play a role in the cell's redox-chemistry Molybdenum must form a prosthetic group known as molybdenum cofactor (Moco) [28]. In eukaryotes, the most prominent Mo-enzymes are nitrate reductase, sulfite oxidase, and the mitochondrial amidoxime reductase, aldehyde oxidase and xanthine dehydrogenase [26]. In detail, xanthine dehydrogenase (XD) catalyzes the terminal two steps of the purine degradation pathway: formation of xanthine from hypoxanthine followed by uric acid from xanthine. This enzyme has been the focus of extensive studies for several reasons: it is a molybdenum containing flavoprotein, because of its interactions with drugs, and for its role in human disorders [29]. A deficiency of XD has been shown to cause xanthinuria which is characterized by hypouricemia and hypouricosuria [30]. Xanthinuria is classified into 2 groups, types I and II and patients with either type tend to develop urinary tract xanthine uroliths due to tissue xanthine deposits [29].

Sulfuration of the molybdenum cofactor of xanthine dehydrogenase (and aldehyde oxidase) is performed by the protein MOCOS which is required for their enzymatic activities [29]. MOCOS is therefore part of a complex pathway and the object of studies at different levels and expression studies associating the protein with genes involved in regulating transcription and apoptosis [31]. In wild silkworms, a MOCOS loss of function mutation was reported showing a mutant phenotype resulting in translucent skin [32]. In humans, mutations in *MOCOS* cause type II xanthinuria (OMIM 613274; Fig. 6). In Japanese Black

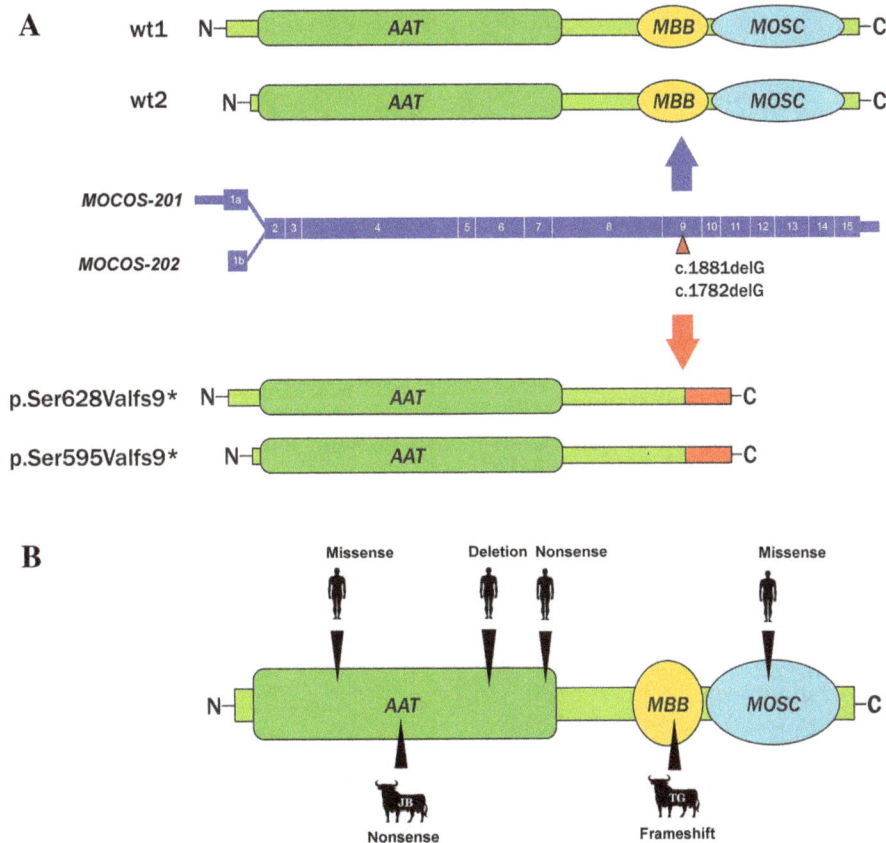

Fig. 6 Summary of known bovine and human MOCOS mutations. **a** The two bovine *MOCOS* transcripts (*MOCOS-201* and *MOCOS-202*) are shown (*center*). The position of the mutation detected in Tyrolean Grey cattle is indicated. The two possible wild type proteins and the relevant domains (AAT: aspartate aminotransferase; MBB: mosc beta barrel; MOSC: MOSC domain) are shown above. Below, the two predicted mutant proteins are shown. Note the predicted mutant truncated proteins lack MBB and MOSC domains. **b** Previously reported mutations causing xanthinuria in human and cattle are displayed (JB: Japanese Black cattle; TG: Tyrolean Grey cattle). The nature of the predicted mutation effect of the protein is shown. The positions of mutated bovine residues are aligned to the length of the human protein

cattle autosomal recessive inherited bovine xanthinuria has been reported which was characterized by elevated xanthine secretion in the urine and lethal growth retardation [14]. Affected cattle had expanded renal tubules containing xanthine calculi ranging from 1 to 3 mm in diameter. A homozygous 3 bp deletion perfectly segregated with the disease, resulting in the loss of the Tyr257 residue (p.Tyr257del) in the bovine MOCOS protein [14]. In humans, two independent patients with classical type II xanthinuria (OMIM 613274) showed a *MOCOS* c.1255C > T nonsense mutation causing a premature stop (p.Arg419*) [33]. A *MOCOS* c.466G > C missense mutation (p.Ala156-Pro) was described in another xanthinuric patient [34]. Finally, a homozygous *MOCOS* c.2326C > T missense mutation (p.Arg776Cys), and a compound heterozygous with a *MOCOS* c.1034insA nonsense mutation (p.Gln347fs32*) was present in a further two xanthinuria patients [32].

In summary, cases of xanthinuria reported in the literature, caused by *MOCOS* mutations follow a recessive inheritance in both bovine and human cases. The xanthinuria occurring in Tyrolean Grey cattle reported in this study is also consistent with recessive inheritance. While it is unclear whether the mutant bovine MOCOS protein is actually expressed in our cases, with 246 amino acids or more than 30% of the normal MOCOS protein missing, it is very unlikely that a mutant protein would fulfill any physiological function. Furthermore, the mutant protein has two functionally important domains missing (Fig. 6). It is therefore more likely that the mutant mRNA is targeted by non-sense-mediated decay, thus the deleterious bovine *MOCOS* variant represents the most likely causative mutation for the renal syndrome.

In a previous case report, the authors diagnosed the renal condition as a renal dysplasia representing a developmental abnormality of the renal tissue. In defense of this diagnosis, it was inferred that the highly disruptive nature of the mutation, the severity of the condition and its probable early onset were responsible for the alteration in the structure of the developing kidney. Other macroscopical features also lead to confusion: e.g. deformity or overgrowth of hooves have been previously reported in cases of renal dysplasia in cattle [6, 10]. In papers describing xanthinuria in cattle, it was not possible to find reports of overgrown hooves [14, 15], but such features could be an indicator of a general sign of renal failure (along with poor growth) and not of a specific disease. The very limited number of cases prevents us from observing in detail how the position and specific nature of the mutation affect the onset and severity of xanthinuria. The two newly reported cases seem to show a milder phenotype compared to the initially observed affected twins. Therefore one could argue that the expressivity of the disease phenotype caused by the *MOCOS* mutation is variable.

Interestingly, there are many phenotypic differences between the initially suspected renal dysplasia and a metabolic disorder like xanthinuria. It was therefore not obvious that the identified mutation would cause the observed kidney development condition, although the detected mutation represents a deleterious loss of function mutation affecting a protein involved in renal metabolism. However, in vertebrates *MOCOS* mutations have always been associated with xanthinuria. Finally, genotyping two additional cases and the urolith analysis, confirmed the suspected diagnosis as xanthinuria.

The third monogenic recessive defect of Tyrolean Grey cattle

Tyrolean Grey cattle (locally known in German as Grauvieh) is a very old, dual-purpose (bred for milk and meat) alpine cattle breed. The population is small with only a few thousand registered cows, predominantly in Austria (Tyrol) and Italy (South Tyrol), and fewer numbers in Switzerland. In the last decade Tyrolean Grey cattle breeders experienced outbreaks of two recessive diseases: degenerative axonopathy [19] and chondrodysplastic dwarfism [20]. For both defects, gene testing to eliminate the disease from the population is still ongoing. This report presents the breeders with a third, although obviously less frequent, genetic defect to take into consideration.

Conclusions

This study reveals the genetic etiology of a very rare early-onset metabolic renal syndrome in Tyrolean Grey cattle. The findings allow targeted selection against a previously unknown or misdiagnosed genetic disorder affecting rearing success.

Additional files

Additional file 1: Homozygous regions detected in the genome of the two animals with renal syndrome (xanthinuria).

Additional file 2: List of private homozygous sequence variants of the sequenced animal located within homozygous IBD regions. Variants with predicted effects on the protein sequence are presented boldface.

Additional file 3: List of private heterozygous variants in annotated genes found in the genome of the sequenced animal.

Abbreviations

Bp: Base pairs; BTA: *Bos taurus* autosome; CLDN16: Claudin 16; FT-IR: Fourier transform infrared spectroscopy; Mb: Megabase pairs; MOCOS: Molybdenum cofactor sulfurase; PCR: Polymerase chain reaction; SNP: Single nucleotide polymorphism

Acknowledgements

The authors wish to thank Nathalie Besuchet-Schmutz and Muriel Fragnière for their invaluable technical assistance. The Next Generation Sequencing Platform of the University of Bern is acknowledged for performing the whole genome re-sequencing experiment and the Interfaculty Bioinformatics Unit of the University of Bern for providing computational infrastructure.

Funding

This research received no specific grant from any funding agency in the public, commercial or non-profit sectors.

Authors' contributions

CP, MB and AG diagnosed the cases and evaluated pedigree information. LM and CD performed the genetic investigations and drafted the manuscript. VJ analyzed the whole genome sequencing data. IDP performed the urolith analysis. CD supervised the genetic project and edited the manuscript. All authors approved the final version of the manuscript.

Competing interests

The authors declare that they have no competing interests.

Consent for publication

Not applicable.

Ethics approval

All animal examinations were conducted according to national and international guidelines for animal welfare. This study was not based on an invasive animal experiment and used naturally occurring cases, therefore there are no associated permit numbers. The samples used were taken from a cattle farm in Italy and the owner agreed that the samples could be used in the study: data was obtained during diagnostic procedures that would have been carried out anyway. As the data is from client-owned cattle that underwent veterinary examinations, there was no "animal experiment" according to the exemptions contemplated by the Italian legislative decree n. 26/2014 (Dir. 2010/63/UE on the protection of animals used for scientific purposes).

Author details

[1]Institute of Genetics, Vetsuisse Faculty, University of Bern, Bern, Switzerland. [2]Servizio Veterinario dell'Azienda Sanitaria dell'Alto Adige, Bozen, Italy. [3]Laboratory on Urolithiasis Research, Department of Medicine, Surgery and Anatomy, Universidad de León, León, Spain. [4]Department of Veterinary Medical Sciences, University of Bologna, Ozzano dell'Emilia, Italy.

References

1. Agerholm JS, Christensen K, Nielsen SS, Flagstad P. Bovine renal lipofuscinosis: prevalence, genetics and impact on milk production and weight at slaughter in Danish cattle. Acta Vet Scand. 2009;51:7.
2. Seifi HA, Karimi K, Movasseghi AR. Renal amyloidosis in cattle - a case report in Iran. Zentralbl Veterinarmed B. 1997;44:631–3.
3. Dunham BM, Anderson WI, Steinberg H, King JM. Renal dysplasia with multiple urogenital and large intestinal anomalies in a calf. Vet Pathol. 1989;26:94–6.
4. Sasaki Y, Kitagawa H, Kitoh K, Okura Y, Suzuki K, Mizukoshi M, Ohba Y, Masegi T. Pathological changes of renal tubular dysplasia in Japanese black cattle. Vet Rec. 2002;150:628–32.
5. Ohba Y, Kitagawa H, Okura Y, Kitoh K, Sasaki Y. Clinical features of renal tubular dysplasia, a new hereditary disease in Japanese Black cattle. Vet Rec. 2001;149:115–8.
6. Castro MB, Szabo MPJ, Ferreira WL, Pereira AA. Renal dysplasia in a Limousin calf. Arquivo Brasileiro de Medicina Veterinária e Zootecnia. 2007;59:517–9.
7. Maxie MG, Newman SJ. Renal dysplasia. In: Kennedy PC, Palmer N, Jubb KVF, editors. Pathology of domestic animals, 5th edition. Amsterdam: Elsevier; 2007. pp 439–42.
8. Aresu L, Zanatta R, Pregel P, Caliari D, Tursi M, Valenza F, Tarducci A. Bilateral juvenile renal dysplasia in a Norwegian Forest Cat. J Feline Med Surg. 2009; 11:326–9.
9. Philbey AW, Mateus A, Bexiga R, Barrett DC, Haining HA, McCandlish IA, Thompson H. Renal dysplasia and nephrosclerosis in calves. Vet Rec. 2009;165:626–30.
10. Hirano T, Kobayashi N, Itoh T, Takasuga A, Nakamaru T, Hirotsune S, Sugimoto Y. Null mutation of PCLN-1/Claudin-16 results in bovine chronic interstitial nephritis. Genome Res. 2000;10:659–63.
11. Ohba Y, Kitagawa H, Kitoh K, Sasaki Y, Takami M, Shinkai Y, Kunieda T. A deletion of the paracellin-1 gene is responsible for renal tubular dysplasia in cattle. Genomics. 2000;68:229–36.
12. Hirano T, Hirotsune S, Sasaki S, Kikuchi T, Sugimoto Y. A new deletion mutation in bovine Claudin-16 (CL-16) deficiency and diagnosis. Anim Genet. 2002;33:118–22.
13. Simon DB, Lu Y, Choate KA, Velazquez H, Al-Sabban E, Praga M, Casari G, Bettinelli A, Colussi G, Rodriguez-Soriano J, McCredie D, Milford D, Sanjad S, Lifton RP. Paracellin-1, a renal tight junction protein required for paracellular Mg2+ resorption. Science. 1999;285:103–6.
14. Watanabe T, Ihara N, Itoh T, Fujita T, Sugimoto Y. Deletion mutation in Drosophila ma-l homologous, putative molybdopterin cofactor sulfurase gene is associated with bovine xanthinuria type II. J Biol Chem. 2000;275: 21789–92.
15. Miranda M, Rigueira L, Suárez ML, Carbajales P, Moure P, Fidalgo LE, Failde D, Vázquez S. Xanthine nephrolithiasis in a galician blond beef calf. J Vet Med Sci. 2010;72:921–3.
16. Testoni S, Mazzariol S, Drögemüller C, Piffer C, Aresu L, Gentile A. Renal dysplasia in grey Alpine breed cattle unrelated to CLDN16 mutations. Vet Rec. 2012;170:22.
17. Purcell S, Neale B, Todd-Brown K, Thomas L, Ferreira MA, Bender D, Maller J, Sklar P, de Bakker PI, Daly MJ, Sham PC. PLINK: a tool set for whole-genome association and population-based linkage analyses. Am J Hum Genet. 2007; 81:559–75.
18. Murgiano L, Wiedemar N, Jagannathan V, Isling LK, Drögemüller C, Agerholm JS. Epidermolysis bullosa in Danish Hereford calves is caused by a deletion in LAMC2 gene. BMC Vet Res. 2015;11:23.
19. Drögemüller C, Reichart U, Seuberlich T, Oevermann A, Baumgartner M, Kühni Boghenbor K, Stoffel MH, Syring C, Meylan M, Müller S, Müller M, Gredler B, Sölkner J, Leeb T. An unusual splice defect in the mitofusin 2 gene (MFN2) is associated with degenerative axonopathy in Tyrolean Grey cattle. PLoS One. 2011;6:e18931.
20. Murgiano L, Jagannathan V, Benazzi C, Bolcato M, Brunetti B, Muscatello LV, Dittmer K, Piffer C, Gentile A, Drögemüller C. Deletion in the EVC2 gene causes chondrodysplastic dwarfism in Tyrolean Grey cattle. PLoS One. 2014; 9:e94861.
21. European Nucleotide Archive. 2016. http://www.ebi.ac.uk/ena/data/view/ PRJEB11962. Accessed 21 Oct 2016.
22. Daetwyler HD, Capitan A, Pausch H, Stothard P, van Binsbergen R, Brøndum RF, Liao X, Djari A, Rodriguez SC, Grohs C, Esquerré D, Bouchez O, Rossignol MN, Klopp C, Rocha D, Fritz S, Eggen A, Bowman PJ, Coote D, Chamberlain AJ, Anderson C, VanTassell CP, Hulsegge I, Goddard ME, Guldbrandtsen B, Lund MS, Veerkamp RF, Boichard DA, Fries R, Hayes BJ. Whole-genome sequencing of 234 bulls facilitates mapping of monogenic and complex traits in cattle. Nat Genet. 2014;46:858–65.
23. Uhler JP, Spåhr H, Farge G, Clavel S, Larsson NG, Falkenberg M, Samuelsson T, Gustafsson CM. The UbL protein UBTD1 stably interacts with the UBE2D family of E2 ubiquitin conjugating enzymes. Biochem Biophys Res Commun. 2014;443:7–12.
24. Vizeacoumar FJ, Arnold R, Vizeacoumar FS, Chandrashekhar M, Buzina A, Young JT, Kwan JH, Sayad A, Mero P, Lawo S, Tanaka H, Brown KR, Baryshnikova A, Mak AB, Fedyshyn Y, Wang Y, Brito GC, Kasimer D, Makhnevych T, Ketela T, Datti A, Babu M, Emili A, Pelletier L, Wrana J, Wainberg Z, Kim PM, Rottapel R, O'Brien CA, Andrews B, Boone C, Moffat J. A negative genetic interaction map in isogenic cancer cell lines reveals cancer cell vulnerabilities. Mol Syst Biol. 2013;9:696.
25. Ensembl Genome Browser. Bovine MOCOS gene entry. 2016. http://www. ensembl.org/Bos_taurus/Gene/Summary?db=core;g=ENSBTAG00000012252; r=24:21198154–21256992. Accessed 21 Oct 2016.
26. Mendel RR. Cell biology of molybdenum. Biofactors. 2009;35:429–34.
27. Mendel RR, Kruse T. Cell biology of molybdenum in plants and humans. Biochim Biophys Acta. 1823;2012:1568–79.
28. Hille R, Nishino T, Bittner R. Molybdenum enzymes in higher organisms. Coord Chem Rev. 2011;255:1179–205.
29. Ichida K, Matsumura T, Sakuma R, Hosoya T, Nishino T. Mutation of human molybdenum cofactor sulfurase gene is responsible for classical xanthinuria type II. Biochem Biophys Res Commun. 2001;282:1194–200.
30. Dent CE, Philpot GR. Xanthinuria, an inborn error (or deviation) of metabolism. Lancet. 1954;266:182–5.

31. Kurzawski M, Dziewanowski K, Safranow K, Drozdzik M. Polymorphism of genes involved in purine metabolism (XDH, AOX1, MOCOS) in kidney transplant recipients receiving azathioprine. Ther Drug Monit. 2012;34:266–74.
32. Fujii T, Ozaki M, Masamoto T, Katsuma S, Abe H, Shimada T. A Bombyx mandarina mutant exhibiting translucent larval skin is controlled by the molybdenum cofactor sulfurase gene. Genes Genet Syst. 2009;84:147–52.
33. Peretz H, Naamati MS, Levartovsky D, Lagziel A, Shani E, Horn I, Shalev H, Landau D. Identification and characterization of the first mutation (Arg776Cys) in the C-terminal domain of the Human Molybdenum Cofactor Sulfurase (HMCS) associated with type II classical xanthinuria. Mol Genet Metab. 2007;91:23–9.
34. Yamamoto T, Moriwaki Y, Takahashi S, Tsutsumi Z, Tuneyoshi K, Matsui K, Cheng J, Hada T. Identification of a new point mutation in the human molybdenum cofactor sulferase gene that is responsible for xanthinuria type II. Metabolism. 2003;52:1501–4.

7

Insemination with border disease virus-infected semen results in seroconversion in cows but not persistent infection in fetuses

Ueli Braun[1]*(ORCID), Fredi Janett[1], Sarah Züblin[1], Michèle von Büren[1], Monika Hilbe[2], Reto Zanoni[3] and Matthias Schweizer[3]

Abstract

Background: This study examined various health variables in cows after artificial insemination with Border disease virus (BDV)-infected semen and the occurrence of persistent infection in ensuing fetuses. Five cows were inseminated (day 0) with BDV-infected semen as well as with semen from a fertile Eringer bull. One cow, inseminated with virus-free semen only, served as a control. Clinical examination, assessment of eating and rumination activities, measurement of intraruminal temperature and leukocyte count were used to monitor the health of the cows. Blood samples were collected at regular intervals for the detection of viral RNA and antibodies against BDV, and the cows were slaughtered on day 56. The uteri, placentae and fetuses were examined macroscopically, histologically, immunohistochemically and by means of molecular methods for the presence of pestiviruses.

Results: The demeanour, eating and rumination activities and intraruminal temperature were not affected by insemination with BDV-infected semen, whereas the total leukocyte and lymphocyte counts dropped transiently and were significantly lower on day 6 than on day 0. Seroconversion occurred by day 28 in the five infected cows but not in the control cow. The uteri, placentae and fetuses had no macroscopic or histological lesions, and immunohistochemical examination and RT-PCR were negative for pestiviruses.

Conclusions: The findings showed that cows inseminated with BDV-infected semen seroconverted and fetuses thus produced were not persistently infected. Transmission of BDV to cattle through infected semen, therefore, seems to be of minor importance.

Keywords: Border disease virus, BDV, Semen, Insemination, Seroconversion, Pestivirus, Cattle, Persistent infection

Background

It is well known that Border disease virus (BDV) can be transmitted from sheep to cattle under natural conditions as well as experimentally [1–9]. Transmission of BDV from a persistently-infected calf to a seronegative heifer in early pregnancy through direct contact was recently reported [10]. Seroconversion in cows was also observed after artificial insemination with semen from a bull persistently infected with BDV even though the cows failed to conceive because of poor semen quality [11]. Twice-daily monitoring showed no abnormalities in demeanour, appetite and rectal temperature of the cows [11] but temperature spikes of short duration may have been missed. Calves had spikes in intraruminal temperature as early as 2 days after intranasal infection with bovine viral diarrhoea (BVD) virus [12]. The mechanism by which BDV affects the leukogram in cattle is not known, but cattle with BVDV infection have leukopenia and lymphopenia [13]. The goal of the study was to determine whether artifical insemination of cows with BDV-infected semen results in persistently infected offspring. The effect of using BDV-infected semen on the health of recipient cows was also investigated by evaluation of general well-being and by determining whether viraemia, pyrexia and/or seroconversion occured.

* Correspondence: ubraun@vetclinics.uzh.ch
[1]Department of Farm Animals, Vetsuisse Faculty, University of Zurich, Zurich, Switzerland
Full list of author information is available at the end of the article

Methods

General procedure

Cows were artificially inseminated using semen from a young bull persistently infected with BDV. Eating and rumination variables were used as proxy for the health status and monitored using a pressure sensor that was incorporated into a halter [14, 15], temperature was measured using an intraruminal temperature-recording bolus, and blood samples were collected to determine the leukocyte count and the presence of BDV virus and antibodies. The cows were slaughtered 56 days after insemination, and the uteri, placentae and fetuses underwent macroscopic and histological examination and molecular testing for BDV.

Animals

Six healthy non-pregnant Swiss Braunvieh cows, 2.8 to 6.2 years (3.8 ± 1.1 years) of age, were used. The cows originated from several farms and had been sold for slaughter because of unsufficient milk production. In all cows, skin biopsy samples tested negative for pestivirus antigen and blood samples were negative for pestivirus antibody. Five cows (nos. 1, 3, 4, 5, 6) served as experimental cows and one (no. 2) served as a control.

Acclimation, oestrus synchronisation, artificial insemination and duration of infection phase

The cows were kept in quarantine during the entire study period and were acclimatised for 10 days. All cows underwent oestrus synchronisation and were artificially inseminated twice 24 h apart (day 0 and day 1 of the infection phase) using semen from a young bull persistently infected with BDV [10, 11]. The genome of the virus causing persistent infection (Sub-Genotype BDSwiss) has since been sequenced [16]. The semen had a high virus titre at 2.51×10^8 $TCID_{50}$ (50% tissue culture infective dose)/ml and 1.44×10^6 $TCID_{50}/10^6$ sperm cells. Because cows inseminated with infected semen failed to become pregnant in an earlier study [11], the cows of the present study were concurrently inseminated with virus-free semen from a bull of the Eringer breed with proven fertility. The goal was to transmit BDV with the semen from the persistently-infected bull and to impregnate the cows with the semen from the fertile bull. The control cow was inseminated with semen from the Eringer bull. The infection phase ended on day 56 when the cows were slaughtered.

Clinical examination and monitoring of eating, rumination and intraruminal temperature

All cows underwent twice-daily clinical examination, which included determination of general demeanour, appetite, rectal temperature, heart rate, respiratory rate, consistency of faeces and the presence of ocular and mucous membrane abnormalities as well as ocular and nasal discharge. To detect possible viraemia-associated reduction in eating and rumination, the cows were fitted with a custom-made halter equipped with a pressure sensor in the noseband (MSR Electronics GmbH, Seuzach, Switzerland) to record jaw movements during eating and rumination [14, 15]. Eating and rumination times, number of regurgitated boluses per day and the number of chewing cycles per bolus were recorded from 5 days before until 15 days after insemination. Body temperature was recorded using an intraruminal temperature bolus (ThermoBolus® Large, San'Phone®, Medria Technologies, Châteaubourg, France). Measurements were made twice hourly and transmitted to a server. The system was programmed to send a text message when temperatures of ≥ 40.5 °C were recorded to initiate blood sampling for the detection of viral RNA.

Haematological and virological examinations

Blood samples were collected every other day starting at day -2 (2 days before the first insemination) until 24 days after insemination for haematological evaluation. Evacuated EDTA tubes (Vacuette, Greiner Bio-One GmbH, Kremsmünster, Oesterreich) were used for blood collection. A haemogram that included a total leukocyte count and leukocyte differential were carried out for each blood sample. The blood samples were examined for viral RNA by means of real-time RT-PCR. RNA from leukocytes was isolated using the QIAamp RNeasy blood mini kit (Qiagen AG, Hombrechtikon, Switzerland), and RT-PCR was done using the Qiagen QuantiTect Probe One Step RT-PCR Kit (Qiagen) according to the manufacturer's instructions. The RT-PCR reactions were run on a thermocycler ABI 7300 (Applied Biosystems, Rotkreuz, Switzerland) with primers in the 5′-untranslated region (5′-UTR) which is used in our Swiss reference laboratory for ruminant pestiviruses and GAPDH as internal control [17]. The amount of viral RNA in the sample was expressed in Ct (cycle threshold) values; values of ≤ 30 were considered positive and values of > 30 up to < 45 as weakly positive. An in-house ELISA [18, 19] was used to detect pestivirus-specific antibodies in serum samples from the cows. This was done twice, 9 days apart, during the acclimation phase and eight times, 7 days apart, during the infection phase (days 7, 14, 21, 28, 35, 42, 49, 56), starting on day 0. Optical density (OD) values $> 30\%$ in relation to a standard serum were considered positive.

Blood samples collected on day 56 with a positive ELISA result underwent a cross-neutralisation test to identify the pestivirus species that induced the antibody response [20]. Briefly, sera were diluted tenfold in Eagles minimal essential medium and inactivated for 30 min at 56 °C. The sera were then further diluted in two-fold steps and incubated in 96-well plates for 1 h with a pre-determined dose of BDV (BDSwiss, R9336/11, isolated from blood of the persistently-infected bull) or with BVD virus (BVDV-1a: R1935/72). Subsequently a suspension of embryonic bovine turbinate cells was added to each well and further incubated for 4 to 5 days, and the cells were then examined for pestivirus using immunoperoxidase staining. A difference in neutralisation titres for BVDV and BDV that was at least four-fold was considered significant, whereas a ratio of lower than 4 was considered as 'indeterminate' [20].

Postmortem examination

All cows were slaughtered in the slaughterhouse of the Faculty on day 56, and the ovaries, uteri and placentae of the cows and skin, multiple bones, brain, heart, lungs, kidneys and intestinal organs of the fetuses were examined macroscopically and histologically. Monoclonal antibodies were used for immunohistochemical examination of fetal skin and organs (cryostat and paraffin sections) and uteri of the cows (paraffin sections) [21]. Cryostat sections were incubated with the BVD-specific antibody Ca3/34-C42 (dilution 1:100; Labor Dr. Bommeli AG, Bern, Switzerland) and the pestivirus-specific antibody C16/1/2 (Institute for Virology, University of Veterinary Medicine Hannover; kindly provided by Sophia Austermann-Busch, European Reference Laboratory for Classical Swine Fever and World Organisation for Animal Health). Paraffin sections were incubated with the pestivirus-specific antibody 15c5 (dilution 1:10,000, E. Dubovi, New York State College of Veterinary Medicine, Cornell University, USA) and the BVDV-specific antibody C42 (dilution 1:400, Prof. Moennig, Institute for Virology Hannover). Primary incubation was followed by incubation with the secondary antibody (DAKO Cytomation, EnVision+™, peroxidase, mouse, K 4001, Zug, Switzerland) and staining with AEC chromogens (DAKO, 3-Amino-9-Ethyl-Carbazole, K 3464). Positive and negative control slides were included in each run.

Samples of fetal thymus and small intestines, fetal and maternal placenta and uterine tissue of non-pregnant cows were examined for viral RNA. Tissue samples were frozen at − 80 °C, cut in small pieces with a scalpel and homogenised using the QIAshredder from the QIAamp RNeasy blood mini Kit (Qiagen) for isolation of total RNA. Virus detection was done by means of RT-PCR analogous to the method described for viral RNA detection in blood.

Pedigree analysis

Fetal tissue samples were subjected to genotyping to determine the sire (Institute for Reproduction of Farm Animals, Schönow, Germany).

Statistical analysis

The program IBM SPSS Statistics 24 (IBM Corporation) was used for analysis, and results were given as frequencies, means and standard deviations. A repeated measures ANOVA with Bonferroni correction and paired t-tests were used to analyse the profiles of eating and rumination variables, intraruminal temperature and white blood cell counts. Differences were considered significant at $P < 0.05$.

Results

Clinical findings

Insemination with BDV-infected semen did not affect the general demeanour of the cows, the duration of eating and rumination (Fig. 1), the number of regurgitated boluses per day or the number of chewing cycles per bolus. The mean intraruminal temperature varied within the reference range and did not differ significantly before (38.6 ± 0.11) and after insemination (38.8 ± 0.09 °C, Fig. 2).

Leukocytes and thrombocytes

The total leukocyte and lymphocyte counts of the experimental cows remained within the reference intervals throughout the study period. They decreased significantly from day 0 to day 6 ($P < 0.05$), after which time they returned to pre-insemination levels (Figs. 3 and 4). The thrombocyte counts did not differ before and after insemination and were in the reference interval in all cows (not shown).

Virus detection in blood and seroconversion

With one exception, viral RNA was not detected in any blood samples before and in the first 24 days after insemination. Cow no. 6 had a weakly positive Ct value of 41.5 on day 24.

All cows were seronegative before insemination and during the first week of the infection phase. Seroconversion with OD values > 30% occurred in all experimental cows two to 4 weeks after insemination (first positive OD values between 41 and 92%; Fig. 5). The OD values further increased to between 111 and 222% by the time of slaughter. The OD of the control cow remained negative between − 7 and + 1%. The SNT was positive for BDV in all experimental cows and negative in the control cow (Table 1). Titres of specific neutralising BDV antibodies varied from 190 to 538. The SNT was negative (cows 3, 5, 6, and control) or very low (cows 1 and 4) for BVDV. The quotient of BDV and BVDV antibody titres ranged from 14 to > 32 in all experimental cows, which were clearly classified as infected with BDV. Neutralising antibodies against BDV and BVDV were not detected in the control cow.

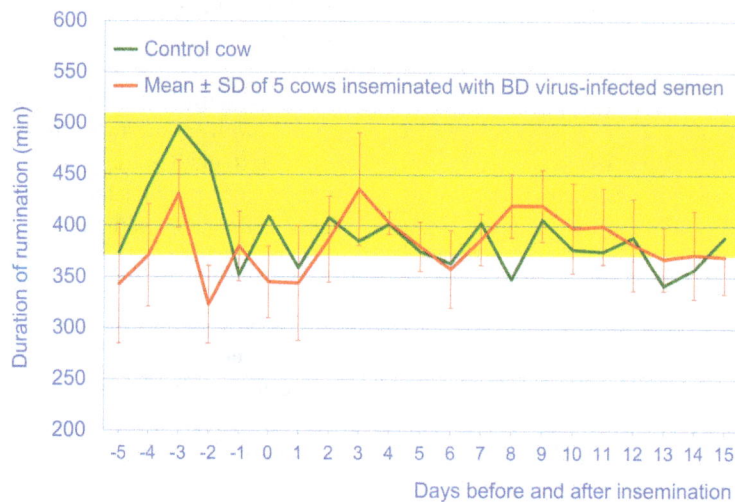

Fig. 1 Daily rumination time from 5 days before until 15 days after insemination in 5 cows artificially inseminated with BDV-infected semen (mean ± SD) and 1 control cow. The yellow bar represents the reference interval (370 to 511 min) established in 300 healthy cows [15]

Examination of uteri placentae, ovaries and fetuses

Four experimental cows (nos. 1, 4, 5, 6) were pregnant and one experimental and the control cow were not. All uteri, placentae, ovaries and fetuses were macroscopically and histologically normal, and the fetal organs and placentae did not yield pestiviral RT-PCR products.

Pedigree analysis

The Eringer bull, the source of the pestivirus-free semen, was identified as the sire of all fetuses.

Discussion

This study confirmed that cows inseminated with semen infected with BDV do not have overt clinical signs of illness [11] even with stringent health monitoring that included twice-hourly intraruminal temperature measurements and continuous recording of eating and rumination activities. The latter are sensitive criteria for the assessment of bovine wellbeing because sick cows usually have reduced rumination times, fewer regurgitated cuds and fewer chewing cycles per cud [22].

The most evident change in the leukogram was a significant decrease in the total leukocyte count on day 6 caused by lymphopenia, which was accompanied by normal neutrophil, eosinophil, basophil and monocyte numbers. Lymphopenia may be a response to stress-induced endogenous corticosteroid secretion [23] but can also occur in the acute phase of infection with viruses, *Ehrlichia*,

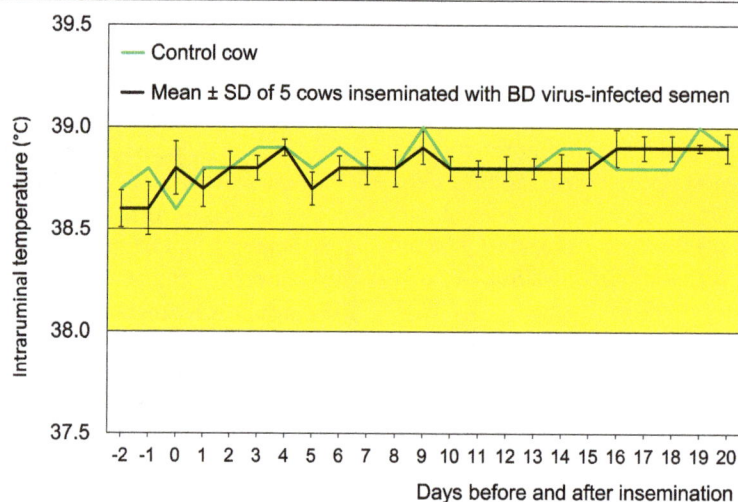

Fig. 2 Intraruminal temperature from 2 days before until 20 days after insemination in 5 cows artificially inseminated with BDV-infected semen (mean ± SD) and 1 control cow. The yellow bar represents the reference interval (38.0 to 39.0 °C) for the rectal temperature [32]

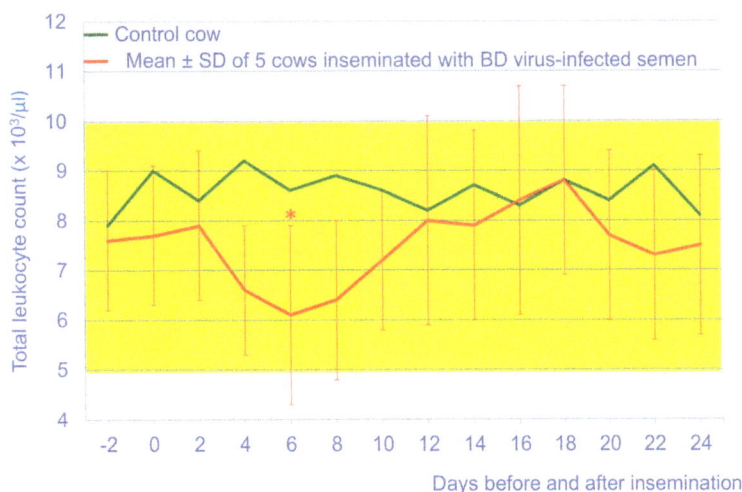

Fig. 3 Total leukocyte count from 2 days before until 24 days after insemination in 5 cows artificially inseminated with BDV-infected semen (mean ± SD) and 1 control cow. The yellow bar represents the reference interval (5 to $10 \times 10^3/\mu l$ blood) [33], * = different from day 0 ($P < 0.05$)

mycoplasma and other microorganisms or with septicaemia [23]. Lambs [24] and pregnant ewes [25] experimentally infected with BDV had significant leukopenia from day 2 to days 6 and 5 post-infection, respectively, with a nadir on day 4 [25]. Differential leukocyte counts were not reported in those studies, but it can be assumed that the leukopenia was attributable to lymphopenia. Leukopenia and lymphopenia were also seen in calves experimentally infected with BVD virus strains of different virulence [13].

Except for one equivocal result, virus was not detected in the present study. Similarly, viraemia was not detected in calves [4, 9] and heifers in early pregnancy [6] housed with sheep persistently infected with BDV. A possible explanation for this is that transient pestivirus infections

are characterised by short-lived and low-level viraemia, which makes detection of viral RNA almost impossible [4]. We believe that the weakly positive Ct value was due to minimal contamination in the laboratory because at the time of detection, the cow had already seroconverted. Seroconversion in cattle infected by sheep [4, 6] or other cattle persistently infected with BDV [9–11] has been reported. Eight heifers in early pregnancy co-housed with nine sheep persistently infected with BDV seroconverted 23 to 28 days after the start of exposure [6], and of nine calves co-housed with two persistently-infected sheep, six seroconverted after 36 to 72 days [10]. Six cows kept with a persistently-infected bull seroconverted after 20 to 40 days [10]. All of five cows had

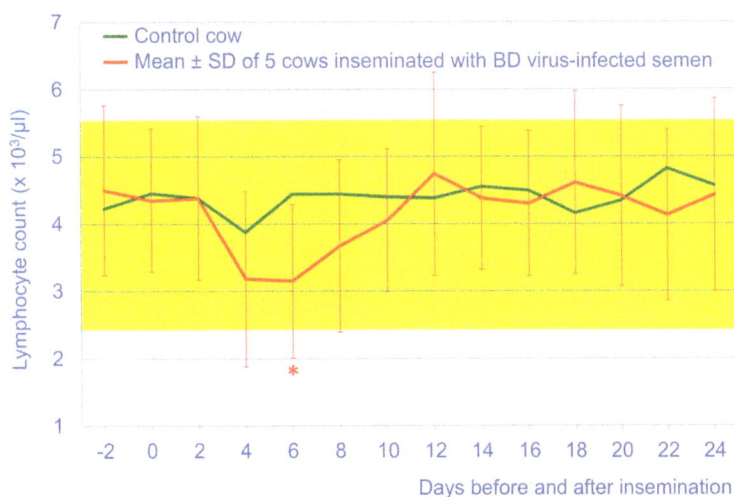

Fig. 4 Lymphocyte count from 2 days before to 24 days after insemination in 5 cows artificially inseminated with BDV-infected semen (mean ± SD) and 1 control cow. The yellow bar represents the reference interval (2.5 to $5.5 \times 10^3/\mu l$ blood) [33], * = different from day 0 ($P < 0.05$)

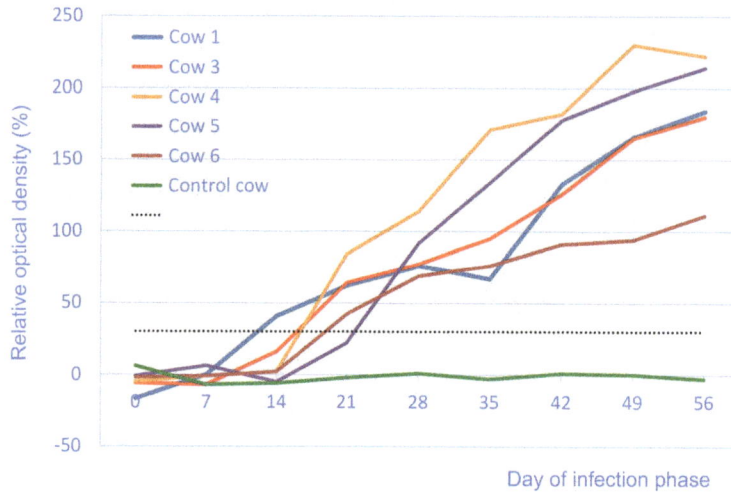

Fig. 5 Relative optical density (OD) in the ELISA for pestivirus antibody in the serum of 5 cows inseminated with BD virus-infected semen and a control cow from day 0 to day 56 of the infection phase expressed as a percentage to the OD of a standard serum. Relative OD values > 30% (dotted line) are defined as positive

seroconverted by day 28 after insemination with BDV-infected semen [11], and the same observation was made in the present study.

In a previous study, three heifers in early pregnancy that were in contact with a calf persistently infected with BDV had persistently-infected fetuses and viral RNA in blood samples [10]. Five of eight heifers co-housed in early pregnancy with persistently-infected sheep aborted infected fetuses and three gave birth to healthy calves [6]. In another experiment, insemination of fertile heifers with BDV-infected semen did not result in pregnancies because of poor semen quality [11]. This problem was circumvented in the present study by using infected semen and virus-free semen, which resulted in pregnancies but, surprisingly, not in infected fetuses. This was in contrast to the results of a study in which pregnant heifers housed with a persistently-infected calf gave birth

to persistently-infected offspring [10]. However, of 61 calves born to a BVDV-infected sire, only two (3.3%) were persistently infected [26]. Assuming a similar infection rate for BDV-infected semen, at least 30 cows would have had to be inseminated to generate one persistently-infected calf. Furthermore, acute BVDV infection only generates persistently-infected calves when it occurs from approximately day 30 to day 120 of pregnancy [27], whereas infection in the first month of pregnancy usually results in loss of pregnancy followed by return to estrus, or in a normal non-infected calf. Some researchers have speculated that the zona pellucida protects the conceptus from virus infection [28] or interferon-tau secreted by the trophoblast in the first 2 to 3 weeks of pregnancy has anti-viral properties [29]. Other factors including selection of a virus variant with different receptor requirements may also play a role in fetal infection [30]. To summarise, the findings of the present as well as earlier studies suggest that close contact with cattle [10] or sheep [4, 6, 20, 31] persistently infected with BDV plays a much bigger role in the pathogenesis of persistent infection than insemination with infected semen.

Table 1 Relative OD values in the Ab-ELISA, SNT titres and quotients of BDV and BVDV SNT titres on day 56 of the infection phase in 5 cows inseminated with BDV-infected semen and in 1 control cow

Cow	OD value (%)	SNT BD virus	SNT BVD virus	Quotient of BD and BVD virus SNT titres
1	184	190	14	14
3	180	226	Negative[a]	≥ 28
4	222	538	17	32
5	214	226	Negative[a]	≥ 28
6	111	190	Negative[a]	≥ 24
Control	-3	Negative[a]	Negative[a]	NA

NA Not applicable
[a] Limit of detection SNT ≤ 8

Conclusions

The findings of the present study showed that artifical insemination using BDV-infected semen led to infection and seroconversion in all cows, but did not result in persistently-infected offspring. This mode of pestivirus infection therefore appears to be unlikely albeit not impossible in a virus-free herd, in contrast to infection through close contact with cattle or sheep persistently infected with BDV.

Abbreviations
ABI: Applied biosystems; AEC: Amino-ethyl-carbazole; BDV: Border disease virus; BVD: Bovine viral diarrhoea; BVDV: Bovine viral diarrhea virus; Ct: Cycle threshold; EDTA: Ethylenediamintetraacetic acid; ELISA: Enzyme-linked immunosorbent assay; GAPDH: Glycerinaldehyde-3-phosphate-dehydrogenase; OD: Optical density; RNA: Ribonucleid acid; RT-PCR: Real-time polymerase chain reaction; SNT: Serum neutralization test; TCID: Tissue culture infective dose

Acknowledgements
The authors thank the technicians of the Medical Laboratory, of the Institute of Veterinary Pathology and of the Institute for Virology and Immunology for the examination of the samples and Hanspeter Müller for taking care of the animals.

Fundings
This study was financed by the University of Zurich, Switzerland.

Authors' contributions
UB initiated, planned and supervised the study and prepared the manuscript together with FJ, MH and MS. SZ and MvB performed the experiments under supervision of UB and FJ as part of their Mastertheses and were involved in the preparation of the manuscript, RZ was responsible for virologic examinations and he was involved in the interpretation and discussion of the results. All authors read and approved the final manuscript.

Competing interests
The authors declare that they have no competing interests.

Author details
[1]Department of Farm Animals, Vetsuisse Faculty, University of Zurich, Zurich, Switzerland. [2]Institute of Veterinary Pathology, Vetsuisse Faculty, University of Zurich, Zurich, Switzerland. [3]Institute for Virology and Immunology, and Department of Diseases and Pathobiology, Vetsuisse Faculty, University of Bern, Bern, Switzerland.

References
1. Carlsson U, Belák K. Border disease virus transmitted to sheep and cattle by a persistently infected ewe: epidemiology and control. Acta Vet Scand. 1994;35:79–88.
2. Becher P, Orlich M, Shannon AD, Horner G, König M, Thiel HJ. Phylogenetic analysis of pestiviruses from domestic and wild ruminants. J Gen Virol. 1997;78:1357–66.
3. Cranwell MP, Otter A, Errington J, Hogg RA, Wakeley P, Sandvik T. Detection of border disease virus in cattle. Vet Rec. 2007;161:211–2.
4. Krametter-Frötscher R, Benetka V, Möstl K, Baumgartner W. Transmission of border disease virus from sheep to calves – a possible risk factor for the Austrian BVD eradication programme in cattle? Wien Tierärztl Mschr. 2008; 95:200–3.
5. Hornberg A, Fernández SR, Vogl C, Vilček S, Matt M, Fink M, Köfer J, Schöpf K. Genetic diversity of pestivirus isolates in cattle from western Austria. Vet Microbiol. 2009;135:205–13.
6. Krametter-Frötscher R, Mason N, Roetzel J, Benetka V, Bago Z, Moestl K, Baumgartner W. Effects of border disease virus (genotype 3) naturally transmitted by persistently infected sheep to pregnant heifers and their progeny. Vet Med. 2010;55:145–53.
7. Strong R, La Rocca SA, Ibata G, Sandvik T. Antigenic and genetic characterisation of border disease viruses isolated from UK cattle. Vet Microbiol. 2010;141:208–15.
8. McFadden AMJ, Tisdall DJ, Hill FI, Otterson P, Pulford D, Peake J, Finnegan CJ, La Rocca SA, Kok-Mun T, Weir AM. The first case of a bull persistently infected with border disease virus in New Zealand. N Z Vet J. 2012;60:290–6.
9. Braun U, Reichle SF, Reichert C, Hässig M, Stalder HP, Bachofen C, Peterhans E. Sheep persistently infected with border disease readily transmit virus to calves seronegative to BVD virus. Vet Microbiol. 2014;168:98–104.
10. Braun U, Hilbe M, Janett F, Hässig M, Zanoni R, Frei S, Schweizer M. Transmission of border disease virus from a persistently infected calf to seronegative heifers in early pregnancy. BMC Vet Res. 2015;11:43.
11. Braun U, Frei S, Schweizer M, Zanoni R, Janett F. Transmission of border disease virus to seronegative cows inseminated with infected semen. Res Vet Sci. 2015;100:297–8.
12. Falkenberg SM, Ridpath J, Vander Ley B, Bauermann FV, Sanchez NCB, Carroll JA. Comparison of temperature fluctuations at multiple anatomical locations in cattle during exposure to bovine viral diarrhea virus. Livest Sci. 2014;164:159–67.
13. Chase CCL, Thakur N, Darweesh MF, Morarie-Kane SE, Rajput MK. Immune response to bovine viral diarrhea virus – looking at newly defined targets. Anim Health Res Rev. 2015;16:4–14.
14. Braun U, Trösch L, Nydegger F, Hässig M. Evaluation of eating and rumination behaviour in cows using a noseband pressure sensor. BMC Vet Res. 2013;9:164.
15. Braun U, Zürcher S, Hässig M. Evaluation of eating and rumination behaviour in 300 cows of three different breeds using a noseband pressure sensor. BMC Vet Res. 2015;11:231.
16. Stalder HP, Marti S, Flückiger F, Renevey N, Hofmann MA, Schweizer M. Complete genome sequences of three border disease virus strains of the same subgenotype, BDSwiss, isolated from sheep, cattle, and pigs in Switzerland. Genome Announc. 2017;5:e01238–17.
17. Schweizer M, Mätzener P, Pfaffen G, Stalder HP, Peterhans E. "self" and "nonself" manipulation of interferon defense during persistent infection: bovine viral diarrhea virus resists alpha/beta interferon without blocking antiviral activity against unrelated viruses replicating in its host cells. J Virol. 2006;80:6926–35.
18. Canal CW, Strasser M, Hertig C, Masuda A, Peterhans E. Detection of antibodies to bovine viral diarrhoea virus (BVDV) and characterization of genomes of BVDV from Brazil. Vet Microbiol. 1998;63:85–97.
19. Bachofen C, Vogt HR, Stalder H, Mathys T, Zanoni R, Hilbe M, Schweizer M, Peterhans E. Persistent infections after natural transmission of bovine viral diarrhoea virus from cattle to goats and among goats. Vet Res. 2013;44:32.
20. Kaiser V, Nebel L, Schüpbach-Regula G, Zanoni RG, Schweizer M. Influence of border disease virus (BDV) on serological surveillance within the bovine virus diarrhea (BVD) eradication program in Switzerland. BMC Vet Res. 2017;13:21.
21. Hilbe M, Arquint A, Schaller P, Zlinszky K, Braun U, Peterhans E, Ehrensperger F. Immunohistochemical diagnosis of persistent infection with bovine viral diarrhea virus (BVDV) on skin biopsies. Schweiz Arch Tierheilk. 2007;149:337–44.
22. Braun U, Tschoner T, Hässig M, Nuss K. Eating and rumination behaviour in cows with traumatic reticuloperitonitis. Schweiz Arch Tierheilk. 2017;159:101–8.
23. Tornquist S, Rigas J. Interpretation of ruminant leukocyte responses. In: Weiss DJ, Wardrop KJ, editors. Schalm's Veterinary Hematology. Ames: Wiley; 2010. p. 307–13.
24. Thabti F, Fronzaroli L, Dlissi E, Guibert JM, Hammami S, Pepin M, Russo P. Experimental model of border disease virus infection in lambs: comparative pathogenicity of pestiviruses isolated in France and Tunisia. Vet Res. 2002; 33:35–45.
25. García-Pérez AL, Minguijón E, Estévez L, Barandika JF, Aduriz G, Juste RA, Hurtado A. Clinical and laboratorial findings in pregnant ewes and their progeny infected with border disease virus (BDV-4 genotype). Res Vet Sci. 2009;86:345–52.
26. Kirkland PD, Mackintosh SG, Moyle A. The outcome of widespread use of semen from a bull persistently infected with pestivirus. Vet Rec. 1994;135:527–9.
27. Brownlie J. The pathogenesis of bovine virus diarrhoea virus infections. Rev Sci Tech Off Int Epiz. 1990;9:43–59.
28. Brock KV, Grooms DL, Givens MD. Reproductive disease and persistent infections. In: Goyal SM, Ridpath JF, editors. Bovine viral diarrhea virus: diagnosis, management and control. Ames: Blackwell Publishing; 2005. p. 145–56.

29. Schweizer M, Peterhans E. Pestiviruses. Annu Rev Anim Biosci. 2014;2:141–63.
30. Swasdipan S, McGowan M, Phillips N, Bielefeldt-Ohmann H. Pathogenesis of transplacental virus infection: pestivirus replication in the placenta and fetus following respiratory infection. Microb Pathog. 2002;32:49–60.
31. Braun U, Bachofen C, Schenk B, Hässig M, Peterhans E. Investigation of border disease and bovine virus diarrhoea in sheep from 76 mixed cattle and sheep farms in eastern Switzerland. Schweiz Arch Tierheilk. 2013;155:293–8.
32. Stöber M. Körpertemperatur. In: Dirksen G, Gründer HD, Stöber M, editors. Die klinische Untersuchung des Rindes. Berlin: Paul Parey; 1990. p. 131–5.
33. Stöber M, Gründer HD. Weisses Blutbild. In: Dirksen G, Gründer HD, Stöber M, editors. Die klinische Untersuchung des Rindes. Berlin: Paul Parey; 1990. p. 208–13.

8

Assisting differential clinical diagnosis of cattle diseases using smartphone-based technology in low resource settings

Tariku Jibat Beyene[1,2]*[iD], Amanuel Eshetu[1], Amina Abdu[1], Etenesh Wondimu[1], Ashenafi Feyisa Beyi[1,3], Takele Beyene Tufa[1], Sami Ibrahim[4] and Crawford W. Revie[5]

Abstract

Background: The recent rise in mobile phone use and increased signal coverage has created opportunities for growth of the *mobile Health* sector in many low resource settings. This pilot study explores the use of a smartphone-based application, *VetAfrica-Ethiopia*, in assisting diagnosis of cattle diseases. We used a modified Delphi protocol to select important diseases and Bayesian algorithms to estimate the related disease probabilities based on various clinical signs being present in Ethiopian cattle.

Results: A total of 928 cases were diagnosed during the study period across three regions of Ethiopia, around 70% of which were covered by diseases included in *VetAfrica-Ethiopia*. Parasitic Gastroenteritis (26%), Blackleg (8.5%), Fasciolosis (8.4%), Pasteurellosis (7.4%), Colibacillosis (6.4%), Lumpy skin disease (5.5%) and CBPP (5.0%) were the most commonly occurring diseases. The highest (84%) and lowest (30%) levels of matching between diagnoses made by student practitioners and *VetAfrica-Ethiopia* were for Babesiosis and Pasteurellosis, respectively. Multiple-variable logistic regression analysis indicated that the putative disease indicated, the practitioner involved, and the level of confidence associated with the prediction made by *VetAfrica-Ethiopia* were major determinants of the likelihood that a diagnostic match would be obtained.

Conclusions: This pilot study demonstrated that the use of such applications can be a valuable means of assisting less experienced animal health professionals in carrying out disease diagnosis which may lead to increased animal productivity through appropriate treatment.

Keywords: Cattle disease, Differential diagnosis, Ethiopia, Smartphone-based application, Bayesian inference

Background

Agriculture is a source of livelihood for an estimated 2.5 billion people and provides employment for 1.3 billion people globally [1, 2]. In developing countries, such as Ethiopia, livestock contribute around 30% of agriculturally related income [3]. Ethiopia has the largest livestock population in Africa, with cattle representing the largest segment at around 56 million animals [4]. Despite the fact that the economic contribution of livestock and their products account for around 20% of the total gross domestic product [5], 45% of the agricultural gross domestic product [5, 6], and directly contribute to the livelihoods of about 65% of Ethiopian families [7], the country has yet to fulfil its potential in this sector [8]. Numerous factors affect livestock production and productivity of which livestock disease is one of the most important [9]. This factor is exacerbated by lack of access to experienced veterinary services and advice, and consequently the mis-diagnosis and incorrect treatment of endemic cattle diseases.

In remote areas of Ethiopia, where most animal health assistants and community animal health workers practice,

* Correspondence: jibattariku@gmail.com
[1]College of Veterinary Medicine and Agriculture, Addis Ababa University, POBox 34, Bishoftu/Debre Zeit, Ethiopia
[2]Business Economics Group, Wageningen University, Hollandseweg 1, 6706 KN Wageningen, The Netherlands
Full list of author information is available at the end of the article

there is little access to continued professional development or to quality reference materials to which such practitioners can refer in cases of multiple tentative diagnoses [10]. Furthermore, under field conditions clinical diagnosis of cattle diseases can be complicated by similarity of clinical presentation [11]. So, where multiple similar signs co-occur, decision support tools can facilitate differential diagnosis [12].

Traditionally, the high burden of disease combined with large rural populations and limited infrastructure, has posed significant challenges for those seeking to tackle animal and human health issues. However, the recent rise in mobile phone usage, increased signal coverage and availability of low cost handsets, has created opportunities for growth in the nascent *mobile Health* sector in many low resource countries [13]. Over the past few years, a variety of *mobile-Health* projects have illustrated that mobile technology can be an appropriate vehicle to deliver medical and agricultural knowledge [14, 15] in a flexible and dynamic manner, as well as for the collection of field-based data [16]. The emergence of a number of software start-ups and technology vendors, such as Microsoft (www.microsoft.com/africa/4afrika/), focussing on Africa as an emerging market [17], illustrate the fact that this is a maturing sector with significant potential.

Previous studies carried out in sub-Saharan Africa have demonstrated that the use of low cost and accessible support tools can aid differential diagnosis and significantly improve the performance of animal health workers [18]. A pilot study in Uganda of a paper-based system not only demonstrated its value to the specific task of differential diagnosis in individual animals but also illustrated the utility of information on clinical signs and disease diagnoses in helping address general epidemiological questions related to syndromic surveillance and proportional disease morbidity [18].

However, further research was required if these early successes were to evolve into operational systems; both in terms of validating the robustness of such systems as well as to demonstrate that these novel diagnostic tools can have significant impacts on the health of cattle and livelihoods in rural communities. In this pilot study we document the introduction of a smartphone-based application, *VetAfrica-Ethiopia*, and explore its potential use in the differential diagnosis in cattle disease.

Methods
Study area and population
This study was conducted in 13 public veterinary clinics in the central (3 clinics), eastern (5 clinics) and southern (5 clinics) regions of Ethiopia shown on Fig. 1. Fifteen final-year veterinary medicine students from the College of Veterinary Medicine and Agriculture of Addis Ababa University took part in the study. The clinics were selected based on their willingness to provide students

supervision, while each student was allocated to a specific veterinary clinic based on the colleges' assignment for final year field clinical practice (fieldwork and case management) in different regions. Three female students were assigned to the same public veterinary clinic in the town in which the College is based (Bishoftu). The students were given basic training on how to use the smartphone app in clinical case management including how to record cases, connect to Internet and ensure that cases had been updated on the Cloud server, as well as to carry out rudimentary trouble-shooting such as restarting pages and editing data entry errors.

Disease selection
The initial selection of cattle diseases to be included in the smartphone-app was based on an interaction, using a modified Delphi protocol [19], with 17 experienced veterinarians from the College of Veterinary Medicine and Agriculture of Addis Ababa University in May 2014. Based on this Delphi exercise, 12 diseases were identified as being of particular significance for Ethiopia in terms of economic importance and/or had a high likelihood of providing a challenge in terms of the interpretation of clinical signs. These diseases were: Anthrax, Babesiosis, Blackleg, Contagious Bovine Pleuropneumonia (CBPP), Colibacillosis, Cowderiosis, Fasciolosis, Pasteurllosis, Parasitic Gastroenteritis (PGE), Rabies, Trypanosomosis and Bovine Tuberculosis.

Development of the *VetAfrica-Ethiopia* smartphone app
Once the diseases had been identified and the related probabilities of various clinical signs determined using simple Naïve Bayes estimation, an application was developed for a smartphone by the commercial partners in the project, Cojengo Ltd. In collaboration with one of the current authors (CWR), Cojengo had in 2014 developed the *VetAfrica* app for use in Uganda and Kenya. A revised version of the app, *VetAfrica-Ethiopia* (VAE), which in addition to being targeted towards appropriate diseases for use in Ethiopia also applied a Naïve Bayes classier to predict disease class, was distributed to the 15 student practitioners on low-cost smartphones.

As part of each student's regular fieldwork they assessed the disease status of cattle which were brought to the veterinary clinic in their district. A complete clinical examination was conducted on sick animals by the student practitioners under the supervision of the veterinarians on duty in the respective clinics. The basic data recorded in the *VetAfrica-Ethiopia* app included: date of examination; location of the case (village); breed, sex and age of the animal; the presenting clinical signs (or absence of same) were recorded and a tentative diagnosis was made by the student practitioner. Based on those signs that were reported to be either present or

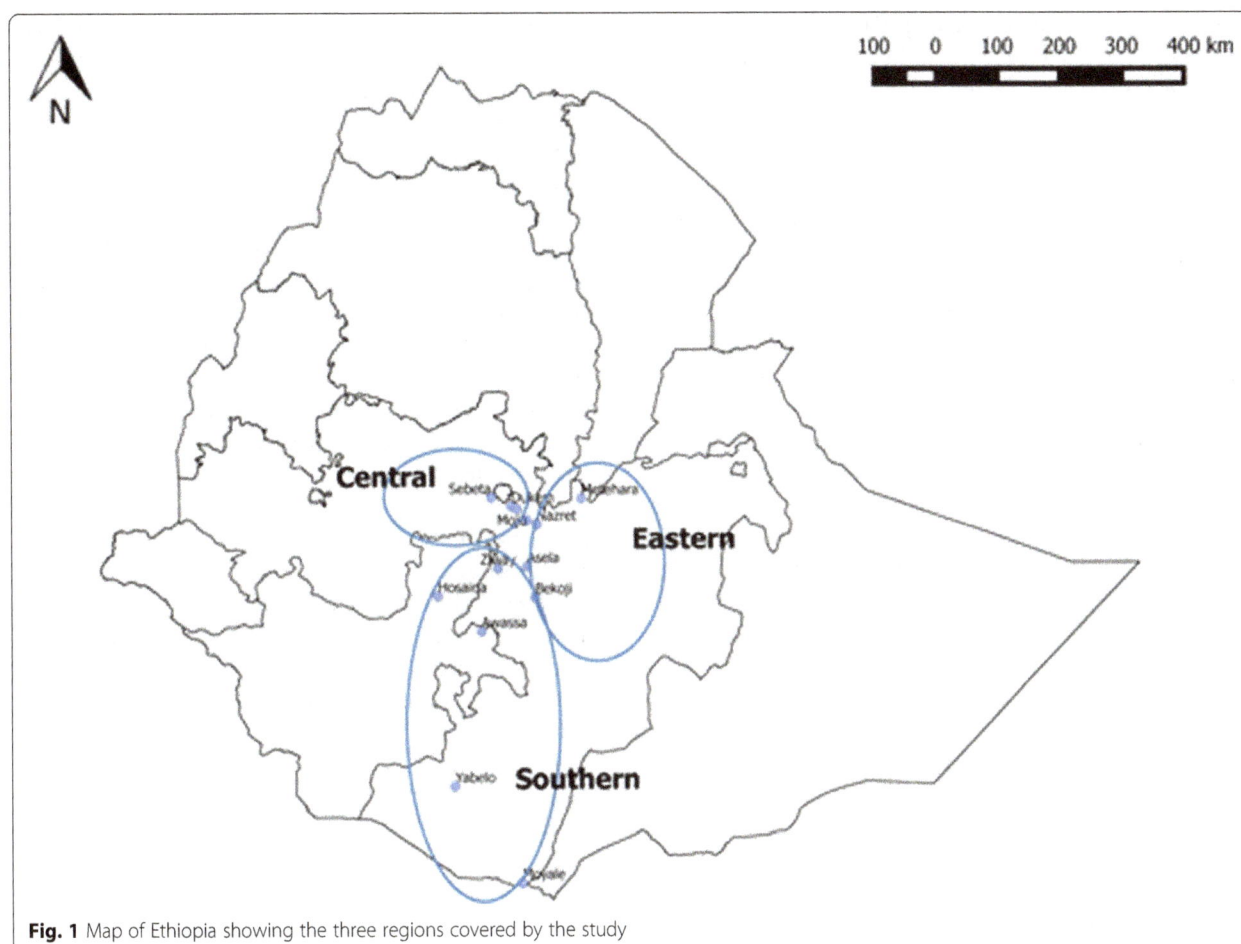

Fig. 1 Map of Ethiopia showing the three regions covered by the study

absent, the *VetAfrica-Ethiopia* app estimates the predicted likelihood for a range of tentative diagnoses in descending order of probability. The student practitioner therefore proposed a diagnosis using the clinical signs and any other relevant case history, while the app gave a range of tentative diagnoses with associated likelihoods. All of these data were then uploaded to a *Cloud*-based server (Microsoft Azure) for further analysis.

Estimation of proportional morbidity

The relative frequencies of a range of diseases from the total number of animals presenting at the public veterinary clinics during the study period, as diagnosed by the student practitioners, were computed to provide an estimate of proportional morbidity.

Evaluation of *VetAfrica-Ethiopia*

An evaluation of the level of agreement between the student practitioner's diagnoses and the prediction(s) of *VetAfrica-Ethiopia* was carried out. First we constructed a simple misclassification matrix to identify the degree of match for each disease. One problem with this simplistic approach to defining a 'match' is that it is sensitive to small changes in the probabilistic estimates associated with the diagnostic outcomes from the *VetAfrica-Ethiopia* app. For example, imagine a case where the student provided a diagnosis of "Blackleg" and *VetAfrica-Ethiopia* provided the following estimates: (Blackleg = 0.42) and (CBPP = 0.39). Strictly speaking this is a "match". However, *VetAfrica-Ethiopia* has also indicated that CBPP is highly likely and we may just have been 'lucky' that Blackleg was shown as 3% more likely. Conversely, what if we had obtained these results in a setting where the student indicated that her diagnosis was "CBPP"; would we not feel that the *VetAfrica-Ethiopia* app was 'almost' right? To deal with these situations we decided to label any case where the second most likely diagnosis suggested by *VetAfrica-Ethiopia* was within 20% of the likelihood value of its primary diagnosis, and where one of these two matched the student's diagnosis, to be a "marginal" case. Thus in the example given above, only when the second diagnosis was less than 0.336 (i.e. 20% lower than 0.42) would we treat Blackleg as a non-marginal case. Clearly in the case where the student stated that the diagnosis was, for example, "PGE" there is no need to consider the 'marginal' condition as both of the leading diagnoses proposed by *VetAfrica-Ethiopia* are incorrect. In the analysis we assess the impact of taking this approach to 'marginal' cases as well as the value of using the remaining 'clear-cut' cases in the regression modelling.

Univariate and multiple-variable logistic regression models were used to explore those variables which could predict the likelihood of there being a 'match' between the diagnosis made by the student practitioner and the diagnosis predicted by the *VetAfrica-Ethiopia* app for all non-marginal cases. A range of variables were screened using a univariate approach to look for candidates that should be preserved in the model; these included: the breed, age, and sex of the animals; the total number of clinical signs provided to *VetAfrica-Ethiopia*; a variable which reflected whether or not the diagnosis under consideration had been included in the *VetAfrica-Ethiopia* app; the diagnosis arrived at by *VetAfrica-Ethiopia*; Town (representing the individual student practitioners); and VAE_Max (the score given by *VetAfrica-Ethiopia* to the most likely diagnosis according to the results of its inference calculations, which was a proxy for the level of certainty the algorithm had in that particular diagnostic outcome). Any variable that had a *p*-value of less than 0.20 was considered for inclusion for the final model. A multiple-variable logistic model was then constructed.

Results

Characteristics of cases reported by student practitioners

The student practitioners reported on a total of 928 cattle cases that visited the veterinary clinics in the three regions. The breakdown of these cases by breed, sex and age group is shown in Table 1. The table indicates that a relatively higher number of animals were examined in the Southern region (around 40% of the total, with 30% in each of the

Table 1 Breakdown of all cases recorded during the study (*N* = 928) by region and in terms of proportions across key variables within these regions

	Central	Eastern	Southern
Total cases	282	279	367
by Sex			
Female	37.6%	41.2%	48.2%
Male	62.4%	58.8%	51.8%
by Breed			
Cross	3.5%	6.5%	7.9%
Exotic	4.6%	11.8%	4.1%
Local	91.8%	81.7%	88.0%
by Age Group			
0-6 months	1.8%	7.5%	13.1%
7-12 months	2.8%	5.7%	13.4%
13-24 months	11.0%	11.8%	10.9%
> 24 months	84.4%	74.9%	62.7%

other two regions). There were significantly more male animals examined (*p* = 0.02) in the Central and East regions, and the majority of cattle belonged to the oldest age category (over 24-months), though the Southern region reported significantly higher proportions (*p* < 0.01) of younger animals. In all regions, over 80% of the cattle presenting were local (zebu) breeds, with limited numbers of exotic and cross-bred animals. There are of course some important associations that are not captured in Table 1. For example, while there were significantly more male animals in total (57% versus 43% female), these proportions were significantly altered for the case of cross-bred (65% female to 35% male) and particularly exotic (79% female to 21% male) animals. This is due to the fact that the imported or cross-bred animals tend to be dairy cattle.

Proportional morbidity

The proportional morbidity (i.e. the relative frequency of each disease from the total of the 928 cases visiting the veterinary clinics during the study period) based on the diagnoses reported by the student practitioners is given in Table 2. In total just over 70 different diseases were diagnosed over the study period. All diseases that had 3 or fewer cases in the study, a total of around 45 separate diseases, are collected into a single category under the label "Other diseases" in Table 2; with the exception of Anthrax for which just a single case was recorded, but this disease has been left in the table as it was included in the *VetAfrica-Ethiopia* app. Around 70% of the cases were associated with a disease that had been identified as important by local veterinary experts and had been included in *VetAfrica-Ethiopia*.

The most common diagnosis, by some margin, was Parasitic Gastrointestinal (PGE) disease with almost 26% of the cases. Other relatively commonly occurring diseases, included: Blackleg (8.5%), Fasciolosis (8.4%), Pasteurellosis (7.4%), Colibacillosis (6.4%), Lumpy skin disease (5.5%) and CBPP (5.0%). Of the top 11 diseases, in terms of proportional morbidity, only Lumpy Skin Disease (LSD), Lungworm and Foot and Mouth Disease (FMD) had not been included in *VetAfrica-Ethiopia*. However, some diseases such as Rabies and Anthrax which were listed as important, perhaps due to their status as notifiable diseases, were found to have proportional morbidities of only 0.5% and 0.1% respectively.

Comparison of diagnoses made by practitioners and *VetAfrica-Ethiopia*

We used a misclassification matrix to explore the level of matching between the diagnosis provided by the student practitioner and that predicted by the *VetAfrica-Ethiopia* app based simply on the highest probability score. This helps to identify those diseases having a high level of discrepancy as shown in Table 3. Accordingly,

Table 2 Summary of proportional morbidity by disease across all cases (N = 928), including an indication as to which diseases were covered by the *VetAfrica-Ethiopia* app

Disease	Proportion	In *VAE*?
PGE	25.8%	Yes
Blackleg	8.5%	Yes
Fasciolosis	8.4%	Yes
Pasteurellosis	7.4%	Yes
Colibacillosis	6.4%	Yes
Lumpy Skin Disease (LSD)	5.5%	No
CBPP	5.0%	Yes
Babesiosis	2.7%	Yes
Lungworm	2.6%	No
Foot and Mouth Disease (FMD)	2.3%	No
Trypanosomiasis	2.2%	Yes
Salmonellosis	1.6%	No
Coccidiosis	1.6%	No
Tuberculosis	1.6%	Yes
Paratuberculosis	1.5%	No
Mastitis	1.4%	No
Actinobacillosis	1.4%	No
Actinomycosis	1.4%	No
Cowdriosis	1.2%	Yes
Pneumonia	0.9%	No
Tick Infestation	0.8%	No
Demodecosis	0.6%	No
Dermatophilosis	0.5%	No
Dermatophytosis	0.5%	No
Rabies	0.5%	Yes
Vesicular Stomatitis	0.5%	No
Anthrax	0.1%	Yes
Other diseases	7.1%	No

considering all 928 cases, Babesiosis was the disease with the highest level of matching; of the 25 cases diagnosed by the student practitioner, 21 were also predicted by the VAE app (84%) with 2 mismatches in each of Fasciolosis and 'Other' disease. Colibacillosis (81%) was the disease with the next highest level of match., In contrast, relatively low levels of matching were observed for Trypanosomiasis (45%) and Pasteurellosis (30%). While providing an outcome of "other" was a valid option from *VetAfrica-Ethiopia*, of the 281 cases for which the student practitioners made this choice, the app only suggested this to be most likely in 118 (42%) of these cases.

There were 89 cases for which the probability score predicted by the *VetAfrica-Ethiopia* app for a second diagnostic outcome was less than 20% lower than the probability score of the highest scoring diagnosis; where at least one of these diagnoses matched that proposed by the student practitioner. As can be seen from Table 4, around 67% of these "marginal" cases would have initially been deemed to represent a 'match' (i.e. the highest scoring diagnosis suggested by *VetAfrica-Ethiopia* was the same as that of the practitioner). In just under 33% of these cases we would initially have deemed there to be 'no match' as the *VetAfrica-Ethiopia* app only had the practitioner's putative diagnosis ranking as a 'close second'. It is thus obvious that discounting these marginal cases in no way inflated the apparent performance of the *VetAfrica-Ethiopia* app, if anything it was likely to slightly under-state the level of agreement. However, in seeking to better understand the factors associated with obtaining the same outcomes between practitioners and the app, it seemed wise to exclude these 89 intrinsically 'confusing' cases.

When considering the 839 (non-marginal) cases that remained, there were some diseases for which the level of match increased and others for which it decreased; for example the level of matching for Babesiosis was

Table 3 Misclassification matrix for all 928 cases, with student practitioner's diagnosis and *VetAfrica-Ethiopia* app prediction shown in vertical columns and horizontal rows respectively. Those where the two diagnoses are in agreement are shown in the shaded main diagonal

Student diagnosis	*VAE* diagnosis Anthrax	Babesi	Blackleg	CBPP	Colibac	Cowdri	Fasciol	Pasteur	PGE	Rabies	Trypano	Tubercu	'Other'	Match
Anthrax			1											0%
Babesiosis		21					2						2	84%
Blackleg			42	2	3	19	1	2	2		3		5	53%
CBPP	1	1	3	30		2		1		1	4		3	65%
Colibacillosis			2		48				6				3	81%
Cowdriosis					1	8					1		1	73%
Fasciolosis	1	1				1	54		14		2		5	69%
Pasteurellosis			1	22	2	5		21	2	3	6	2	5	30%
PGE		1	3		13	1	13		151	4	1	1	51	63%
Rabies						1			1	3				60%
Trypanosomiasis		1			1				1		9		8	45%
Tuberculosis				2				3				9	1	60%
'Other'	16	6	34	4	20	12	5	3	18	17	25	3	118	42%

Table 4 Distribution of 'marginal' cases by disease (n = 89)

	Initially categorised as a match?	
Disease	No	Yes
Babesiosis	–	2
Blackleg	1	2
CBPP	3	10
Colibacillosis	–	2
Fasciolosis	5	1
Pasteurellosis	4	9
PGE	11	11
Trypanosomiasis	–	1
Tuberculosis	–	1
Other diseases	5	21
Total	29	60

90% (up from 84%) while for Colibacillosis it was 70% (down from 81%). However, the graphical summary in Fig. 2 illustrates that the broad level of agreement across these cases was in line with that seen when considering all 928 animals. (There are in fact only 838 cases summarised in Fig. 2 and in the regression analyses that follow, as the single case of Anthrax in the study has also been excluded.)

Determinants of diagnosis matching between practitioners and VetAfrica-Ethiopia

Univariate logistic analyses were run, using the non-marginal 838 cases, to assess the potential of the available variables to predict diagnostic matching. This demonstrated no significant association (even at the $p < 0.20$ level) for the variables Age and Sex of the animal, or for the total number of clinical signs noted (S_Count). However, the Breed, Region, Town, User_Diag, In_VAE and VAE_Max variables were all found to be significant (p < 0.20) candidate predictors of a match between the student practitioner and the diagnosis made by the VetAfrica-Ethiopia app (Table 5).

The candidate variables were included in a multi-variable logistic model, with the exception of Region, as each Town (where a student practitioner was based) was situated in a specific geographical region and therefore Region could not be included along with Town. It was also found the Breed, and In_VAE were no longer significant, with the variability in the latter variable being mostly captured by the diagnosis being suggested (User_Diag). This resulted in a final model as summarised in Table 6.

As can be seen from Table 6, the likelihood of a match was significantly associated with the disease diagnosed; with Babesiosis (the diagnosis with the highest likelihood of a match) acting as the reference category. There were four disease options (including the catch-all 'Other' category) for which the likelihood of gaining a match were significantly lower. The value of VAE_Max was also noted as being a highly significant contributor to the likelihood of obtaining a matched outcome. The impact of this variable is illustrated by the lowess curves shown in Fig. 3, which indicate that as the value of the score

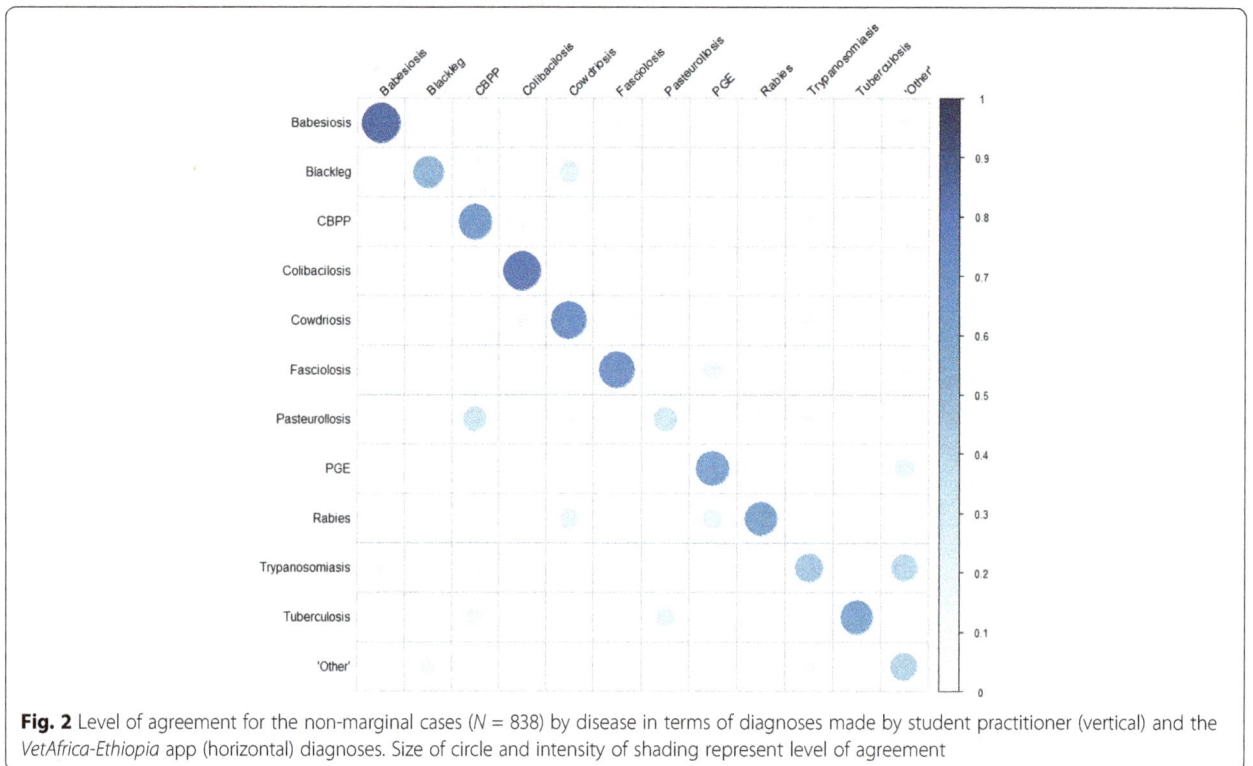

Fig. 2 Level of agreement for the non-marginal cases (N = 838) by disease in terms of diagnoses made by student practitioner (vertical) and the VetAfrica-Ethiopia app (horizontal) diagnoses. Size of circle and intensity of shading represent level of agreement

Table 5 Significance of each predictor in a simple univariate logistic model to predict the likelihood of a match ($N = 838$)

Variable Name	Explanation	P-value
Age	Age of animal	0.43
Sex	Sex of animal	0.86
Breed	Breed of animal	0.07
S_Count	Number of signs provided for this case	0.72
Region	Region of the country from which case came	0.00
Town	Town within which the student practitioner was working	0.00
User_Diag	The diagnosis provided by the student practitioner	0.00
In_VAE	Was diagnosis listed as a possible outcome within the VAE app?	0.00
VAE_Max	The actual maximum probability score associated with the diagnosis predicted to be the most likely match by the VAE app	0.00

Table 6 Summary of multivariable logistic model output for variables predicting a match between the diagnosis provided by the student practitioner and the VAE app ($N = 838$)

Matched	Coef.	Std. Err.	z	P > |z|	95% Confidence interval	
User_Diag						
Blackleg	−1.48	0.63	−2.34	0.02	−2.72	−0.24
CBPP	−1.02	0.70	−1.47	0.14	−2.39	0.34
Colibacillosis	−0.37	0.68	−0.54	0.59	−1.69	0.96
Cowdriosis	−0.76	0.92	−0.83	0.41	−2.56	1.04
Fasciolosis	−0.53	0.65	−0.81	0.42	−1.80	0.75
PGE	−0.95	0.60	−1.57	0.12	−2.13	0.23
Pasteurellosis	−3.26	0.69	−4.72	0.00	−4.61	−1.90
Rabies	−0.50	1.14	−0.44	0.66	−2.74	1.73
Trypanosomiasis	−1.52	0.75	−2.02	0.04	−2.99	−0.05
Tuberculosis	−1.02	0.83	−1.23	0.22	−2.64	0.60
Other	−2.05	0.60	−3.43	0.00	−3.23	−0.88
VAE_Max	1.31	0.49	2.69	0.01	0.35	2.26
Town						
HOS	−1.73	0.48	−3.58	0.00	−2.68	−0.78
ASS	−0.42	0.55	−0.77	0.44	−1.49	0.65
BI1	−0.80	0.48	−1.69	0.09	−1.73	0.13
BI2	−1.18	0.48	−2.48	0.01	−2.12	−0.25
BI3	−2.80	0.56	−5.00	0.00	−3.90	−1.70
BOK	−0.25	0.46	−0.55	0.58	−1.15	0.65
DUK	−2.72	0.54	−5.03	0.00	−3.78	−1.66
HWS	−0.97	0.53	−1.84	0.07	−2.00	0.07
MAT	−2.01	0.56	−3.56	0.00	−3.12	−0.90
MOD	−1.34	0.50	−2.65	0.01	−2.33	−0.35
MOY	−1.59	0.48	−3.34	0.00	−2.52	−0.66
SEB	−1.03	0.50	−2.04	0.04	−2.02	−0.04
YAB	−1.86	0.49	−3.83	0.00	−2.82	−0.91
ZIW	−1.77	0.44	−4.08	0.00	−2.63	−0.92
_cons	2.10	0.80	2.62	0.01	0.53	3.66

LR chi^2(26) = 204; Log likelihood = −475
Prob > chi^2 = 0.00; Pseudo R^2 = 0.18

associated with the most likely diagnosis (VAE_Max) increases, the likelihood of a match also increases, but with varying slope across diagnoses. For instance, the likelihood of matching does not vary a great deal for the case of Babesiosis (light red), while it does so in the case of Pasteurellosis (light green). The outputs in Fig. 3 are based on a scenario that assumes the student practitioner under consideration was from ADA, the best performing Town in terms of gaining a match. The output in Table 6 indicates that Town (i.e. the student involved) was significantly associated with the chance of a match; with 10 students have significantly less likelihood of obtaining a match than was the case for ADA (the referent student/Town). The full model, including the variables Diagnosis, Town and VAE_Max resulted in an AUC value of 0.77.

Discussion

In this pilot study we compared the level of match between the *VetAfrica-Ethiopia* app and a set of student practitioners' diagnoses. It was not possible to confirm cases by laboratory investigation as many clinics were far from suitable laboratories and such investigation would have been highly resource intensive. In the absence of a definitive identification of the disease causing agent, which is assumed to be the gold standard in many cases [20], we still wished to explore the likelihood that the app had made the 'correct' diagnosis. Strictly we did not make the assumption that the student practitioners had always suggested the correct diagnosis, but rather assessed how often the student practitioner's diagnosis and that provided by the *VetAfrica-Ethiopia* app were in agreement. To reduce potential bias in situations with a high degree of uncertainty, we introduced the concept of marginal cases.

It must be acknowledged that this was a pilot study and that as such the results should only be extrapolated with caution. For example, due to financial and logistic constraints we chose to use final year veterinary medicine students to evaluate the use of the *VetAfrica-Ethiopia* app in the field. We are aware that collecting data from students during their clinical rotation is not the same as observing an operational veterinarian in the field. In addition, in many such rural settings diagnostic assistance is provided by animal health technicians or community animal health workers; as such it would be valuable to assess the utility of this type of tool for those communities. There was some facility to capture more

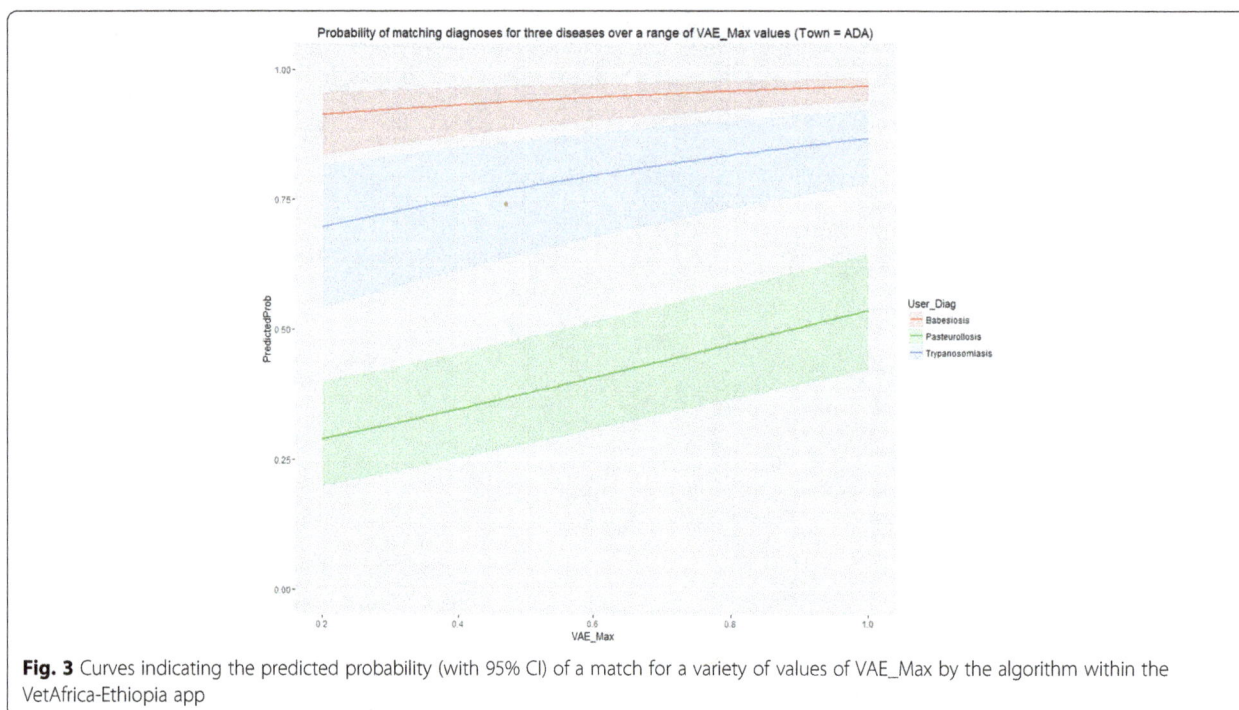

Fig. 3 Curves indicating the predicted probability (with 95% CI) of a match for a variety of values of VAE_Max by the algorithm within the VetAfrica-Ethiopia app

general case information within the app, including information of treatment options and notes as to whether samples were collected for laboratory analyses. However, these aspects of *VetAfrica-Ethiopia* functionality were not evaluated within this study. As such a complete consideration of how this type of technology might impact on the full spectrum issues involved in cattle disease management is only partially informed by this work. However, we do believe that the more limited goal set out here, that of assessing diagnostic capability, can be meaningfully summarised.

Misclassification analyses of the 12 identified diseases revealed that the highest level of matching occurred for the disease Babesiosis followed by Colibacillosis, while the lowest level of accuracy was seen in the case of Pasteurellosis. This may be due to the fact that diseases with the highest level of matching such as Babesiosis and Colibacillosis are less challenging in terms of the interpretation of clinical signs and tend to be well characterised. However, diseases with low levels of matching such as Pasteurellosis and Fasciolosis tend to have less clearly distinguished clinical signs or have similarities in terms of their clinical presentation to other diseases [21].

In this study we also found that the likelihood of a diagnostic match between the student practitioner and *VetAfrica-Ethiopia* depended on the particular student making the diagnosis. This could reflect differing degrees of diagnostic competence between students or their ability to use the mobile phone application. In addition each student was under the supervision of a more experienced veterinarian in the clinic to which they were

assigned, and some of the student-to-student variation could be due to a differential level of consultation with these supervisors. As their assignment to the various clinics was not random, bias may also have been introduced due to location-specific factors such as variable local mentoring or differential exposure to disease. In addition, the score given by *VetAfrica-Ethiopia* to the most likely diagnosis strongly influenced the chance of there being a match. This suggested that in future versions of the app, some minimum acceptable threshold of score should be set below which the app would state that the diagnosis was "inconclusive" or required additional clinical data, rather than providing putative diagnoses with low likelihoods.

As noted above, a limitation of this study was the inability to confirm the disease in each case based on a laboratory diagnosis. However, as we were attempting to cover a wide geographic area, it was not logistically feasible to access appropriate laboratories. Even if this had been feasible there would likely have been significant uncertainties associated with the diagnostic sensitivity and specificity of results for a number of these diseases [22]. In addition, if we assume a 'null' model, i.e. that cases fall on the leading 'matching' diagonal merely by chance, our expectation of obtaining a match would be just under 8%; as reported, the actual level of matching observed was more than 7 times this level. However, to aid with more complete validation in future, the authors have arranged workshops with veterinary experts from which one of the key outputs will be a set of 'control' cases with defined outcomes that will be available to be used to assess the performance of revised editions of *VetAfrica*.

The 12 cattle diseases of particular significance in terms of trade and economic importance or that were challenging in terms of the interpretation of clinical signs, as identified by experts during the Delphi exercise, were also found to be among the top listed diseases targeted for control by the veterinary services of Ethiopia [23]. The clinical signs chosen for inclusion in the app included those shared among many of the diseases [21]. The assignment of final year veterinary student practitioners to different veterinary clinics during their clinical rotation provided an opportunity to test this relatively low cost technology in a wider agro-ecology with a diverse distribution of livestock diseases. This demonstrated the potential for wider dissemination to veterinary professionals in remote areas. While all of the diseases selected by the experts were of some significance for Ethiopian cattle they are obviously not all equally prevalent. This was a simplifying assumption of this initial instantiation of *VetAfrica-Ethiopia* which may have affected the effectiveness of the diagnostic algorithm. This decision was made due to the absence of detailed disease prevalence estimates in cattle within the areas in which the app was tested. However, the algorithm has been set up to work equally well with any matrix of prior prevalence values as such those estimated as part of the proportional morbidity calculations made here could immediately be used, as could any other data- or model- driven estimates of more accurate priors for disease prevalence.

As far as signs, rather than disease distribution, are concerned the Naïve Bayes approach makes the assumption that each sign is independent with respect to its association with the diseases under consideration. However, the Bayesian Belief Network (BBN) that underlies the algorithmic approach here [24, 25] allows this assumption to be relaxed while maintaining mathematical integrity. Their main challenge, and the reason that the simpler Naïve Bayes assumptions were adopted here in the first instance, lies in estimating the non-independent (conditional) probabilities – which tend to grow in a combinatoric fashion. However, a range of data-driven learning algorithms such as tree-augmented Naïve Bayes (TAN) are available to help made such challenges tractable [26].

The student practitioners were given no instruction to select specific types of case so we assume that the breakdown is broadly reflective of the cases brought to public veterinary clinics seeking diagnostic and treatment services. Male cattle were more commonly seen in these veterinary clinic visits, perhaps reflecting the higher perceived value of males in terms of draft power and market sales value [27]. We found the large majority of the cattle presenting to be of local breed and that over two thirds of the animals were in the over 24-month category. These findings are in line with reports from the Central Statistical Authority of Ethiopia stating about 98.7% of cattle kept in Ethiopia are of indigenous breed and 65.6% are over 3 years of age [4]. Indeed the fact that the proportions of exotic and cross-bred cattle present in our study was well over the 1.3% that might be expected from the census data, may indicate the higher value of such cattle to the farmer and/or their higher level of susceptibility to endemic disease.

The intentions in quantifying proportional morbidity in this study were twofold: to characterise the distribution of disease among the animals examined and to assess the proportion of cases that were covered by those diseases included in the *VetAfrica-Ethiopia* app. The 12 diseases covered by *VetAfrica-Ethiopia* captured around 70% of the putative diagnoses made by the practitioners for cases they attended throughout the study period. The diseases most commonly diagnosed by practitioners were helminthiases: parasitic gastroenteritis (PGE) and bacterial diseases: Blackleg, Pasteurellosis and Colibacillosis. Lumpy skin disease, Lungworm and FMD were found to be diseases that presented relatively more frequently than some which were initially included in *VetAfrica-Ethiopia*, such as Rabies and Anthrax. While these cases were not confirmed by laboratory tests, the disease profile was consistent with the endemic diseases reported to the national veterinary services as published by the Ministry of Agriculture in the animal health year book [23]. They are also reflected in the range of veterinary drugs sold in veterinary drug shops in Ethiopia as authorised by the drug administration and control authority of Ethiopia [28]. However, as the reports included in the current study were based on cattle cases presented to public veterinary clinics they might not be entirely in line with the true proportional morbidity of disease occurrence. It has been reported elsewhere that some cases will be taken to traditional healers, rarely they may be seen by private veterinarians, while some will be left untreated [29–31]. The *VetAfrica-Ethiopia* developers, with inputs from veterinary experts, have now added the three most commonly occurring diseases not included in the app (LSD, lungworm and FMD) so that a set of 15 specific diseases have likelihood estimate values, which would cover just over 80% of the cases seen in this pilot study. One great advantage of a Cloud-based system where cases are available in near real-time is that an alert could easily be set up to let data administrators know when a given disease (currently not included in the app) had been recorded more than some minimum threshold of cases in a given time period. This could then initiate the required work with veterinary experts to explore the inclusion of this additional disease into the app.

The recent increase in availability of low cost mobile handsets and network coverage has created opportunities for applications in the health sector, especially in low resource settings [32, 33]. However, to date these have been predominantly focused in the domain of human health with only limited application to the challenges of diseases in animal populations [34]. Although such smartphone-based applications have several advantages in resource-limited settings, there are also limitations. These include the non-comprehensiveness of the disease list that we constructed at the beginning of the study. In this regard, three additional diseases which accounted for almost 10% of the total cases have been included in a revised version of the app. The diagnostic prediction of the app was also found to have limited accuracy for some diseases. To address this, the Bayesian learning aspect of the diagnostic algorithm supports ongoing modifications in sign-disease weightings. In addition, the fact that the score associated with the most likely diagnosis in VAE strongly influenced the likelihood of practitioner agreement, led to the adoption of a 'minimum acceptable likelihood threshold' below which no diagnosis will be suggested by VAE. Despite increased telecoms coverage in Ethiopia in recent years, the technology is still not entirely reliable and there were some instance where delays were experienced in uploading data to the Cloud. This was also in part due to some technical limitations of the Windows Mobile platform used during the study; to help address these, the revised version of VAE is available on Android devices which provides a more seamless interaction between off-line and on-line data access.

This pilot study has demonstrated the potential use of a smartphone-based application for animal disease diagnosis. Correctly diagnosing cattle diseases is known to be a key constraint on animal production efficiency in developing countries. To the authors' knowledge this is the first attempt to evaluate such an approach in a resource-limited setting. It also seems likely that such an approach would have great potential in other constrained sectors of veterinary service provision such as disease surveillance.

Conclusions

In conclusion, we have demonstrated that smartphone-based applications can be used by animal health professionals and have several advantages in resource-limited settings. In this pilot study we evaluated the performance of the *VetAfrica-Ethiopia* smartphone-based application based on the level of match between student practitioners' diagnoses and the app's predictions. The main findings of this study indicated that an acceptable overall level of matching could be achieved and that the major determinants of such matching were the disease being diagnosed, the diagnostic ability of the student practitioner and the level of certainty the *VetAfrica-Ethiopia* app assigned to the most likely diagnosis. It was shown that the higher this predicted likelihood, the more likely there would be a matching diagnosis, which

led to suggestions for design changes in the way that outputs from the algorithm were presented to the user. However, these likelihood values did vary according to disease and it should be noted that some important cattle diseases are currently not captured by the app. In addition the results from a pilot study involving 15 final year veterinary students should be treated with some caution, when drawing broader implication regarding wide-scale adoption. We have begun to explore the potential of smartphone applications such as *VetAfrica-Ethiopia* in providing assistance to less experienced animal health professionals. Further research, involving more definitive case outcomes, is required to fully access improvements in disease diagnosis and the provision of the most appropriate treatment advice; ultimately leading to an increase in animal productivity. We have also illustrated that the careful evaluation of such approaches can lead to better and ultimately more sustainable solutions.

Acknowledgements

We wish to thank district veterinary officers and clinicians as well as all of the 2015 graduating veterinary students from Addis Ababa University College of Veterinary Medicine and Agriculture who participated in the study.

Funding

This work was supported partly by funding from the International Development Research Centre (IDRC/CRDI)/Canada - Grant Number: 107,718-00020599-009.

Authors' contributions

TJB SI and CWR conceived and designed the experiments:, TJB AE AA EW AFB and TBT performed the experiments, TJB and CWR analysed the data, TJB AE AA EW AFB TBT and CWR wrote the paper. All authors read and approved the final manuscript.

Consent for publication

Not applicable.

Competing interests

The authors declare that they have no competing interests.

Author details

[1]College of Veterinary Medicine and Agriculture, Addis Ababa University, POBox 34, Bishoftu/Debre Zeit, Ethiopia. [2]Business Economics Group, Wageningen University, Hollandseweg 1, 6706 KN Wageningen, The Netherlands. [3]Department of Animal Sciences, University of Florida, Gainesville, FL, USA. [4]Cojengo Ltd, Glasgow, UK. [5]Department of Health Management, Atlantic Veterinary College, University of Prince Edward Island, Charlottetown, PEI, Canada.

Reference

1. World Bank. Agriculture for development. World development report 2008. Washington, DC: The World Bank; 2007. http://siteresources.worldbank.org/INTWDR2008/Resources/WDR_00_book.pdf.
2. FAO. State of food and agriculture (SOFA). Livestock in the balance. Rome: FAO; 2009. http://www.fao.org/docrep/012/i0680e/i0680e.pdf.
3. World Bank. Minding the stock: bringing public policy to bear on livestock sector development, Report no. 44010-GLB. Washington, DC: The World Bank, Agriculture and Rural Development Department; 2009. http://

siteresources.worldbank.org/INTARD/Resources/FinalMindingtheStock.pdf.

4. CSA. Report on livestock and livestock characteristics, agricultural sample survey 2014/15. Addis ababa: Federal Democratic Republic of Ethiopia Central Statistical Agency; 2015. http://www.csa.gov.et/newcsaweb/images/general/news/livestock%20report%202005%20ec_2012_13.pdf.

5. FAO. Food And Agriculture Organization Ethiopia Country Programming Framework. 2012-2015. Food and Agriculture Organization of the united nations office of the fao representative In Ethiopia To AU And ECA- Addis Ababa, December, 2011. 2011. http://www.fao.org/3/a-aq402e.pdf.

6. Behnke R. The contribution of livestock to the economies of IGAD member states: study findings, application of the methodology in Ethiopia and recommendations for further work. IGAD LPI working paper 02–10. Addis Ababa: IGAD Livestock Policy Initiative; 2010. http://www.fao.org/fileadmin/user_upload/drought/docs/IGAD%20LPI%20WP%2002-10.pdf.

7. Solomon A, Workalemahu A, Jabbar MA, Ahmed MM, Hurissa B. Livestock marketing in Ethiopia: a review of structure, performance and development initiatives. Socio-economics and policy research working paper 52. Nairobi: ILRI (International Livestock Research Institute); 2003. p. 35. http://www.fao.org/fileadmin/templates/agphome/images/iclsd/documents/wk2_c5_gerard.pdf.

8. Dorosh P, Rashid S. Food and agriculture in Ethiopia: progress and policy challenge. Washington, DC: IFPRI Issue Brief 74; 2013. http://www.upenn.edu/pennpress/book/209.html.

9. Sejian V, Naqvi, SMK, Ezeji T, Lakritz LR. Environmental stress and amelioration in livestock production. Springer-Verlag Berlin Heidelberg: Springer; 2012. doi:10.1007/978-3-642-29205-7.

10. Graham TW, Turk J, McDermott J, Brown C. Preparing veterinarians for work in resource-poor settings. J Am Vet Med Assoc. 2013;243(11):1523–8. doi:10.2460/javma.243.11.1523.

11. Dwinger R, Agyemang K, Kaufmann J, Grieve A, Bah M. Effects of trypanosome and helminth infections on health and production parameters of village N'Dama cattle in the cambia. Vet Parasitol. 1994;54(4):353–65.

12. Magona J, Walubengo J, Olaho-Mukani W, Revie CW, Jonsson N, Eisler M. A Delphi survey on expert opinion on key signs for clinical diagnosis of bovine trypanosomosis, tick-borne diseases and helminthoses. Bull Anim Health Prod Afr. 2004;52(3):130–40.

13. Bhatnagar S. Using ICT to improve governance and service delivery to the poor. Governance Dev Asia. 2015:296. doi:10.4337/9781784715571.00022.

14. WHO. Compendium of new and emerging health technologies. Geneva: World Health Organization - Technical Report Series; 2011.

15. Kaewkungwal J, Apidechkul T, Jandee K, Khamsiriwatchara A, Lawpoolsri S, Sawang S, Krongrungroj S. Application of mobile Technology for Improving Expanded Program on immunization among highland minority and stateless populations in northern Thailand border. JMIR mHealth and uHealth. 2015;3(1). doi:10.2196/mhealth.3704.

16. Lwin KK, Murayama Y. Web-based GIS system for real-time field data collection using personal mobile phone. J Geogr Inf Syst. 2011;3(04):382. doi:10.4236/jgis.2011.34037.

17. Norris P. Digital divide: Civic engagement, information poverty, and the Internet worldwide. Cambridge: Cambridge University Press; 2001.

18. Eisler MC, Magona JW, Revie CW. Diagnosis of cattle diseases endemic to sub-Saharan Africa: evaluating a low cost decision support tool in use by veterinary personnel. PLoS One. 2012;7(7). doi:10.1371/journal.pone.004068).

19. Gustafson L, Klotins K, Tomlinson S, Karreman G, Cameron A, Wagner B, Remmenga M, Bruneau N, Scott A. Combining surveillance and expert evidence of viral hemorrhagic septicemia freedom: a decision science approach. Prev Vet Med. 2010;94(1-2):140–53. doi:10.1016/j.prevetmed.2009.11.021.

20. Gardner IA, Hietala S, Boyce WM. Validity of using serological tests for diagnosis of diseases in wild animals. OIE Revue Scientifique et Technique. 1996;15(1):323–35.

21. Radostits OM, Gay C, Hinchcliff KW, Constable PD. A textbook of the diseases of cattle, horses, sheep, pigs and goats. Veterinary medicine. 10th ed. London: Saunders; 2007. p. 2045–50. doi:10.1016/S0737-0806(00)80409-8.

22. Pai NPVC, Denkinger C, Engel N, Pai M. Point-of-care testing for infectious diseases: diversity, complexity, and barriers in low- and middle-income countries. PLoS Med. 2012;9(9):e1001306. doi:10.1371/journal.pmed.1001306.

23. MOA. Animal health yearbook (2009/10). Ethiopia: Ministry of Agriculture Animal and Plant Health Regulatory Directorate; 2010.

24. Heckerman D. A Tutorial on Learning with Bayesian Networks. In: Holmes DE, Jain LC, editors. Innovations in Bayesian Networks. Studies in

Computational Intelligence, vol 156. Berlin, Heidelberg: Springer; 2008.

25. Seidel M, Breslin C, Christley RM, Gettinby G, Reid SWJ, Revie CW. Comparing diagnoses from expert systems and human experts. Agric Syst. 2003;76(2):527–38.

26. Friedman N, Geiger D, Goldszmidt M. Bayesian network classifiers. Mach Learn. 1997;29(2):131–62.

27. Kassie G, Abdulai A, Wollny C, Ayalew W, Dessie T, Tibbo M, Haile A, Okeyo AM. Implicit prices of indigenous cattle traits in central Ethiopia: Application of revealed and stated preference approaches. ILRI Research Report 26. Nairobi: ILRI; 2011.

28. DACA. List of veterinary drugs for Ethiopia. Addis Ababa: Drug Administration and Control Authority (DACA) of Ethiopia; 2002.

29. Mesfin T, Obsa T. Ethiopian traditional veterinary practices and their possible contribution to animal production and management. Revue scientifique et technique (International Office of Epizootics). 1994;13(2):417–24.

30. Tolossa K, Debela E, Athanasiadou S, Tolera A, Ganga G, Houdijk JG. Ethno-medicinal study of plants used for treatment of human and livestock ailments by traditional healers in south Omo, southern Ethiopia. J Ethnobiol Ethnomed. 2013;9(1):32. doi:10.1186/1746-4269-9-32.

31. Yigezu Y, Haile DB, Ayen WY. Ethnoveterinary medicines in four districts of Jimma zone, Ethiopia: cross sectional survey for plant species and mode of use. BMC Vet Res. 2014;10(1):76. doi:10.1186/1746-6148-10-76.

32. Hall CS, Fottrell E, Wilkinson S, Byass P. Assessing the impact of mHealth interventions in low-and middle-income countries–what has been shown to work? Glob Health Action. 2014;7. doi:10.3402/gha.v7.25606.

33. Qiang CZ, Yamamichi M, Hausma V, Altman D, Unit IS. Mobile applications for the health sector. Washington, DC: World Bank; 2011.

34. Robertson C, Sawford K, Daniel SLA, Nelson TA, Stephen C. Mobile phone–based infectious disease surveillance system, Sri Lanka. Emerg Infect Dis. 2010;16(10):1524–31. doi:10.3201/eid1610.100249.

9

Distinct correlations between lipogenic gene expression and fatty acid composition of subcutaneous fat among cattle breeds

David Gamarra[1], Noelia Aldai[2*], Aisaku Arakawa[3], Luis Javier R. Barron[2], Andrés López-Oceja[1], Marian M. de Pancorbo[1] and Masaaki Taniguchi[3*] (iD)

Abstract

Background: The fatty acid (FA) composition of adipose tissue influences the nutritional quality of meat products. The unsaturation level of FAs is determined by fatty acid desaturases such as stearoyl-CoA desaturases (SCDs), which are under control of the transcription factor sterol regulatory element-binding protein (SREBP). Differences in *SCD* genotype may thus confer variations in lipid metabolism and FA content among cattle breeds. This study investigated correlations between FA composition and lipogenic gene expression levels in the subcutaneous adipose tissue of beef cattle breeds of different gender from the Basque region of northern Spain. Pirenaica is the most important beef cattle breed in northern Spain, while Salers cattle and Holstein-Friesian cull cows are also an integral part of the regional beef supply.

Results: Pirenaica heifers showed higher monounsaturated FA (MUFA) and conjugated linoleic acid (CLA) contents in subcutaneous adipose tissue than other breeds ($P < 0.001$). Alternatively, Salers bulls produced the highest oleic acid content, followed by Pirenaica heifers ($P < 0.001$). There was substantial variability in *SCD* gene expression among breeds, consistent with these differences in MUFA and CLA content. Correlations between *SCD1* expression and most FA desaturation indexes (DIs) were positive in Salers ($P < 0.05$) and Pirenaica bulls, while, in general, *SCD5* expression showed few significant correlations with DIs. There was a significant linear correlation between *SCD1* and *SRBEP1* in all breeds, suggesting strong regulation of *SCD1* expression by *SRBEP1*. Pirenaica heifers showed a stronger correlation between *SCD1* and *SREBP1* than Pirenaica bulls. We also observed a opposite relationship between *SCD1* and *SCD5* expression levels and opposite associations of isoform expression levels with the Δ9 desaturation indexes.

Conclusions: These results suggest that the relationships between FA composition and lipogenic gene expression are influenced by breed and sex. The opposite relationship between *SCD* isoforms suggests a compensatory regulation of total SCD activity, while opposite relationships between *SCD* isoforms and desaturation indexes, specially 9c-14:1 DI, previously reported as an indicator of SCD activity, may reflect distinct activities of SCD1 and SCD5 in regulation of FA content. These findings may be useful for beef/dairy breeding and feeding programs to supply nutritionally favorable products.

Keywords: Cattle, Desaturation index, Fatty acid composition, Gene expression, Lipid metabolism, *SCD*, Subcutaneous adipose tissue, *SREBP*, Δ9-desaturase

---

* Correspondence: noelia.aldai@ehu.eus; masaakit@affrc.go.jp
[2]Lactiker Research Group; Lascaray Research Center, University of the Basque Country (UPV/EHU), 01006 Vitoria-Gasteiz, Spain
[3]Animal Genome Unit, Institute of Livestock and Grassland Science, National Agriculture and Food Research Organization (NARO), Tsukuba 305-0901, Japan
Full list of author information is available at the end of the article

Background

In recent years, consumers have expressed growing concern regarding the amount and types of dietary fat due to reported deleterious health effects of saturated and *trans* fatty acids (FAs) [1]. On the other hand, monounsaturated FAs (MUFAs) and polyunsaturated FAs (PUFAs) are recognized as beneficial for human health [2]. The FA composition of meat influences the lipid melting point [3], and an increase in the ratio of MUFA to saturated FA (SFA) increases fat softness, thereby improving palatability. Thus, enhancing MUFA content improves both the quality and nutritional value of animal products [4].

Pirenaica is the most important beef cattle breed raised in the Basque region of northern Spain and is highly appreciated both for its value as a genetic resource as well as for the production system that has developed around it. In addition to Pirenaica, Salers, a rustic cattle breed used for beef production, has grown in importance due to its ready adaptability to local management and environmental conditions [5]. Finally, Holstein-Friesian, primarily as cull dairy cows, are also an integral part of the regional beef supply chain.

There have been a number of studies investigating associations between lipogenic enzyme genotype and FA composition in cattle. In the subcutaneous and intramuscular fat depots of beef cattle, the majority of SFA conversion to MUFA is catalyzed by stearoyl-CoA desaturase (SCD, EC 1.14.19.1 or $\Delta9$-desaturase) [6]. In addition, SCD enzyme can also catalyze the conversion of substrates like vaccenic acid (11 *t*-18:1) to its corresponding conjugated linoleic acid (CLA) isomer (9*c*,11 *t*-18:2 or rumenic acid). The association of *SCD*1 genotype with FA composition has been previously investigated in Japanese Black [7], Canadian Holstein [8], Fleckvieh [9] and crossbred cattle [10]. In addition to regulation of FA profile by *SCD1*, the novel $\Delta9$-desaturase isoform *SCD5*, previously found in humans, has also been identified in cattle which shares 65% identity at the amino acid level [11]. Further, a relationship between genetic polymorphisms in *SCD5* and the ratio of SFA to unsaturated FA (UFA) has been reported in Holstein milk fat [12]. It thus appears that both bovine isoforms *SCD*1 and *SCD5* contribute to FA composition. Therefore, the mechanism by which *SCD* isoforms are activated is a major determinant of FA composition and of great interest to breeders.

Da Costa et al. [13] reported a correlation between *SCD1* expression and FA composition of subcutaneous fat in Portuguese cattle, whereas the expression of *SCD5* and its relation to the subcutaneous FA profile was not investigated. The *SCD1* gene is controlled by the key transcription factor *SREBP1* [14]. In Japanese Black beef cattle, *SREBP1* polymorphisms have been associated

with FA composition [15]. Alternatively, transcriptional regulation of bovine *SCD5* remains unclear, although a recent study using human choriocarcinoma trophoblastic cells (JEG3) reported that *SREBP1* can bind to the *SCD5* promoter [16].

The complex associations between the biochemical pathways regulating fat content and genetic variability of lipogenic genes are not yet fully understood in European cattle breeds, although recent studies have begun to elucidate these relationships in a specific genetic background of Japanese Black cattle [17]. The objectives of the present study were to investigate the expression levels of three key genes controlling $\Delta9$-desaturated FA content, *SCD1*, *SCD5*, and *SREBP1*, and their associations with the FA composition of subcutaneous adipose tissue from the major commercial cattle breeds produced in northern Spain, Pirenaica, Salers, and Holstein-Friesian. Based on the findings of this study, we discuss how these associations may give information on the mechanisms of the differences in meat quality among these cattle commercial types.

Methods

Sample collection

In the present study, cattle commercial types typically destined for meat production in northern Spain (Basque region) were examined. Sample collection was designed according to data from the Bovine Identification Document and inferred relationships (parentage and sibships) computed from 29 microsatellites (Software Colony 2.0. 6.2) [18]. Neither parentage nor maternal half-sibs were observed, and paternal half-sibs were maintained at low frequencies (Pirenaica, 0.009; Salers 0.013; Holstein-Friesian, 0.019). A total of 100 subcutaneous adipose tissue samples were collected from pure breed cattle (13 Salers bulls, 37 Pirenaica bulls, 29 Pirenaica heifers, and 21 Holstein-Friesian) slaughtered in a local commercial abattoir (Urkaiko S. Coop., Zestoa, Gipuzkoa, Spain) during 12 days over 5 weeks in June and July 2014. Animals came from different farms [19].

Backfat samples were obtained from the left half carcass between the 5-6th ribs and stored in plastic bags with the air removed for FA analysis or preserved in RNAlater™ (Ambion, Austin, TX) for RNA analysis. All samples were transported to the laboratory in insulated coolers and stored at − 80 °C until analysis.

Salers and Pirenaica were yearling calves with similar age (average of 12.9 ± 1.4 months), while Holstein-Friesian were cull cows (70.0 ± 19.43 months) at slaughter, which are regular ages of commercial types used for beef production in the region. Hot carcasses of Salers and Pirenaica commercial bulls were of similar weight (average of 325 ± 38.4 kg) while carcasses of Pirenaica heifers and Holstein-Friesian cows were markedly lighter (291 ± 59.6 and 253 ± 33.1 kg, respectively).

In the abattoir, conformation and degree of fat cover of each carcass were recorded. European regulations were followed for carcass classification at 24 h post-mortem [20] including the EUROP scale for conformation and a 1-to-5 scale for fat cover scoring. Each level of both scales was divided in 3 sub-levels and transformed to a numerical scale ranging from 1 to 15, with 15 being the best conformation and the thickest fat cover.

Fatty acid composition

A 50 mg sample of subcutaneous fat tissue was weighed, freeze-dried, and directly methylated with sodium methoxide (0.5 N methanolic base, Supelco Inc., Bellefonte, PA, USA) [21]. For quantitation, 1 mL of internal standard (23:0 methyl ester) was added prior to methylation, and FA methyl esters were analyzed by gas chromatography with flame ionization detection (GC/FID) using two complementary 100 m columns (SP-2560 [22] and SLB-IL111 [23]) and following the conditions and details reported in [24]. Main FA groups and potential Δ9 substrates, products, and inhibitors (10 t,12c-18:2; [25]) have been determined for the individual FAs measured in this study. From the potential substrates (14:0, 15:0, 16:0, 17:0, 18:0, 19:0, 20:0, 6-8 t-18:1, 11 t-18:1, 12 t-18:1, 13 t/14 t-18:1 and 15c-18:1) and products (9c-14:1, 9c-15:1, 9c-16:1, 9c-17:1, 9c-18:1, 9c-19:1, 9c-20:1, 7 t,9c-18:2, 9c,11 t-18:2, 9c,12 t-18:2, 9c,13 t-18:2, 9c,15c-18:2), individual desaturation indexes were calculated by the following formula:

$$\text{Desaturation index (DI)} = [\text{product}]/([\text{substrate}]$$

$$+[\text{product}]).$$

Total DI (sum of all individual DIs) was also computed for each commercial type, while minor products and substrates (i.e., 11 t,15c-18:2 & 9c,11 t,15c-18:3 [26]) or below quantification limits were not considered in the present study.

RNA extraction and quantitative real-time PCR

A 100 mg sample of frozen subcutaneous fat tissue was disrupted and simultaneously homogenized to fine powder with a mortar and pestle under liquid N_2. Total RNA was extracted using the RNeasy Lipid Tissue kit (Qiagen Inc., Valencia, CA, USA) following the manufacturer's instructions. An additional DNase digestion step was performed to remove any contaminating genomic DNA. Concentration and quality of the extracted RNA were assayed by measuring the 260 nm and 280 nm absorbance using a NanoDrop ND-1000 Spectrophotometer (Peqlab, Erlangen, Germany). Absorbance ratios

(260/280) of all preparations were at least 1.8. Integrity of RNA was checked by denaturing agarose gel electrophoresis. Aliquots of RNA were stored at − 80 °C and dehydrated in RNAstable 96-Well Plates (Biomatrica, San Diego, CA, USA) for long-term storage. Reverse transcription was performed in a 30 μL final reaction volume containing 250 ng total RNA, 3.3 μL RNase/DNase-free water, 5 μL of 5 × RT buffer, 1.5 μL dNTPs, 0.8 μL RNAase inhibitor, 0.8 μL random primer, and 0.8 μL high efficient ReverTra Ace reverse transcriptase (TOYOBO, Osaka, Japan). Cycle parameters were 30 °C for 10 min, 42 °C for 20 min, 99 °C for 5 min, and 4 °C for 5 min. Custom TaqMan Assays (Applied Biosystems, Foster City, CA, USA) were conducted to measure the relative expression levels of bovine *SCD1*, *SCD5* and *SREBP1* using the primers and FAM/TAMRA probes reported in Table 1. Each candidate gene was amplified in multiplex with an internal control (18S rRNA Endogenous Control VIC/TAMRA Probe, Primer Limited) by the co-application reverse transcription method (Co-RT) [27]. This multiplexing approach guarantees the same conditions (thus equal amplification efficiency) and same reverse transcriptase activity for both genes, thereby yielding better normalization and reproducibility. The reaction mixture included primers (10 μM each), FAM-labeled probe (10 μM), 0.6 μL of 18S RNA Endogenous Standard containing VIC-labeled probe and limited primers, and 2 × TaqMan Gene Expression Master Mix (7.5 μL) (Applied Biosystems). Real-time PCR was performed in triplicate using the ABI Prism 7500 Sequence Detection System (Applied Biosystems, Foster City, CA, USA) with a standard two-step cycling program of 40 cycles at 95 °C for 15 s and 60 °C for 1 min. The average of the gene expression levels was used for further analyses. PCR efficiency was monitored by the increase in absolute fluorescence [28], mainly because this allows PCR efficiency calculation for individual samples/reactions and prevents problems arising from the use of standard curves. Raw data were obtained from the ABI Prism 7500 SDS software v1.4, exported in Rn format, and imported to LinRegPCR (Heart Failure Research Center, Amsterdam, the Netherlands). LinRegPCR determines baseline fluorescence, sets a window of linearity for each amplicon, and calculates the PCR efficiency (E) per sample and amplicons using an iterative algorithm. In this study, efficiencies were over 90% for all samples and correlation coefficients were higher than 0.99.

The comparative threshold cycle method (ΔCt) was employed to calculate relative gene expression based on the following formula:

$$\Delta Ct = \left(Ct_{\text{target gene}} - Ct_{\text{18S rRNA gene}}\right).$$

Table 1 Primer sequences, product sizes, and annealing temperatures of bovine genes analyzed by RT-PCR

Gene symbol [GenBank accession]	Primer sequence (5' - 3')	Product (bp)	Annealing temperature
SCD1	P: CCTCTGGAACATCACCAGCTTCTCGGC	106	60
[NM_173959]	F: GCTGTCAAAGAAAAGGGTTCCAC		
	R: AGCACAACAACAGGACACCAG		
SCD5	P: CAGAACCCGCTCGTCACCCTGGG	82	60
[NM_001076945]	F: CCCTATGACAAGCACATCAGCC		
	R: GATGGTAGTTATGGAAACCTTCACC		
SREBP1	P: CAGCCCCAGTCCTGGATCAGCCGA	83	60
[NM_001113302]	F: CTTGGAGCGAGCACTGAATTG		
	R: GGGCATCTGAGAACTCCTTGTC		

P = probe, F forward, R reverse

Statistical analysis

Statistical analysis was conducted using IBM SPSS Statistics 22 for Windows (SPSS Inc., IBM Corporation, NY, USA). First of all, data was checked for normality and homoscedasticity. Then, the following general linear model $Y_{ijk} = \mu + CT_i + A_j + HCW_k + e_{ijk}$ was used for analysis of variance (ANOVA), including commercial type (CT; Salers bulls, Pirenaica bulls, Pirenaica heifers, Holstein-Friesian cows) as fixed effect and age at slaughter (A) and hot carcass weight (HCW) as covariates. The effect of sire was also checked but not included in the model as it was statistically not significant. LSD post hoc test was applied for multiple comparison of means among commercial breeds studied.

Simple linear regression analyses were also performed to investigate the relationship between genes (gene-gene) for each commercial type studied.

Finally, partial Pearson correlations coefficients adjusted for A and HCW were computed to determine the associations among gene expression (ΔCt) and FA (Δ9 DIs) data.

Three significant figures were used to express the data, and significance was declared at $P < 0.05$.

Results

Carcass traits and fatty acid composition

Pirenaica heifer carcasses showed the highest fat cover, while those from Pirenaica bulls and Holstein-Friesian cows were lower, and carcasses from Salers showed an intermediate degree of fat cover ($P < 0.001$). In terms of FA composition, several significant differences in specific SFA species were found among commercial types (i.e., 14:0, 16:0, 19:0, 20:0 SFAs), but there were no significant differences in total SFA content. Pirenaica heifers exhibited the highest content of cis- and trans-MUFAs, and Salers bulls had higher cis-MUFA content than Pirenaica bulls ($P < 0.001$; Table 2). Accordingly, Pirenaica heifers showed the highest contents of 9c-14:1 and 9c-16:1, while 9c-17:1 and 9c-18:1 were the highest in Pirenaica heifers but also in Salers bulls. Additionally, Pirenaica

heifers exhibited the highest content of individual trans-18:1 isomers, while vaccenic acid (11 t-18:1) and trans-12-octadecenoic acid (12 t-18:1) contents did not differ among commercial types. The total CLA content was highest in Pirenaica heifers ($P < 0.01$). However, no significant differences were observed in rumenic acid (9c,11 t-18:2), the major CLA isomer. The second major CLA isomer (7 t,9c-18:2), other non-conjugated dienes (6-8 t-18:1and 13 t/14 t-18:1), and potential products of Δ9-desaturation (9c,12 t-18:2, 9c,13 t-18:2 and 9c,15c-18:2) were significantly higher in Pirenaica heifers than in the other commercial types. In contrast, n-6 PUFA content was similar in Pirenaica heifers and Pirenaica bulls, and significantly lower in both compared to Salers bulls. Finally, the content of 10 t,12c-CLA, reported as an inhibitor of Δ9-desaturase, was higher in fat tissues of Pirenaica heifers than in other commercial types ($P < 0.001$; Table 2).

Gene expression

The relative mRNA expression levels of the lipogenic genes SREBP1, SCD1, and SCD5 were similar in Pirenaica bulls and heifers (Fig. 1). Overall, SCD1 expression was higher than SREBP1 expression ($P < 0.001$) and SCD5 expression ($P < 0.001$) in all commercial types, with average–ΔCt values of – 7.91, – 13.4, and – 17.2, respectively (Fig. 1). Differences among breeds were observed for each gene. The mRNA expression of SCD1 was significantly higher in Salers (– 7.36) and Pirenaica cattle (average –ΔCt value of – 6.10) than Holstein-Friesian cows (– 13.8) ($P < 0.001$). In contrast, SCD5 mRNA expression was lowest in Pirenaica bulls and heifers (average of – 17.8) among commercial types, highest in Holstein-Friesians cows (– 15.3), and at intermediate expression levels in Salers bulls (– 17.1; $P < 0.001$). In addition, expression of SREBP1 mRNA was higher in Pirenaica bulls and heifers (average of – 12.73) than in the other commercial types (average – 14.8; $P < 0.001$).

Table 2 Comparisons of fatty acid composition (mg/g of subcutaneous fat) and carcass parameters among commercial types

	Commercial type				
	Salers bulls ($n = 13$)	Pirenaica bulls ($n = 37$)	Pirenaica heifers ($n = 29$)	Holstein-Friesian cows ($n = 21$)	p-value
Conformation	8.45 ± 0.37[b]	10.9 ± 0.3[a]	11.0 ± 0.3[a]	2.02 ± 0.64[c]	< 0.001
Fatness	5.79 ± 0.46[b]	4.52 ± 0.33[c]	7.47 ± 0.33[a]	1.96 ± 0.81[d]	< 0.001
14:0 [s1]	31.6 ± 2.4[ab]	30.5 ± 1.7[b]	35.6 ± 1.8[a]	16.0 ± 4.2[c]	< 0.001
15:0 [s2]	4.71 ± 0.33	4.07 ± 0.23	3.99 ± 0.24	3.13 ± 0.57	0.111
16:0 [s3]	230 ± 11[ab]	217 ± 8[b]	246 ± 8[a]	170 ± 20[bc]	0.002
17:0 [s4]	9.01 ± 0.71	7.48 ± 0.50	8.08 ± 0.51	5.54 ± 1.22	0.057
18:0 [s5]	131 ± 11	119 ± 8	98.0 ± 8.4	129 ± 20	0.059
19:0 [s6]	0.565 ± 0.074[a]	0.595 ± 0.052[a]	0.38 ± 0.05[b]	0.708 ± 0.129[a]	0.005
20:0 [s7]	0.907 ± 0.119[ab]	0.816 ± 0.084[b]	0.468 ± 0.085[c]	1.34 ± 0.21[a]	< 0.001
9c-14:1 [p1]	8.05 ± 1.27[b]	7.80 ± 0.90[b]	12.4 ± 0.9[a]	1.46 ± 2.21[c]	< 0.001
9c-15:1 [p2]	0.208 ± 0.028	0.183 ± 0.020	0.206 ± 0.020	0.15 ± 0.05	0.572
9c-16:1 [p3]	36.7 ± 3.7[b]	33.5 ± 2.6[b]	45.3 ± 2.7[a]	9.74 ± 6.51[c]	< 0.001
9c-17:1 [p4]	6.73 ± 0.44[a]	5.24 ± 0.31[b]	6.96 ± 0.32[a]	2.15 ± 0.77[c]	< 0.001
9c-18:1 [p5]	308 ± 14[a]	261 ± 10[b]	333 ± 10[a]	196 ± 24[c]	< 0.001
9c-19:1 [p6]	0.987 ± 0.055[a]	0.789 ± 0.039[b]	0.801 ± 0.040[b]	0.795 ± 0.096[ab]	0.004
9c-20:1 [p7]	0.726 ± 0.075	0.614 ± 0.053	0.728 ± 0.054	0.777 ± 0.130	0.231
6-8 t-18:1 [s8]	3.14 ± 0.38[bc]	3.66 ± 0.27[ab]	4.12 ± 0.28[a]	1.83 ± 0.67[c]	0.010
11 t-18:1 [s9]	10.3 ± 2.0	12.1 ± 1.4	8.20 ± 1.47	6.15 ± 3.55	0.162
12 t-18:1 [s10]	2.26 ± 0.28	2.56 ± 0.20	2.71 ± 0.21	1.59 ± 0.50	0.174
13 t/14 t-18:1 [s11]	4.37 ± 0.51[b]	5.23 ± 0.36[ab]	5.80 ± 0.37[a]	3.67 ± 0.89[ab]	0.029
15c-18:1 [s12]	1.08 ± 0.16[c]	1.38 ± 0.11[b]	2.07 ± 0.11[a]	0.596 ± 0.271[c]	< 0.001
7 t,9c-18:2 [p8]	0.813 ± 0.116[bc]	0.845 ± 0.082[b]	1.25 ± 0.08[a]	0.295 ± 0.201[c]	< 0.001
9c,11 t-18:2 [p9]	3.11 ± 0.53	3.26 ± 0.376	3.53 ± 0.38	1.80 ± 0.92	0.453
9c,12 t-18:2 [p10]	0.520 ± 0.067[b]	0.608 ± 0.048[b]	0.854 ± 0.048[a]	0.312 ± 0.118[b]	< 0.001
9c,13 t-18:2 [p11]	0.963 ± 0.131[b]	1.07 ± 0.09[b]	1.61 ± 0.09[a]	0.567 ± 0.229[b]	< 0.001
9c,15c-18:2 [p12]	0.461 ± 0.044[b]	0.341 ± 0.031[c]	0.587 ± 0.032[a]	0.266 ± 0.077[bc]	< 0.001
10 t,12c-18:2 [i]	0.221 ± 0.041[b]	0.195 ± 0.029[b]	0.324 ± 0.030[a]	0.099 ± 0.072[b]	0.001
SFA	410 ± 21	382 ± 15	395 ± 15	328 ± 37	0.285
MUFA	427 ± 18[b]	379 ± 13[c]	489 ± 13[a]	245 ± 32[d]	< 0.001
cis-MUFA	385 ± 18[b]	331 ± 13[c]	430 ± 13[a]	221 ± 32[d]	< 0.001
trans-MUFA	42.2 ± 5.2[bc]	48.3 ± 3.6[b]	58.9 ± 3.7[a]	23.5 ± 9.0[c]	0.001
CLA	4.92 ± 0.56[bc]	5.22 ± 0.40[b]	6.42 ± 0.40[a]	2.68 ± 0.98[c]	0.002
PUFA	29.9 ± 2.1[a]	24.3 ± 1.5[bc]	24.4 ± 1.5[b]	14.9 ± 3.7[c]	0.009
n-6	27.7 ± 2.0[a]	22.2 ± 1.4[b]	22.1 ± 1.4[b]	12.8 ± 3.4[c]	0.005
n-3	2.09 ± 0.19	2.05 ± 0.14	2.15 ± 0.14	2.04 ± 0.33	0.942

Least square means ± standard deviations
t trans, *c* cis, *SFA* saturated fatty acids, *MUFA* monounsaturated fatty acids, *CLA* conjugated linoleic acid, *PUFA* polyunsaturated fatty acids
[s] substrate; [p] product; [1-12]Same superscript numbers indicate the substrate-product pairs
[i] inhibitor of SCD enzyme [25]
[a,b,c,d] Values within a row with different superscripts differ significantly at $P < 0.05$
SFA = 10:0 + 12:0 + 13:0 + 14:0 + 15:0 + 16:0 + 17:0 + 18:0 + 19:0 + 20:0 + 21:0 + 22:0 + 23:0 + 24:0
MUFA = 9c-14:1 + 9c-15:1 + 7c-16:1 + 9c-16:1 + 10c-16:1 + 11c-16:1 + 12c-16:1 + 13c-16:1 + 5c-17:1 + 7c-17:1 + 9c-17:1 + 9c-18:1 + 11c-18:1 + 12c-18:1 + 13c-18:1 + 14c-18:1 + 15c-18:1 + 16c-18:1 + 9c-19:1 + 11c-19:1 + 13c-19:1 + 9c-20:1 + 11c-20:1 + 6 t/7 t-16:1 + 8 t-16:1 + 9 t-16:1 + 10 t-16:1 + 11 t/12 t-16:1 + 4 t-18:1 + 5 t-18:1 + 6-8 t-18:1 + 9 t-18:1 + 10 t-18:1 + 11 t-18:1 + 12 t-18:1 + 13 t/14 t-18:1 + 15 t-18:1 + 16 t-18:1
CLA = 9c,11 t-18:2 + 7 t,9c-18:2 + 8c,10 t-18:2 + 9 t,11c-18:2 + 11c,13 t-18:2 + 10 t,12c-18:2 + 11 t,13c-18:2 + other t,t-18:2
PUFA = 18:2n-6 + 18:3n-6 + 20:2n-6 + 20:3n-6 + 20:4n-6 + 22:4n-6 + 18:3n-3 + 18:4n-3 + 20:5n-3 + 22:5n-3 + 22:6n-3 + 20:3n-9

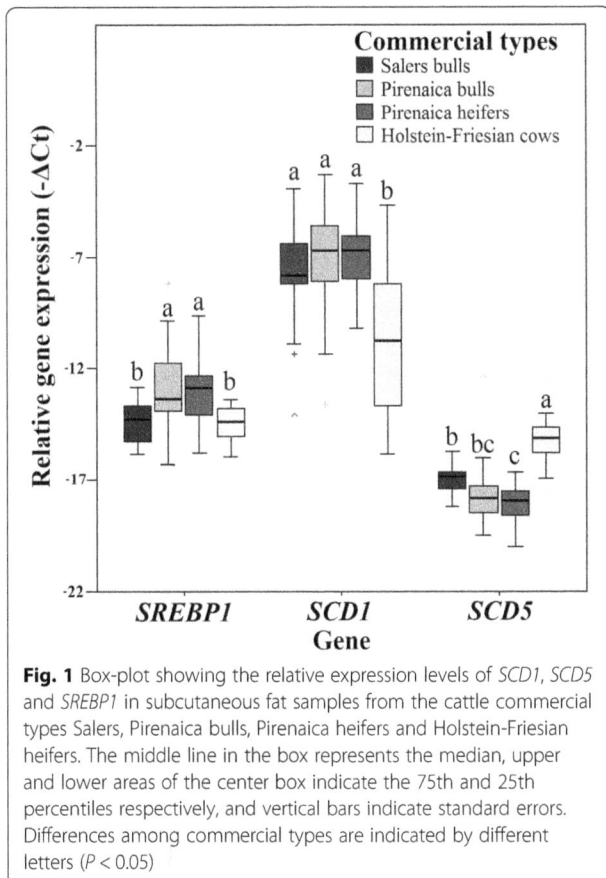

Fig. 1 Box-plot showing the relative expression levels of *SCD1*, *SCD5* and *SREBP1* in subcutaneous fat samples from the cattle commercial types Salers, Pirenaica bulls, Pirenaica heifers and Holstein-Friesian heifers. The middle line in the box represents the median, upper and lower areas of the center box indicate the 75th and 25th percentiles respectively, and vertical bars indicate standard errors. Differences among commercial types are indicated by different letters ($P < 0.05$)

Relationships among gene expression and fatty acid composition data

Significant correlations were observed between studied gene pairs in all commercial types, with particularly strong correlation between *SCD1* and *SREBP1* (Fig. 2a). Pirenaica heifers showed the highest regression coefficient between *SCD1* and *SREBP1* among the commercial types ($R^2 = 0.491$; $P < 0.001$). Salers bulls and Holstein-Friesian cows also showed relatively high regression coefficients between *SCD1* and *SREBP1* ($R^2 = 0.385$; $P = 0.024$ and $R^2 = 0.395$; $P = 0.002$, respectively), while Pirenaica bulls showed the lowest values ($R^2 = 0.239$; $P = 0.002$). A positive correlation between *SCD5* and *SREBP1* (Fig. 2b) was observed in Pirenaica bulls ($R^2 = 0.114$; $P = 0.040$) and Holstein-Friesian cows ($R^2 = 0.213$; $P = 0.035$), while in Salers bulls and Pirenaica heifers was not ($P > 0.05$). No significant correlations were observed between *SCD5* and *SCD1* gene expression except for Holstein-Friesian cows ($R^2 = 0.266$, $P = 0.017$; Fig. 2c).

In all commercial types, *SREBP1* expression was positively correlated with the DI of most FA species, and correlations were significant for 9c-15:1 and 7 t,9c-18:2 in Pirenaica bulls and 9c-15:1 and 9c,12 t-18:2 in Pirenaica heifers ($P < 0.05$; Fig. 3a). In general, Salers bulls

showed the highest positive correlations ($R > 0.65$) between *SCD1* expression and DIs for 9c-16:1, 9c-17:1, 9c-18:1, 9c-20:1, 7 t,9c-18:2 and 9c,12 t-18:2 (Fig. 3b). Pirenaica bulls also showed significant positive correlations between *SCD1* expression and DIs for 9c-17:1, 9c,13 t-18:2, and 9c,15c-18:2 DIs ($P < 0.05$), while Pirenaica heifers did not ($P > 0.05$). In contrast to *SREBP1* and *SCD1*, there were few significant correlations between *SCD5* and DIs among commercial types (Fig. 3c). A negative correlation was observed between *SCD5* and 9c,12 t-18:2 DI in Salers and 9c-14:1 DI in Pirenaica heifers ($P < 0.05$). Total DI was positively correlated with *SCD1* in Salers ($R > 0.65$, $P < 0.05$) and Pirenaica bulls ($R > 0.35$, $P < 0.05$), but negatively correlated with *SCD5* in Salers bulls ($R > 0.60$, $P < 0.05$).

Discussion

Fat deposition and the FA composition of fat depots are controlled by a complex regulatory system including lipogenesis and lipolysis pathways. Adipose tissue is the main site for the storage of excess energy in the form of triacylglycerols, with the Δ9-desaturase product oleic acid (9c-18:1) being the predominant FA [29]. Therefore, Δ9-desaturase activity is critical for triglyceride storage in adipose tissue. While several pathways are involved in regulating FA composition, FAs produced from the precursors acetate and NADH, from the hydrolysis of triacylglycerols, and produced and deposited as rumen biohydrogenation metabolites can act as substrates for Δ9-desaturase. Adipose tissue develops in inter- and intra-muscular depots and both have a major impact on the quality and palatability of commercial beef. There is evidence for differential gene expression profiles in these two fat depots [30]. In this regard, the present study aimed to evaluate the regulation of *SCD* and *SREBP1*, genes strongly affecting the FA composition of subcutaneous adipose tissue, in three genetically diverse bovine breeds commercialized in the Basque region of northern Spain; Pirenaica, Salers and Holstein-Friesian [5, 31, 32]. Expression of *SCD1* did not differ significantly between Pirenaica bulls and heifers or among young cattle of Salers and Pirenaica. This may be partially explained by a similar feeding regimen, typically including concentrates, when meat production is the final purpose (Fig. 1). However, the content of Δ9 products, such as *cis*-MUFA, was higher in Salers bulls and Pirenaica heifers than corresponding bulls (Table 2). The Salers bulls and Pirenaica heifers, together with Holstein-Friesian cows, showed stronger correlations between *SCD1* and *SREBP1* compared to Pirenaica bulls (Fig. 2). This suggests that, in Pirenaica breed, the FA composition is affected by the lipogenic gene regulation in a sex-dependent manner. Similarly, in a crossbred study, heifers exhibited higher *SCD1* mRNA levels and higher MUFA content than bulls

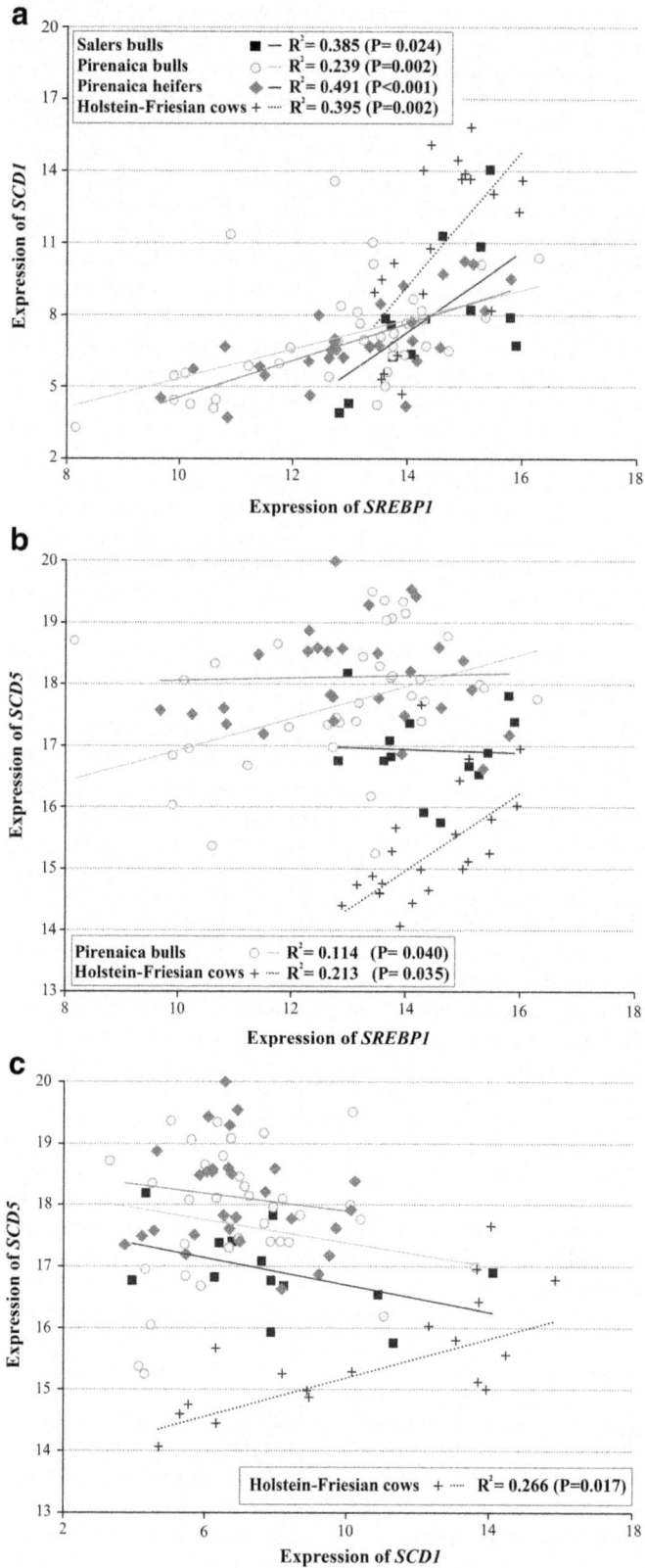

Fig. 2 Estimated linear regression equations between (**a**) *SCD1* and *SREBP1*, (**b**) *SCD5* and *SREBP1*, and (**c**) *SCD5* and *SCD1*

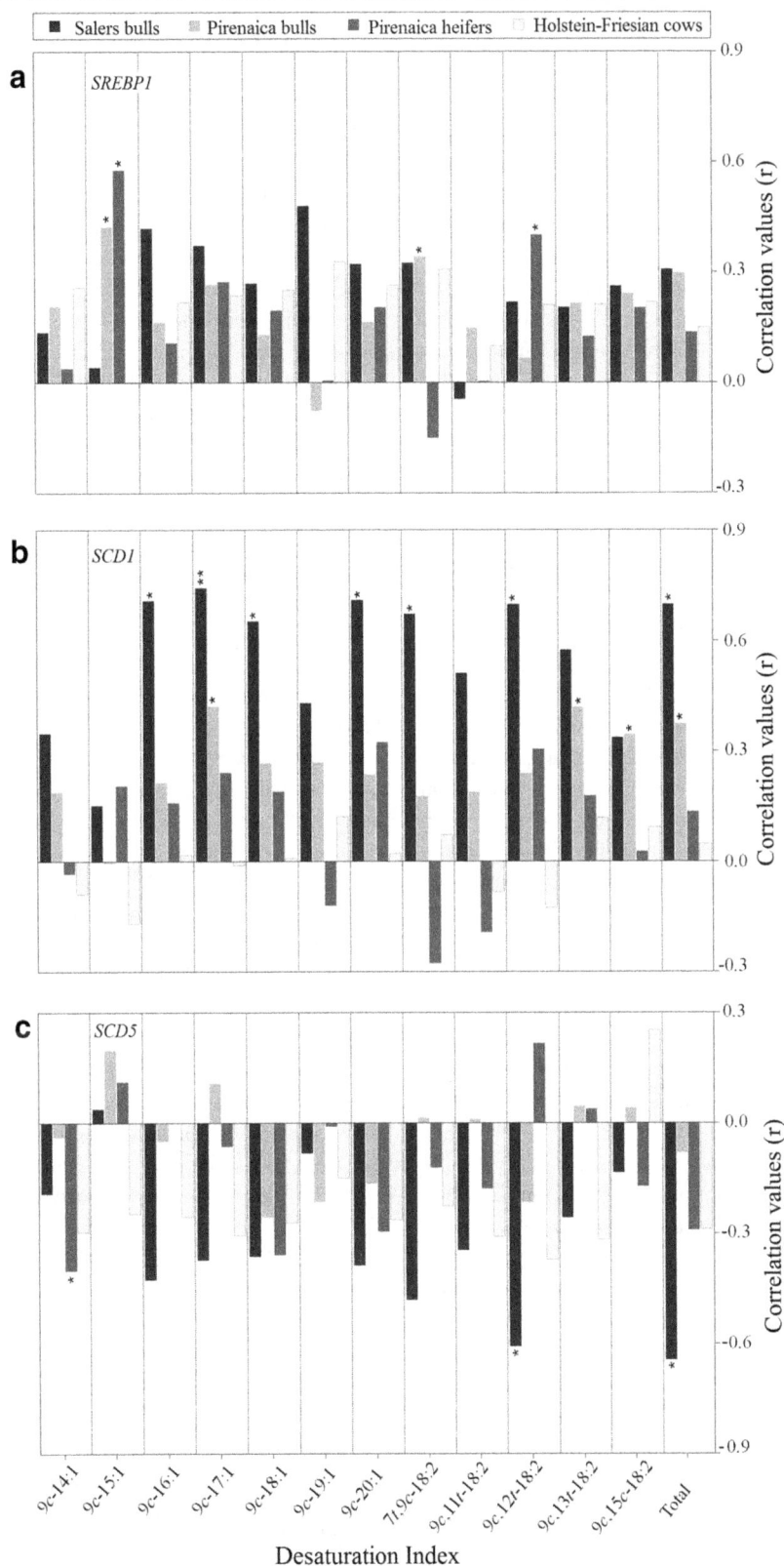

Fig. 3 Partial correlations controlling for age and HCW between gene expression of *SREBP1* (**a**), *SCD1* (**b**), *SCD5* (**c**) and desaturation indexes calculated from fatty acid composition data of cattle commercial types. *P < 0.05, **P < 0.01. Total is sum of all individual DIs. Desaturation indexes were calculated as [SCD product]/([SCD substrate] + [SCD product])

in subcutaneous adipose tissue [33], and a possible effect of sex hormones on enzymatic systems affecting lipid metabolism has been suggested [34]. Indeed, the growth hormone, sexually differentiated in mammals, seems to increase *SREBP*1 and *SCD1* gene expression in females [35]. Alternatively, age and diet have been demonstrated to influence adipocyte development in Pirenaica bulls [36]. Hence, the activation of *SCD1* due to a potentially higher concentrate consumption [19, 37] agrees with the greater total MUFA content of Salers and Pirenaica, while higher MUFA content in Pirenaica heifers than bulls seems to be more sex-dependent (Table 2). The greater variability in *SCD1* expression within Holstein-Friesian cows compared to young Pirenaica and Salers (Fig. 1) could be related to the less homogeneous diet and older age of these animals. Nevertheless, the generally lower *SCD1* expression observed in Holstein-Friesian cows was also reported in other mature culled cows [4], in which linoleic acid (18:2n-6) was suggested as the primary agent depressing *SCD* gene expression in adipose tissue [38].

We detected variability in *SCD5* mRNA expression levels among breeds ($P < 0.01$) and generally greater expression of *SCD1* relative to *SCD5* in all breeds. Lengi and Corl (2007) [11] also reported over 40-fold greater expression of *SCD1* compared to *SCD5* in adipose tissue of bulls (albeit with unspecified feeding). Variation among breeds was observed in the expression of both *SCD* isoforms, especially between beef and dairy cattle breeds (Salers and Pirenaica vs. Holstein-Friesian), suggesting that even if FA differences are generally small, there may still be differences in the underlying lipogenic gene expression or enzyme profile [39].

These differences in *SCD1* and *SCD5* expression levels (Fig. 1) also suggest that *SCD5* expression is more breed dependent than *SCD1* expression. However, it is also possible that *SCD5* expression is more sensitive than *SCD1* expression to other environmental factors (i.e., feeding) that differ among commercial types. Our results also revealed a potential opposite association between *SCD* isoforms within each breed. In general, this opposite correlation of DIs with *SCD5* and *SCD*1 expression levels suggests that regulatory factors that upregulate *SCD*1 also downregulate *SCD5* (and vice versa). However, since both *SCD* isoforms are expressed in adipose tissue, both may contribute to the maintenance of desaturation. In contrast to Pirenaica and Salers, this opposite association between *SCD1* and *SCD5* was not observed in Holstein-Friesian cows, a breed selected intensively for dairy production and most often utilized for beef production as cull cows at no specific age. This opposite pattern was more evident when the Holstein-Friesian sample was stratified by age (data not shown). Furthermore, Holstein-Friesian cows may exhibit side effects of dairy selection that differentially affect the genetic

sequences containing *SCD* genes, thereby influencing transcriptional regulation. Moreover, variability in the regulatory DNA sequences of *SCD* genes may confer differences in gene expression and physiological changes that could also explain the different correlation patterns with DIs observed among commercial types. 9*c*-14:1 DI has been reported as the best indicator of overall SCD enzyme activity by Corl et al. (2002) [40]. Feedstuffs are normally devoid of 9*c*-14:1 and, therefore, this FA is produced by de novo FA synthesis. In mammary gland, significant correlation between 9*c*-14:1 DI and *SCD1* expression was observed, whereas 9*c*-14:1 DI and *SCD5* correlation was not [41]. We did not observe significant correlation between 9*c*-14:1 DI and *SCD1*, similarly to a previous study in intestinal adipose tissue, skeletal muscle or mammary gland [42]. However, we observed that 9*c*-14:1 DI and *SCD5* were negatively correlated in subcutaneous adipose tissue of Pirenaica heifers ($P < 0.05$). Thus, these results suggest that *SCD* gene expression may directly affect 9*c*-14:1 content, but 9*c*-14:1 DI correlation with *SCD5* and *SCD1* might be breed and tissue specific as well.

As previously reported by Horton et al. [14], *SCD1* and *SREBP1* appear to be directly related as there was a significant linear association between these two genes in all commercial types studied (Fig. 2a). Differences in slope and coefficient of determination (R^2) values, however, revealed variability in this relationship among commercial types. A previous study suggested that the FA synthesis pathway is regulated in a coordinated manner by the SREBP family of membrane-bound transcription factors, and regulation of *SCD1* by *SREBP1* via the SRE binding site of *SCD1* has been demonstrated [43].

Significant correlation between *SCD5* and *SREBP1* specifically observed in Pirenaica bulls and Holstein-Friesian cows suggest that *SCD5* expression may be more variable among commercial types than *SCD1* expression, possibly due to differences in regulation by *SREBP1*. For example, according to Lengi and Corl (2012) [16], the early growth response protein 2 (*EGR2*) and *SREBP1* may bind to the same DNA site of the bovine *SCD5* promoter. They observed that expression of *EGR2* or *SREBP1* did not increase endogenous *SCD5* mRNA expression but did activate a truncated bovine *SCD5* promoter luciferase reporter constructs in human JEG3 cells. Therefore, they attributed the lack of increase in *SCD5* expression to the presence of additional negative-regulation sites in this gene. In our case, the absence of significant differences in other commercial types could be due to breed or other environmental factors that could, in part, modulate these putative negative-regulation sites.

Among these commercial types of the Basque region, correlations between lipogenic genes (*SCD1* and *SREBP1*) and calculated DIs were stronger in Salers bulls than Pirenaica bulls and heifers (Fig. 3). In Holstein-Friesian cows, correlations were not significant (Fig. 3). However, *SREBP1* and *SCD1* correlations with DIs became significant ($P < 0.05$) when computed without age and HCW as covariates (data not shown). Moreover, correlations between *SCD1* expression and DIs were slightly higher in younger Holstein-Friesians, while *SCD5* correlations with DIs were higher in older Holstein-Friesians. This suggests an effect of age on gene expression–DI correlations in Holstein-Friesian cows. In addition to 16:1 and 18:1 [42], positive correlations between *SCD1* and calculated DIs were observed for other FA species, suggesting desaturase activity also targets minor FAs of subcutaneous fat. In this regard, DIs and MUFA content were more susceptible to the expression of lipogenic genes in Pirenaica heifers than bulls. Furthermore, the effect of lipogenic gene expression on DIs was stronger in Salers than Pirenaica heifers. Our findings are supported by a previous study [13] suggesting that the FA composition of subcutaneous adipose tissue is mainly dependent on genetic background, which may in turn indicate inter-breed differences in lipid metabolism. The effect of breed appears to be more strongly associated with *SREBP1* expression level than *SCD1* or *SCD5* expression level (Fig. 1), whereas the underlying regulation of *SCD1* and *SCD5* could be responsible for inter-breed differences in DIs and FA profiles.

We also report an opposite effect of *SCD* isoforms on certain DI values, especially in Salers bulls. This stronger pattern may stem for a more homogeneous production system (Salers breeder, personal communication) that may reduce the influences of extraneous factors. Positive correlations between DIs and *SCD*1 (Fig. 2b) and contrasting negative correlations between DIs and *SCD*5 (Fig. 2c) are likely due to genetic compensation. Lower expression of one *SCD* isoform could well be compensated for by upregulation of the other isoform (Fig. 1). This compensation theory was previously suggested in *Caenorhabditis elegans* [44]. The reciprocal expression observed between different isoforms and the underlying epigenetic processes require further investigation.

The CLA isomer 10 *t*,12*c*-18:2 was examined because it was previously described as an important inhibitor of *SCD1* in dairy cattle [25]. In our study, although Pirenaica heifers accumulated the highest amounts of 10 *t*,12*c*-18:2 in subcutaneous adipose tissue (Table 2), no significant correlation was observed between 10 *t*,12*c*-18:2 and lipogenic gene expression (data not shown). Nevertheless, both isoforms may be differently regulated. In contrast to *SCD1*, which tends to be reduced by 10 *t*,12*c*-18:2 [45], *SCD5* appears to be more stable due to lack of an N-terminal PEST sequence for degradation [11]. However, further research is needed to establish relationships among DIs and SCD isoform

mRNA expression levels, and to clarify the effects of 10 *t*,12*c*-18:2 on bovine adipose and muscle tissues. Analysis of lipogenic gene expression changes with dietary treatment in ruminant species as well as promoter sequencing would provide valuable insight into the regulation of these genes and their impact on the synthesis of MUFAs and PUFAs.

Conclusion

The present study suggests that the differences in subcutaneous fat FA composition among bovine commercial types of the Basque region are related to genetic variability in lipogenic gene expression. The expression of lipogenic genes in Salers bulls showed clear effects on desaturation indexes and FA composition. All breeds show a strong correlation between *SREBP1* and *SCD1* expression. In addition, distinct correlations between *SCD* isoforms and DIs suggest a novel genetic compensation mechanism between *SCD1* and *SCD5* that warrants further investigation.

Abbreviations

c: *Cis*; CLA: Conjugated linoleic acid; DI: Desaturation index; FA: Fatty acid; HCW: Hot carcass weight; MUFA: Monounsaturated fatty acids; PUFA: Polyunsaturated fatty acids; SCD: Stearoyl-CoA desaturase; SFA: Saturated fatty acids; SREBP: Sterol regulatory element-binding protein; *t*: *Trans*; UFA: Unsaturated fatty acids

Acknowledgements

Help and advice provided by Dr. Mikawa is very much appreciated. Technical support provided by SGIker (DNA Bank Service of UPV/EHU) and abattoir staff from Arakai-Urkaiko is also acknowledged.
The authors would like to thank Enago (www.enago.jp) for the English language review.

Funding

Department of Economic Development & Competitiveness of the Basque Government supported the doctoral fellowship of D.G. and A.L.
The Spanish Ministry of Economy & Competitiveness and the UPV/EHU for 'Ramón y Cajal (RYC-2011-08593)' research contract of N.A.
This work was supported by grants from the Dept. of Education, Universities & Research of the Basque Government (IT766–13 & IT833–13).

Authors' contributions

DG analyzed the overall experimental data used in this study and wrote the manuscript. NA analyzed and interpreted fatty acid composition in adipose tissue of cattle. AA and LJRB performed statistical analyses. MT analyzed and interpreted lipogenic gene expression data. ALO and MMP participated in discussion. All authors contributed to drafting the manuscript and gave final approval of the version to be published.

Authors' information

Masaaki Taniguchi: Molecular genetics analyses on meat quality of beef and pork using high throughput technologies such as single nucleotide polymorphisms (SNP) array and gene expression microarray.
Noelia Aldai: Lipids in animal science: factors influencing the composition, strategies to improve the nutritional quality of meat, and comprehensive analytical methods that demonstrate their value.

Competing interests

The authors declare that they have no competing interests.

Author details

[1]Biomics Research Group, University of the Basque Country (UPV/EHU), 01006 Vitoria-Gasteiz, Spain. [2]Lactiker Research Group; Lascaray Research Center, University of the Basque Country (UPV/EHU), 01006 Vitoria-Gasteiz, Spain. [3]Animal Genome Unit, Institute of Livestock and Grassland Science, National Agriculture and Food Research Organization (NARO), Tsukuba 305-0901, Japan.

References

1. World Health Organization. Diet, nutrition and the prevention of chronic diseases: report of a joint WHO/FAO expert consultation. Geneva: WHO Technical Report Series; 2003. p. 916.
2. Burlingame B, Nishida C, Uauy R, Weisell R. Fats and fatty acids in human nutrition. Joint FAO/WHO expert consultation. Ann Nutr Metab. 2009;55:5–7.
3. Wood JD, Richardson RI, Nute GR, Fisher AV, Campo MM, Kasapidou E, et al. Effects of fatty acids on meat quality: a review. Meat Sci. 2004;66:21–32.
4. Brooks MA, Choi CW, Lunt DK, Kawachi H, Smith SB. Subcutaneous and intramuscular adipose tissue stearoyl-coenzyme a desaturase gene expression and fatty acid composition in calf- and yearling-fed Angus steers. J Anim Sci. 2011;89:2556–70.
5. Gamarra D, Lopez-Oceja A, de Pancorbo M. Genetic characterization and founder effect analysis of recently introduced Salers cattle breed population. Animal. 2017;11:24–32.
6. St John LC, Lunt DK, Smith SB. Fatty acid elongation and desaturation enzyme activities of bovine liver and subcutaneous adipose tissue microsomes. J Anim Sci. 1991;69:1064–73.
7. Taniguchi M, Mannen H, Oyama K, Shimakura Y, Oka A, Watanabe H, Kojima T, Komatsu M, Harper GS, Tsuji S. Differences in stearoyl-CoA desaturase mRNA levels between Japanese black and Holstein cattle. Livest Prod Sci. 2004;87:215–20.
8. Kgwatalala PM, Ibeagha-Awemu EM, Hayes JF, Zhao X. Single nucleotide polymorphisms in the open reading frame of the stearoyl-CoA desaturase gene and resulting genetic variants in Canadian Holstein and Jersey cows. DNA Seq. 2007;18:357–62.
9. Barton L, Kott T, Bures D, Rehak D, Zahradkova R, Kottova B. The polymorphisms of stearoyl-CoA desaturase (SCD1) and sterol regulatory element binding protein-1 (SREBP-1) genes and their association with the fatty acid profile of muscle and subcutaneous fat in Fleckvieh bulls. Meat Sci. 2010;85:15–20.
10. Li C, Aldai N, Vinsky M, Dugan ME, McAllister TA. Association analyses of single nucleotide polymorphisms in bovine stearoyl-CoA desaturase and fatty acid synthase genes with fatty acid composition in commercial cross-bred beef steers. Anim Genet. 2012;43:93–7.
11. Lengi AJ, Corl BA. Identification and characterization of a novel bovine stearoyl-CoA desaturase isoform with homology to human SCD5. Lipids. 2007;42:499–508.
12. Rincon G, Islas-Trejo A, Castillo AR, Bauman DE, German BJ, Medrano JF. Polymorphisms in genes in the SREBP1 signalling pathway and SCD are associated with milk fatty acid composition in Holstein cattle. J Dairy Res. 2012;79:66–75.
13. da Costa AS, Pires VM, Fontes CM, Prates JA. Expression of genes controlling fat deposition in two genetically diverse beef cattle breeds fed high or low silage diets. BMC Vet Res. 2013;9:118.
14. Horton JD, Goldstein JL, Brown MS. SREBPs: activators of the complete program of cholesterol and fatty acid synthesis in the liver. J Clin Invest. 2002;109:1125–31.
15. Hoashi S, Ashida N, Ohsaki H, Utsugi T, Sasazaki S, Taniguchi M, Oyama K, Mukai F, Mannen H. Genotype of bovine sterol regulatory element binding protein-1 (SREBP-1) is associated with fatty acid composition in Japanese black cattle. Mamm Genome. 2007;18:880–6.
16. Lengi AJ, Corl BA. Regulation of the bovine SCD5 promoter by EGR2 and SREBP1. Biochem Biophys Res Commun. 2012;421:375–9.
17. Widmann P, Nuernberg K, Kuehn C, Weikard R. Association of an ACSL1 gene variant with polyunsaturated fatty acids in bovine skeletal muscle. BMC Genet. 2011;12:96.
18. Jones OR, Wang J. COLONY: a program for parentage and sibship inference from multilocus genotype data. Mol Ecol Resour. 2010;10:551–5.
19. Aurtenetxe M, Belaunzaran X, Bravo-Lamas L, Gamarra D, Barron LJR, Aldai N. Caracterización comercial y nutricional de la grasa subcutánea de terneros y vacas de desvieje sacrificados en la Comunidad Autónoma del País Vasco. Archivos de Zootecnia. 2017;66:435–42.
20. The Commission of the European Communities. Commission regulation (EEC) no 103/2006 of 20 January 2006 adopting additional provisions for the application of the Community scale for the classification of carcases of adult bovine animals. Off J Eur Communities. 2006;49:6–8.
21. Kramer JK, Fellner V, Dugan ME, Sauer FD, Mossoba MM, Yurawecz MP. Evaluating acid and base catalysts in the methylation of milk and rumen fatty acids with special emphasis on conjugated dienes and total trans fatty acids. Lipids. 1997;32:1219–28.
22. Kramer JK, Hernandez M, Cruz-Hernandez C, Kraft J, Dugan ME. Combining results of two GC separations partly achieves except CLA isomers of milk fat as demonstrated using ag-ion SPE fractionation. Lipids. 2008;43:259–73.
23. Delmonte P, Fardin-Kia AR, Kramer JKG, Mossoba MM, Sidisky L, Tyburczy C, Rader JI. Evaluation of highly polar ionic liquid gas chromatographic column for the determination of the fatty acids in milk fat. J Chromatogr A. 2012;1233:137–46.
24. Bravo-Lamas L, Barron LJ, Kramer JK, Etaio I, Aldai N. Characterization of the fatty acid composition of lamb commercially available in northern Spain: emphasis on the trans-18:1 and CLA content and profile. Meat Sci. 2016;117:108–16.
25. Baumgard LH, Matitashvili E, Corl BA, Dwyer DA, Bauman DE. Trans-10, Cis-12 conjugated linoleic acid decreases Lipogenic rates and expression of genes involved in milk lipid synthesis in dairy cows. J Dairy Sci. 2002;85:2155–63.
26. Destaillats F, Trottier JP, Galvez JMG, Angers P. Analysis of alpha-linolenic acid biohydrogenation intermediates in milk fat with emphasis on conjugated linolenic acids. J Dairy Sci. 2005;88:3231–9.
27. Zhu LJ, Altmann SW. mRNA and 18S-RNA coapplication-reverse transcription for quantitative gene expression analysis. Anal Biochem. 2005;345:102–9.
28. Ramakers C, Ruijter JM, Lekanne Deprez RH, Moorman AFM. Assumption-free analysis of quantitative real-time polymerase chain reaction (PCR) data. Neurosci Lett. 2003;339:62–6.
29. Mauvoisin D, Mounier C. Hormonal and nutritional regulation of SCD1 gene expression. Biochimie. 2011;93:78–86.
30. Bong JJ, Cho KK, Baik M. Comparison of gene expression profiling between bovine subcutaneous and intramuscular adipose tissues by serial analysis of gene expression. Cell Biol Int. 2010;34:125–33.
31. Martín-Burriel I, Rodellar C, Cañon J, Cortés O, Dunner S, Landi V, Martínez-Martínez A, Gama LT, Ginja C, Penedo MC, Sanz A, Zaragoza P, Delgado JV. Genetic diversity, structure, and breed relationships in Iberian cattle. J Anim Sci. 2011;89:893–906.
32. Lopez-Oceja A, Muro-Verde A, Gamarra D, Cardoso S, de Pancorbo MM. New Q lineage found in bovine (Bos taurus) of Iberian peninsula. Mitochondrial DNA. 2016;27:3597–601.
33. Barton L, Bureš D, Kott T, Rehák D. Effect of sex and age on bovine muscle and adipose fatty acid composition and stearoyl-CoA desaturase mRNA expression. Meat Sci. 2011;89:444–50.
34. Zhang Y-Y, Zan L-S, Wang H-B, Xin Y-P, Adoligbe CM, Ujan JA. Effect of sex on meat quality characteristics of Qinchuan cattle. African J Biotechnol. 2010;9:4504–9.
35. Ameen C, Linden D, Larsson BM, Mode A, Holmang A, Oscarsson J. Effects of gender and GH secretory pattern on sterol regulatory element-binding protein-1c and its target genes in rat liver. Am J Physiol Endocrinol Metab. 2004;287:E1039–48.
36. Soret B, Mendizabal A, Arana A, Alfonso L. Expression of genes involved in adipogenesis and lipid metabolism in subcutaneous adipose tissue and longissimus muscle in low-marbled Pirenaica beef cattle. Animal. 2016;1:1–31.
37. Smith S, Gill C, Lunt D, Brooks M. Regulation of fat and fatty acid composition in beef cattle. Asian-Australasian. J Anim Sci. 2009;22:1225–33.
38. Smith SB. Functional development of Stearoyl-CoA desaturase gene expression in livestock species. In: Ntambi J, editor. Stearoyl-CoA desaturase genes lipid Metab. New York: Springer; 2013. p. 141–59.
39. De Smet S, Raes K, Demeyer D. Meat fatty acid composition as affected by fatness and genetic factors: a review. Anim Res. 2004;53:81–8.

40. Corl BA, Baumgard LH, Griinari JM, Delmonte P, Morehouse KM, Yurawecz MP, Bauman DE. Trans-7, cis-9 CLA is synthesized endogenously by Δ9-desaturase in dairy cows. Lipids. 2002;37:681–8.

41. Jacobs AA, van Baal J, Smits MA, Taweel HZ, Hendriks WH, van Vuuren AM, Dijkstra J. Effects of feeding rapeseed oil, soybean oil, or linseed oil on stearoyl-CoA desaturase expression in the mammary gland of dairy cows. J Dairy Sci. 2011;94:874–87.

42. Rezamand P, Watts JS, Yavah KM, Mosley EE, Ma L, Corl BA, McGuire MA. Relationship between stearoyl-CoA desaturase 1 gene expression, relative protein abundance, and its fatty acid products in bovine tissues. J Dairy Res. 2014;81:333–9.

43. Tabor DE, Kim JB, Spiegelman BM, Edwards PA. Identification of conserved cis elements and transcription factors required for sterol regulated transcription of stearoyl CoA desaturase 1 and 2. J Biol Chem. 1999;274:20603–10.

44. Brock TJ, Browse J, Watts JL. Genetic regulation of unsaturated fatty acid composition in C. Elegans. PLoS Genet. 2006;2:e108.

45. Gervais R, McFadden JW, Lengi AJ, Corl BA, Chouinard PY. Effects of intravenous infusion of trans-10, cis-12 18:2 on mammary lipid metabolism in lactating dairy cows. J Dairy Sci. 2009;92:5167–77.

10

Epidemiological and economic consequences of purchasing livestock infected with *Mycobacterium avium* subsp. *paratuberculosis*

Carsten Kirkeby[1][*] ⓘ, Kaare Græsbøll[1,2], Søren Saxmose Nielsen[3], Nils Toft[1] and Tariq Halasa[1]

Abstract

Background: Paratuberculosis (PTB) is a chronic disease which may lead to reduced milk yield, lower animal welfare and death in cattle. The causative agent is *Mycobacterium avium* subsp. *paratuberculosis* (MAP). The economic consequences are particularly important incentives in the control and eradication of the infection. One strategy to control PTB in a herd is to purchase animals from farms with a low risk of MAP infection. We wanted to investigate the epidemiological and economic consequences of buying livestock from different supplier farms of low, medium or high risk, as well as farms with unknown status. We also wanted to estimate the probability of spontaneous fadeout if the farmer of an initially MAP-free herd bought a specified number of infected animals in a single year, or continually bought infected animals. This was achieved through simulation modeling, and the effects of consistently introducing one, five or ten infected animals annually into an initially infection-free herd was also modeled.

Results: Our findings show that once infected, a farm can relatively safely purchase animals from other low and medium-risk farms without experiencing an increase in the prevalence, highlighting the importance of certification programmes. Furthermore, farms free of MAP are highly susceptible and cannot purchase more than a small number of animals per year without having a high risk of being infected. The probability of spontaneous fadeout after 10 years was 82% when introducing a single infected animal into an initially MAP-free herd. When purchasing ten infected animals, this probability was 46%. The continual purchase of infected animals resulted in very low probabilities of spontaneous fadeout.

Conclusions: We demonstrated that MAP-free farms can purchase a small number of animals, preferably from certified farms, each year and still remain free of MAP. Already infected farms have little risk of increasing the prevalence on a farm when purchasing animals from other farms.

Background

Paratuberculosis (PTB) is a chronic infection caused by *Mycobacterium avium* ssp. *paratuberculosis* (MAP), occurring in dairy cattle. Infections can be latent for many years [1], but infected cows can shed large amounts of MAP [2]. After the latent state, animals can become clinical with fatal diarrhea. Symptoms in infected cows include reduced milk yield and weight loss [3, 4], the

economic implications of which may motivate farmers to reduce the prevalence. Furthermore, reduced animal health and welfare, and the potential benefits of being able to certify a herd free of infection can also serve as motivators for farmers to lower the prevalence [5].

Around half of the Danish dairy herds are open herds, and the farmers purchase cattle (often pregnant heifers) either on a regular or irregular basis. This is similar to other countries like the USA, with 34% closed herds [6] and 47% in the Netherlands [7]. Farmers purchase cattle for many reasons, e.g. to improve the genetic potential of the herd, to increase the herd size or to restore the

* Correspondence: ckir@vet.dtu.dk
[1]National Veterinary Institute, Technical University of Denmark, Kemitorvet, bygning 204, 2800 Kgs. Lyngby, Denmark
Full list of author information is available at the end of the article

herd size following a disease outbreak. Approximately half of the Danish open herds use a single supplier. Using fewer suppliers may potentially reduce the risk of infection if the chosen supplier has a low prevalence. In contrast, the risk of purchasing PTB-infected cattle can be diluted by using more than one supplier, if the prevalence is low among suppliers. A certification system, as introduced in some countries, can help the farmer to reduce the risk of buying infected animals [5, 8–12].

There are approximately 3000 Danish dairy herds, with an average herd size of about 200 cows. The number of herds has reduced from approximately 12,000 in the past 20 years, while the number of cows has remained relatively stable, thus greatly increasing the herd-size and the need for purchase of livestock. Denmark has a voluntary control programme for MAP where participating herds are required to test all lactating cows four times per year, or use a risk-based test strategy where cows are tested one time per year unless they have a positive test [13]. Testing is done using a milk ELISA (IDvet, Graebels, France), and a number of recommendations are associated to the test-positive animals [14]. About 25% of the Danish farms are enrolled in the MAP control programme. Cows that have two positive tests are regarded as high-risk animals and hence farmers are recommended to cull these animals. Cows with variable results are still considered potentially infectious, but farmers are not recommended to cull these cows. The repeated testing increases the global test sensitivity [14]. Furthermore, herds participating in the control programme for MAP are advised on how to optimize the hygiene on the farm in order to break transmission routes, e.g. cleaning of calving areas and avoiding feeding calves with colostrum and waste milk from infected cows [15]. The between-herd prevalence of MAP has been estimated to be >80%, and it has been estimated that more than 75% of the herds enrolled in the MAP control programme were infected [14]. The within-herd prevalence of MAP infection in Danish dairy herds has also been investigated for herds that were not enrolled in the control programme: The median within-herd prevalence was estimated to be 5.6% of cows in the herds, but the prevalence in one herd was 45% [15]. Danish dairy cattle farms can be certified with a given MAP risk score [5, 8].

Many sellers have a low probability of infection in the cattle, and it might be considered safe to purchase from a farm with unknown risk. Where farmers use multiple suppliers and only occasionally purchase an infected animal, the probability of infection might be negligible. It is important to quantify the epidemiological and economic impact of cattle purchases on the risk of MAP introduction and spread within a dairy cattle herd. This is influenced by the MAP prevalence within the buying and selling herds, the number and frequency of cattle purchases and the number of suppliers. Knowledge of these risks would enable the provision of recommendations to help limit MAP introduction and/or spread within dairy cattle herds. Furthermore, MAP-free herds needing to purchase animals can be assisted in maintaining their free status. Therefore, our first aim was to assess the impact of purchasing cattle from herds with low, medium or high risk, or from farms with unknown status (random risk). We wanted to investigate these scenarios for both a MAP-free herd and a herd with median prevalence.

Secondly, we wanted to estimate the probability of infection fadeout after 10 years in an open, infection-free herd, where one or more infected animals were introduced at the beginning of the simulations, and to estimate the probability of acquiring a persistent infection in these circumstances.

In this study, we investigated the impact of purchase strategies by simulating combinations of situations for a dairy farm, namely: purchase from single vs. multiple suppliers; purchase of one, five or ten animals per year, and purchase from random farms (applying no MAP control measures) or farms with low, medium or high probability of infection. We also simulated an infection-free herd where the farmer bought one, five or ten animals either yearly, in a single year, or continually over 10 years. We then evaluated the resulting prevalence and probability of fadeout for all simulations.

Methods

We used the iCull PTB simulation model, which is a bio-economic model that simulates a dairy farm using single-day time steps [15]. The model simulates a farm with 200 dairy cows, corresponding to a medium-sized Danish dairy herd. The model is a mechanistic, stochastic simulation model that simulated the individual cows in great detail, e.g. with individual lactation curves and SCC curves based on data. The iCull PTB model runs in daily time steps. Animals enter the herd as calves, are reared as heifers and later milked as cows. The farmer takes weekly decisions on which cows to cull if there are more than 200 cows present in the herd, thus keeping the number of cows stable [15]. In this study, we simulated that the farmer purchased one, five or ten pregnant heifers per year. We simulated both a situation where the farmer had a single supplier of animals, and a situation where the farmer bought from multiple suppliers (see description below). Within the model, cows feed, lactate and are inseminated and dried off as in a real herd. In the model, two thirds of culled animals are culled involuntarily due to diseases (such as lameness and mastitis) or acute injuries. This means that the farmer can make decisions for one third of the culled animals, thus resembling a real farm.

The model simulated spread of MAP between animals through the environment: bacteria are shed by infected animals every day. Accumulated bacteria followed a survival curve estimated from data. The probability of infection from the environment is based on the amount of bacteria present in the farm section where each animal is located. Additionally, animals can be infected in utero, through colostrum and waste milk [15]. If no new animals are infected from the environment or other transmission routes, disease fadeout occurs. We here defined fadeout as a permanent situation where no new animals were infected during the simulations. However, MAP bacteria can still be present in the environment for some time without infecting new animals.

The purchases were evenly spread over the year. We simulated purchases of pregnant heifers from other farms, which is a common practice in Denmark and other countries such as France [16]. Purchased pregnant heifers were introduced into the heifer section of the housing. The number of days remaining before calving was randomly drawn from a uniform distribution between 42 and 280 days after insemination. In this study, we assume that the risk of infection from an animal bought from a supplier herd is the same as the prevalence in the supplier herd, resembling that the pregnant heifer came from another farm. In Denmark, it is common for farmers to trade animals directly or through a cattle market. The probability of infection in the purchased animals was modeled according to each scenario (see below). All other properties of the purchased heifers were generated from the same distributions used in the model by [15]. In this study, we simulated for 10 years and repeated the stochastic simulations 500 times, which was previously found to be appropriate for convergence [15]. In this study, no control actions against the spread of MAP were simulated.

Open herd scenarios

We simulated a herd with a stable herd size of 200 cows and a steady prevalence of either 0% or 5.6%. A 10-year period was chosen for this study, though it should be noted that an increase in infection prevalence or an introduction of MAP would have consequences spanning more than 10 years if no control or eradication measures were implemented. All purchased animals were pregnant heifers, as this is common practice in Denmark.

The model simulates a reduced milk production in the subclinical phase [15]. If there are infected animals in the herd, they are culled when they are detected, preventing the farmer from automatically culling the lowest producing animals. Thus the model captures that low

producing cows are kept in the herd for longer time than normal. The economic calculations did not include salary for personnel or expenses such as machines and housing. Neither did we include changes in the feed conversion ratio of infected animals or potential consequences for trade. The model simulates other expenses like insemination (16.1 EUR), feed (0.133 EUR per feed unit) and the destruction of dead animals (64.8 EUR) [15]. The price of a pregnant heifer was set to EUR 1275 [17]. Each milk-ELISA cost 5.3 EUR [15].

In the simulations, the farmer could buy cattle from random herds with unknown probability of infection, or from herds certified with a low, medium or high probability of infection. We used the true within-herd prevalence in Danish herds based on [15] as the probability that a purchased animal from a given farm was infected. Therefore, in the scenarios where random suppliers were used, the probability of infection in the purchased animals was sampled from the empirical distribution of true prevalence found in 102 tested Danish herds. We also simulated three risk levels when purchasing animals: low risk, medium risk and high risk. Low risk animals had 0% to 5% probability of being infected at purchase. Medium risk animals had 5% to 15% probability of being infected, and high risk animals had 15% to 45% probability of being infected at purchase. These intervals were chosen based on the empirical distribution of within-herd prevalence in Denmark [15]. For each purchase, the probability of infection from the herd of origin was drawn from a uniform distribution between the numbers given for the risk levels described above. Whether the purchased animal was infected or not was then decided using a binomial distribution based on this probability. This resembles that a farmer purchase from a farm with a given risk level where the animal has a probability of being infected. We simulated both scenarios where the farmer purchased from a single supplier and from multiple suppliers. If multiple suppliers were used, the probability of infection in each purchased cow was redrawn from the respective distribution (described above) for every purchase.

The number of cattle purchased annually is likely to have an impact on the prevalence and thus on the economic output. We therefore simulated scenarios where farmers purchased one, five or ten animals per year. We also evaluated the impact of using single or multiple suppliers. For this purpose, we used a dataset with the number of suppliers for 19,056 Danish dairy herds registered between 01 March 2014 and 28 February 2015. Of these herds, 19,015 used fewer than 50 suppliers in that year (Fig. 1). In the scenario where multiple suppliers were used, we sampled from the empirical distribution of the number of suppliers for each farm. However, we limited the number to a maximum of 50 suppliers since

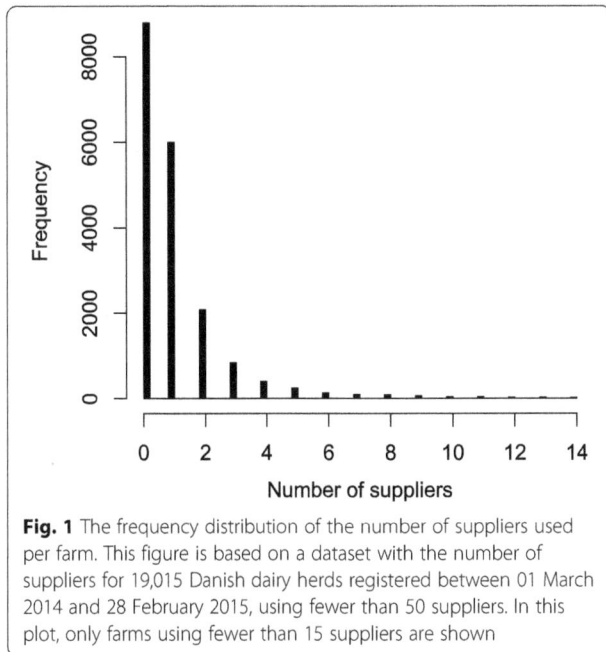

Fig. 1 The frequency distribution of the number of suppliers used per farm. This figure is based on a dataset with the number of suppliers for 19,015 Danish dairy herds registered between 01 March 2014 and 28 February 2015, using fewer than 50 suppliers. In this plot, only farms using fewer than 15 suppliers are shown

we presumed that farms with more than 50 suppliers were not dairy farms. In practice, however, the maximum number of possible suppliers was ten because in this study the farmer bought either one, five or ten animals per year.

We simulated all combinations of scenarios with random, low, medium or high-risk purchases of one, five or ten heifers per year from single and multiple suppliers. All simulations included a burn-in period of 3 years and thereafter we simulated for 10 years.

Infection fadeout
Infection fadeout is when the farm becomes free of MAP infections. It is important to estimate the probability of infection fadeout in order to make informed decisions on factors such as the implementation of disease control actions. In order to estimate the probability of infection fadeout in a herd, we simulated an initially disease-free herd under different scenarios where MAP infection was introduced. We first wanted to estimate the probability of infection fadeout in a farm without MAP, where the farmer bought one to ten infected animals in the first year only. These simulations were run for 10 years in order to estimate the probability of infection fadeout over time. The probability of infection fadeout was calculated from 500 simulations.

We then wanted to estimate the probability of disease fadeout when the farmer of an initially MAP-free herd consistently bought one to ten animals per year. This was achieved by running simulations where the farmer bought a fixed number of animals per year with a 100% probability of infection. We then calculated what

percentage of the 500 simulations resulted in a fadeout of disease after 1 to 10 simulated years.

Model updates
We used the iCull PTB model version 9.1 in this study [15]. This model included monthly milk testing in order to observe the milk yield for each cow, which was updated from an earlier version that used quarterly testing. The observed milk production is affected by the infection status [18] and as a result, the farmer would be more likely to cull infected cows with a lower milk yield. Therefore, we readjusted the force of infection and the infection probability from colostrum and waste milk in the model (using the procedure as described in [15]), in order to obtain a steady prevalence in the herd. Therefore, a fundamental presumption is that the prevalence in the simulated herd is steady without any control actions.

We also updated the model with a standard lactation fitted to every cow so that the farmer could compare the observed milk yield with the expected milk yield. This allowed for a better estimate of the milk yield level, thus helping the farmer to prioritize cows with lower production for culling.

Results
The number of suppliers for each farm is shown in Fig. 1. Of the 19,056 farms, 51% had no suppliers, meaning that they were closed herds, and 49% had one or more suppliers.

Open herd scenarios
In the scenarios with an initial prevalence of 5.6% and where the farmer purchased animals from a low-risk farm, the herd prevalence reached 2% to 4% after 10 simulated years. When farmers purchased animals from a random farm, it resulted in 3% to 4% prevalence (Fig. 2). In these two scenarios, there was no pronounced difference between using single or multiple suppliers. In general, buying ten animals per year marginally reduced the prevalence in these two scenarios, when compared to buying one or five animals. When purchasing animals from medium and high-risk farms, the prevalence increased markedly with an increasing number of heifers bought per year, resulting in the median herd prevalence of 5% and 10%, respectively.

The economic output in the scenarios with an initial prevalence of 5.6% ended up fairly similar in all simulations (Fig. 3).

In all the scenarios with initially MAP-free herds, the resulting prevalences are clearly affected by the number of heifers purchased annually (Fig. 4). In the scenarios where the farmer purchased animals with a random risk, the resulting median prevalence was between 1% and

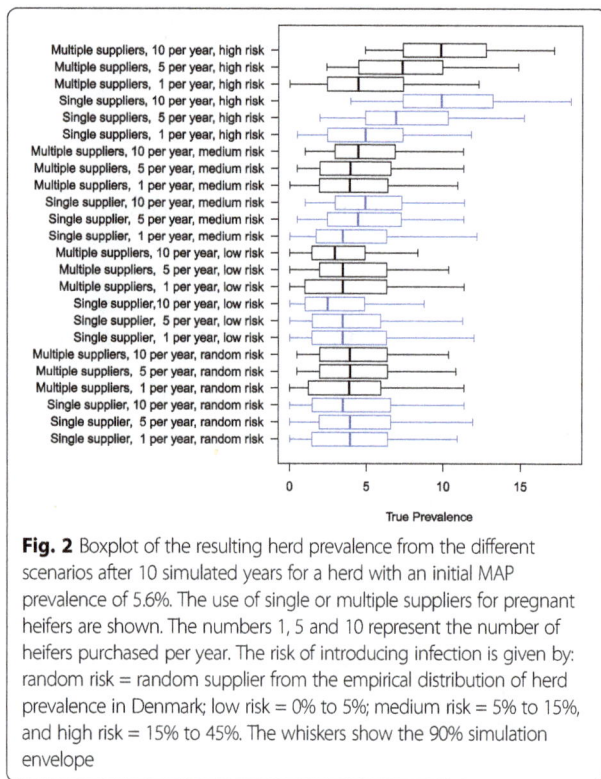

Fig. 2 Boxplot of the resulting herd prevalence from the different scenarios after 10 simulated years for a herd with an initial MAP prevalence of 5.6%. The use of single or multiple suppliers for pregnant heifers are shown. The numbers 1, 5 and 10 represent the number of heifers purchased per year. The risk of introducing infection is given by: random risk = random supplier from the empirical distribution of herd prevalence in Denmark; low risk = 0% to 5%; medium risk = 5% to 15%, and high risk = 15% to 45%. The whiskers show the 90% simulation envelope

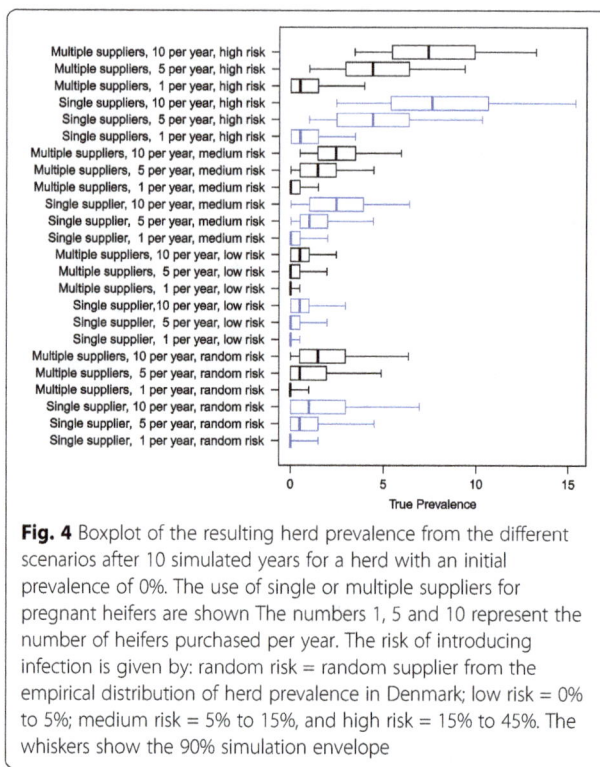

Fig. 4 Boxplot of the resulting herd prevalence from the different scenarios after 10 simulated years for a herd with an initial prevalence of 0%. The use of single or multiple suppliers for pregnant heifers are shown The numbers 1, 5 and 10 represent the number of heifers purchased per year. The risk of introducing infection is given by: random risk = random supplier from the empirical distribution of herd prevalence in Denmark; low risk = 0% to 5%; medium risk = 5% to 15%, and high risk = 15% to 45%. The whiskers show the 90% simulation envelope

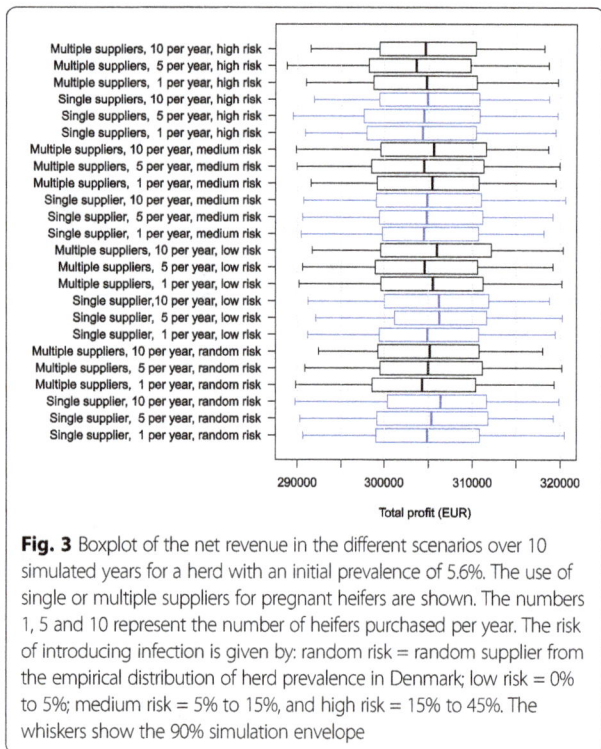

Fig. 3 Boxplot of the net revenue in the different scenarios over 10 simulated years for a herd with an initial prevalence of 5.6%. The use of single or multiple suppliers for pregnant heifers are shown. The numbers 1, 5 and 10 represent the number of heifers purchased per year. The risk of introducing infection is given by: random risk = random supplier from the empirical distribution of herd prevalence in Denmark; low risk = 0% to 5%; medium risk = 5% to 15%, and high risk = 15% to 45%. The whiskers show the 90% simulation envelope

2%. However, if the farmer purchased only one animal from a random-risk farm per year, the median was zero and the upper 90% simulation envelope reached approximately 2%. Furthermore, 79% of the simulations resulted in a prevalence of zero (data not shown). Likewise, if the farmer bought one animal per year from a low-risk farm, 93.4% of the simulations resulted in a prevalence of zero (data not shown).

Using low-risk suppliers resulted in lower herd prevalence than the random-risk scenarios, as was expected (Fig. 4). Purchasing ten low-risk animals per year resulted in a herd prevalence of up to 3% (90% simulation envelope).

Purchasing animals from medium-risk farms resulted in a median herd prevalence of zero when purchasing one animal per year, and a median herd prevalence of 2% when purchasing ten animals per year. When purchasing animals from high-risk suppliers, the resulting median prevalence ranged between 1% (when buying one animal per year) and 8% (when buying ten animals per year). Buying ten high-risk animals per year caused the upper 90% simulation envelope to increase to 15%.

The economic output in the scenarios with an initial freedom from MAP was also found to be very similar in all simulations (Fig. 5).

Infection fadeout

In the simulations where the farmer purchased a number of infected animals in the first year, the probability

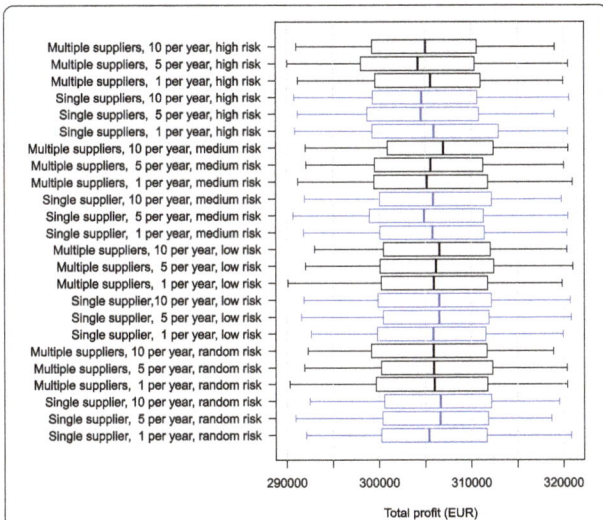

Fig. 5 Boxplot of the net revenue in the different scenarios over 10 simulated years for a herd that was initially free of MAP. The use of single or multiple suppliers for pregnant heifers are shown. The numbers 1, 5 and 10 represent the number of heifers purchased per year. The risk of introducing infection is given by: random risk = random supplier from the empirical distribution of herd prevalence in Denmark; low risk = 0% to 5%; medium risk = 5% to 15%, and high risk = 15% to 45%. The whiskers show the 90% simulation envelope

of a fadeout decreased with the number of infected animals purchased (Table 1). For instance, when the farmer bought one infected animal, the probability of infection fadeout was 58% in the first year and 82% after 10 years. If the farmer bought ten infected animals in the first year, the probability of infection fadeout was only 16% in the first year and 46% after 10 years.

In the scenario where the farmer continually purchased infected animals, the probability of infection fadeout also decreased with an increasing number of infected animals purchased (Table 2). If the farmer bought one infected animal per year, there was a 99.6% probability of infection fadeout in the first year. This high percentage is influenced by the fact that the purchased heifers have not yet given birth so have not been tested, and are therefore not included in the true prevalence. As a result, the prevalence will be low before they start milking. If the farmer bought six or more animals every year, the probability of fadeout was zero in all simulated years (Table 2).

Discussion

In this study, we found that purchasing animals generally increased the herd prevalence proportionally to the probability of infection of the purchased animals and the amount of animals bought per year. However, it was still possible to buy a small number of animals from low or medium-risk farms each year and maintain a low prevalence. Furthermore, in herds that were initially MAP free, the prevalence was still zero after 10 years (within the 90% simulation envelope) if the farmer bought one heifer with random risk per year. This highlights the advantage of a certification system for farms selling livestock, if herds can be reliably divided into risk groups [8, 9, 12].

If a farmer bought a single animal from a low-risk farm per year, the simulations suggested that 93.4% of cases did not result in an infected herd. This means that in this scenario, there is only 6.6% risk of MAP infection on the farm over 10 years, again highlighting the importance of a certification programme where the farmer can actively choose to buy animals from herds with a low

Table 1 The consequence of a single purchase of infected animals

| Simulated years | Number of animals purchased in the first year | | | | | | | | | |
	1	2	3	4	5	6	7	8	9	10
1	58	48	41	36	32	27	22	16	17	16
2	68	58	51	47	39	34	29	24	24	21
3	72	62	57	54	44	40	35	27	27	24
4	74	66	59	54	51	42	38	31	31	27
5	78	69	64	57	54	47	43	36	36	32
6	79	70	68	59	57	48	46	38	39	36
7	80	73	70	65	61	51	48	41	41	38
8	80	74	70	67	63	54	51	44	44	42
9	82	76	72	67	64	55	55	46	47	45
10	82	77	73	69	66	56	56	49	49	46

The probability (%) of fadeout after 1 to 10 simulated years when the farmer buys one to ten infected animals in the first year. For instance, in the first year, the probability of fadeout is lower when 10 animals are purchased, than when only one animal is purchased. The probabilities are calculated from 500 simulations

Table 2 The consequence of a multiple purchases of infected animals

| Simulated years | Number of animals purchased per year | | | | | |
	1	2	3	4	5	6
1	99.6	35.2	2.8	3.2	1.4	0.0
2	15.8	2.6	0.0	0.0	0.0	0.0
3	7.0	0.4	0.0	0.0	0.0	0.0
4	3.2	0.0	0.2	0.0	0.0	0.0
5	2.2	0.0	0.0	0.0	0.0	0.0
6	0.6	0.0	0.0	0.0	0.0	0.0
7	0.4	0.0	0.0	0.0	0.0	0.0
8	0.2	0.0	0.0	0.0	0.0	0.0
9	0.2	0.0	0.0	0.0	0.0	0.0
10	0.8	0.0	0.0	0.0	0.0	0.0

The probability of disease fadeout (calculated as percentages from 500 simulations) after 1 to 10 simulated years when buying one to six animals per year. When buying more than six animals per year, the probability of infection fadeout was always zero risk of infection over farms with a higher risk of infection.

The economic consequences did not appear to differ to a large degree, although the variation within the simulations was fairly large. As previously described in Kirkeby et al. [15], the economic consequences of having MAP within a farm are often small, so these results were as expected. In herds with an initial median prevalence of 5.6%, the purchase of heifers from low-risk farms lowered the upper 75% and 95% percentiles of simulated prevalence. This indicates that, to a certain extent, purchasing low-risk heifers can aid disease control on a farm. In these simulations, we also found that the prevalence did not change considerably even when the farmer bought animals from medium-risk suppliers. This is probably because purchased heifers do not normally come into contact with the highly susceptible calves on the farm. The risk of vertical transmission in the model is 39% and most calves born from infected dams are therefore not infected [19].

We found no noteworthy difference in prevalence between farmers using single and multiple suppliers, when these suppliers had the same risk level. We expected that using multiple suppliers would increase the risk of purchasing animals with PTB and thus introduce MAP into the herd. However, if there is only a small risk of pathogen spread within the farm from purchased animals, then a larger number of infected animals are required to increase the prevalence on the farm.

The prevalence in the simulated scenarios without fadeout (shown in Figs. 6 and 7) is steadily increasing and has not yet reached a steady state. This underlines the slowly evolving nature of PTB. We chose to look at a 10-year timespan in this study, but it is evident that the impact of disease would continue over a much longer period. The model assumes that the same management strategy is used throughout the simulated years.

A previous study [20] simulated a closed herd with endemic MAP infection, and found that the probability of infection fadeout decreased with the herd size. They used a median prevalence of 20%, which is much higher than in this model.

We found that farmers that bought one infected animal per year had a 99.6% chance of having a disease-free herd after the first year (Table 2). However, in the second year they had only a 15% chance of being MAP-free (which would also imply that the purchased and infected animal was culled). If the farmer bought two infected animals per year, he would have only a 35% chance of a MAP-free herd after the second year. The low risk of infection after the first year can be partly attributed to the fact that the true prevalence is only calculated for cows, and does not include infected (purchased) heifers or calves. Furthermore, these estimates include the purchased infected animals, so the probability of a fadeout is also affected by the probability of the purchased animals being slaughtered.

A simulation study by Pouillot et al. [21] showed that in a dairy herd with 40 cows, the maximum fadeout probability of 20% was reached after 10 years. Others [22, 23] estimated that the introduction of one infected animal into a MAP-free herd with 114 cows resulted in disease fadeout in 66% of the simulations. Marcé et al. [22] found that when one animal is purchased per year over 10 years (0.66^{10}), the chance of fadeout is reduced to 2%. In the study by Lu et al. [24], it was found that an initially infection-free herd with 140 cows had a 42% probability of infection fadeout 10 years after the introduction of a single infected animal. In the present iCull model, this probability was higher at 82%. This may be the effect of a relatively low force of infection in the

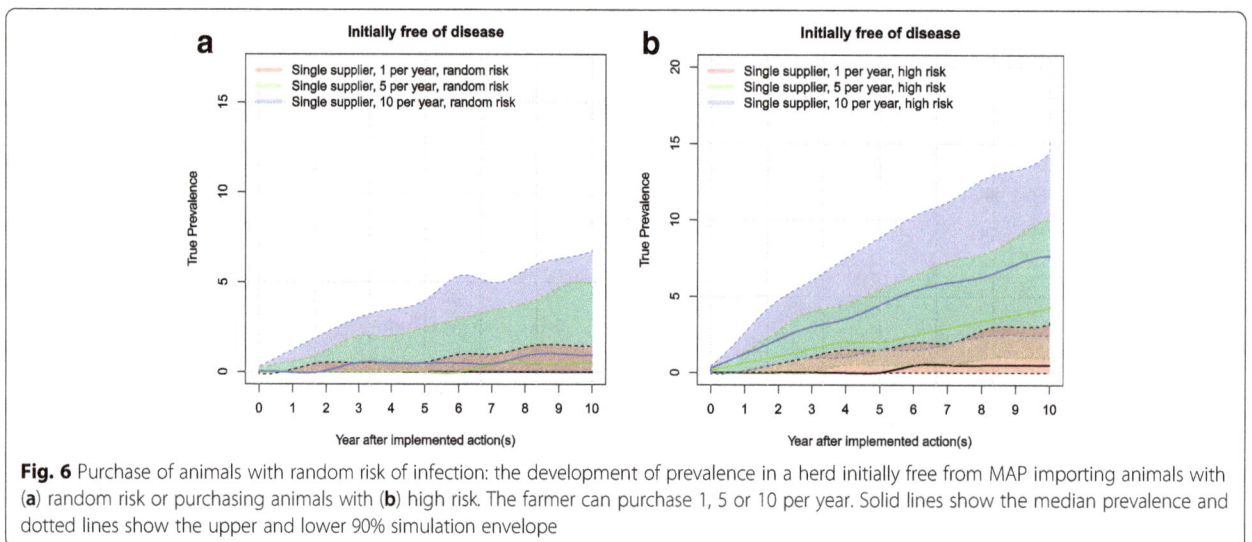

Fig. 6 Purchase of animals with random risk of infection: the development of prevalence in a herd initially free from MAP importing animals with (**a**) random risk or purchasing animals with (**b**) high risk. The farmer can purchase 1, 5 or 10 per year. Solid lines show the median prevalence and dotted lines show the upper and lower 90% simulation envelope

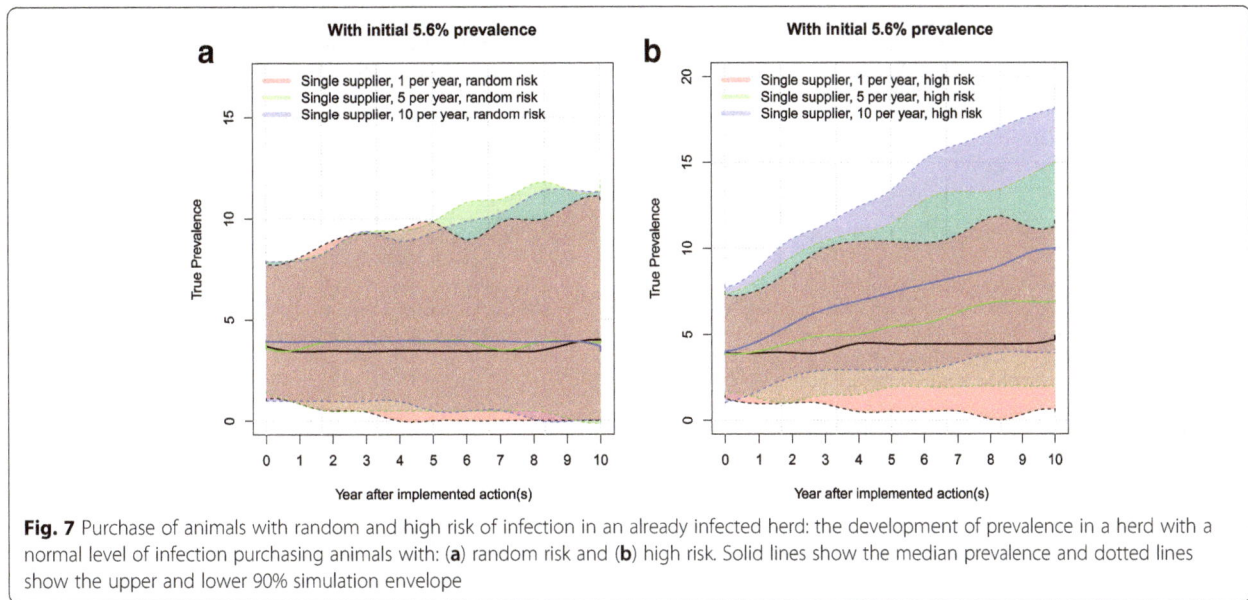

Fig. 7 Purchase of animals with random and high risk of infection in an already infected herd: the development of prevalence in a herd with a normal level of infection purchasing animals with: (**a**) random risk and (**b**) high risk. Solid lines show the median prevalence and dotted lines show the upper and lower 90% simulation envelope

iCull model, which was introduced to calibrate the model to reflect the prevalence in Danish herds.

No control actions were simulated since we chose to look at the general effect of purchasing animals into the herd. Clearly, implementing control actions would increase the probability of disease fadeout, but this is beyond the scope of this study. The impact of control actions on MAP spread within Danish dairy cattle herds was recently examined by [15].

In this model, we included contamination between farm sections. This mechanism might not be relevant on an infected farm, but on an uninfected farm it could potentially transfer MAP from a newly imported heifer to the calf section. The risk of infection is higher for calves due to a decreasing susceptibility with age, and therefore in this case, contamination between farm sections would be critical [25]. The low probabilities of a disease fadeout highlight the importance of buying uninfected animals.

We here assumed that farmers bought pregnant heifers. Therefore our results are limited to this situation. Furthermore, the simulated farm is specifically built to mimic a Danish dairy herd. Consequently our results are not directly applicable to other scenarios where the farm configuration differs a lot from such a farm. The probability of fadeout in this study is dependent on the survival rate of MAP. If, for instance, another strain of MAP has a higher survival rate in the environment, it would take longer time for MAP to fadeout.

Conclusions

To conclude, the simulations showed that it is possible for MAP-free farms to import a small number of

animals each year and still remain free of MAP. If the farmer continually imports animals from certified herds with a known low risk of infection, the actual risk of infection in their herd is very low. In contrast, if the farmer buys animals from a random source, there is a substantial risk of acquiring infection in the herd.

Acknowledgements
The authors wish to thank Jørgen Nielsen from SEGES for providing data on the number of suppliers for Danish herds.

Funding
This project was funded by the Green Development and Demonstration Programme (GUDP) under the Danish Directorate for Food, Fisheries and Agriculture, grant no. 34009–13 0596.

Authors' contributions
All authors participated in planning the study. CK conducted the simulation analyses and wrote the first draft of the manuscript. All authors participated in writing the manuscript and approved the final version.

Consent for publication
(Not applicable)

Competing interests
The authors declare that they have no competing interests.

Author details
[1]National Veterinary Institute, Technical University of Denmark, Kemitorvet, bygning 204, 2800 Kgs. Lyngby, Denmark. [2]DTU Compute, Section for

Dynamical Systems, Department of Applied Mathematics and Computer Science, Technical University of Denmark, Richard Petersens Plads, Bygning 324, 2800 Kgs. Lyngby, Denmark. [3]Department of Large Animal Sciences, Section for Animal Welfare and DiseaseControl, University of Copenhagen, Grønnegaardsvej 8, 1870 Frb. C, København, Denmark.

References

1. Sweeney RW. Transmission of paratuberculosis. The veterinary clinics of North America. Food Animal Practice. 1996;12:305–12.

2. Whittington RJ, Sergeant ESG. Progress towards understanding the spread, detection and control of *Mycobacterium avium* subsp. *paratuberculosis* in animal populations. Aust Vet J. 2001;79:267–78.

3. Nielsen SS, Krogh MA, Enevoldsen C. Time to the occurrence of a decline in milk production in cows with various paratuberculosis antibody profiles. J Dairy Sci. 2009;92:149–55.

4. Kudahl AB, Nielsen SS. Effect of paratuberculosis on slaughter weight and slaughter value of dairy cows. J Dairy Sci. 2009;92:4340–6.

5. Nielsen SS. Dairy farmers' reasons for participation in the Danish control programme on bovine paratuberculosis. Prev Vet Med. 2011;98:279–83.

6. Hassan L, Mohammed HO, McDonough PL. Farm-management and milking practices associated with the presence of *Listeria monocytogenes* in New York state dairy herds. Prev. Vet. Med. 2001;51:63–73.

7. Schaik GV, Schukken YH, Nielen M, Dijkhuizen AA, Barkema HW, Benedictus G. Probability of and risk factors for introduction of infectious diseases into Dutch SPF dairy farms: a cohort study. Prev. Vet. Med. 2002;54:279–89.

8. Krogh K, Nielsen SS, 2012. Lessons learned on control of paratuberculosis in Denmark. Proceedings of the 3rd ParaTB Forum, 4 February 2012, Sydney, Australia, p 5–9.

9. Whipple D. National paratuberculosis certification program. In United States Animals Health Association Proceedings: Report of the 97th Annual Meeting of the United Stated Animal Health Association. Las Vegas. 1993. October, pages 23–29.

10. Kennedy DJ, Allworth MB. Progress in national control and assurance programs for bovine Johne's disease in Australia. Vet Microbiol. 2000;77:443–51.

11. Muskens J, Zijderveld FV, Eger A, Bakker D. Evaluation of the long-term immune response in cattle after vaccination against paratuberculosis in two Dutch dairy herds. Vet Microbiol. 2002;86:269–78.

12. Dufour B, Pouillot R, Durand B. A cost/benefit study of paratuberculosis certification in French cattle herds. Vet Res. 2004;35:69–81.

13. Kirkeby C, Græsbøll K, Nielsen SS, Christiansen LE, Toft N, Halasa T. Evaluation of adaptive test strategies for control and eradication of paratuberculosis within dairy cattle herds. SVEPM 2017 conference proceedings. 2017; 111–118.

14. Verdugo C, Toft N, Nielsen SS. Within- and between-herd prevalence variation of *Mycobacterium avium* subsp. *paratuberculosis* infection among control programme herds in Denmark (2011 – 2013). Prev. Vet. Med. 2015;121:282–7.

15. Kirkeby C, Græsbøll K, Nielsen SS, Christiansen LE, Toft N, Rattenborg E, et al. Simulating the epidemiological and economic impact of Paratuberculosis control actions in dairy cattle. Frontiers in veterinary science. 2016;3

16. Ezanno P, Fourichon C, Beaudeau F, Seegers H. Between-herd movements of cattle as a tool for evaluating the risk of introducing infected animals. Anim Res. 2006;55:189–208.

17. Ettema JF, Østergaard S. Short communication: economics of sex-biased milk production. J Dairy Sci. 2015;98:1078–81.

18. Græsbøll K, Nielsen SS, Halasa TH, Christiansen LE, Kirkeby C, ,Toft, N. Association of MAP specific ELISA-responses and productive parameters in 367 Danish dairy farms. In Proceedings of the 12th International Colloquium on Paratuberculosis, Parma, Italy, June 22–26, 2014; pages 139–141.

19. Whittington RJ, Windsor PA. In utero infection of cattle with *Mycobacterium avium* subsp. *paratuberculosis*: a critical review and meta-analysis. Vet. J. 2009;179:60–9.

20. Lu Z, Schukken YH, Smith RL, Gröhn YT. Stochastic simulations of a multi-group compartmental model for Johne's disease on US dairy herds with test-based culling intervention. J Theoret Biol. 2010;264:1190–201.

21. Pouillot R, Dufour B, Durand B. A deterministic and stochastic simulation model for intra-herd *paratuberculosis* transmission. Vet Res. 2004;35:53–68.

22. Marcé C, Ezanno P, Seegers H, Pfeiffer DU, Fourichon C. Predicting fadeout versus persistence of paratuberculosis in a dairy cattle herd for management and control purposes: a modelling study. Vet Res. 2011;42:36.

23. More SJ, Cameron AR, Strain S, Cashman W, Ezanno P, Kenny K, et al. Evaluation of testing strategies to identify infected animals at a single round of testing within dairy herds known to be infected with *Mycobacterium avium* Ssp. *paratuberculosis*. J. Dairy Sci. 2015;98:5194–210.

24. Lu Z, Schukken YH, Smith RL, Gröhn YT. Using vaccination to prevent the invasion of *Mycobacterium avium* subsp. *paratuberculosis* in dairy herds: a stochastic simulation study. Prev. Vet. Med. 2013;110:335–45.

25. Marcé C, Seegers H, Pfeiffer DU, Fourichon C. The influence of contact structure on disease transmission in a dairy herd using paratuberculosis as an example. In Society for Veterinary Epidemiology and Preventive Medicine. Proceedings, Nantes, France, 24–26 March, 2010; pages 135–144.

11

Molecular identification of *Mycobacterium bovis* from cattle and human host in Mali: expanded genetic diversity

Mamadou Diallo[1†], Bassirou Diarra[2*†], Moumine Sanogo[2], Antieme C. G. Togo[2], Anou M. Somboro[2], Mariam H. Diallo[2], Bréhima Traoré[2], Mamoudou Maiga[2], Younoussa Koné[1], Karim Tounkara[1], Yeya dit Sadio Sarro[2], Bocar Baya[2], Drissa Goita[2], Hamadoun Kassambara[2], Bindongo P. P. Dembélé[2], Sophia Siddiqui[3], Robert L. Murphy[4], Sounkalo Dao[2], Souleymane Diallo[2], Anatole Tounkara[2ˆ] and Mamadou Niang[1]

Abstract

Background: Bovine tuberculosis (BTB) is a contagious, debilitating human and animal disease caused by *Mycobacterium bovis*, a member of the *Mycobacterium tuberculosis* complex. The study objective were to estimate the frequency of BTB, examine genetic diversity of the *M. bovis* population in cattle from five regions in Mali and to determine whether *M. bovis* is involved in active tuberculosis (TB) in humans. Samples from suspected lesions on cattle at the slaughterhouses were collected. Mycobacterial smear, culture confirmation, and spoligotyping were used for diagnosis and species identification. *Mycobacterium* DNA from TB patients was spoligotyped to identify *M. bovis*.

Results: In total, 675 cattle have been examined for lesions in the five regions of Mali. Out of 675 cattle, 79 specimens presented lesions and then examined for the presence of *M. bovis*. Thus, 19 (24.1 %) were identified as *M. bovis*; eight (10.1 %) were non-tuberculous *Mycobacterium* (NTM). Nineteen spoligotype patterns were identified among 79 samples with five novel patterns. One case of *M. bovis* (spoligotype pattern SB0300) was identified among 67 TB patients.

Conclusion: This study estimates a relatively true proportion of BTB in the regions of Mali and reveals new spoligotype patterns.

Keywords: Bovine tuberculosis, Frequency, Spoligotyping, Mali

Background

Bovine tuberculosis (BTB) is a contagious, debilitating human and animal disease. It is caused by *Mycobacterium bovis* a member of the *M. tuberculosis* complex (MTBc), which also includes *M. tuberculosis*, *M. africanum*, *M. microti*, *M. canettii*, *M. caprae*, *M. pinnipedii* and the new discovered strain *M.mungi* [1]. Cattle are considered as natural reservoir for *M. bovis*.

In Mali, West Africa, the total number of cattle is estimated to be more than 9,438,000, and the livestock sector occupies an important place in the development of the economy [2]. In terms of the national economy, the livestock sector contribution to gross domestic product (GDP) is estimated to be 10 % and its contribution to export earnings was 40 billion CFA ($83 million United States of America Dollar: USD) in 1995, which represents 17.5 % of total exports, occupying third place after cotton and gold [3]. The economic contributions from cattle have steadily declined in recent years, and were only 28 billion CFA ($58 million USD) in 1999 representing 7 % of total exports [4].

The monitoring of BTB is based partially on screening by tuberculin skin test of live animals and also on the testing of suspected lesions found at slaughterhouses. Bovine tuberculosis is a zoonotic disease caused by *M. bovis*, which is transmitted, from animals to humans by ingestion of raw milk, contaminated meat or aerosol [5]. *Mycobacterium bovis* infection in humans often presents as extra pulmonary tuberculosis, especially in children

* Correspondence: bdiarra@icermali.org
†Equal contributors
ˆDeceased
2SEREFO, University of Sciences, Techniques and Technologies of Bamako (USTTB), Point-G, Bamako, Mali

[5]. The possible role of *M. bovis* in human tuberculosis infection is not well known in Mali because tuberculosis in humans due to *M. bovis* is clinically indistinguishable from that of *M. tuberculosis.*

In many resource-limited settings, the diagnosis of tuberculosis is based only on smear microscopy which cannot differentiate *Mycobacterium* species or the different strains of MTBc. Bovine tuberculosis cannot be specifically diagnosed without confirmation using additional microbiological methods. New molecular methods such Mycobacterial Interspersed Repetitive Units (MIRU), Variable Number of Tandem Repeats (VNTR) and restriction fragment length polymorphism (IS6110 RFLP) techniques [6] can determine the importance and factors influencing the transmission of BTB among cattle. Among these methods, spoligotyping technique is most often used for molecular typing of *Mycobacterium* strains [7]. This technique has the advantage to confirm the diagnosis of tuberculosis and also to classify subspecies of *Mycobacterium tuberculosis* complex (MTBc) [7]. Because of this method, *M. bovis* was isolated in countries where mycobacterial isolates from humans were fully typed [7].

In Mali, there has been an increase in the number and variety of cattle, and as well as in the number of tuberculin skin test-positive cases. Several studies on *M. bovis* population focused only in the capital city of Bamako have been performed excluding other regions of Mali [8, 9]. The goal of this study was first to estimate the apparent prevalence of BTB, second to examine the genetic diversity of strains of *M. bovis* obtained in five distinct regions of Mali from infected cattle, using spoligotyping technique. Lastly, we examined whether *M. bovis* is involved in active TB in the human host.

Methods
Slaughterhouses and specimens collection
The regions of Ségou, Mopti, Sikasso, Kayes and Bamako were selected for this study. These five regions are home to more than 80 % of the livestock of Mali [4]. In each region there is a slaughterhouse maintained by a trained veterinarian.

We conducted this study from January 2008 to December 2010. During that period, 675 cattle were examined for the presence of lesions in the regions of Kayes (75), Sikasso (160), Ségou (80), Mopti (140) and Bamako (220). At the slaughterhouses, specimen collection was based on suspicion of bovine tuberculosis lesions after slaughtering by meat inspection [10]. Suspected lesions were lymph nodes and or necrotic nodules. They were size different and were containing yellow, green or tan pus. Tissue specimens were collected from affected organs such as lymph nodes, lungs, chest, and liver as described by Cosins [9]. Upon collection, tissue specimens

were transported on ice to the Central Veterinary Laboratory of Bamako (Laboratoire Central Vétérinaire de Bamako) and kept at −80 °C before testing. Each specimen was uniquely identified by collection date, sample number and region of origin.

Specimens processing, smear microscopy and culture
All the laboratory work was performed in the BSL-3 facility within the Human Immunodeficiency Virus and Tuberculosis (HIV-TB) Research and Training Center (SEREFO Laboratories) at the University of Sciences, Techniques and Technologies of Bamako (USTTB).

Tissue specimens were washed with sterile saline solution (Remel Inc., Lenexa, KS 66215, USA). Specimens were then cut into small pieces using the Tissue Grinder (Precision Disposable Tissue Grinder, Covidien, Mansfield, MD, USA). Thereafter, specimens were digested and decontaminated using the standard N-Acetyl-L-Cysteine/4 % NaOH solution (Alpha Tec System. Inc., Vancouver, USA), concentrated by high speed centrifugation (4500 rpm or 3000 g, Eppendorf centrifuge) and inoculated on both liquid (Mycobacterium Growth Incubator Tube, BBL™ MGIT™ Becton Dickinson, Sparks MD, USA), and solid (Middlebrook 7H11 Agar and Selective 7H11 Agar) media. Simultaneously, an aliquot of concentrated specimen was prepared for Auramine-Rhodamine staining (BBL™ Becton Dickinson, Sparks MD, USA) for microscopic examination (Olympus, Olympus Corporation, Tokyo, Japan).

Identification of mycobacterial isolates
Positive cultures were confirmed by smear using Kinyoun (Remel Inc. Lenexa, KS 66215, US) stain acid fast method. The mycobacterial infection of specimens was based on a combination of the initial fluorescent acid-fast microscopy result, colonial morphology of growth in culture, and confirmation by Kinyoun staining. Suspected tuberculosis cases were confirmed by spoligotyping technique.

In order to test whether there is transmission of *M. bovis* in human hosts, we examined the spoligotyping results for 67 new human TB cases screened at the University Teaching Hospital of Point G and the six-TB care units located in the health referral centers of Bamako between 2006 and 2010 from SEREFO database.

Molecular strain typing for genetic diversity study within M. bovis
All Kinyoun-positive colonial growth suggestive of *M. bovis* underwent molecular strain typing. Spoligotyping technique was performed using a commercially available kit (Isogen Life Science, Netherlands) by following the manufacturer's instructions to amplify the whole direct repeat region (DR) within the *M. tuberculosis* genome.

The amplified biotin labeled PCR product was then hybridized to 43 ordered synthetic DNA oligonucleotide spacers bound to a nylon membrane. After hybridization, the membrane was developed by incubating with a streptavidin peroxidase conjugate, which binds to the biotin label on the PCR product. Next, the membrane was washed and incubated with Enhanced Chemiluminescent (ECL) Detection Reagents. This reaction results in the emission of light, and is detected by exposing a light sensitive film to the membrane. Comparison and naming of the patterns were done through global *M. bovis* database (http://www.Mbovis.org).

Statistical analysis

The proportion of each strain type with 95 % confidence interval (CI) and their geographical region were compared using the χ^2 test or Fisher's exact test. The study included all the suspected cases which were seen at each slaughterhouse during our visit. A descriptive analysis was performed by evaluating the number of isolates determined to be *M. bovis cases* by spoligotyping technique. Thereafter similar spoligotypes were grouped in a same cluster. The relationship of clusters to geographical origin of animal was also evaluated. At the same these patters were compared to the *M.bovis* case which was isolated from human during one of our previous study on human tuberculosis.

Results

During the study period, in total 675 cattle were examined in the 4 regions and the district of Bamako where the study was conducted (Fig. 1).

Among the 675 cattle screened between January 2008 and December 2010, 79 have presented lesions leading to a rate of 11.7 % [95 % CI 9.3–14.4] (Table 1). The regions of Kayes and Mopti have presented the highest rate of lesions respectively 14.7 % and 13.6 %. In contrast, Sikasso has the lower rate with 7.5 % (Table 2). The general frequency of *M. bovis* infection was 2.81 % with the highest prevalence in Bamako (8/675) with

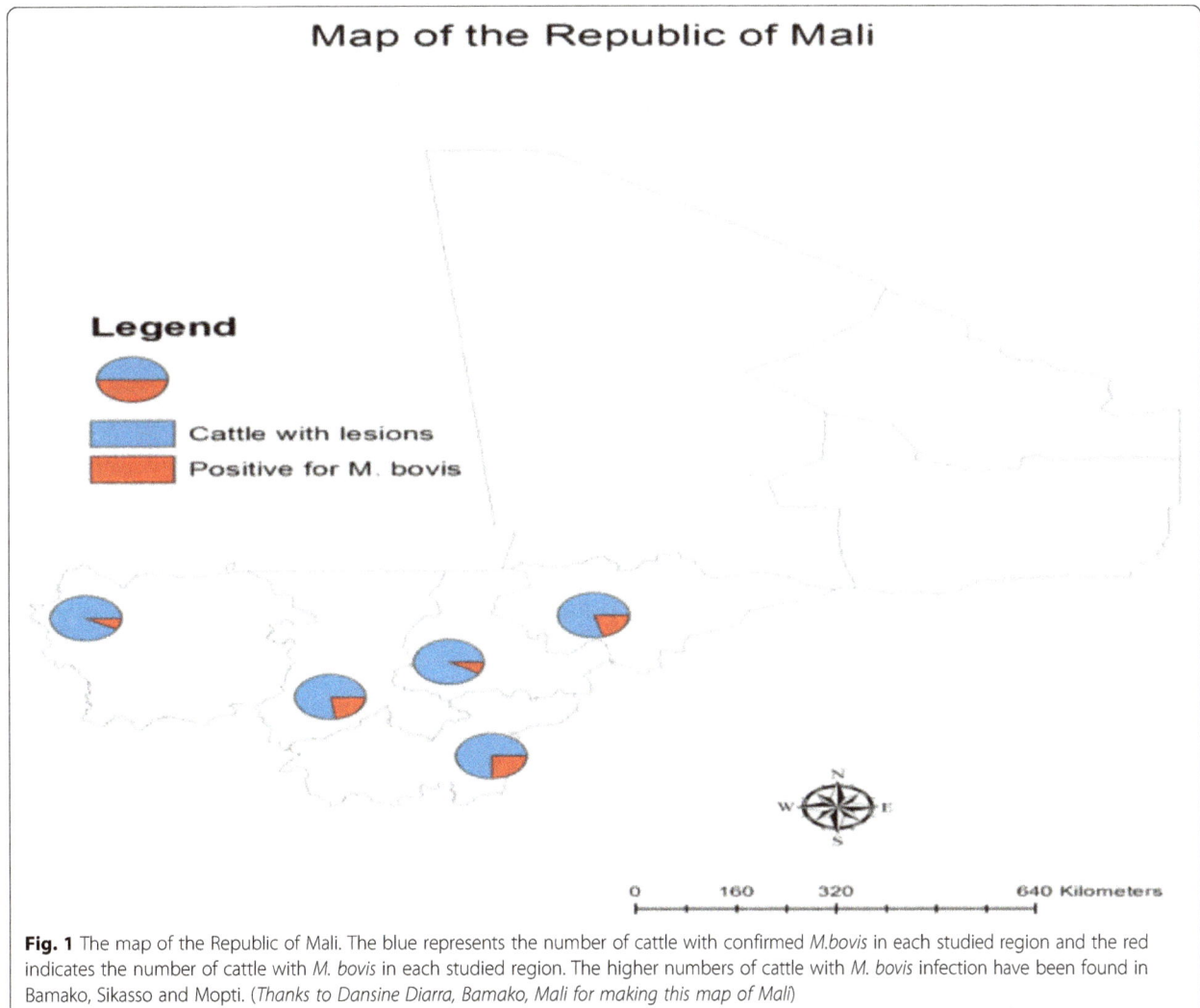

Fig. 1 The map of the Republic of Mali. The blue represents the number of cattle with confirmed *M.bovis* in each studied region and the red indicates the number of cattle with *M. bovis* in each studied region. The higher numbers of cattle with *M. bovis* infection have been found in Bamako, Sikasso and Mopti. (*Thanks to Dansine Diarra, Bamako, Mali for making this map of Mali*)

Table 1 Tissue sample number, region of provenance and prevalence using liquid media technique

Sample provenance	Number of examined cattle	Number of cattle with lesions	Percentage of cattle with lesion	Number of positive *M. bovis*	Percentage of positive (microscopy)	Percentage of positive (liquid media culture)
Kayes	75	11	14.7 %	1	1/11 (9.1 %)	1/11 (9.1 %)
Sikasso	160	12	7.5 %	4	4/12 (33.3 %)	4/12 (33.3 %)
Ségou	80	10	12.5 %	1	0/10 (0 %)	1/10 (10 %)
Mopti	140	19	13.6 %	5	4/19 (21.55)	5/19 (26.3 %)
Bamako	220	27	12.3 %	8	7/27 (25.9 %)	8/27 (29.6 %)
Total	675	79	11.70	19	16/79 (20.25 %)	19/79 (24.05 %)

Among the 675 cattle screened, 79 have presented lesions leading to a rate of 11.70 %. The regions of Kayes and Mopti have presented the highest rate of lesions respectively 14.7 % and 13.6 %. In contrast, Sikasso has the lower rate with 7.5 %. The general prevalence of the infection was 2.81 % with the highest prevalence in Bamako (8/675) with 1.18 % and the lowest was observed in Kayes and Sikasso (1/675) with 0.15 %. Tests performed on the 79 lesions obtained, the number of infected lesions with *M. bovis* observed was higher in Sikasso, Bamako and Mopti with respectively 33.3, 29.6 and 26.3 %. This table shows that 34.2 % of cattle were investigated in Bamako followed by Mopti with 24.1 % and Ségou has the lowest number of cattle 12.6 %. This was no statistically significant difference between the number of specimens collected in each site ($p = 0.47$)

1.18 % and the lowest was observed in Kayes and Sikasso (1/675) with 0.15 % (Table 1). Tests performed on the 79 lesions obtained, the number of infected lesions with *M. bovis* was observed in Sikasso, Bamako and Mopti with respectively 33.3, 29.6 and 26.3 % (Table 2). The mean age of the cattle was 4 years old, and the majority were female.

Using the culture technique, 19 specimens out of 79 were confirmed as BTB (24.1 %) (Table 3). Eight (10.1 %) cultures positive for acid-fast bacilli were found to be non-tuberculous mycobacteria (NTM) strains, while 23 samples (29.1 %) were contaminated with other bacteria or yeast (Table 1). The sequencing of the 16sRNA to identify the species of those bacteria was not performed.

Spoligotype patterns were assessed during the study. The absence of spacers 3, 9, 16 and 39–43 was used to distinguish *M. bovis* [7]. Our data showed that all 19 specimens investigated were identified as BTB and lacked these spacers (Table 4). Comparison of the patterns with the global *M. bovis* database revealed five spoligotypes that had not previously been reported (SB2262, SB2263, SB2264, SBXXXX mopti and SBXXXX bamako) in Mali and elsewhere (Table 4). Those five were subsequently registered in the *M. bovis* database in http://www.mbovis.org.

There was a disparity in the distribution of spoligotypes among the five sites. Among the newly identified spoligotype patterns, the SB2262 was the most frequent with 36.84 % and found only in Bamako. The spoligotype pattern SB1275 was present only in Sikasso (in the south of the country at the Côte D'Ivoire border) as well as the SB0944, which represented 15.78 % of all the patterns (Table 4). The SB0300 was widely spread throughout the country (Bamako, Kayes, Ségou and Mopti) except Sikasso; it represented 31.57 % of all the patterns (Table 4). Also, a unique pattern was observed in Mopti, which is the SBXXX mopti. The characteristic of this pattern is that, the spacers 4, 5, 22 and 23 are lacking (Table 3). In addition, the lack of the spacer 32 led to the spoligotype pattern SBXXXX bamako and found only in Bamako. In addition, we found that spoligotype patterns SB2262 and SB2263 lacked both spacers 6 and 21 (Figs. 2 and 3). These strains were found only in Bamako and Mopti.

Among the 67 specimens obtained from individual human subjects infected with pulmonary tuberculosis, only one had confirmed *M. bovis*. This case was also a pulmonary tuberculosis infection. In comparison to the spoligotypes from cattle, the spoligotype pattern obtained from human host was similar to the pattern observed in

Table 2 Proportion of infection per region

Sample provenance	Number of examined cattle (%)	Number of Positive (*M. bovis*) by microscopy and (%)	Number of Positive (*M. bovis*) by liquid media culture (%)
Kayes	75 (11.1)	1/75 (1.33 %)	1/75 (1.33 %)
Sikasso	160 (23.7)	4/160 (2.5 %)	4/160 (2.5 %)
Segou	80 (11.9)	0/80 (0 %)	1/80 (1.25 %)
Mopti	140 (20.7)	4/140 (2.86 %)	5/140 (3.57 %)
Bamako	220 (32.6)	7/220 (3.18 %)	8/220 (3.63 %)
Total	675	16/675 (2.37 %)	19/675 (2.81 %)

The estimated prevalence (stated) as proportion of infection of *M. bovis* was 2.37 % (16/675) by microscopy and 2.81 % (19/675) using the liquid media culture. The gold standard test is liquid media culture. There is disparity among regions, the highest prevalence (culture) was in Mopti (3.57 %) and Bamako (3,63 %) and the lowest was in Segou (1.25 %)

Table 3 Culture results of tissue specimens from Malian cattle. Specimens were digested, decontaminated, centrifuged at high speed and inoculated onto two types of media

Tissue specimens culture results	Number (N)	Percentage (%)
Positive (M. bovis)	19	24.1
Positive (NTM)	8	10.1
Negative	29	36.7
Contaminated with other bacteria	23	29.1
Total	79	100

Using the culture system, 27 of 79 were positive; one can notice that 19 were M. bovis (24.1 %) while eight were Non Tuberculous Mycobacteria (10.1 %). The contamination with other species of bacteria has been estimated

the cattle *M. bovis* pattern SB0300. Both lack the spacers 6 and 30 (Table 4). The patient with the SB0300 spoligotype pattern of *M. bovis* had developed active tuberculosis with characteristic symptoms including weight loss, cough and fever. The serological testing for HIV was negative.

Discussion

The estimated frequency of lesion was 11.70 % (79/675) with a disparity among regions. The highest lesion rate was observed in Kayes in the Western part of the country with 14.7 % whereas it was two times less in Sikasso in the Southern region with 7.5 % (Table 2). In contrast, the infection rate of *M. bovis* presents another profile; the overall frequency was 2.37 % (16/675) with the microscopy and 2.81 % (19/675) with the use of liquid media culture (Table 5). The highest frequency of *M. bovis* infection (by culture) was observed in Bamako and Mopti with respectively 3.63 % (8/220) and 3.57 % (5/140). The lowest frequency was found in Ségou with 1.25 % and in Kayes (1.33 %) where the lesion rate was higher (Table 1). There is then an indication that all the lesions were not due to *M. bovis* infection. Mopti and Bamako are two important places for livestock, Mopti is an excellent place in terms of cattle breeding, and Bamako is the big market for big consumption. Thus, the risk seems

higher based on those data, it is likely that out of 100 cattle there is risk that at least two could be infected with *M. bovis*.

Sixty-three percent of specimens from tuberculosis lesions were culture-positive. This proportion is similar to the frequency observed in Chad, which was 65 % as reported by Diguimbaye-Djaibé & al. in 2006 [11]. The smear positivity was only one fifth (20.3 %) of the specimens, which was low and confirms the lack of sensitivity of smear microscopy for tuberculosis diagnosis in this setting.

The frequency of BTB varies from one geographical zone to another depending on the population density of cattle. An example is Tanzania, where the prevalence of BTB (by using a single tuberculin skin test) was 14 % in the Southern Highlands [12], and 0.2 % in the Lake Victoria zone [13]. This difference was explained partially by the density and number of cattle in the region.

Tuberculosis in cattle has been demonstrated to be prevalent in sedentary breeding style, where cattle are used for milk production as these conditions allow better transmission of *M. bovis* infection between animals [14, 15]. In Mali, the prevalence of tuberculin-skin test positive cattle in Sikasso and Ségou was respectively 20 % and 64.3 % in different herds,1 % and 2.8 % in individual cattle, and the same observation was seen in the suburban area of Bamako in 2003 [14, 16].

We further found that 10.1 % of specimens were infected with non-tuberculous mycobacteria (NTM) and 29.1 % with bacteria or yeast. Almost one third (36.7 %) of the specimens was culture negative, a proportion higher than what was seen in Chad (27.3 %) [11]. The isolation of NTM from tissue samples may represent true infections, co-infections or colonization. However, we do not know from our data the importance and infectiousness of those isolates. In view of the lack of pasteurization of milk, these data do suggest that, cattle may be one source of NTM transmission.

Table 4 Spoligotype patterns revealed during the study in Mali and their frequency

SB number	Frequency	Percent	Pattern	Region
SB2262[a]	2	10.52	1101101101111100111101111111101111111100000	Bamako
SB2263[a]	3	15.78	1101101101111101111011111111101101111100000	Bamako
SB2264[a]	1	5.26	1101101101111101111111111111101011111100000	Bamako
SB0300	6	31.57	1101101101111101111111111111101111111100000	Kayes, Bamako, Ségou, Mopti
SB1275	1	5.26	1101111101111101111011111111101111111100000	Sikasso
SBXXXX	1	5.26	1100011101111101111100111111101111111100000	Mopti
SBXXXX	2	10.52	1101101101111101111111111111101011111100000	Bamako
SB0944	3	15.78	1101111101111101111111111111101111111100000	Sikasso
Total	19	100		

[a]Three new patterns have been identified, SB2262, SB2263 and SB2264 and are found only in Bamako. The SB1275 and SB0944 are found in Sikasso. In contrast, the SB0300 is found in all sites except Sikasso. The frequency of the SB2262 pattern is more frequent in Bamako with 36.84 % followed by the SB0300 pattern with 31.57 % in the four of the five study sites

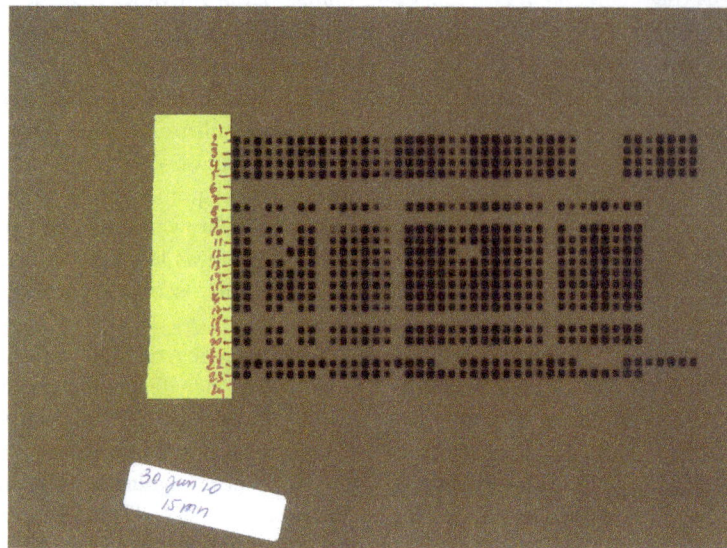

Fig. 2 Spoligotyping test run on 30 Jun 2010 (*30 Jun 10*). 1 = Buffer. 2 = MAL02020101. 3 = MAL02020201. 4 = MAL02020501. 5 = MAL02020801. 6 = LCV-1. 7 = LCV-7. 8 = LCV-9. 9 = LCV-10. 10 = AFS-2. 11 = MMP-1. 12 = MMP-2. 13 = MMP-3. 14 = SKSS-1. 15 = SKSS-3. 16 = SKSS-4. 17 = SKKL-1. 18 = SKKL-2. 19 = SgSg-2. 20 = Mate-1. 21 = Negative control. 22 = Positive control-1 (H37Rv). 23 = Positive control-2 (*M.bovis*). 24 = Buffer. Numbers represent the patients Laboratory unique ID or Tissues Identification numbers

Our study was limited to the south of the country, and the cattle were from the urban areas of each region. The transmission profiles in rural areas of the regions may be different if we were able to include cattle from these areas. Thus, a cross sectional with larger sample size and sampling combining slaughterhouses, rural and urban areas from all the country may reveal the true country profile of *M.bovis* strains circulating in Mali. Another limitation was that our focus was mainly for identification of *M.bovis* and we didn't focus on identifying other mycobacteria species. Despite these limitations, the study revealed that the Bamako region has the highest frequency of BTB and showed a wide variety of *M. bovis* strains (SB2262, SB2263, SB2264, and SB0300). Only the SB0300 was previously reported [8]. This frequency is higher than the one reported by Sidibé in 2003 [14].

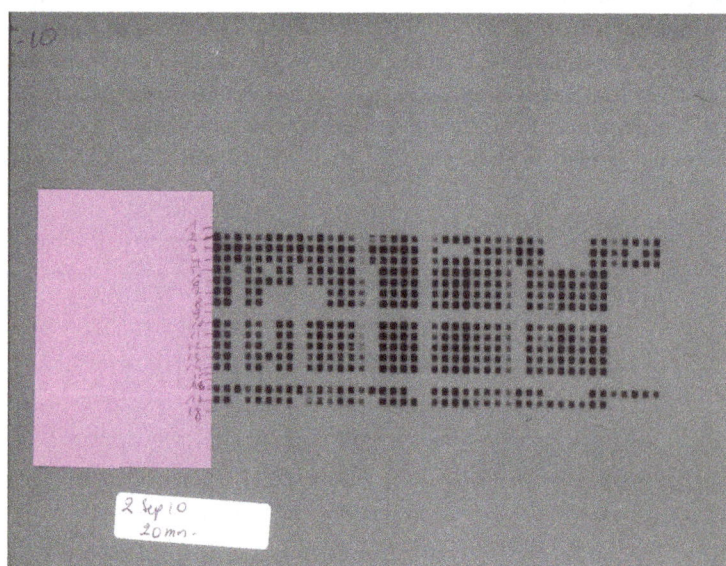

Fig. 3 Spoligotyping test run on 2 September 2010 (*2 Sep 10*). 1 = Buffer. 2 = MAL02020601. 3 = MAL02020901. 4 = MAL020201101. 5 = Bko-2. 6 = SKSS-2. 7 = AFS-1. 8 = AFS-3. 9 = LCV-7. 10 = MMP-2. 11 = Bko-4. 12 = Bko-1. 13 = SKSS-2. 14 = Bko-3. 15 = Negative control. 16 = Positive control-1 (H37Rv). 17 = Positive control-2 (*M.bovis*). 18 = Buffer. Numbers represent the patients Laboratory unique ID or Tissues Identification numbers

Table 5 Results of the smear microscopy from tissue pellets

Microscopy (AFB-Smear)	Number (N)	Percentage (%)
Positive	16	20.3
Negative	63	79.7
Total	79	100

By indirect Auramine-Rhodamine (smear from tissue pellets), 20.3 % of the samples were positive by microscopy

Such high prevalence and diversity in Bamako, the capital city of Mali, may be due to the semi-intensive breeding style used in this region to improve milk and meat production. The high demand for meat in the capital city of Bamako increases the different varieties of cattle, the biggest market for livestock in Mali [14]. These data are in line with previous studies that showed that imported cattle are sensitive to different types of diseases in general and bovine tuberculosis in particular [14]. Another explanation may be the high number of tissue samples collected from Bamako region in comparison to the other regions (27/79), although this difference was not significant ($p = 0.47$). Bamako was followed by the Mopti region, which has the highest number of cattle in Mali and is an excellent geographical breeding zone because of its diverse climate [14]. We discovered a wide variation in *M. bovis* strains with five new spoligotype patterns identified. However, the particular distribution of the spoligotype SB0944 in Sikasso has to be investigated further. In fact, this region in the south of the country has received funding to breed the *N'dama race*, which is Trypano- tolerant [17]. In addition, all the strains observed in Sikasso had the spacer 6 which was absent in all strains studied so far in Mali [8]. This may explain the clustering of this spoligotype in Sikasso and we believe that strain was imported through the intense breeding related to this N'dama project. We also have identified seven cattle infected with *M. bovis* that are lacking the spacers 6 and 21. This is the first time those spoligotypes have been observed in Mali. In the previous study done in the Bamako abattoir, Muller & al. [8], have identified spoligotypes lacking spacer 6 in Bamako. We observed that five cattle were infected with strains of *M. bovis* lacking simultaneously the spacers 6 and 21, which may indicate that the genome is flexible enough to allow creation of new spoligotypes that in turn helps to increase the antigenic repertoire within *M. bovis*. The cluster number six was the most prevalent and was seen in four regions out of five [8].

All our examined specimens showed an absence of spacer 30 (Fig. 2), which confirmed the results obtained in 2007 by Muller et al. [8]. However, Muller found the presence of the spacer 30 among seven strains out of 20 from Bamako abattoir [18]. Taken together, *M. bovis* strains from Cameroun, Nigeria, Chad and Mali examined by spoligotyping technique showed similar spoligotype patterns that lacked the spacer 30 [18]. This may suggest that strains from those countries are phylogenetically similar [18]. The reason may be a specific deletion within a chromosomal DNA (Af 1) [18]. The shortest sequence through the Direct Repeat region was observed in the SBXXXX Mopti. In addition to known missing spacers, there are the spacers 4, 5, 22 and 23 which were also missing. The deletion of those spacers which are contiguous may have a rich AT content which is prone for deletion. During this study, *M. caprae*, which is known to be associated tuberculosis infection in sheep and goats, was not observed [19].

The use of spoligotyping for accurate diagnosis is important for assessing the fight against this zoonosis [20]. In this study, spoligotyping was used as a diagnostic and confirmatory tool, but also yielded important insight into the epidemiology and characteristic of *M. bovis* in Mali and revealed that more spoligotyping patterns are emerging. Another important impact of this study concerns the neighbor countries of Mali, since a large percentage of meat from cattle in Mali is consumed in those countries, and the mobility of livestock in the vast region of Sahel which could potentially lead to further transmission of BTB [4].

In this study, we identified one case of *M. bovis* infection in humans for a crude estimated prevalence of 1.5 % (1 out 67). This zoonotic TB was found in several countries in Africa with a median of 2.8 % (range 0–37.7 %). Our prevalence falls in that range and below the prevalence observed in Ethiopia, Nigeria and Tanzania [12, 21, 22]. We assume that the proportion of human cases of *M. bovis* depends on the efficacy of disease control such as regular milk pasteurization, and slaughterhouse meat inspection [10, 23]. In Mali, little is known about the transmission of *M. bovis* to, or between humans. *Mycobacterium bovis* can be easily transmitted through ingestion of unpasteurized milk, or undercooked meat and can cause pulmonary tuberculosis mainly in children [5, 12] among workers from farms and slaughterhouses [24]. Although we have not performed a proper study on risk factors analysis, but we may speculate that those means are the common route of *M. bovis* among our patient.

Conclusion

M. bovis is present in Mali and accounts for approximately one quarter of suspected lesions from infected cattle in the tested regions. Our approach by the use of spoligotyping is a logical step for a better understanding of different mycobacterial species circulating in the country. In order to further characterize the frequency of BTB in Mali and to estimate human transmission rates of NTM as well as BTB, more extensive testing is necessary. The presence of confirmed *M. bovis* in one

Malian subject suggests that there may be a lack of effective disease control which includes regular milk pasteurization and slaughterhouse meat inspection.

Abbreviations

BTB, bovine tuberculosis; DR, direct repeat; ECL, enhanced chemiluminescent; GDP, Growth Domestic Product; HIV, human immunodeficiency virus; IS6110 RFLP, restriction fragment length polymorphism; MIRU, Mycobacterial Interspersed Repetitive Units; MTBc, Mycobacterium tuberculosis complex; NTM, non-tuberculous mycobacteria; TB, tuberculosis; USD, Dollar of United States of America; USTTB, University of Sciences, Techniques and Technologies of Bamako; VNTR, variable number of tandem repeats

Acknowledgments

The authors express special thanks to all the slaughterhouses technicians of the different regions, for their collaboration during the study period, and also to all the staff of the Laboratoire Central Vétérinaire de Bamako. They also express our special thanks to Prof. Ousmane Koita for his assistance and guidance in this manuscript, thanks to Janice A Washington for her assistance with the manuscript, to Dansine Diarra for making the map of Mali, to Bakary Konaté and Oumar Mangané for their assistance and help for samples collection and transportation, and Michael Polis from the intramural program of the US National Institute of Allergy and Infectious Diseases, National Institutes of Health.

Funding

This study was funded by World Bank through Programme Maladie Infectieuse (PMI PMI-23/LCV) and le programme d'appui aux services agricoles et aux organisations paysannes (PASAOP) and by the Division of Clinical Research (DCR) of NIAID/NIH. This work was also partially supported by NIH grants 5D43TW007995 and 5UM1AI69471.

Author s'contributions

MD conceived and developed the protocol. BD, MN, SDa, SDi, RLM, SS, YK, BK, OM, KT, helped in the protocol development. MN, SS, SDi, and RLM led the study. BD, MS, MHD, BT, MM, AMS, ACGT, YSS, DG, BB, HK, BPPD, were instrumental in data generation and laboratory assays. BD, MD, BT, YSS, SDi. and MN collated, reviewed and interpreted data and contributed to the writing of the manuscript. MD, YK, BK, OM, KT, reviewed and selected cattle for the study. YSS did all statistical analysis and figures. BD and MD wrote the first draft. SS, SDa, SDi, RLM, and MC reviewed the drafts. SS, SDi, SDa, MN, RLM, and AT provided administrative and scientific oversight for this work. All authors had access to the data, read and approved the final manuscript.

Authors' information

Dr Mamadou Diallo is a senior researcher at the laboratoire central veterinaire (Bamako, Mali), his research is aimed at understanding the zoonotic diseases.
Dr Bassirou Diarra is a senior scientist at SEREFO, specialist of tropical medicine; head of BSL-3 laboratory and his work is related to Human TB.

Competing interests

The authors declare that they have no competing interests.

Consent for publication

Not applicable.

Ethics approval and consent to participate

The protocol for spoligotyping of *M. tuberculosis* DNA was approved by the Ethical committees of Northwestern University of Chicago, IL, USA, and the USTTB, Mali. The study performed on human subjects was approved by the "Comite d'éthique de la Faculté de Médecine, de pharmacie et d'odontostomatologie" of the University of sciences, techniques and technologies of Bamako and the institutional review board (IRB) of the National Institute of Allergy and Infectious Diseases, National Institutes of Health (NIAID/NIH), USA. The protocol for the cattle

study was also approved by the ethic committee of the "laboratoire central vétérinaire de Bamako", Mali.
Before collecting tissue samples, verbal consent was obtained from the livestock owners for samples to be taken and used.

Author details

[1]Laboratoire Central Vétérinaire, Bamako, Mali. [2]SEREFO, University of Sciences, Techniques and Technologies of Bamako (USTTB), Point-G, Bamako, Mali. [3]Division of Clinical Research, NIH, Bethesda, USA. [4]Northwestern University, Chicago, IL, USA.

References

1. Alexander KA, et al. Novel Mycobacterium tuberculosis complex pathogen, *M. mungi*. Emerg Infect Dis. 2010;16(8):1296–9.
2. Ministère de l'élevage et de la Pêche RM. Direction Nationale de l'appui au Monde Rural, Rapport d'activité 2000. 2000.
3. Ministère de l'Economie et des Finances RM. Direction Nationale de la Statistique, Rapport d'activités 1997. 1997.
4. Ministère de l'élevage et de la Pêche RM. Politique Nationale de Développement de l'élevage au Mali, Rapport d'activité 2003. 2003.
5. Dankner WM, Davis CE. Mycobacterium bovis as a significant cause of tuberculosis in children residing along the United States-Mexico border in the Baja California region. Pediatrics. 2000;105, E79.
6. Barnes PF, Cave MD. Molecular epidemiology of tuberculosis. N Eng J Med. 2003;349:1149–56.
7. Kamerbeek J, et al. Simultaneous detection and strain differentiation of Mycobacterium tuberculosis for diagnosis and epidemiology. J Clin Microbiol. 1997;35(4):907–14.
8. Muller B, Steiner B, Bonfoh B, Fane A, Smith NH, Zinsstag J. Molecular characterization of *M. bovis* isolated from cattle slaughtered at the Bamako abattoir in Mali. BMC Vet Res. 2008;4:26.
9. Cousins DV. Mycobacterium bovis infection and control in domestic livestock. Rev Sci Tech. 2001;20:71–85.
10. Cosivi O, et al. Zoonotic tuberculosis due to Mycobacterium bovis in developing countries. Emerg Infect Dis. 1998;4:59–70.
11. Diguimbaye-Djaibe C, et al. Mycobacterium bovis isolates from tuberculous lesions in Chadian zebu carcasses. Emerg Infect Dis. 2006;12:769–71.
12. Kazwala RR, et al. Isolation of Mycobacterium bovis from human cases of cervical adenitis in Tanzania: a cause for concern? Int J Tuberc Lung Dis. 2001;5:87–91.
13. Jiwa SF, et al. Bovine tuberculosis in the Lake Victoria zone of Tanzania and its possible consequences for human health in the HIV/AIDS era. Vet Res Commun. 1997;21:533–9.
14. Sidibe SS NAD, Fane A, Doumbia RM, Sidibe CK, Kante S, Mangane O, Konate B, Kone AZ, Maiga MS, Fofana M. Tuberculose bovine au Mali: Résultats d'une enquête épidémiologiques dans les élevages laitiers de la zone périurbaine du district de Bamako. Revue Elev Med Vet Pays Trop. 2003;56:115–20.
15. Cosivi O, et al. Epidemiology of Mycobacterium bovis infection in animals and humans, with particular reference to Africa. Rev Sci Tech. 1995;14:733–46.
16. Mamadou, D. Etude de la prévalence de la tuberculose bovine au Mali et caractérisation des souches de mycobactéries isolées, étude pilote. Rapport de projet PMI-23/LCV 2011.
17. International Livestock Research institute (ILRI). http://agtr.ilri.cgiar.org/index. php?option=.
18. Müller B, Hilty M, Berg S, Garcia-Pelayo MC, Dale J, Boschiroli ML, Cadmus S, Ngandolo BN, Godreuil S, Diguimbaye-Djaibé C, et al. African 1, an epidemiologically important clonal complex of *Mycobacterium bovis* dominant in Mali, Nigeria, Cameroon, and Chad. J Bacteriol. 2009; 191:1951–60.
19. Aranaz A, Liebana E, Gomez-Mampaso E, Galan JC, Cousins D, Ortega A, Blazquez J, Baquero F, Mateos A, Suarez G, Dominguez L. *Mycobacterium tuberculosis* subsp. caprae subsp. nov.: a taxonomic study of a new member of the Mycobacterium tuberculosis complex isolated from goats in Spain. Int J Syst Bacteriol. 1999;49:1263–73.
20. Mazurek GH, et al. Chromosomal DNA fingerprint patterns produced with IS6110 as strain-specific markers for epidemiologic study of tuberculosis. J Clin Microbiol. 1991;29:2030–3.

21. Alemayehu R, Girmay M, Gobena A. Bovine tuberculosis is more prevalent in cattle owned by farmers with active tuberculosis in Central Ethiopia. Vet J. 2008;178:119–25.
22. Mawak J, Gomwalk N, Bello C, Kandakai-Olukemi Y. Human pulmonary infections with bovine and environment (atypical) mycobacteria in Jos, Nigeria. Ghana Med J. 2006;40:132–6.
23. Ayele WY, Neil SD, Zinsstad J, Weiss MG, Pavlik I. Bovine tuberculosis: an old disease but a new threat to Africa. Int J Tuberc Lung Dis. 2004;8:924–37.
24. Daborn CJ, Grange JM, Kazwala RR. The bovine tuberculosis cycle–an African perspective. Soc Appl Bacteriol Symp Ser. 1996;25:27S–32.

12

Development and validation of a protocol to identify and recruit participants into a large scale study on liver fluke in cattle

Catherine M. McCann[1,5]* , Helen E. Clough[1,2], Matthew Baylis[3,4] and Diana J. L. Williams[1]

Abstract

Background: Liver fluke infection caused by the parasite *Fasciola hepatica* is a major cause of production losses to the cattle industry in the UK. To investigate farm-level risk factors for fluke infection, a randomised method to recruit an appropriate number of herds from a defined geographical area into the study was required. The approach and hurdles that were encountered in designing and implementing this study are described. The county of Shropshire, England, was selected for the study because of the variation between farms in exposure to fluke infection observed in an earlier study.

Results: From a sampling list of 569 holdings in Shropshire randomly drawn from the RADAR cattle population dataset, 396 (69.6%) holdings were successfully contacted by telephone and asked if they would be interested in taking part in the study. Of 296 farmers who agreed to receive information packs by post, 195 (65.9%) agreed to take part in the study. Over the period October 2014 – April 2015 visits were made to 100 dairy and 95 non-dairy herds. During the farm visits 40 faecal samples +/− bulk-tank milk samples were collected and a questionnaire administered. Composite faecal samples were analysed for the presence of *F. hepatica* eggs by sedimentation and bulk tank milk samples were tested with an antibody ELISA for *F. hepatica*. Forty-five (49%) of non-dairy herds were positive for liver fluke infection as determined by the finding of one or more fluke eggs, while 36 (36%) dairy herds had fluke positive faecal samples and 41 (41%) dairy herds were positive for *F. hepatica* antibody. Eighty-seven (45.8%) farmers said that they monitored their cattle for liver fluke infection and 118 (62.1%) reported that they used flukicide drugs in their cattle.

Conclusions: Using a protocol of contacting farmers directly by telephone and subsequently sending information by post, 79% of the target sample size was successfully recruited into the study. A dataset of farm-specific information on possible risk factors for liver fluke infection and corresponding liver-fluke infection status was generated for the development of statistical models to identify risk factors for liver fluke infection at the farm-level.

Keywords: *Fasciola hepatica*, Cattle, Farmers, Holdings, Shropshire, Fasciolicide

Background

Liver fluke infection, caused by the trematode parasite *Fasciola* spp., has an economic impact on livestock production worldwide as a result of both morbidity and mortality. In the UK, where the predominant host species are cattle and sheep, liver fluke infection costs the agricultural industry in the region of £300 million per year due to production losses. Acute clinical disease or sudden death due to liver fluke, a common feature of disease in sheep, is rare in cattle. In dairy cattle liver fluke has been associated with reduced milk production [1, 2], reduced milk fat content and increased calving interval [3]. In high yielding dairy cows, an increase in *F. hepatica* exposure from the 25th to the 75th percentile was associated with a 15% decrease in milk yield [4]. In beef cattle the impact of liver fluke infection on growth rate and weight gain has been less

* Correspondence: Catherine.Mccann@liverpool.ac.uk;
cathy_mccann@hotmail.com
[1]Department of Infection Biology, Institute of Infection and Global Health, University of Liverpool, Liverpool, UK
[5]Present address: Epidemiology Research Unit, Scotland's Rural College (SRUC), An Lòchran, Inverness Campus, Inverness IV2 5NA, UK
Full list of author information is available at the end of the article

easy to demonstrate [5]. Carcasses from cattle infected with fluke had lower cold weight, lower conformation score and lower fat content than carcasses free of liver fluke in a study of abattoir data in Scotland [6]. Milk yield in dairy cattle or carcass weight of slaughtered beef cattle are reported to decrease on average by 3–5% or 0.5–0.7% respectively [7].

Recent estimates of the prevalence of liver fluke infection in dairy herds in the UK are 80% in 2012 in England, Wales and Scotland [4]; 48 and 86% in 2006–2007 in England and Wales respectively [8]; 72 and 84% in 2005 in England and Wales respectively [9]; and 61–65% in 2011–2013 in Northern Ireland [10]. Traditionally in the UK, most cases of fasciolosis are found in wetter, western areas which provide ideal climatic conditions; however in recent years fasciolosis has emerged in other areas including East Anglia and parts of Scotland [11, 12]. Seasonal risk forecasts that provided an approximation of the potential impact of climate change on fasciolosis in the UK predict an overall long term increase in prevalence of infection in all regions of the UK with an expectation of spatio-temporal variation in risk levels [13].

Liver fluke has an indirect life-cycle involving a snail intermediate host. Principle definitive hosts of *F. hepatica* in the UK are cattle and sheep, however wild herbivorous animals such as deer, rabbits and hares may act as reservoir hosts. Liver fluke have an indirect life cycle involving a snail intermediate host. Undifferentiated fluke eggs are passed out in the faeces of infected animals. The egg hatches to release a miracidium which enters the snail. The main intermediate host is *Galba truncatula*, the amphibious dwarf pond snail. Following further development in the snail, several hundred cercariae are released which then encyst on the pasture. Cattle, sheep and other herbivores become infected when they ingest contaminated herbage, the metacercariae hatch, the newly excysted juveniles burrow through the gut wall and migrate into the liver. [18]. In the presence of suitable definitive hosts, development of *F. hepatica* is dependent primarily on there being suitable environmental conditions for the snail hosts and the intra-molluscan and free-living stages of the parasite. Rainfall and temperature are critical to the development of both the parasite and snail. Other determinants such as soil properties [14], vegetation and altitude [15] and farm management factors [16] may also influence *F. hepatica* transmission. In a recent study of farm management and environmental risk factors for *F. hepatica* exposure in high yielding dairy herds in Great Britain, higher rainfall, grazing boggy pasture, presence of beef cattle on farm, access to a stream or pond and smaller herd size were all associated with an increased rate of exposure [4].

Control of fasciolosis is aimed at reducing the prevalence of disease to permit economic livestock production,

as it is unlikely that infection will be eradicated [17]. There are currently no commercial vaccines; hence control is based on the use of anthelminthic drugs to reduce disease and sub-clinical economic losses and the rate of contamination of pastures by reducing fluke egg output [18], together with pasture management, including drainage to reduce the survival of free-living stages and to prevent the establishment of snail populations [19]. There are a number of medicines available to treat cattle for fluke (fasciolicides); these vary in terms of whether they target the immature or adult stages of the parasite and their milk and meat withhold periods. The drug of choice for treatment of fasciolosis is the benzamine derivative triclabendazole, because it is effective against both adult and juvenile fluke [20, 21]; however resistance to triclabendazole has been reported in a number of countries, [21–30] which is a cause for concern especially as triclabendazole is also the drug of choice to treat human fasciolosis. Triclabendazole may only be used at the start of the dry period in dairy animals because of its long withhold period, e.g. milk for human consumption can only be taken from 50 days after treatment with Fasinex® 240 (Elanco Animal Health) [31] In dairy herds with year-round calving, treatment for fluke is likely to be done throughout the year. In the UK, the majority of cattle herds are housed through the winter months; once housed, cattle should no longer be at risk of infection. Hence farmers often treat cattle at a time after housing, to ensure that any fluke present will be killed.

If farmers are able to accurately assess the risk of fasciolosis in their specific location they will be able to make informed decisions on treatment and prevention of disease. Such targeted control programmes should not only result in reduced use of fasciolicides but also lead to better disease control. In the UK, the National Animal Disease Information Service (NADIS) produces a parasite forecast for 10 regions in Great Britain and Ireland (http://www.nadis.org.uk/). This provides farmers with a forecast of the risk of fluke infection on a regular basis, however the coarse spatial resolution of the forecasts do not provide farmers with an accurate assessment of risk of disease to the livestock in their farm.

In an earlier study we developed linear regression models, using a combination of environmental, soil and climatic variables, to describe the observed pattern of exposure to *F. hepatica* in England and Wales using UK postcode area (PCA) to define a spatial unit [32]. The area of these PCAs varied from approximately 150–6000 km^2, with a mean of 2000 km^2. Whilst the model explained over 73% of the spatial variation in exposure it did not account for variation of exposure within PCAs. This paper describes the methodology used to recruit nearly 200 dairy and beef enterprises within a small spatial area, to identify more precisely farm-level determinants of *F. hepatica* infection that may explain the variation in prevalence of fluke infection between farms exposed to the same climatic variables. We describe the processes used to identify, contact and recruit farms into the study; once farm proprietors had agreed to participate, visits were made to the farms during which samples were collected to be analysed for evidence of *F. hepatica* infection and a questionnaire administered to collect farm demography, herd health and land and herd management information. Whilst the main aim of this paper is to describe a framework for a truly random recruitment of participants into a study, we also include preliminary data on the prevalence of *F. hepatica* infection in cattle herds in a major farming area of England.

Methods

Study population

To estimate the prevalence of exposure to *F. hepatica* in the study area 124 dairy and 124 beef (non-dairy) herds were required for a study with 7.5% precision and 95% level of confidence, based on an assumed herd prevalence of exposure to *F. hepatica* of 76%. Under a data confidentiality agreement with the Animal and Plant Health Agency (APHA), data on all agricultural holdings in England with dairy and/or beef herds were obtained from the Rapid Analysis and Detection of Animal-related Risk (RADAR) cattle population dataset [33]. This database combines the lists held by the Agricultural Census, the Animal Movements Licensing System and the British Cattle Movements Service. Data provided on agricultural holdings included their County Parish Holding (CPH) number, the farm name and address, the Ordnance Survey Easting and Northing and the number of cattle of each breed registered at each holding on 1 May 2014. Most holdings were represented by one farm; however some are used by more than one farm.

Cattle farms in the county of Shropshire in England were chosen as the study population because this population contains large numbers of both dairy and beef farms, and exposure to fluke infection varies from negative to high, based on our 2006/7 survey. A sampling frame of agricultural holdings identified as being located in Shropshire by their postal address, was constructed from the RADAR cattle dataset. The sampling frame comprised 1970 holdings each with between 1 and 2929 cattle registered. Using internet search engines (Google.com and bing.com) phone numbers were found for 1129 holdings; the name of the farm proprietor was also obtained for many of the farms. Two lists comprising 576 holdings with 50 or more non-dairy and 405 holdings with 50 or more dairy breed type cattle were created for a random selection of holdings to be contacted. Based on experience from similar studies, and assuming a participation rate of 40% amongst farmers contacted, a sampling list of 310

beef and 310 dairy herds was required. Random numbers generated in MS Excel were used to select 310 holdings with cattle of each type. Some holdings were selected in both categories; this resulted in a final list of 569 holdings; 300 holdings remained unselected. The list of holdings was randomised to ensure recruitment in a random order to reduce the risk of bias in the time of contacting holdings and subsequently visiting farms and collecting samples.

Farmer recruitment

Farmers were invited to take part in the study by telephone in the order of the random recruitment list. Between one and seven attempts were made to contact farmers from 15 October 2014–31 March 2015; when attempts had been made to call all the farms the recruiter returned to the beginning of the list. At the beginning of the phone call, the recruiter asked to speak to the proprietor of the farm. Using a prepared script the recruiter then provided information about the study and offered to send further information by post. When possible, eligibility of the farm was determined during the phone call. Inclusion criteria for the study were that there were at least 40 cattle aged at least one year, the cattle had not been treated with an anthelminthic for liver fluke in the previous 12 weeks and the cattle had grazed on pasture on the farm during 2014. The 12-week restriction was not applied to dairy herds with year-round calving which were likely to have year-round liver fluke treatment regimes. If the proprietor was not available and a suitable time to call back could not be arranged, agreement to send the information by post was made. If the farmer was unwilling to receive the information pack, or for any other reason the study information was not sent, the reason was recorded. Herds were classified as dairy or non-dairy (beef suckler, stores/finishers and dairy replacements) based on the details provided during the phone call.

An information pack comprising a Participant Information Sheet, Consent Form and personalised letter was sent to farmers within a week of the recruitment phone call. The Participant Information Sheet provided full details of the study including its objectives, the funding agencies, and what the farmers would be required to do. Farmers were advised that they would be contacted again shortly to be asked if they would be interested in taking part.

One to 3 weeks later, farmers were re-contacted by phone, again using a prepared script, to ask them if they were interested in taking part in the study; if they were, a farm visit was arranged. Again, reasons for not agreeing to a visit were recorded. A maximum of nine attempts were made to re-contact farmers by telephone.

All telephone calls and farm visits were made by the same research scientist. Recruitment was made on a rolling pattern with the aim of obtaining agreement from 10 to 16 farmers per week and to visit 8–10 farms each week. A number of farms located in counties that bordered Shropshire were included in the study – all farms in the RADAR dataset with 'Shropshire' in the postal address were included however some were spatially located outside the county border. The study farms were located in an area of approximately 4800 km^2.

Description of questionnaire

The questionnaire was piloted on five farms outside the study area. Initially two versions of the questionnaire were used, one for beef and one for dairy farms. After the pilot, a number of adjustments were made including the change to a single version for all herd types.

The final 12 page questionnaire was divided into three sections which covered details about the farm demography and herd health, the herd, pastures and farm management, and cattle production and fertility. Both closed and open-ended questions were used. Questions were designed to determine whether there was a history of liver fluke infection in the herd and to find out if any flukicide drugs were used on the farm and if so, which drugs were used, frequency of treatment and which classes of cattle (and sheep) were treated.

Farm visit

Signed consent forms were either returned by mail to the study team at the University of Liverpool or collected during the visit.

During the farm visit, which lasted between 1 and 2 h, the questionnaire was completed during a face to face interview with the farmer or their representative. Due to time constraints for six farms the questionnaire was left with the farmer to complete and return by mail to the study team at the University of Liverpool. Three of these six farmers failed to return the questionnaire by post. Interviewees identified areas used for cattle grazing during the 2014 grazing season on farm maps or by providing parcel numbers of the areas grazed. On some farms, photographs were taken of habitat that could potentially harbour *G. truncatula*.

During farm visits, 40 separate faecal samples were collected from individual faecal pats on the pasture or the floor of the shed if cattle were housed. In non-dairy herds, faecal samples were collected from the main cattle group, i.e. from suckler cows, store cattle, fattener cattle or replacement heifers. If the main group of cattle was made up of less than 40 animals, samples were collected from adult cattle, where available. On dairy farms, faecal samples were collected from the cows that were in milk on the day of the visit; additionally, a milk sample was collected from each bulk tank. A preservative bronopolnatamyin (MSI, Nottingham) was added to the milk samples at the time of collection.

Determination of *F. hepatica* infection status
Faecal egg counts

The faecal samples were stored at 4 °C until analysis. For the faecal analysis, the Herdsure® protocol [34] was followed. This uses composite faeces samples to establish the liver fluke infection status of a herd. In brief, 5 g of faeces was taken from each faecal sample and pooled into four 50 g composites for each farm. Each composite sample was mixed with water and a full faecal egg count carried out using the sedimentation technique [35]. By testing four pools of 10 samples there is a 95% confidence level of detecting one positive animal if the within herd prevalence is at least 20%. *F. hepatica* eggs per gram of faeces was calculated for each composite sample. Farms were classified as positive if one or more fluke eggs were detected in at least one of the four composite samples.

Bulk tank milk samples

F. hepatica antibody levels in the bulk milk tank (BMT) samples were determined using an *F. hepatica* excretory/secretory (ES) antigen specific enzyme-linked immunosorbent assay (ELISA) as previously described [8]. Results were expressed as a percent positivity (PP), which is the optical density (OD) reading of test sample divided by the OD of a positive control, times 100. If the test sample is more strongly positive than the control, the PP will exceed 100%. Herds classed as positive were categorised into low positive (LP) ($27 \leq$ PP-value < 50); medium positive (MP) ($50 \leq$ PP-value < 100) and high positive (HP) (PP-value ≥ 100). The sensitivity and specificity of the ELISA to detect herds in which more than 25% of the cows are infected are 96% (95% CI 89–100%) and 80% (95% CI 66–94%), respectively [8].

All farmers were provided with written results for their farms.

Data analysis

An MS Access database was constructed to hold all data generated relating to farmer recruitment, including telephone call records, for the selected holdings.

The questionnaires were checked manually for inconsistencies and then entered into a database (Epi Info™ version 7.0). Descriptive statistics were estimated and comparisons between the dairy and non-dairy farms performed using STATA (StataCorp. 2015. Stata Statistical Software: Release 14. College Station, TX: StataCorp LP). A Geographical Information System (GIS) (ArcGIS version 10.1) was constructed using data layers of the liver fluke infection prevalence levels at the study farms.

To determine whether there is a statistically significant association between (a) total cattle; (b) grass acreage and farm type (dairy or non-dairy), given that both outcomes are counts the first model of choice is a Poisson log-linear model. The Poisson model assumes that the mean and the variance of the outcome data are the same, however in the case of both outcomes the variance of the data is considerably larger than the mean (mean = 246.9; variance = 29,877.2 for total cattle; mean = 48.1, variance 752.1 for grass acreage). This effect is consistent with most farms having smaller numbers of cattle/acres of grass, and fewer farms having very large numbers of cattle/acres of grass. An ad-hoc way of handling this is to fit the model using quasi-likelihood (a "quasi-poisson model"), which allows the dispersion parameter (which is fixed at 1 in the standard Poisson model) to vary, and to be estimated from the data (in the presence of over-dispersion the estimated dispersion parameter will be greater than 1). The significance of coefficients in a Poisson model which does not allow for over-dispersion will be over-stated, and the effect of allowing for over-dispersion will be to make the estimated coefficient standard errors larger, thereby reducing their significance in the model. We hence fit a quasi-Poisson model to each of the outcomes with farm type (coded 1 = dairy, 0 = non-dairy) as a single covariate.

The Kulldorff spatial scan statistic was used to test whether liver fluke infected farms were randomly distributed within the study area and if not, to identify significant spatial anomalies [36]. Analysis was performed using the Bernoulli model implemented in version 9.4 of the SaTScan software (https://www.satscan.org/). This programme creates circular windows that are moved systematically throughout the geographic space to identify significant anomalies in the spatial distribution of infection. The windows are centred on each of the farms; the maximum window size, to be specified by the user, was defined here as 50% of the farms (i.e. the largest possible cluster would encompass 50% of the farms). For each location and size of the scanning window, SaTScan performs a likelihood ratio test to evaluate whether infection is more prevalent inside than outside that given circular window. Separate analyses were performed for (1) dairy farms with positive BTM samples, (2) dairy farms with positive faecal samples, and (3) non-dairy farms with positive faecal samples. P values were determined by Monte Carlo replications of the data set; a 5% significance level was adopted [37].

Results

Recruitment

Nine hundred and seventy-four phone calls, lasting a total of 1528 min were made to recruit farmers to the study (Table 1). A minimum of 1 min was recorded for every phone call which was answered, had an answer phone message or was not answered. It was not possible to make telephone calls to 28 farms because the telephone number was not valid. Of the 569 farms selected,

Table 1 Summary of the number of attempted telephone calls made to recruit farms to the study

Number of calls	Number (%) of farms	
	All farms	Farms successfully contacted
0	28 (4.9)[a]	–
1	281 (49.4)	246 (62.1)
2	166 (29.2)	106 (26.8)
3	60 (10.5)	27 (6.8)
4	13 (2.3)	8 (2.0)
5	5 (0.9)	3 (0.8)
6	8 (1.4)	5 (1.3)
7	8 (1.4)	1 (0.3)
Total	569	396

[a]Telephone numbers used were not valid

contact was made with 396 (69.60%); 296 of those contacted agreed to receive further information about the study. It was not possible to contact 173 (30.40%) farms in the list of selected farms for a variety of reasons (Table 2). Due to the working pattern of many farmers, phone calls were generally made between 9 and 10.30 am, 12–13.30 pm and after 4 pm unless arrangements were made to speak to farmers at other times. Twenty-seven percent of the recruitment telephone calls where a farmer was successfully contacted were made between 5 and 6 pm. Phone calls made between 12 and 1 pm and 4–5 pm each resulted in 12% of the total successful recruitment phone calls.

Figure 1 provides a flow chart of the recruitment process.

Seven hundred and eighty five follow-up phone calls, lasting 1328 min were made to the 296 farmers who had been sent information packs (Table 3); 195 (65.54%) farmers agreed to take part in the study and a farm visit was arranged. It was not possible to re-contact 13 farmers. The reasons for non-participation in the study are provided in Table 4. Overall 56% of farmers who had been

Table 2 Summary of farms which could not be contacted - results of phone calls made

Telephone call result	Number of farms (%)
Number not valid/not connected	28 (16.2)
Wrong number – telephone call answered but not correct for farm	19 (11)
No answer	26 (15)
Answerphone	85 (49.1)
Call answered, unable to speak to proprietor and failed to make contact with farm again	15 (8.7)
Total	173

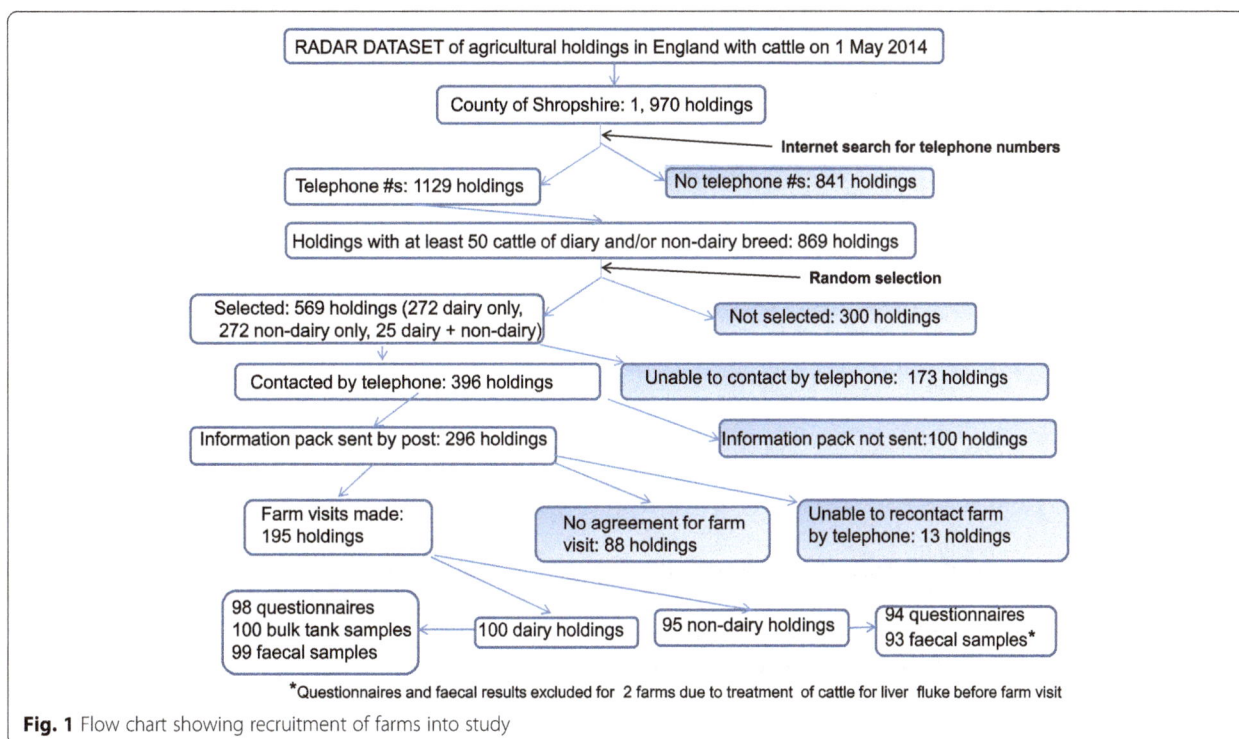

Fig. 1 Flow chart showing recruitment of farms into study

contacted but did not take part were unwilling either because they were too busy or not interested. Other reasons for non–participation included farms not fulfilling inclusion criteria for the study or ill health of the farmer.

Visits were made to 195 farms between 29 October 2014 and 30 April 2015. The farms that participated in the study comprised 100 dairy herds and 95 non-dairy (75 beef suckler, 14 stores/finishers and six dairy replacements). A total of 192 questionnaires were completed. During the visits made to two beef suckler farms it was revealed that the cattle had been treated with fasciolicides within the previous 12 weeks. Faecal samples were collected and analysed for these two farms but the results were excluded from any further analysis. Faecal samples from one dairy and two non-dairy farms were not available for analysis.

Questionnaires

Farms with a milking herd were classed as dairy farms; all other farms were classed as non-dairy farms. Table 5 provides a summary of the farm size and number of each class of cattle and sheep on each enterprise.

The outputs from the quasi-Poisson model are displayed in Table 6. The positive sign of the dairy/non-dairy term in each case suggests that dairy farms have more cattle and more grass acreage than non-dairy farms. The degree of statistical significance of these findings is assessed via analysis of deviance of the fitted models, using an F-test which is appropriate when quasi-likelihood is used. The analysis of deviance confirms that there is a statistically significant positive association between the total number of cattle and the type of farm $(F = 38.02, p \approx 0)$ but the association between total grass acreage, though positive, is not statistically significant $(F = 1.503, p = 0.222)$.

Table 3 Summary of the number of attempted follow-up phone calls made to farms that had been sent recruitment packs

Number of calls	Number of farms (%)
0[a]	8 (2.7)
1	87 (29.4)
2	73 (24.7)
3	56 (18.9)
4	28 (9.5)
5	19 (6.4)
6	10 (3.4)
7	7 (2.4)
8	4 (1.4)
9	4 (1.4)
Totals	296

[a]The farm visit was arranged at the time of the initial phone call for eight farms

Table 4 Reasons provided by farmers for non-participation in study

Reason for non-participation	Recruitment Stage		Total farms (%)
	Initial phone call	Follow-up phone call	
Cattle do not go out	13	2	15 (8)
No cattle/not farming	16	0	16 (8.5)
Not interested or did not want to take part	27	39	66 (35.1)
Too busy/no time	11	29	40 (21.3)
Cattle treated with fasciolicides within the last 90 days	8	8	16 (8.5)
Not enough cattle (< 50)	12	4	16 (8.5)
Retired/ill health/death	11	5	16 (8.5)
Otherwise unsuitable	2	1	3 (1.6)
Totals	100	88	188

Evidence of liver fluke in herds and use of fasciolicides as reported in questionnaires

Eighty-seven farmers of the 190 farmers who completed the questionnaire (45.8%) said that they monitored their cattle for liver fluke infection. Sixty-one (62.2%) and 26 (28.3%) of dairy and non-dairy farmers respectively said they monitored their cattle for fluke using one or more methods. Table 7 summarises the results reported by farmers of tests carried out to monitor fluke infection on the farms. Some farmers reported that they suspected that liver fluke infection was present but did not have any definitive evidence of fluke infection in their cattle. On 91 (47.9%) farms, farmers reported either recent, past (more than two years ago) or suspected fluke infection (Table 8). Fasciolicide use was reported in cattle and sheep, and in cattle only, on 135 (71.1%) and 118 (62.1%) farms respectively (Table 9).

Prevalence of *F. hepatica*

Bulk tank milk samples were analysed for 100 herds and faecal analysis results were available for 190 herds. In the dairy herds 41 (41, 95% CI 31.4–50.6%) of 100 BTM samples tested positive for antibodies to *F. hepatica* (Table 10). Across all herds, 81 (42.6, 95% CI 35.6–49.7%) of 190 composite faecal samples were positive for *F. hepatica* eggs (Table 11). The prevalence of faecal positivity was greater in non-dairy herds compared to dairy herds: 45 (49.5, 95% CI 39.2–59.7%) of 91 and 36 (36.4, 95% CI 26.9–45.8%) of 99 composite faecal samples respectively. Faecal positivity rates each month ranged

Table 5 Summary of farms size (acres) and numbers of cattle and sheep according to class present on the study farms as reported by farmers. Only farms with at least one animal are included for each class

Variable	Dairy				Non-dairy				All farms			
	No. of farms	Mean	Min	Max	No. of farms	Mean	Min	Max	No. of farms	Mean	Min	Max
Total acres	95	346.5	40	5000	91	370.2	27	1600	186[a]	358.1	27	5000
Grassland acres	98	251.1	30	1236	90	221	27	900	188[b]	236.7	27	1236
Dairy cows	98	180.8	45	500	2	21.5	9	34	100	177.6	9	500
Beef cows	3	9	2	17	74	554.6	4	320	77	52.8	2	320
Dairy heifers	88	59.1	5	260	6	162.8	27	280	94	66.7	5	280
Beef heifers	1	4	4	4	61	13.2	1	80	62	13.5	1	80
Calves	90	62.8	2	250	78	41.4	1	310	168	52.9	1	310
Fatteners and stores	32	68.1	5	300	75	91.1	1	468	107	84.2	1	468
Bulls	73	1.8	1	16	64	1.9	1	8	137	1.9	1	16
Total cattle	98	315.4	68	886	92	173.9	25	986	190	246.9	25	986
Sheep	18	246.8	2	2000	64	396.4	3	1900	82	363.6	2	2000
Overwinter sheep	47	323.2	30	1500	25	262.4	50	1000	72	302.1	30	1500

[a]Data not provided by four farms
[b]Data not provided by two farms

Table 6 Output for the quasi-Poisson model to each of the outcomes (a) total cattle and (b) total acreage with farm type (coded 1 = dairy, 0 = non-dairy) as a single covariate

Model	Coefficient	Standard error
Total cattle		
Intercept	5.16	0.08
Dairy/non-dairy	0.59	0.10
Total acreage		
Intercept	5.40	0.08
Dairy/non-dairy	0.13	0.10

from 34.5% (January 2015) to 50% (December 2014) of herds tested; bulk tank milk positivity rates each month ranged from 28% (April 2015) to 59% (November 2014) (Fig. 2).

Figures 3 and 4 show the spatial distribution and results for the sampled farms in the county of Shropshire. Figure 3 shows that most of the dairy herds are located in the centre and north of the county. Figure 4, which includes all the sampled herds, demonstrates that *F. hepatica* infection appears to be spread throughout the county with infected and uninfected herds located in close proximity to each other.

Spatial clustering

Spatial scan statistics revealed no significant anomalies in the spatial distribution of liver fluke infection on dairy or non-dairy farms.

Discussion

The primary purpose of this paper is to document the methodology used to recruit farms to a study to identify risk factors for liver fluke infection on dairy and beef farms. Many researchers have recruited farmers by contacting them by post with varying results. The RADAR dataset from which we selected our sample provided limited information on the

farms; most importantly we did not know whether the farms were still in operation and whether the enterprises were primarily beef or dairy. By contacting everyone by telephone we were able to engage directly with farmers to determine the eligibility of their farms for the study and gauge their interest. Further, by visiting the farms we were able to conduct the questionnaire by face to face interview and provide a diagnostic test to determine the *F. hepatica* status of their herds. Whilst a farmer can collect and send bulk tank milk samples to a laboratory for analysis, the same cannot be said for faecal samples using the exact sampling strategy required for this study.

Only 78.6% of the target sample was successfully recruited. The RADAR dataset from which the names and addresses of cattle farms in Shropshire were obtained did not included contact telephone numbers for the farms, and hence it was necessary to use internet search engines to find the numbers with no guarantee that the telephone numbers were still in service or correct. In addition we were unable to find telephone numbers for 841 of the farms listed for Shropshire in the RADAR list. We found that some of the numbers used to contact farms were incorrect or invalid and also that a number of the farms we contacted were no longer in operation. We have no knowledge of whether the farms for which we were unable to find a telephone number were still in operation. We believe that these issues are likely to be of a random nature and should not result in selection bias. We were unable to contact 30.4% of the farms in the sampling list. With more time, the target sample size could have been achieved by randomly selecting farms from the remaining 300 farms from the list of 869 holdings from which the farms had been selected. This would have required extending the study period; however in the spring time (April 2015) farmers were found to be increasingly occupied with farm activities (e.g. ploughing) and had less time available to accommodate the farm visits despite

Table 7 Methods used to monitor herds for liver fluke and farmer-reported results of tests used

Method		Dairy herds (n = 98)		Non-dairy herds (n = 92)		All herds (n = 190)	
		Number (%)	Positive (%)	Number (%)	Positive (%)	Number (%)	Positive (%)
F. hepatica Bulk tank milk ELISA	Yes	48 (49)	28 (58.3)	NA[a]	–	–	–
	No	50 (51)					
Liver inspection at abattoir	Yes	32 (32.7)	24 (75.0)	23 (25.0)	16 (69.6)	55 (28.9)	40 (72.7)
	No	36 (36.7)	–	14 (15.2)	–	50 (26.3)	–
	Don't know	2 (2.0)	–	0 (0.0)	–	2 (1.1)	–
	NA	28 (28.6)	–	55 (59.8)	–	83 (43.7)	–
F. hepatica faecal egg count analysis	Yes	8 (8.2)	4 (50.0)	6 (6.5)	2 (33.3%)	14 (7.4)	6 (42.86)
	No	90 (91.8)	–	86 (93.5)	–	176 (92.6)	–

[a]NA not applicable

Table 8 Summary of farmers' reports of liver fluke infection on their farms

	Dairy herds ($n = 98$)	Non-dairy herds ($n = 92$)	All herds ($n = 190$)
	Number (%)	Number (%)	Number (%)
Evidence of fluke in cattle in last two years	34 (34.7)	19 (20.7)	53 (27.9)
Evidence of fluke in cattle more than two years ago but not in last two years	15 (15.3)	6 (6.5)	21 (11.1)
Suspect fluke (Clinical signs +/− diagnosed in sheep)	8 (8.2)	9 (9.8)	17 (8.9)
No fluke reported on farm	41 (41.8)	58 (63.0)	99 (52.1)

Table 10 Results of *F. hepatica* faecal egg count tests according to *F. hepatica* enzyme-linked immunosorbent assay (ELISA) result on dairy farms

BTM ELISA (PP-value)	BTM result category	Faecal result			Total (%)
		Negative	Positive	Not available	
< 27	Negative	45	14	0	59 (59.0)
27–49	Low positive	16	10	1	27 (27.0)
50–99	Medium positive	1	11	0	12 (12.0)
≥ 100	High positive	1	1	0	2 (2.0)
					100

the longer day length. Also collecting faecal samples was found to be more difficult in the open fields when the cattle were at grass compared to when the cattle were housed so the period of the study was not extended.

Achieving the target recruitment of 124 each of dairy and non-dairy farmers was further complicated because many farms had been included in both the dairy and non-dairy sampling lists. This discrepancy was generally a result of farms of each type often having stores or fatteners of dairy and or beef breeds, for example approximately one-third of the dairy farms kept such animals. Of the farmers who did not participate in the study, 44% were because of inability, rather than unwillingness, to do so. Forty-seven of the farms that did not participate were not eligible to take part by reason of there being no cattle or too few cattle or the cattle did not graze outside. On about 8% of farms contacted the cattle had been treated with a fasciolicide in the last 12 weeks.

Farmers were recruited into this study by telephone. The only information available on the farms recruited

into this study was the farm address, location and number and types of cattle as provided in the RADAR dataset. By contacting farms by telephone it was possible to find out whether a farm, listed in the RADAR dataset, was actively operating and whether it fulfilled the criteria to take part in the study. In addition, by contacting farmers by telephone it was possible to obtain information on why farmers who fulfilled the study criteria were unable to participate. This method of recruitment was very time-consuming. Recruitment telephone calls took 25 h, in addition follow-up calls were made, many of which were made outside of normal working hours. The internet search for the telephone numbers took approximately 1 week. However, the personal contact with the farmers resulted in successfully recruiting 195 farms – 65% of the farms that had been sent information packs. Recruitment methods such as contacting farmers in the first instance by mail would have been less time consuming in the first instance, however no information would be obtained on farms from which there had been no response unless letters were returned as undelivered.

This study did offer farmers the incentive of having their herd tested for liver fluke; however it also required farmers to give up 1–2 h of their time to accommodate the visit of the research scientist. It is not possible to determine whether there was any response bias associated with non-participation because the only information available on the non-participating eligible farms was the

Table 9 Summary of reported use of fasciolicide drugs on farms

		Dairy ($n = 98$)	Non-dairy ($n = 92$)	Total ($n = 190$)
		Number (%)	Number (%)	Number (%)
Treat cattle or sheep with fasciolicide within the last two years	Yes	63 (64.3)	72 (78.3)	135 (71.1)
	No	34 (34.7	19 (20.7)	53 (27.9)
	Don't know	1 (1.0)	1 (1.1)	2 (1.1)
Treat cattle with fasciolicide	Yes	60 (61.2)	58 (63.0)	118 (62.1)
	2 or more years ago	6 (6.1)	10 (10.9)	16 (8.4)
	No	31 (31.6)	23 (25.0)	54 (28.4)
	Don't know	1 (1.0)	1 (1.1)	2 (1.1)

Table 11 Results of *F. hepatica* faecal egg count tests in dairy and non-dairy herds

	Farm type					
	Dairy		Non-dairy		All herds	
	n	%	n	%	n	%
Negative	63	63.6	46	50.6	109	57.4
Positive	36	36.4	45	49.5	81	42.6
Total	99	100	91	100	190	100

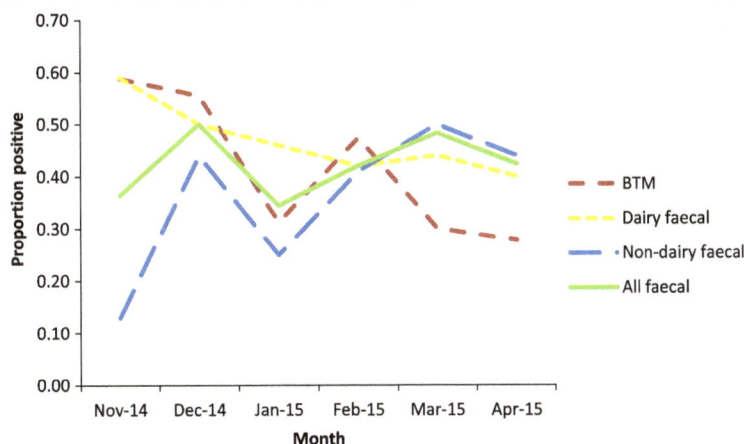

Fig. 2 Proportion of farms with positive results in the i. *F. hepatica* sedimentation test for composite faecal samples ii. *Fasciola*-specific bulk tank milk (BTM) ELISA each month (November 2014 – April 2015)

numbers of cattle as provided by the RADAR dataset. The RADAR database provides a comprehensive repository of data on cattle farms; however the data were often found to be out of date, as a number of farms were no longer operating. Similar issues of RADAR data accuracy have been reported in other studies, for example in a study to determine bluetongue virus vaccine uptake from a random sample of 1866 ruminant holdings, using the RADAR data as a sample frame, 823 questionnaires were returned of which 49 were not valid because they had the wrong address [38].

Other studies on fluke have contacted farms by post, but in all these cases the farmer cohort was under the terms of their milk contract required to participate in research projects funded by the contracting company [4], or used an advisory service [39, 40] hence these were not truly randomised studies. Four hundred and 50 dairy farmers were invited to participate in a survey on liver fluke in Ireland and were sent a questionnaire and a bulk tank milk sampling kit [39]. Contacted farmers either participated in Teagasc (Irish Agriculture and Food Development Authority) discussion groups or were selected by Teagasc dairy advisors, which is a probable explanation for the high response rate of 82%. In a postal survey of Irish dairy farms to obtain information on parasite control practices, 96% of the 312 farmers who were sent the questionnaire responded. The study farms were randomly selected from the Irish Cattle Breeding Federation database and all were members of HerdPlus®, a management decision support tool for dairy farmers [40]. Participation in the study was entirely voluntary and non-incentivised. In a observational study on liver fluke of high-yielding dairy farms who were contracted to supply milk to a major-supermarket chain, 58% of 606 farmers who were sent a questionnaire by post

either completed a paper or online version [4]. Reasons for non-completion of the questionnaire were not obtained, however all the farms contacted were operating as the researchers obtained bulk tank milk samples for all of them from National Milk Laboratories. In these three studies, enough information was known about the farms to believe that they were operating and that recruitment by post would result in successful contact with the farms concerned.

In this study, 189 questionnaires were completed during the farm visit; however only three of the six farmers who were asked to return the questionnaire by mail returned them. This may be due to a lack of interest in the study, but is unlikely to introduce bias because of the small number of farms involved.

In this study, by choosing to contact farms by telephone using landline telephone numbers that were obtained using internet search engines, farms were excluded from participating if they could not be contacted using the number found. Some of these farms may have no longer been in operation. However, it is also possible that, with the increasing popularity of mobile phones, some farmers may no longer use landline telephones. Of 64% of the farms in the sampling list that could not be contacted, the telephone was either not answered or there was an answerphone; however failure to contact farms was unlikely to be a cause of non-response bias. In addition, randomising the order in which farms were visited should have prevented any sampling bias in the results according to type, size and management system of the farms.

Fluke eggs were found in nearly 50% of the non-dairy herds. Only 28% of the non-dairy farmers reported that they monitored their cattle for fluke infection with 37% of these reporting that they either had evidence of fluke in their herds or suspected that their cattle were

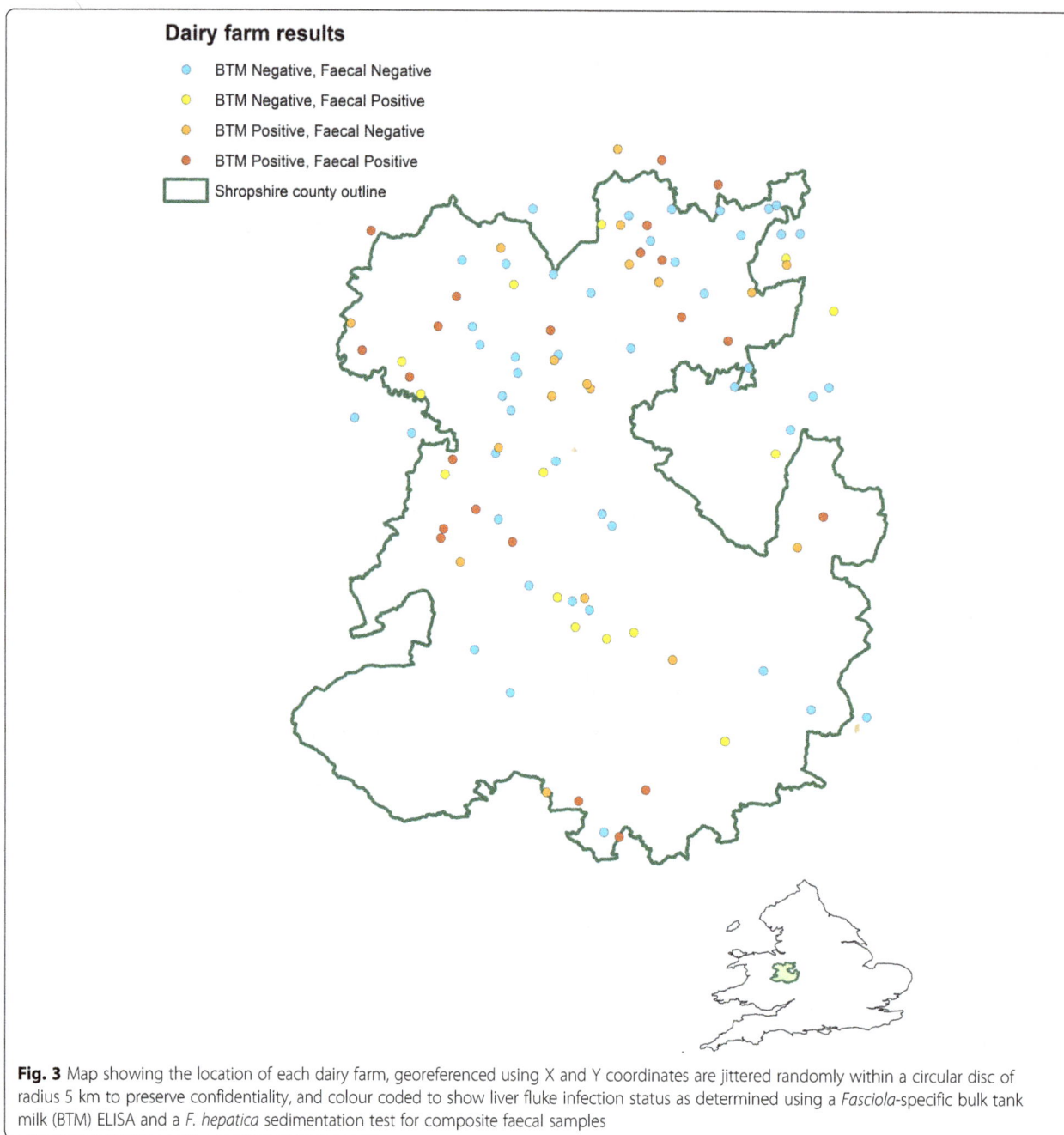

Fig. 3 Map showing the location of each dairy farm, georeferenced using X and Y coordinates are jittered randomly within a circular disc of radius 5 km to preserve confidentiality, and colour coded to show liver fluke infection status as determined using a *Fasciola*-specific bulk tank milk (BTM) ELISA and a *F. hepatica* sedimentation test for composite faecal samples

infected. However 63% of non-dairy farmers reported using fasciolicide; it would appear that many farmers use drugs without knowing if their animals are infected.

In dairy herds where 62% of farmers reported that they monitored their herds for fluke infection by one or more method, fluke eggs were found in 36% of herds and 41% of herds were positive by the BTM *F. hepatica* ELISA. The reported use of fasciolicides in 61% of dairy herds was similar to that reported in non-dairy herds although the timing of treatment was different.

The majority of dairy herds sampled had year round calving and in 50% of herds, cows were treated at the beginning of the dry period which may not coincide with when they are most at risk of infection. The BTM antibody ELISA detects liver fluke antibodies that are present in milk and is a good screening test for lactating dairy cows that contribute their milk to the BTM sample. Antibodies may persist 4–10 weeks after treatment; hence the ELISA cannot distinguish between current and recent exposure [41]. Discordant

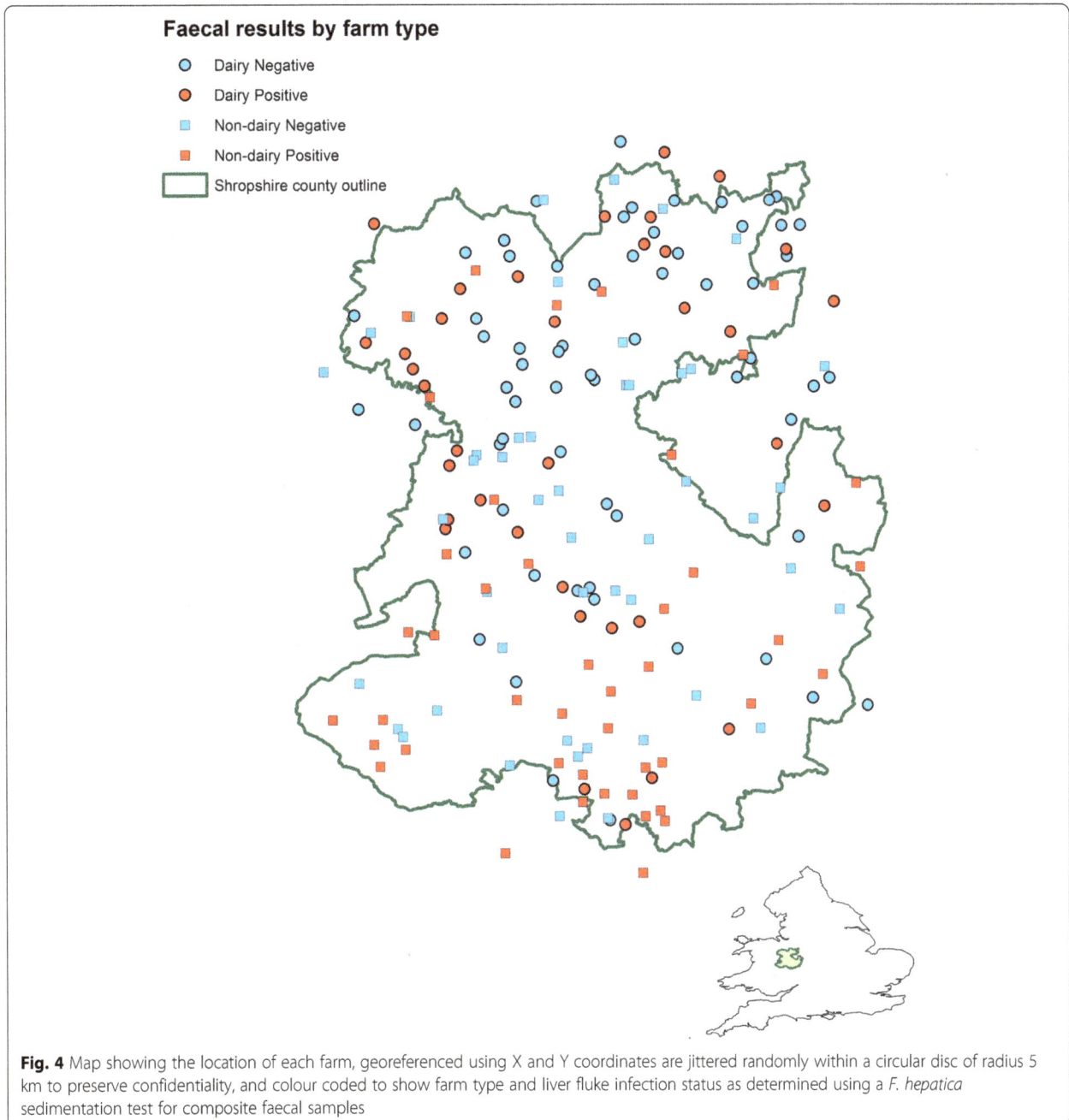

Fig. 4 Map showing the location of each farm, georeferenced using X and Y coordinates are jittered randomly within a circular disc of radius 5 km to preserve confidentiality, and colour coded to show farm type and liver fluke infection status as determined using a *F. hepatica* sedimentation test for composite faecal samples

results were found between the *F. hepatica* BTM antibody ELISA and the faecal analyses in 32 herds. Eighteen herds were positive by the BTM ELISA but negative for fluke eggs. There are several possible explanations for this result. Firstly whilst the specificity of the faecal egg count test is high, the sensitivity is low, 30–70% depending on method and study area [7]. In this study we used a composite test which adds a layer of complexity. Also, in herds with year-round calving and routine dry cow treatment for fluke, there is likely to be a continuous entry of newly calved uninfected cows that may still be antibody positive, thus resulting in a positive BTM ELISA and negative faecal egg count. Also *F. hepatica* antibodies are produced in the serum 2–4 weeks after infection hence an ELISA that detects antibody in the milk can detect early, pre patent infection whilst a faecal egg count will only detect patent infection.

The finding of 14 dairy herds which were negative by the BTM and positive for faecal eggs may be explained by the characteristics of the BTM ELISA used in the study. A positive BTM ELISA result identifies herds in which least 25% of cows have been exposed which means that some of the cows from herds with a within-herd prevalence of less than 25% may be infected. The method of collection of faecal samples for the study was random, hence samples from infected cows was probable in some herds which had low levels of infection. The composite method for faecal samples is designed to detect a positive herd with 95% certainty whilst the BMT ELISA is designed to detect herds with more than 25% of cattle infected.

In this study the prevalence of fluke infection was 55% in dairy herds by BTM ELISA and/or a composite faecal egg count test, while the reported use of fasciolicide drugs was over 60%. Whilst previous data are not available for the herds in this study, other researchers have found that in endemic regions of Spain, Switzerland and Belgium, where repeated surveys were carried out, there was little change in prevalence of infection despite regular fasciolicide treatment [7].

The prevalence of fluke as determined by the composite faecal egg count test in the non-dairy herds was 49%. The detection of fluke eggs in faeces requires a patent infection – egg laying commences 10–12 weeks after ingestion of metacercariae. In the UK fasciolosis is a seasonal disease, development of the parasite in the intermediate host occurs between May and October if the weather conditions are appropriate, and cattle acquire infection in the autumn. In this study, farms were recruited and visited over a seven month period from October 2014 – April 2015. It is likely that some of the cattle which were sampled in October or November may have become infected with metacercariae in the autumn and may have tested negative for liver fluke eggs because the infection was not patent at the time of sample collection.

In this study we did not find any spatial pattern in the spread of negative and positive results, this is important because the farms were situated in a small area and would have had similar climates.

Conclusions

In this study we successfully recruited 79% of the target sample of farms into a study of liver fluke in cattle. Whilst recruitment by telephone was time consuming, by engaging directly with farmers we were able to determine why farmers were able/not able to participate. The information collected from each farm, together with environmental data, is being used to develop models to identify risk factors associated with different levels of infection at the farm level.

Abbreviations
APHA: Animal and Plant Health Agency; BMT: Bulk milk tank; CI: Confidence interval; CPH: County Parish Holding; ELISA: Enzyme-linked immunosorbent assay; ES: Excretory/secretory; GIS: Geographical Information System; HP: High positive; LP: Low positive; MP: Medium positive; NADIS: National Animal Disease Information Service; OD: Optical density; PCA: Postcode area; PP: Percent positivity; RADAR: Rapid Analysis and Detection of Animal-related Risk; Teagasc: Irish Agriculture and Food Development Authority

Acknowledgements
We would like to thank Catherine Hartley, John Graham-Brown, Catherine Glover, Buddhini Bandara Athauda and Alice Balard for laboratory support and Jennifer Carney for administrative support. This study would not have been possible without the support and cooperation of the cattle farmers in Shropshire who gave up their time to take part in the study.

Funding
The authors would like to acknowledge financial support of the BBSRC (BBSRC BB/K015591/1), the Agricultural and Horticultural Development Board, Quality Meat Scotland, Hybu Cig Cymru and Agrisearch Northern Ireland. The funding bodies had no role in the design of the study and collection, analysis and interpretation of data and in writing the manuscript.

Authors' contributions
The study was conceived and designed by DW, CM and MB; HC provided statistical advice on study design and analysis. The manuscript was written by CM and commented on by all authors. All authors have read and approved the manuscript.

Competing interests
The authors declare that they have no competing interests.

Author details
[1]Department of Infection Biology, Institute of Infection and Global Health, University of Liverpool, Liverpool, UK. [2]Department of Public Health and Policy, Institute of Psychology, Health and Society, University of Liverpool, Liverpool, UK. [3]National Institute of Health Research Health Protection Research Unit in Emerging and Zoonotic Infections, University of Liverpool, Liverpool, UK. [4]Department of Epidemiology and Population Health, Institute of Infection and Global Health, University of Liverpool, Liverpool, UK. [5]Present address: Epidemiology Research Unit, Scotland's Rural College (SRUC), An Lòchran, Inverness Campus, Inverness IV2 5NA, UK.

References
1. Mezo M, Gonzalez-Warleta M, Castro-Hermida JA, Muino L, Ubeira FM. Association between anti-*F. hepatica* antibody levels in milk and production losses in dairy cows. Vet Parasitol. 2011;180(3–4):237–42.
2. Charlier J, Van der Voort M, Hogeveen H, Vercruysse J. ParaCalc(R)- a novel tool to evaluate the economic importance of worm infections on the dairy farm. Vet Parasitol. 2012;184(2–4):204–11.
3. Charlier J, Duchateau L, Claerebout E, Williams D, Vercruysse J. Associations between anti-*Fasciola hepatica* antibody levels in bulk-tank milk samples and production parameters in dairy herds. Prev Vet Med. 2007;78(1):57–66.
4. Howell A, Baylis M, Smith R, Pinchbeck G, Williams D. Epidemiology and impact of *Fasciola hepatica* exposure in high-yielding dairy herds. Prev Vet Med. 2015;121(1–2):41–8.
5. Cawdery MJ, Conway A. Production effects of the liver fluke, *Fasciola hepatica*, on beef cattle. Vet Rec. 1971;89(24):641–3.
6. Sanchez-Vazquez MJ, Lewis FI. Investigating the impact of fasciolosis on cattle carcase performance. Vet Parasitol. 2013;193(1–3):307–11.
7. Charlier J, Vercruysse J, Morgan E, van Dijk J, Williams DJ. Recent advances in the diagnosis, impact on production and prediction of *Fasciola hepatica* in cattle. Parasitology. 2014;141(3):326–35.
8. Salimi-Bejestani MR, Daniel RG, Felstead SM, Cripps PJ, Mahmoody H, Williams DJ. Prevalence of *Fasciola hepatica* in dairy herds in England and Wales measured with an ELISA applied to bulk-tank milk. Vet Rec. 2005; 156(23):729–31.
9. McCann CM, Baylis M, Williams DJ. Seroprevalence and spatial distribution of *Fasciola hepatica*-infected dairy herds in England and Wales. Vet Rec. 2010;166(20):612–7.
10. Byrne AW, McBride S, Lahuerta-Marin A, Guelbenzu M, McNair J, Skuce RA, McDowell SW. Liver fluke (*Fasciola hepatica*) infection in cattle in Northern Ireland: a large-scale epidemiological investigation utilising surveillance data. Parasit Vectors. 2016;9:209.
11. Pritchard GC, Forbes AB, Williams DJ, Salimi-Bejestani MR, Daniel RG. Emergence of fasciolosis in cattle in east Anglia. Vet Rec. 2005;157(19):578–82.

12. Kenyon F, Sargison ND, Skuce PJ, Jackson F. Sheep helminth parasitic disease in south eastern Scotland arising as a possible consequence of climate change. Vet Parasitol. 2009;163(4):293–7.

13. Fox NJ, White PC, McClean CJ, Marion G, Evans A, Hutchings MR. Predicting impacts of climate change on *Fasciola hepatica* risk. PLoS One. 2011;6(1):e16126.

14. Malone JB. Biology-based mapping of vector-borne parasites by geographic information systems and remote sensing. Parassitologia. 2005;47(1):27–50.

15. Fuentes MV. Proposal of a geographic information system for modeling zoonotic fasciolosis transmission in the Andes. Parasitol Latinoam. 2004;59:51–5.

16. Bennema SC, Ducheyne E, Vercruysse J, Claerebout E, Hendrickx G, Charlier J. Relative importance of management, meteorological and environmental factors in the spatial distribution of *Fasciola hepatica* in dairy cattle in a temperate climate zone. Int J Parasitol. 2011;41(2):225–33.

17. Brunsdon RV. Principles of helminth control. Vet Parasitol. 1980;1:185–215.

18. Williams DJ, Howell A, Graham-Brown J, Kamaludeen J, Smith D. Liver fluke – an overview for practitioners. 2015. http://www.cattleparasites.org.uk/app/uploads/2018/04/Liver-fluke-an-overview-for-practitioners.pdf. Accessed 17 Nov 2017.

19. Sargison ND, Scott PR. Diagnosis and economic consequences of triclabendazole resistance in *Fasciola hepatica* in a sheep flock in south-East Scotland. Vet Rec. 2011;168(6):159–64.

20. Boray JC, Crowfoot PD, Strong MB, Allison JR, Schellenbaum M, Von Orelli M, Sarasin G. Treatment of immature and mature *Fasciola hepatica* infections in sheep with triclabendazole. Vet Rec. 1983;113(14):315–7.

21. Fairweather I. Triclabendazole: new skills to unravel an old(ish) enigma. J Helminthol. 2005;79(3):227–34.

22. Overend DJ, Bowen FL. Resistance of *Fasciola hepatica* to triclabendazole. Aust Vet J. 1995;72(7):275–6.

23. Mitchell GB, Maris L, Bonniwell MA. Triclabendazole-resistant liver fluke in Scottish sheep. Vet Rec. 1998;143(14):399.

24. Moll L, Gaasenbeek CP, Vellema P, Borgsteede FH. Resistance of *Fasciola hepatica* against triclabendazole in cattle and sheep in the Netherlands. Vet Parasitol. 2000;91(1–2):153–8.

25. Gaasenbeck CP, Moll L, Cornelissen JB, Vellema P, Borgsteede FH. An experimental study on triclabendazole resistance of *Fasciola hepatica* in sheep. Vet Parasitol. 2001;95:37–43.

26. Alvarez-Sanchez MA, Mainar-Jaime RC, Perez-Garcia J, Rojo-Vazquez FA. Resistance of *Fasciola hepatica* to triclabendazole and albendazole in sheep in Spain. Vet Rec. 2006;159(13):424–5.

27. Brennan GP, Fairweather I, Trudgett A, Hoey E, McCoy MCM, Meaney M, Robinson M, McFerran N, Ryan L, Lanusse C, Mottier L, Alvarez L, Solana H, Virkel G, Brophy PM. Understanding triclabendazole resistance. Exp Mol Pathol. 2007;82(2):104–9.

28. Daniel R, van Dijk J, Jenkins T, Akca A, Mearns R, Williams DJ. Composite faecal egg count reduction test to detect resistance to triclabendazole in *Fasciola hepatica*. Vet Rec. 2012;171(6):153.

29. Brockwell YM, Elliott TP, Anderson GR, Stanton R, Spithill TW, Sangster NC. Confirmation of *Fasciola hepatica* resistant to triclabendazole in naturally infected beef and dairy cattle. Int J Parasitol Drugs Drug Resist. 2014;4(1):48–54.

30. Hanna RE, McMahon C, Ellison S, Edgar HW, Kajugu PE, Gordon A, Irwin D, Barley JP, Malone FE, Brennan GP, Fairweather I. *Fasciola hepatica*: a comparative survey of adult fluke resistance to triclabendazole, nitroxynil and closantel on selected upland and lowland sheep farms in Northern Ireland using faecal egg counting, coproantigen ELISA testing and fluke histology. Vet Parasitol. 2015;207(1–2):34–43.

31. NOAH, National Office for animal health. NOAH Compendium. 2017. http://www.noahcompendium.co.uk/. Accessed 20 May 2017.

32. McCann CM, Baylis M, Williams DJ. The development of linear regression models using environmental variables to explain the spatial distribution of *Fasciola hepatica* infection in dairy herds in England and Wales. Int J Parasitol. 2010;40(9):1021–8.

33. Paiba GA, Roberts SR, Houston CW, Williams EC, Smith LH, Gibbens JC, Holdship S, Lysons R. UK surveillance: provision of quality assured information from combined datasets. Prev Vet Med. 2007;81(1–3):117–34.

34. Animal Health and Veterinary Laboratories Agency (AHVLA), Protocol for liver fluke infection in cattle herds. Herd® Catt Health Improv Serv, 2009 Version 6, 40–52.

35. MAFF. Ministry of Agriculture, Fisheries and Food (MAFF). Manual of Veterinary Parasitological Laboratory Techniques. 1986.

36. Kulldorff M, Nagarwalla N. Spatial disease clusters: detection and inference. Stat Med. 1995;14:799–819.

37. de Souza EA, da Silva-Nunes M, Malafronte Rdos S, Muniz PT, Cardoso MA, Ferreira MU. Prevalence and spatial distribution of intestinal parasitic infections in a rural Amazonian settlement, acre state, Brazil. Cad Saude Publica. 2007;23:427–34.

38. Webb CR, Floyd T, Brien S, Oura CA, Wood JL. Bluetongue serotype 8 vaccine coverage in northern and South-Eastern England in 2008. Vet Rec. 2011;168(16):428.

39. Selemetas N, Phelan P, O'Kiely P, de Waal T. The effects of farm management practices on liver fluke prevalence and the current internal parasite control measures employed on Irish dairy farms. Vet Parasitol. 2015;207(3–4):228–40.

40. Bloemhoff Y, Danaher M, Andrew F, Morgan E, Mulcahy G, Power C, Sayers R. Parasite control practices on pasture-based dairy farms in the Republic of Ireland. Vet Parasitol. 2014;204(3–4):352–63.

41. Salimi-Bejestani MR, McGarry JW, Felstead S, Ortiz P, Akca A, Williams DJ. Development of an antibody-detection ELISA for *Fasciola hepatica* and its evaluation against a commercially available test. Res Vet Sci. 2005;78(2):177–81.

13

Molecular study on *Pasteurella multocida* and *Mannheimia granulomatis* from Kenyan Camels (*Camelus dromedarius*)

Ilona V. Gluecks[1†], Astrid Bethe[2†], Mario Younan[3†] and Christa Ewers[4*]

Abstract

Background: Outbreaks of a Haemorrhagic Septicaemia (HS) like disease causing large mortalities in camels (*Camelus dromedarius*) in Asia and in Africa have been reported since 1890. Yet the aetiology of this condition remains elusive. This study is the first to apply state of the art molecular methods to shed light on the nasopharyngeal carrier state of Pasteurellaceae in camels. The study focused on HS causing *Pasteurella multocida* capsular types B and E. Other Pasteurellaceae, implicated in common respiratory infections of animals, were also investigated.

Methods: In 2007 and 2008, 388 nasopharyngeal swabs were collected at 12 locations in North Kenya from 246 clinically healthy camels in 81 herds that had been affected by HS-like disease. Swabs were used to cultivate bacteria on blood agar and to extract DNA for subsequent PCR analysis targeting *P. multocida* and *Mannheimia*-specific gene sequences.

Results: Forty-five samples were positive for *P. multocida* genes *kmt* and *psl* and for the *P. multocida* Haemorrhagic Septicaemia (HS) specific sequences KTSP61/KTT72 but lacked HS-associated capsular type B and E genes *capB* and *capE*. This indicates circulation of HS strains in camels that lack established capsular types. Sequence analysis of the partial 16S rRNA gene identified 17 nasal swab isolates as 99% identical with *Mannheimia granulomatis*, demonstrating a hitherto unrecognised active carrier state for *M. granulomatis* or a closely related *Mannheimia* sp. in camels.

Conclusions: The findings of this study provide evidence for the presence of acapsular *P. multocida* or of hitherto unknown capsular types of *P. multocida* in camels, closely related to *P. multocida* strains causing HS in bovines. Further isolations and molecular studies of camelid *P. multocida* from healthy carriers and from HS-like disease in camels are necessary to provide conclusive answers. This paper is the first report on the isolation of *M. granulomatis* or a closely related new *Mannheimia* species from camelids.

Keywords: Pasteurellaceae, *Pasteurella multocida*, *Mannheimia*, Camels, *Camelus dromedarius*, Haemorrhagic Septicaemia

Background

Pasteurella multocida capsular types B and E are the specific cause of seasonal outbreaks of Haemorrhagic Septicaemia (HS) in tropical cattle and buffaloes [1–3]. The veterinary literature provides a long record of an HS-like disease causing significant mortality in camels [4–8]. Yet the aetiology of HS in camels remains elusive [1, 3, 9]. Previous reports on HS-like disease in camels often failed to isolate the pathogen or provided only limited information on phenotypical characteristics of the isolates [9]. Attempts to infect camels with highly virulent *Pasteurella multocida* capsular type B strains produced only very mild clinical symptoms that resolved completely within 3 days [9].

A respiratory disease in Ethiopian camels caused by *Pasteurella (Mannheimia) haemolytica* has been described by Bekele [10]. This pilot study is the first of its' kind to use state of the art molecular methods for investigating Pasteurellaceae in camels.

* Correspondence: Christa.Ewers@vetmed.uni-giessen.de
†Equal contributors
4Institute of Hygiene and Infectious Diseases of Animals, Justus-Liebig University Giessen, Giessen, Germany

Methods

In 2007 and 2008, 388 nasopharyngeal swabs were collected at 12 locations in North Kenya from 246 clinically healthy camels in 81 herds that had reportedly been affected by outbreaks of HS-like disease. Flocculated swabs (FLOQSwabs®) were used for DNA extraction while swabs in Amies transport medium (Sterilin®) were used for standard bacteriological investigation. Bacteriological examination and extraction of DNA was carried out at a laboratory in Nairobi (Analabs Ltd.). DNA eluates from 341 swabs and 19 isolated *Pasteurella*-like cultures were transferred to the Institute for Microbiology and Epizootics at Freie Universität Berlin (IMT/FUB) in Germany. At the IMT/FUB, DNA was extracted from 19 culture isolates, two of which did not belong to the Pasteurellaceae based on PCR results. Seventeen isolates were subsequently selected for biochemical testing to differentiate *M. granulomatis* from other *Mannheimia* spp. according to Ewers et al. (2004) [11]. Three hundred five samples with positive reaction in the 16S rRNA gene PCR, as well as, heat denatured DNA from bacterial cultures were investigated for the presence of *P. multocida*- and *M. haemolytica*- specific DNA sequences. For *P. multocida* the specific sequences tested were *kmt*, *psl*, and KTSP61/KTT72. Samples that proofed positive in one of these PCRs were further tested for capsular genes *capA*, *capB*, *capD*, *capE*, and *capF*, and for virulence-associated genes *toxA*, *ptfA*, *pfhA* *oma87*, *ompH*, *hgbA*, *hgbB*, *exbB/tonB*, *tbpA*, *nanB*, and *nanH*. To detect pathogenic *Mannheimia* spp., we screened for the leukotoxin (*lktA*) and the outer membrane protein (*pomA*) genes. All PCRs, except for the *P. multocida* capsular gene PCR, which was performed as a multiplex PCR, were conducted as single PCRs according to previously published protocols [12–18].

Results

The species *P. multocida* could not be isolated from 312 nasal swabs cultured on Blood Agar. Of the 305 DNA eluates containing sufficient quantities of DNA to undergo molecular characterization, 60 were positive for at least one of the *P. multocida* species-specific sequences tested by PCR. Forty-five samples gave a positive result for the HS- associated sequence KTSP61/KTT72, known to be present in *P. multocida* capsular type B strains. Neither *capB* nor *capE* gene, encoding HS-associated capsular types B and E, were detected in the sample material. Blast search of 18 KTSP61/KTT72 sequences [19] matched with high scores (98.8–99%) to nucleotide sequences submitted under GenBank accession numbers AF016260 (*P. multocida* unknown protein 1 gene, partial cds and unknown protein 2 gene, complete cds), AY948545.1 (*P. multocida* HS-B specific genomic sequence) and AJ421513.1 (*P. multocida* DNA fragment

specific for HS). *CapA* and *capD* genes encoding for *P. multocida* capsular types A and D were identified in five and in one sample, respectively. Screening for adhesion-related genes *ptfA* and *pfhA* was positive in 15 and 2 samples, respectively. Outer membrane protein gene *ompH* was detected in 10, *oma87* in 14 samples. While *toxA*, encoding the dermonecrotoxin, was not found in any sample, iron acquisition-related genes were present as follows: *hgbA* (*n* = 12), *hgbB* (*n* = 11), *exbB/tonB* (*n* = 1). Neuraminidase encoding genes *nanB* and *nanH* were present in 9 and 11 samples, respectively.

The biochemical profiles of the 17 *Mannheimia* spp. isolates characterised phenotypically were inconsistent and differed from the biochemical profile described for *M. granulomatis* [20], namely Sorbitol: pos.; α-Fucosidase: neg.; β-Galactosidase: variable. Of the 305 eluates tested, none was positive for *Mannheimia* spp. associated sequences *lktA* or *pomA*. According to the 16S rRNA sequence analysis performed on 19 cultures from the nasopharynx of healthy camels, 17 sequences showed closest relatedness (98–99%) to sequences from *M. granulomatis*, previously classified as *P. granulomatis*, Bisgaard taxon 20 and *P. haemolytica* biogroup 3 J [21, 22].

Discussion

This study investigated the active (nasopharyngeal) carrier state for Pasteurellaceae in Kenyan camels. Importance of the latent carrier state in the epidemiology of Pasteurellosis was reiterated by Dziva et al. (2008) [23].

The fact that no *P. multocida* capsular type B or E specific DNA sequences were identified in this study may indicate the presence of non-capsulated strains or the emergence of a hitherto unknown capsular type of *P. multocida* in camels. According to the OIE Terrestrial Manual [24] vaccines against Haemorrhagic Septicaemia in cattle and buffaloes must be based on local isolates that represent the prevalent serotype; seed cultures for the production of HS vaccines should contain capsulated organisms. Based on the results of this study the indiscriminate use of vaccines based on HS causing *P. multocida* isolates from cattle and buffaloes cannot be recommended for the prevention of HS-like disease in camels. - This is the first molecular study to confirm the presence of *P. multocida* capsular types A and D specific DNA sequences in the nasopharynx of healthy Kenyan carrier camels, albeit at low frequency. Both capsular types have been reported in camels previously [25] based on phenotypic characterisation. Failure to culture *P. multocida* in this study is possibly related to the fastidiousness of the species which does not withstand cold chain transport of several days [26].

The *lktA* or *pomA* sequences do not occur regularly in all *Mannheimia* species or strains [27, 28], hence negative findings do not rule out presence of *Mannheimia* spp., but *lktA*-negative *M. haemolytica* strains are reported to be less virulent. In this study the most common Pasteurel-laceae species cultured from the nasopharynx of healthy carrier camels in North Kenya and identified by 16S rRNA sequence analysis was a *Mannheimia* sp. with 98% to 99% sequence identity to *M. granulomatis*. Only more recently has *M. granulomatis* been recognised as a significant pathogen in domestic and wild ruminants [29–31]. Pheno-typic characterization of Pasteurellaceae species is of lim-ited value [32] and published phenotypic characteristics for *M. granulomatis* are based on a limited number of bo-vine, leprine and deer strains [2, 21, 22, 30]. Hence it is to be expected that biochemical reactions of Kenyan camelid *M. granulomatis* strains differ from those described for *M. granulomatis* under http://www.bacterio.net/mannhei mia.html/. Comparison of our results with a previous communication on involvement of *P. (M.) haemolytica* in respiratory disease in Ethiopian camels [10] is also limited, because the methodology used there would not have per-mitted a differentiation between *M. granulomatis, M. (Pasteurella) haemolytica* and other *Mannheima* spp. The possibility that this respiratory pathogen isolated from Ethiopian camels [10] may in fact have been *M. granulo-matis* or a new *Mannheimia* species very closely related to *M. granulomatis* cannot entirely be ruled out.

Conclusions

This study has documented the carrier state for acapsu-lar *P. multocida* or unknown capsular types of *P. multo-cida*, closely related to the *P. multocida* strains causing HS in cattle and buffaloes, in healthy camels. At the same time the study found no evidence for the presence of *P. multocida* capsular types B and E or their specific DNA sequences in healthy camels in North Kenya. Further isolations and molecular studies of camelid *P. multocida* from healthy carriers and from HS-like disease in camels are necessary to provide conclusive answers. To our best knowledge this is the first report on the isolation of *M. granulomatis* or a closely related new *Mannheimia* species from camelids.

Abbreviations
DNA: Deoxyribonucleic acid; HS: Haemorrhagic Septicaemia; IMT/ FUB: Institute for Microbiology and Epizootics at Freie Universität Berlin; PCR: Polymerase Chain Reaction; RNA: Ribonucleic acid; rRNA: Ribosomal ribonucleic acid

Acknowledgements
We would like to thank Ines Diehl for excellent technical assistance.

Funding
This work was supported by the Wellcome Trust (Grant number 081073/Z/06/Z).

Authors' contributions
All authors conceived and planned the study. IG and MY performed the sampling, DNA extraction, primary cultures and organized the transfer of samples to Germany. AB and CE designed the laboratory and molecular experiments. All authors contributed on the draft of the manuscript. All authors read and approved the final manuscript.

Competing interests
The authors declare that they have no competing interests.

Author details
[1]Vétérinaires sans Frontières Suisse, Nairobi, Kenya. [2]Institute of Microbiology and Epizootics, Centre for Infection Medicine, Free University Berlin, Berlin, Germany. [3]Vétérinaires sans Frontières Germany, Nairobi, Kenya. [4]Institute of Hygiene and Infectious Diseases of Animals, Justus-Liebig University Giessen, Giessen, Germany.

References
1. Bain RVS, De Alwis MCL, Carter GR, Gupta BK. Haemorrhagic Septicaemia. Rome: FAO Animal Production and Health Paper 33; 1982. p. 54.
2. Songer JG, Post KW. The Genera *Mannheimia* and *Pasteurella*. In: Songer JG, Post KW, editors. Veterinary Microbiology, Bacterial and Fungal Agents of Animal Disease, Chapter 23. USA: Elsevier Inc; 2005. p. 434.
3. Office International Epizootics (OIE). Report of the 2nd Meeting of the OIE ad hoc Group on Disease of Camelids. Paris, 3–5 May 2010.
4. Steel JH. A Manual of the Diseases of the Camel and of his Management and Uses. Madras: Indian Veterinary Manuals III, The Lawrence Asylum Press (printed by GW Taylor); 1890. p. 54–5.
5. Leese AS. The Camel – a treatise on the one-humped camel in health and disease. Stanford, Lincolnshire, UK: Haynes & Son; 1927. p. 270–2.
6. Masia R. Les maladies microbiennes du dromadaire et leur importance en Afrique du Nord. Thése pour le Doctorat Vétérinaire, L'École Nationale Vétérinaire d'Alfort, Imprimerie R.Foulon, Paris; 1953. p. 33–34.
7. McGrane JJ, Higgins A. (1986) Infectious Diseases of the Camel, Viruses Bacteria and Fungi, Pasteurellosis. In: Higgins A, editor. . Baillière Tindall: The Camel in Health and Disease; 1986. p. 103.
8. Schwartz HJ, Dioli M. The one-humped camel in Eastern Africa - A pictorial guide to disease, health care and management. Berlin, Germany: Verlag Josef Margraf; 1992.
9. Wernery U, Kinne J, Schuster RK. 1.1.8 Pasteurellosis. In: Camelid Infectious Disorders. Paris, France: World Organisation for Animal Health OIE; 2014. p. 58–65.
10. Bekele T. Studies on the respiratory disease 'sonbobe' in camels in the eastern lowlands of Ethiopia. Trop Anim Health Prod. 1999;31(6):333–45.
11. Ewers C, Lübke-Becker A, Wieler LH. *Mannheimia haemolytica* and the pathogenesis of pneumonic pasteurellosis. Berl Munch Tierarztl Wochenschr. 2004;3-4:97–115.
12. Doughty SW, Ruffolo CG, Adler B. The type 4 fimbrial subunit gene of *Pasteurella multocida*. Vet Microbiol. 2000;72:79–90.
13. Ewers C, Luebke-Becker A, Bethe A, Kiessling S, Filter M, Wieler LH. Virulence genotype of *Pasteurella multocida* strains isolated from different hosts with various disease status. Vet Microbiol. 2006;114:304–17.
14. Kasten RW, Hansen LM, Hinojoza J, Bieber D, Ruehl WW, Hirsh D. *Pasteurella multocida* produces a protein with homology to the P6 outer membrane protein of *Haemophilus influenzae*. Infect Immun. 1995;63:989–93.
15. Townsend KM, Frost AJ, Lee CW, Papadimitriou JM, Dawkins HJ. Development of PCR assays for species- and type-specific identification of *Pasteurella multocida* isolates. J Clin Microbiol. 1998;36:1096–100.
16. Townsend KM, Hanh TX, O'Boyle D, Wilkie I, Phan TT, Wijewardana TG, Trung NT, Frost AJ. PCR detection and analysis of *Pasteurella multocida* from the tonsils of slaughtered pigs in Vietnam. Vet Microbiol. 2000;72:69–78.
17. Guenther S, Schierack P, Grobbel M, Lubke-Becker A, Wieler LH, Ewers C. Real-time PCR assay for the detection of species of the genus *Mannheimia*. J Microbiol Methods. 2008;75:75–80.
18. Townsend KM, Boyce JD, Chung JY, Frost AJ, Adler B. Genetic organization of *Pasteurella multocida cap* Loci and development of a multiplex capsular PCR typing system. J Clin Microbiol. 2001;39:924.929.
19. http://blast.ncbi.nlm.nih.gov/Blast.cgi (Accessed on 18th March, 2017).
20. http://www.bacterio.cict.fr (Accessed on 18th March, 2017).
21. Angen O, Mutters R, Caugant DA, Olsen JE, Bisgaard M. Taxonomic relationships of the [*Pasteurella*] *haemolytica* complex as evaluated by DNA-

DNA hybridizations and 16S rRNA sequencing with proposal of *Mannheimia haemolytica* gen. nov., comb. nov., *Mannheimia granulomatis* comb. nov., *Mannheimia glucosida* sp. nov., *Mannheimia ruminalis* sp. nov. and *Mannheimia varigena* sp. nov. Int J Syst Bacteriol. 1999;49:67–86.

22. Angen O, Quirie M, Donachie W, Bisgaard M. Investigations on the species specificity of *Mannheimia (Pasteurella) haemolytica* serotyping. Vet Microbiol. 1999;65:283–90.

23. Dziva F, Muhairwa AP, Bisgaard M, Christensen H. Diagnostic and typing options for investigating diseases associated with *Pasteurella multocida*. Vet Microbiol. 2008;128(1–2):1–22.

24. World Organisation for Animal Health – WOAH/OIE. Manual of Diagnostic Tests and Vaccines for Terrestrial Animals (mammals, birds and bees). 2012; 7th edition, Volume 1, Chapter 2.4.12. - Haemorrhagic septicaemia, 732–744.

25. Seleim RS, Amal RT, Sahar RM, Nada H, Gobran RA. ELISA and other tests in the diagnosis of *Pasteurella multocida* infection in camels. Presented at Deutscher Tropentag, International Research on Food Security, Natural Resource Management and Rural Development Georg-August-Universität Göttingen, October 8–10 2003; posted on 'health & life sciences' May 2003 http://www.priory.com/vet/camel.htm.

26. Younan M. Biochemical characterization and identification of capsular antigen of ovine Pasteurella strains of different geographical provenance (Syria / South Germany) and of a collection of bovine Pasteurella haemolytica isolates. Veterinary Faculty - Free University Berlin, 1988.

27. Omaleki L, Browning GF, Barber SR, Allen JL, Srikumaran S, Markham PF. Sequence diversity, cytotoxicity and antigenic similarities of the leukotoxin of isolates of *Mannheimia* species from mastitis in domestic sheep. Vet Microbiol. 2014;174(1–2):172–9.

28. Shanthalingam S, Goldy A, Bavananthasivam J, Subramaniam R, Batra SA, Kugadas A, Raghavan B, Dassanayake RP, Jennings-Gaines JE, Killion HJ, Edwards WH, Ramsey JM, Anderson NJ, Wolff PL, Mansfield K, Bruning D, Srikumaran S. PCR assay detects *Mannheimia haemolytica* in culture-negative pneumonic lung tissues of bighorn sheep (*Ovis canadensis*) from outbreaks in the western USA, 2009-2010. J Wildl Dis. 2014;50(1):1–10.

29. Riet-Correa F, Ladeira SL, Andrade GB, Carter GR. Lechiguana (focal proliferative fibrogranulomatous panniculitis) in cattle. Vet Res Comm. 2000;24(8):557–72.

30. Blackall PJ, Bisgaard M, Stephens CP. Phenotypic characterisation of Australian sheep and cattle isolates of *Mannheimia haemolytica*, *Mannheimia granulomatis* and *Mannheimia varigena*. Australian Vet J. 2002;80(1–2):87–91.

31. Bojesen AM. Larsen J, Pedersen AG, M¨orner T, Mattson R & Bisgaard M. Identification of a novel *Mannheimia granulomatis* lineage from lesions in roe deer (Capreolus capreolus). J Wildl Dis. 2007;43:345–52.

32. Koendgen S, Leider M, Lankester F, Bethe A, Lubke-Becker A, Fabian H, Leendertz EC. *Pasteurella multocida* Involved in Respiratory Disease of Wild Chimpanzees. PLoS One. 2011;6(9):e24236.

14

Development of an ovine efferent mammary lymphatic cannulation model with minimal tissue damage

Hung-Hsun Yen*, Elizabeth Washington, Wayne Kimpton, Evan Hallein, Joanne Allen, Silk Yu Lin and Stuart Barber* (ID)

Abstract

Background: Two mammary lymphatic cannulation models in sheep have been described with minimal use in the past 50 years. The purpose of this study was to investigate a new surgical technique to allow long term monitoring of mammary lymph flow and composition from the mammary glands, with rapid ewe recovery and minimal complications post-surgery.

Results: We developed a modified methodology for cannulating the efferent mammary lymphatic from the mammary lymph node with minimum tissue damage. Compared to the previous models, our method required only a small incision on the aponeurosis of the external abdominal oblique muscles and thus reduced the difficulties in suturing the aponeurosis. It allowed for lymph collection and assessment for at least one week post-surgery with concurrent milk collection.

Conclusion: This method allows for good ewe recovery post-surgery and in vivo sampling of efferent mammary lymph from the mammary lymph nodes in real-time and comparison with milk parameters.

Keywords: Lymphatic, Cannulation, Ewe, Model, Mammary, Ovine

Background

An improved understanding of immunological responses in the mammalian udder during disease processes requires an ability to collect lymph from the infected gland over multiple time points, with the collection method causing minimal interference with the animal. One method to achieve this is via surgical cannulation of either afferent lymphatic ducts that lead to mammary lymph nodes, also called supra-mammary lymph nodes, or from efferent lymphatic ducts that drain lymph from these lymph nodes [1–3].

Surgical cannulation of the lymphatic vessels provides an approach for harvesting lymph draining the target tissues for biomedical research in many species including rats [4], humans [5], sheep [2], goats [6], cattle [7], mice [8] and dogs [9]. In sheep, a number of lymphatic catheterisation models for accessing pseudo-afferent or efferent lymph from different lymphatic vessels have been described, including the hepatic lymphatic [2], the efferent lymphatic of the mammary lymph nodes [2, 6], the prefemoral lymphatic [10], the popliteal lymphatic draining the lower hind limb [11], the efferent duct of the prescapular (superficial cervical) lymph nodes [12], intestinal lymphatic draining the small intestine [2, 13], the tracheal trunks draining oro-nasal regions [14, 15], the thoracic duct with thoracotomy [2] and the thoracic duct without thoracotomy [16]. These ovine cannulation models are useful tools in immunological research, with the ability to study in vivo, long-term pathogen and host interplay over time [17].

Lymphatic cannulation models such as the prefemoral model have been broadly applied to many studies, whereas mammary lymphatic cannulation has not, with only a few published reports of its use [1, 3, 18] since the publication by Lascelles and Morris (1961) more than 50 years ago [2] and none following the alternative method by Linzell (1960) [6]. One of the reasons for the limited application

* Correspondence: hyen@unimelb.edu.au; srbarber@unimelb.edu.au
Faculty of Veterinary and Agricultural Sciences, The University of Melbourne, Parkville, VIC 3010, Australia

of the model of Lascelles and Morris (1961) is that the surgical procedures require incising a large portion of the aponeuroses of the external and internal abdominal oblique muscles and rejoining the aponeuroses, which requires advanced surgical skills. Indeed, the authors mentioned, "It is important to cut through the aponeurosis of the external and internal oblique abdominal muscles carefully as this tissue is difficult to suture. Unless the incision is closed correctly, the underlying peritoneum is exposed and the peritoneal contents may herniate." [2].

The development of an improved mammary cannulation model will facilitate investigation of sheep diseases and the use of sheep as a model for investigation of diseases in other species, such as cattle and humans. One example of a potential use for disease investigation is mastitis, as this is a relatively common disease with a variety of aetiological agents [19]. It is particularly important in farmed ruminants as a source of reduced animal welfare, production and profitability [20, 21]. The development of RNA-seq [22, 23] and recent findings of the presence of exosomes in various body fluids [24, 25] has expanded the potential benefit of this type of model to enable long-term RNA profiling and exosome monitoring in lymph. Here, we report the development of a modified mammary lymphatic cannulation model of sheep.

Methods
Animals
The cadavers of three sheep were collected after they had been euthanised at completion of animal trials not associated with this study at the Faculty of Veterinary and Agricultural Sciences animal facility. Dissections were performed on the cadavers to approach the aponeurosis of the external abdominal oblique muscles and identify the external pudendal vessels medial to the aponeurosis. The use of cadavers to develop a new approach to the mammary lymphatic duct prior to surgery on live animals was undertaken to reduce the use of live animals. The use of live ewes for mammary cannulation in this experiment was covered under University of Melbourne Faculty of Veterinary and Agricultural Science Animal Ethics Committee (AEC) number 1312857.2 and for prefemoral cannulation under University of Melbourne Faculty of Veterinary Science AEC approval number 251683.

Mammary cannulation
Four lactating East Friesian cross ewes, aged 2.2–2.3 years were purchased from a commercial dairy. Three ewes were on their second lactation and one on her first lactation. The ewes had been lactating between 94 and 127 days prior to arrival at the animal house and averaged 1.05 litres milk produced per day post lambing. The ewes were transported to the animal house and fed on a mixed lucerne and oaten chaff ration (50:50) *ad*

libitum for the first four days. They were then fed a mixture of this chaff and manufactured sheep pellets (Rumevite, Ridley Corporation, VIC, Australia) for the duration of the trial. They were milked by hand twice daily after arrival and acclimatised for a minimum of seven days before surgery. Prior to each milking the teat ends were disinfected with 70% ethanol soaked swabs. Milk volume was determined and 30 mL of milk was set aside for cell count and component analysis following the first few squirts of milk. An 18 mg bronopol tablet (Broad spectrum microtabs, Advanced Instruments) was added to the 30 mL tube to allow samples to be sent for analysis weekly with milk refrigerated within 1 h of collection. Measurements of milk cell count and components (fat, protein, solids-not-fat (SNF) & lactose) for each udder half were performed at Dairy Technical Services (North Melbourne, VIC, Australia) using a CombiFoss 5000 with standard FOSS reagents and technique, using International Dairy Federation standards.

Prefemoral cannulation
An additional three ewes of matched ages to those with mammary lymphatic cannulation had prefemoral efferent lymphatic ducts cannulated in order to compare the lymphocyte outputs and subsets between the two lymphatic circulations. These ewes were housed and fed under the same conditions as the ewes with mammary cannulations.

General surgical procedures
On the day prior to surgery for mammary cannulation, ewes were fasted overnight and provided water *ad libitum* until the time of surgery. Anaesthesia was induced by intravenous injection of 1.0–1.5 mg thiopentone sodium (Boehringer Ingelheim, Australia) in 20–30 mL distilled water per sheep and then maintained with isoflurane (1.5–2.5%) and oxygen following intubation. The general surgical procedures and method for securing the bottles for lymph collection has been described in previous publications [14, 16]. We used clear vinyl cannulae (internal diameter 0.58 mm; external diameter 0.96 mm, Dural Plastics, Australia) coated with bioactive heparin (CBAS, Carmeda AB, Stockholm, Sweden) for all lymphatic cannulation surgeries. The bottles for lymph collections were secured on an animal with two tubular elastic net bandages (size 6, Surgifix, Australia). Each ewe was administered one injection of Temgesic (2.2 mg/kg) intramuscularly per day for the first two days postsurgery. Following cannulation, ewes were maintained on *ad libitum* feed and water in individual pens in sight of other ewes.

The presurgery and anaesthetic protocol for ewes undergoing prefemoral surgery was as per mammary cannulation surgery. The method for cannulating the prefemoral efferent lymphatic ducts of sheep was first described by Hall [10]. Briefly, an incision of about 10 cm was made through the skin and cutaneous muscle from the tuber coxae along the anterior border of the thigh. Subcutaneous fat was divided by blunt dissection then the anterior border of the tensor fascia was retracted to expose the circumflex iliac blood vessels and the associated prefemoral efferent duct. The detailed procedure used in cannulating this duct was identical to that used for the mammary efferent lymphatic duct as described later in this paper.

Collection and analysis of mammary and prefemoral lymph

Lymph was collected twice daily in sterile 100 or 250 mL polypropylene collection bottles (Plastilab, Kartell Labware, Noviglio Italy) containing 1000 or 2000 IU of heparin (Pfizer). The bottle was fixed to netting surrounding the sheep's abdomen to avoid inadvertent removal of the bottle or tube. The bottle was tethered to the inner tubular netting bandage using strings and held between the two tubular bandages. At each lymph collection the bottle was removed and replaced with a clean, sterile bottle containing heparin. To change the bottles, the cannula was removed from a small opening in the cap of the bottle before untying the bottle. The fresh bottle was secured to the netting with the same strings. The free end of the cannula was disinfected with 0.5% w/v Hibitane in 70% v/v alcohol before inserting it into the bottle through the small opening of the cap and sealed with adhesive tape. The total volume of lymph collected for each duct was measured and the average rate of lymph flow was determined.

Cells from a 50 μl sample of lymph were counted using a model Z1 Coulter Particle Counter (Beckman Coulter, USA). Cells in lymph were then washed 3 times in PBS containing 2% BSA, 0.4% EDTA and 0.1% azide (FACS wash) and stained for flow cytometric analysis of lymphocyte subsets. Monoclonal antibodies (mAb) against the T cell subsets CD4 (44–38), CD8 (38–65) and γδTCR (86D) were obtained from Dr Scheerlinck (Centre for Animal Biotechnology, The University of Melbourne) and have been described previously [26–28]. They were used as cell culture supernatants and detected with PE-conjugated sheep anti-mouse immunoglobulin (Ig) (Chemicon, Australia).

Cells were analysed fresh on a FACSCalibur Cytometer equipped with argon and red diode lasers (BD Immunocytometry Systems, USA). The instrument was calibrated with Calibrite Beads (BD Biosciences) and samples were collected and analysed using CellQuest Pro software (BD). Forward and side scatter were used to exclude dead cells.

Results

Baseline milk parameters of ewes before and after mammary cannulation

For the first six days after arrival in the animal house prior to surgery, the average half-day milk production for a single gland from the ewes ranged from 51.5–180 millilitres. Milk was not collected from ewes in the evening following surgery. After surgery, the average half-day milk volume had an obvious drop in the first 3–4 days post-surgery in all sheep and then started to increase. At 4–7 days post-surgery, production levels of two ewes came back to similar quantities of that before surgery, but the milk production of the other two ewes remained at lower levels (Fig. 1). The quantities of milk collected were however adequate for milk quality tests.

We also monitored the amounts of fats, proteins, lactose and SNF in the milk before and after surgery. Daily fluctuations and individual differences on the percentages of these components in the milk were observed, but no obvious changes were noticed post-surgery. The corresponding range in percentages of fats, proteins, lactose and SNF in the milk were 6.54–10.02%, 5.86–8.4%, 4.36–5.1% and 12.03–13.38% before surgery and 7.04–10.13%, 5.92–8.57%, 4.53–5.17%, and 11.72–13.96% respectively post-surgery.

Establishment of a modified mammary lymphatic cannulation

Prior to live animal surgery, examination of cadaver anatomy showed that efferent mammary lymphatics coursed with the external pudendal vessels in parallel underneath the aponeurosis of the external abdominal oblique muscle entering the abdomen through the inguinal canal. The mammary lymphatics could have multiple branches and could be either cranial or caudal to the external pudendal vessels (Fig. 2). For live animal surgery a ventro-cranial to dorso-caudal orientated skin

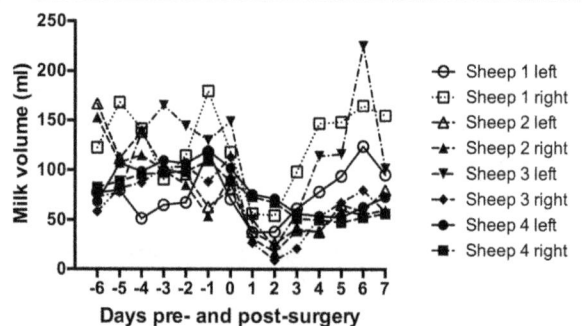

Fig. 1 Milk production (half day) before and after mammary lymphatic cannulation surgery. The capacity for milk production in four ewes following mammary lymphatic cannulation surgery was monitored. Data represents the mean of milk production of each gland calculated from the twice-daily collections

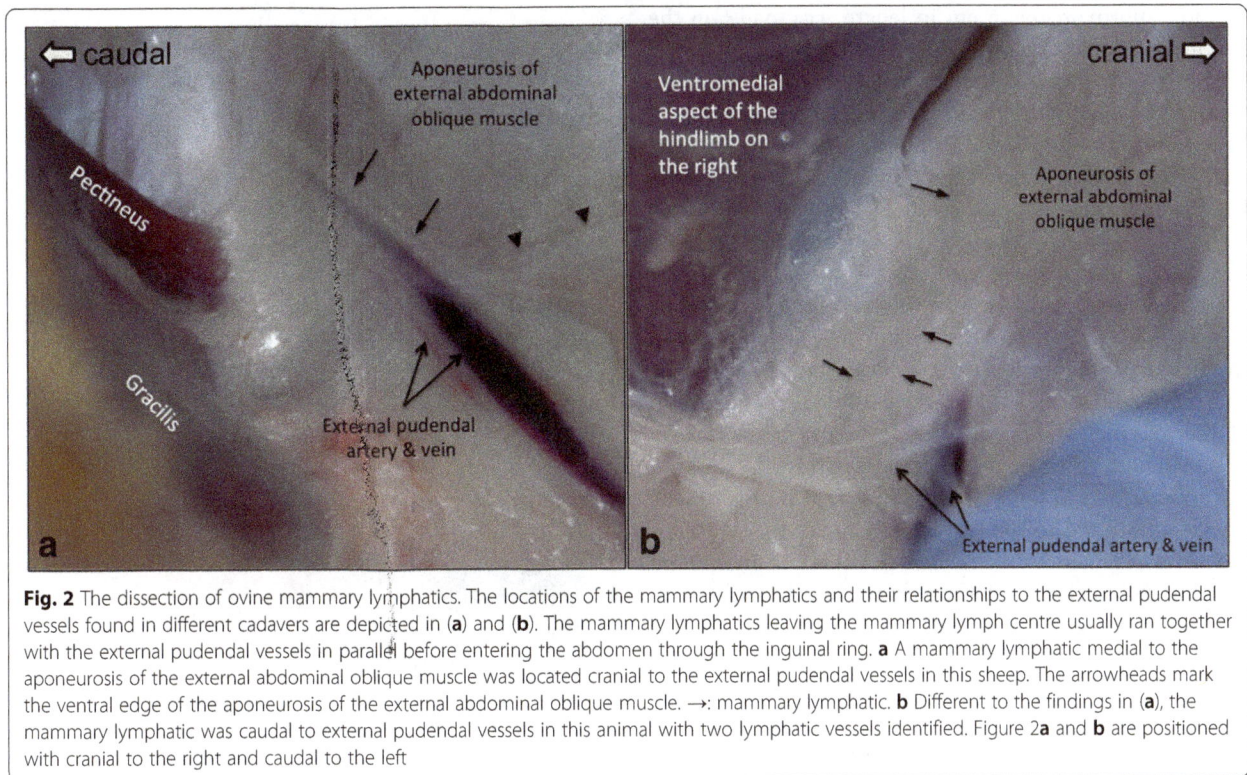

Fig. 2 The dissection of ovine mammary lymphatics. The locations of the mammary lymphatics and their relationships to the external pudendal vessels found in different cadavers are depicted in (**a**) and (**b**). The mammary lymphatics leaving the mammary lymph centre usually ran together with the external pudendal vessels in parallel before entering the abdomen through the inguinal ring. **a** A mammary lymphatic medial to the aponeurosis of the external abdominal oblique muscle was located cranial to the external pudendal vessels in this sheep. The arrowheads mark the ventral edge of the aponeurosis of the external abdominal oblique muscle. →: mammary lymphatic. **b** Different to the findings in (**a**), the mammary lymphatic was caudal to external pudendal vessels in this animal with two lymphatic vessels identified. Figure 2**a** and **b** are positioned with cranial to the right and caudal to the left

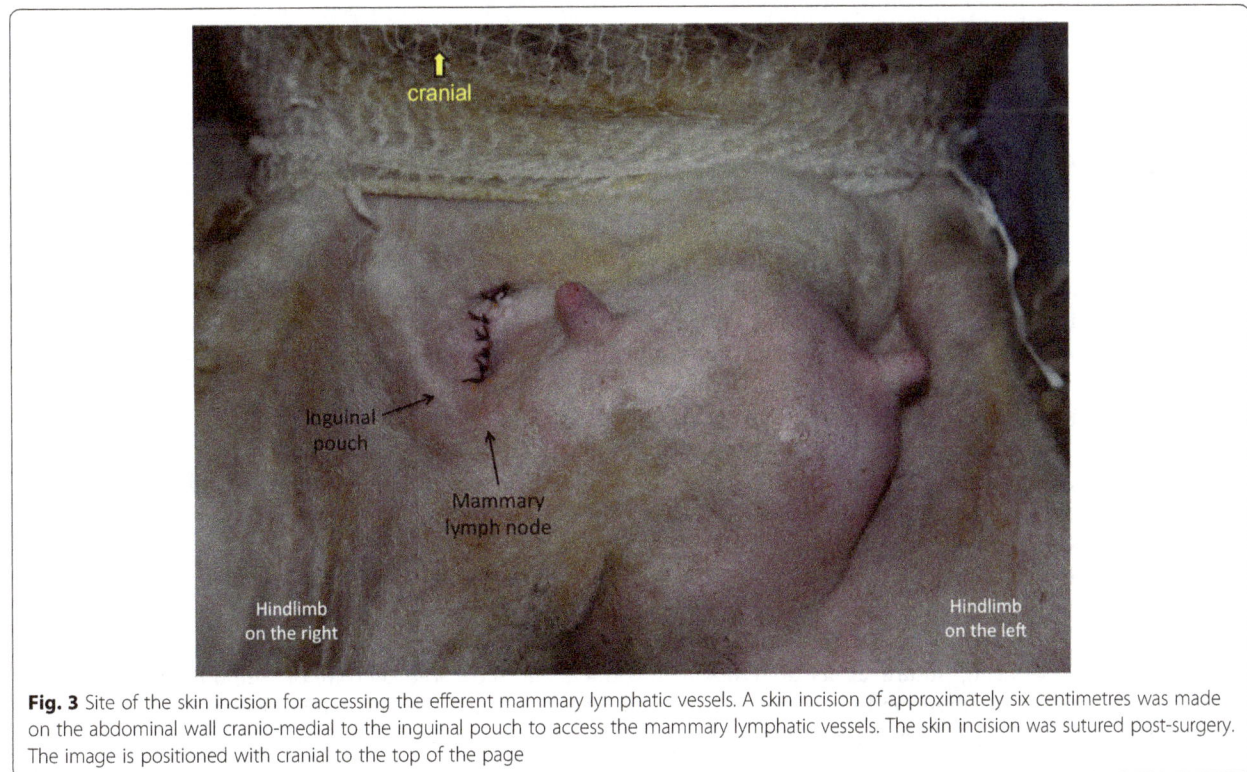

Fig. 3 Site of the skin incision for accessing the efferent mammary lymphatic vessels. A skin incision of approximately six centimetres was made on the abdominal wall cranio-medial to the inguinal pouch to access the mammary lymphatic vessels. The skin incision was sutured post-surgery. The image is positioned with cranial to the top of the page

incision approximately 6 cm in length was made on the abdominal wall, cranio-medial to the inguinal pouch (Fig. 3). After skin incision, blunt dissection was performed to penetrate the subcutaneous fat and the superficial fasciae to approach the aponeuroses of the internal and external abdominal oblique muscles. After identifying the aponeuroses, a self-retained retractor was placed in the skin opening to generate an operating area. The lymphatics under the aponeurosis of the external abdominal oblique muscles were then identified as shown in a cadaver in Fig. 2. After identifying all branches of the mammary lymphatics, the external pudendal vessels and the lymphatics were detached from the aponeurosis of the external abdominal muscles using blunt dissection. By cutting through the caudoventral insertion of the aponeurosis of the external abdominal muscle to the fasciae connected to the rectus abdominal muscle, more space was created to access the segments of mammary lymphatics adjacent to the mammary lymph nodes. The surgical field was expanded by placing two fingers through the skin incision to increase the space to approach the lymphatic vessels caudal to the external pudendal vein (Fig. 4). Similar to the cadaver image in Fig. 2, multiple branches of similarly sized mammary lymphatic vessels were found during surgery (Fig. 4), while in other surgeries one mammary lymphatic vessel was tightly attached to the external pudendal vein (Fig. 5). This lymphatic vessel was the largest lymphatic branch in this surgery with a smaller lymphatic vessel next to the vein. We found the strong attachment of the largest mammary lymphatic branch to the external pudendal vein in two surgeries. It was necessary to identify all lymphatic branches and ligate them, with the largest lymphatic selected for cannula insertion.

The bevelled end of a cannula was placed beneath the aponeurosis of the external abdominal oblique muscle through a small stab incision at the dorsal part of the aponeurosis before inserting it into the lymphatic. The basic technique for inserting a cannula into a lymphatic has been described in previous publications [14–16]. Briefly, the procedures of cannula insertion into a lymphatic as shown in Fig. 6 were: 1. To place two pre sutures around the lymphatic upstream from its ligation, 2. To make a cut in the lymphatic using a pair of corneal scissors and then to insert the cannula into the lymphatic, 3. To secure the cannula in the lymphatic with the preplaced sutures.

After cannula insertion into the mammary lymphatic, the free end of the cannula was threaded through the skin near the wing of the ilium. The cannula was secured using a purse-string suture at its skin opening ventral to the wing of the ilium and craniodorsal to the pre-femoral lymph node(s) after exteriorising its free end. An additional suture was made to secure the

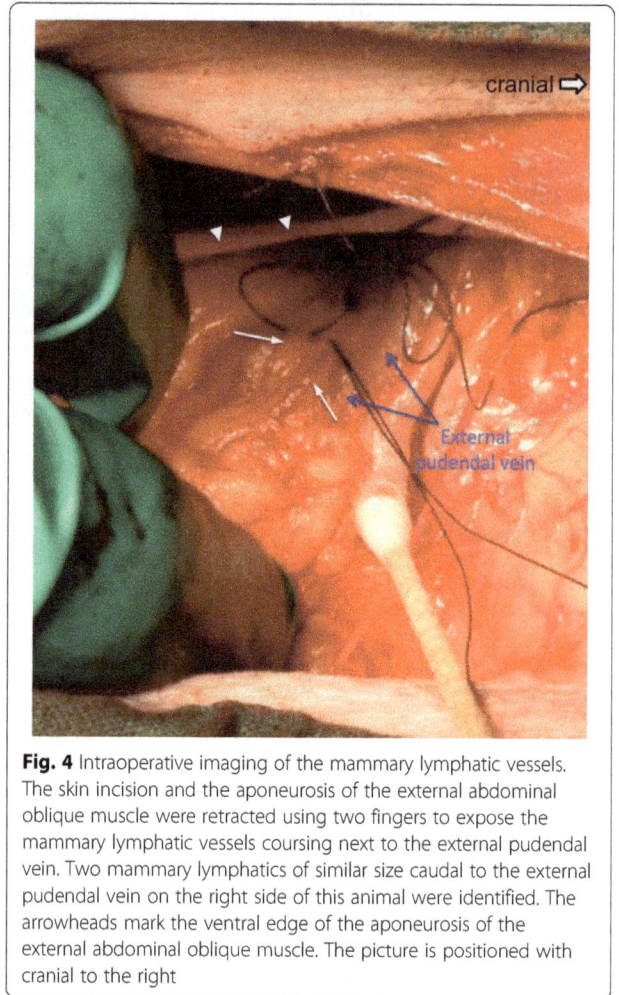

Fig. 4 Intraoperative imaging of the mammary lymphatic vessels. The skin incision and the aponeurosis of the external abdominal oblique muscle were retracted using two fingers to expose the mammary lymphatic vessels coursing next to the external pudendal vein. Two mammary lymphatics of similar size caudal to the external pudendal vein on the right side of this animal were identified. The arrowheads mark the ventral edge of the aponeurosis of the external abdominal oblique muscle. The picture is positioned with cranial to the right

cannula on the skin. Following surgery, the success of mammary lymphatic cannulation was confirmed in all ewes by the presence of blue dye in the cannula following an injection of 1 mL (0.5 g patent blue violet in 10 mL PBS) into the mammary tissues dorso-cranial to the teat and lymph dripping from the cannula. At the end of this study incisions on all ewes were healing well with no evidence of swelling and ewes were enrolled in a further study that required lymph collection.

Prefemoral lymph cannulation
A single prefemoral efferent lymphatic duct was successfully cannulated in three additional non-lactating ewes.

Cell profiles and lymph volume of mammary and prefemoral lymph
We collected lymph from all eight cannulated mammary ducts and measured the rate of lymph flow, the cell concentrations and the cell output per hour. The presence of a low number of red blood cells was noted in the lymph samples for up to 4–5 days post-surgery. Lymph

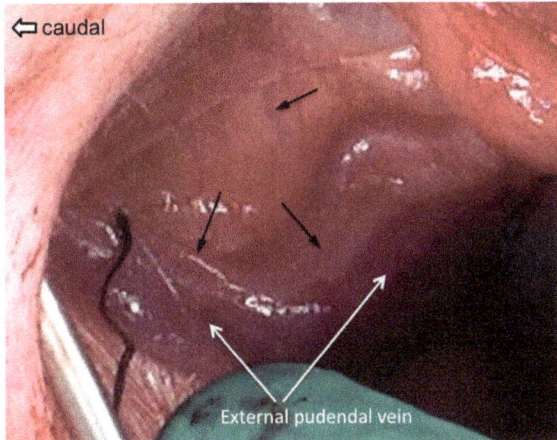

Fig. 5 Intraoperative image showing adjoined mammary efferent lymphatic and external pudendal vein. A major mammary lymphatic ligated by silk suture was found to be adjacent to the external pudendal vein. The fibro-connective tissues strongly attached the lymphatic to the wall of the vein. A smaller mammary lymphatic running next to the vein was also found. The picture is positioned with cranial to the right. →: mammary lymphatics

flowed well in four ducts with the flow rate ranging between 1.58 and 5.72 mL/h (daily volume, 37.8–205.7 mL) and the individual cell concentrations and outputs are shown in Fig. 7a and c respectively. Three ducts had slow, but continuous flow rates ranging from 0.09–1.07 mL/h. One cannulation completely blocked at day one post-surgery (Sheep 2 left side).

The flow of the three prefemoral efferent ducts ranged from 3–10 mL/h and the individual cells concentrations and outputs are shown in Fig. 7e and g respectively with the mean concentrations and outputs in Fig. 7f and h respectively. The drainage area of one prefemoral duct (Sheep 6) showed higher values than the other two cannulated sheep.

More than 97% of cells in the lymph samples were small lymphocytes, with the remainder large or blast-like lymphocytes. The phenotypes of small lymphocytes in mammary efferent lymph and those in the prefemoral efferent lymph of similar aged ewes are shown in Table 1. Over 80% of the lymphocytes in both the mammary and prefemoral efferent lymph were T cells (86% and 82% in mammary and prefemoral efferent lymph, respectively) with CD4+ T cells comprising 60% and 67% of the total T cells in mammary and prefemoral lymph respectively (Table 1).

Discussion

The surgical procedures developed in this study provide a successful and less invasive approach for cannulating the efferent lymphatic vessels draining the mammary lymph nodes and the mammary glands. From the reports by Lascelles and Morris (1961) and the images in

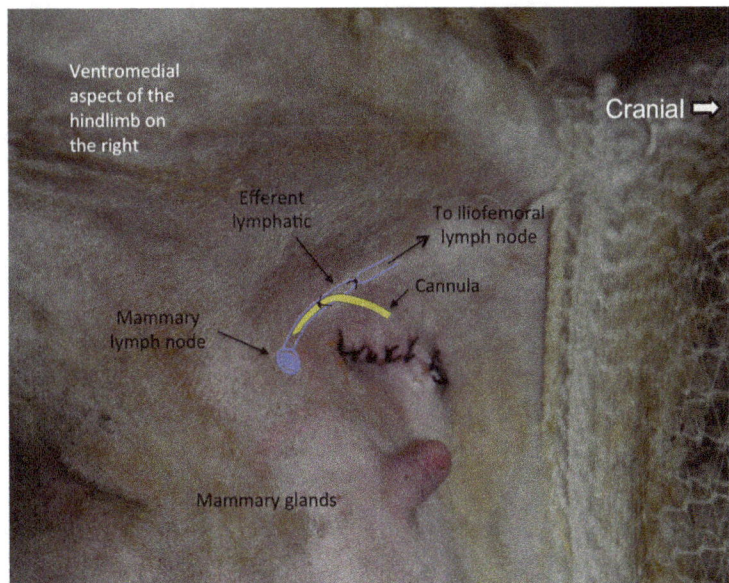

Fig. 6 A schematic view of efferent mammary lymphatic cannulation. The sketch indicates the cannula placement location in the efferent mammary lymphatic vessel relative to the locations of the lymphatic ligation and a mammary lymph node adjacent to the right mammary gland. A segment of the efferent mammary lymphatic vessel near to the mammary lymph node was ligated using suture. The external pudendal vessels and lymphatics can show curved segments in the deep inguinal region, not demonstrated on this diagram. The efferent lymphatic vessels of the mammary lymph nodes (on both sides) enter the iliofemoral lymph node(s). The cannula was inserted into the lymphatic upstream from the ligation and secured with ligations (only one ligation is shown in this schematic picture). The sketch is superimposed over a photograph of a ewe post-surgery. The image is positioned with cranial to the right

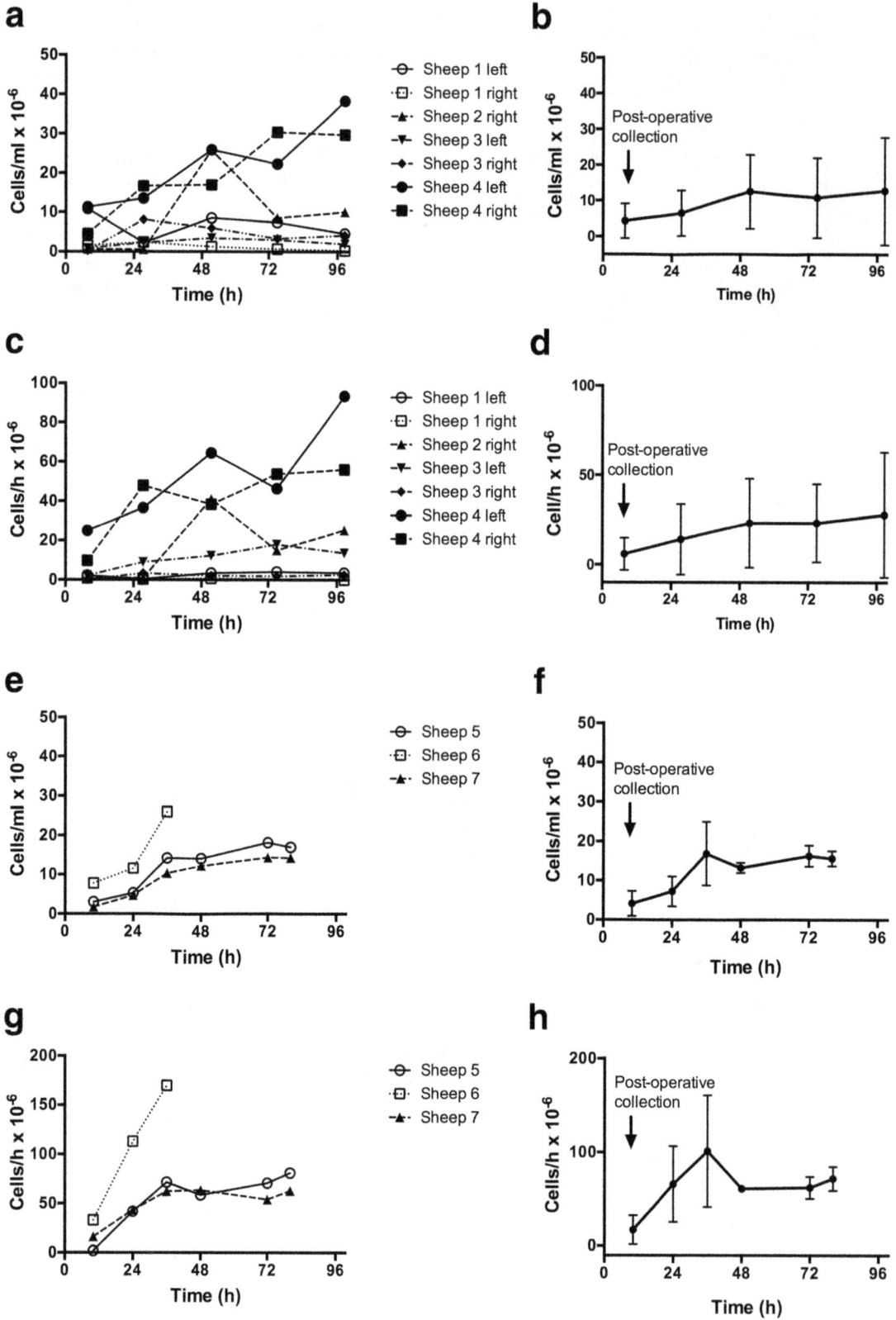

Fig. 7 (See legend on next page.)

Fig. 7 Lymphocyte concentration and output in efferent lymph from mammary and prefemoral lymph nodes. Comparison of cell concentrations and cell outputs in efferent lymph from mammary lymph nodes and prefemoral lymph nodes. **a**. Individual mammary cell concentrations. **b**. Mean mammary cell concentration ± standard deviation (SD). **c**. Individual mammary cell outputs. **d**. Mean mammary cell output ± SD. **e**. Individual prefemoral cell concentrations. **f**. Mean prefemoral cell concentration ± SD. **g**. Individual prefemoral cell outputs. **h**. Mean prefemoral cell output ± SD

Fig. 2, it is clear that locating and accessing the mammary lymphatic vessel(s) directly from the dorsal portion of the aponeurosis of the external abdominal oblique muscle is easier since there are generally less adipose tissues in that region [2]. However, it is technically demanding and time-consuming to suture the aponeurosis and the sutured aponeurosis is likely to induce tangles of the cannula, in particular at the cannula's insertion end in the lymphatic. The aponeurosis is a thin smooth sheet of fibro-connective tissue with specific orientations and arrangements of the tissue fibres. Damages of the aponeurosis can cause uneven healing of the connective tissues, resulting in an irregular shape of the aponeurosis. The strong pressure from the abdominal cavity on the damaged aponeurosis could alter its structure and shape even if it is properly sutured. Compared to the procedures by Lascelles and Morris (1961), to identify and access the segments of the efferent mammary lymphatics using our approach is more time-consuming since the efferent lymphatics course in the rich adipose tissues in the inguinal regions. However, to follow the segments of mammary efferent lymphatic vessels underneath the aponeurosis upstream, lymphatics in the inguinal regions can be located. To secure the success of the surgery and study, it is critical to confirm that all branches of the efferent mammary lymphatic vessels are identified and ligated. To find all lymphatic vessels in the adipose tissues in the inguinal region is the most challenging and time-consuming part of the surgery described in our study. However, our surgical procedures caused minimal damage to the aponeurosis of the external abdominal oblique muscle and the local tissues and are technically easier to perform compared to previous techniques. We suggest that the method described in

this manuscript make the mammary lymphatic cannulation model more successful, combined with good animal recovery post-surgery.

Similar to that reported previously by Lascelles and Morris [1], we found variations in lymph flow volumes among different cannulated lymphatics. However, the total amount of lymph collected in our sheep each day (37.8–205.7 mL/day) are lower than that (450–900 mL/day) reported by Lascelles and Morris but similar to Watson and Davies [18]. Lascelles and Morris found that the stage of lactation, amount of milk production as well as the ewe's activities correlated to lymph production. Our ewes were kept in metabolic cages with less exercise, were late in lactation and did not have lambs suckling them. This may explain why we harvested comparatively less lymph from the ewes in our study. A single cannula on one ewe blocked soon post-surgery in this study, however from our previous experience in other lymphatic cannulation surgeries this is a relatively common risk that lymph clots can completely block a cannula.

The cell concentrations in mammary and prefemoral efferent lymph were similar at around 10×10^6/mL. The outputs of the prefemoral ducts (around $50–100 \times 10^6$/ mL) were higher than the mammary ducts (around $30–40 \times 10^6$/mL), though there was wide variation in both. The cell output from a non-stimulated peripheral lymph node reflects the size of the node and hence blood flow, with the majority of lymphocytes derived from blood rather than afferent lymphatics [29, 30]. Both the prefemoral node, which weighs 2–3 g in an adult sheep, and the popliteal node, which weighs 1–2 g, have efferent outputs of around 50×10^6/h [31]. The larger 3-6 g prescapular (superficial cervical) node has efferent outputs around 150×10^6/mL [32]. It is difficult to relate the mammary node output directly to that of a peripheral node when the relative contribution of afferent lymph from the mammary gland is not also measured, especially as blood flow to the gland itself increases markedly during lactation [33].

The lymphocyte subsets in mammary efferent lymph were similar to prefemoral efferent lymph, indicating no difference between the mammary/mucosal and skin-draining efferent circulations in healthy sheep and typical of other non-stimulated adult efferent lymphatic ducts such as those draining the popliteal and prescapular lymph nodes [31, 32].

Table 1 Phenotypes of lymphocytes in mammary and prefemoral efferent lymph

Subset	Mammary efferent lymph	Prefemoral efferent lymph
	Mean % ± SD	Mean % ± SD
CD4	51 ± 14	55 ± 7
CD8	15 ± 2	16 ± 6
γδ TCR	20 ± 9	11 ± 6

Cells were stained with monoclonal antibodies and analysed by flow cytometry. Data represents the phenotypes of lymphocytes from seven mammary efferent lymphatic ducts (lymph not available from one duct that blocked) and three prefemoral efferent lymphatic ducts. Cell subsets were compared between compartments using unpaired Mann–Whitney tests and no significant difference was found at $p \leq 0.05$

The ewes' milk production following surgery is a key feature in the establishment of this lymphatic cannulation model. To maximise the applications of the mammary lymphatic models, it is necessary to confirm that samples from the lactiferous passages and alveoli and the lymph draining the inter-alveolar tissues can be harvested at the same time for comparative data analysis. The findings in our study verified that it is practical to collect and analyse the components in the milk following surgery. We suggest allowing the sheep to recover for at least 4–5 days post-surgery for milk levels and red blood cells in lymph to return to normal before receiving any further experimental treatments. Red blood cells may appear in efferent lymph in very small numbers due to the damage of capillaries that exist in the wall of the lymphatic vessels [11]. This may explain why we observed the presence of RBCs in lymph. In general, there should be no red blood cells in pure ovine efferent lymph. In addition to simply detecting the changes in milk samples in traditional mammary disease studies, the ability to monitor the responses in the lymph draining the gland can bring additional understanding of disease. In future work investigating the immune-biology of mammary disease and normal function, cannulation of the efferent duct could be combined with cannulation of the afferent mammary lymphatics to provide even more information from this model. Dendritic cells are present in ovine milk [34] and circulate to the mammary lymph node in afferent lymph, so by adding mammary afferent lymph [18, 35] to our model, we could obtain immune cells from three different compartments including antigen-presenting cells from the lactiferous passages (alveoli and canals), dendritic cells from the mammary inter-alveolar connective tissues and effector cells from the efferent mammary lymph of the same sheep. This further modification of the mammary lymphatic cannulation model would provide a powerful tool to examine the responses to disease.

Conclusions

In conclusion, this improved cannulation technique enabled lymphocyte subset monitoring from ewes in late lactation for at least eight days following surgery and will be useful as a model to further study mammary disease and mucosal immunity. This model may also have significant application for monitoring vaccination or antibiotic performance at the level of the mammary gland.

Abbreviations
SNF: solids-not-fat

Acknowledgements
The authors thank Mr Bob Geyer and animal house staff for their care of experimental animals. The authors also thank Dr Jean-Pierre Scheerlinck for providing some surgical instruments and reagents for this study.

Funding
This study was funded by an Early Career Researcher grant from the University of Melbourne and by Australian sheep producers and the Australian Government through Meat & Livestock Australia (MLA) Limited. Neither of these organisations were involved in the design, interpretation or reporting of study results.

Authors' contributions
All authors contributed to writing and reviewing the manuscript for this study. HHY performed all mammary surgeries and SB performed all anaesthetic induction and monitoring. WK performed surgery for prefemoral cannulation. EW performed FACS analysis on both prefemoral and mammary cannulations. SB and EH were responsible for animal maintenance with other tasks shared amongst the authors. All authors read and approved the final manuscript.

Competing interests
The authors declare that they have no competing interests.

Consent for publication
Not applicable.

Ethics approval
Cadavers of three sheep were used to develop the initial surgical approach for this study. These cadavers were obtained from sheep from other AEC approved research trials. Use of ewes for mammary cannulation in this experiment, was covered under University of Melbourne Faculty of Veterinary and Agricultural Science Animal Ethics Committee number 1312857.2 with prefemoral lymph cannulation under University of Melbourne Faculty of Veterinary Science AEC approval number 251683. The studies were carried out in accordance with the approved guidelines.

References
1. Lascelles AK, Morris B. The flow and composition of lymph from the mammary gland in merino sheep. Q J Exp Physiol Cogn Med Sci. 1961;46: 206–15.
2. Lascelles AK, Morris B. Surgical techniques for the collection of lymph from unanaesthetized sheep. Q J Exp Physiol Cogn Med Sci. 1961;46:199–205.
3. McKeever DJ, Reid HW. The response of the supramammary lymph node of the sheep to secondary infection with orf virus. Vet Microbiol. 1987; 14(1):3–13.
4. Bollman JL, Cain JC, Grindlay JH. Techniques for the collection of lymph from the liver, small intestine, or thoracic duct of the rat. J Lab Clin Med. 1948;33(10):1349–52.
5. Bierman HR, Byron Jr RL, Kelly KH, Gilfillan RS, White LP, Freeman NE, Petrakis NL. The characteristics of thoracic duct lymph in man. J Clin Investig. 1953;32(7):637–49.
6. Linzell JL. The flow and composition of mammary gland lymph. J Physiol. 1960;153:510–21.
7. Hartmann PE, Lascelles AK. The flow and lipoid composition of thoracic duct lymph in the grazing cow. J Physiol. 1966;184(1):193–202.
8. Ionac M. One technique, two approaches, and results: thoracic duct cannulation in small laboratory animals. Microsurgery. 2003;23(3):239–45.
9. Uhley H, Leeds SE, Sampson JJ, Friedman M. A technic for collection of right duct lymph flow in unanesthetized dogs. Proc Soc Exp Biol Med. 1963; 112:684–5.
10. Hall JG. A method for collecting lymph from the prefemoral lymph node of unanaesthetised sheep. Q J Exp Physiol Cogn Med Sci. 1967;52(2):200–5.
11. Hall JG, Morris B. The output of cells in lymph from the popliteal node of sheep. Q J Exp Physiol Cogn Med Sci. 1962;47:360–9.
12. Glover DJ, Hall JG. A method for the collection of lymph from the prescapular lymph node of unanaethetized sheep. Lab Anim. 1976;10(10): 403–8.
13. Hein WR, Barber T, Cole SA, Morrison L, Pernthaner A. Long-term collection and characterization of afferent lymph from the ovine small intestine. J Immunol Methods. 2004;293(1–2):153–68.

14. Yen HH, Scheerlinck JP, Gekas S, Sutton P. A sheep cannulation model for evaluation of nasal vaccine delivery. Methods. 2006;38(2):117–23.

15. Schwartz-Cornil I, Epardaud M, Bonneau M. Cervical duct cannulation in sheep for collection of afferent lymph dendritic cells from head tissues. Nat Protoc. 2006;1(2):874–9.

16. Yen HH, Wee JL, Snibson KJ, Scheerlinck JP. Thoracic duct cannulation without thoracotomy in sheep: A method for accessing efferent lymph from the lung. Vet Immunol Immunopathol. 2009;129(1–2):76–81.

17. Hein WR, Griebel PJ. A road less travelled: large animal models in immunological research. Nat Rev Immunol. 2003;3(1):79–84.

18. Watson DL, Davies HI. Immunophysiological activity of supramammary lymph nodes of the ewe during the very early phase of staphylococcal mastitis. Res Vet Sci. 1985;39(1):52–8.

19. Contreras GA, Rodriguez JM. Mastitis: comparative etiology and epidemiology. J Mammary Gland Biol Neoplasia. 2011;16(4):339–56.

20. Hogeveen H, Huijps K, Lam TJ. Economic aspects of mastitis: new developments. N Z Vet J. 2011;59(1):16–23.

21. Bergonier D, de Cremoux R, Rupp R, Lagriffoul G, Berthelot X. Mastitis of dairy small ruminants. Vet Res. 2003;34(5):689–716.

22. Wang Z, Gerstein M, Snyder M. RNA-Seq: a revolutionary tool for transcriptomics. Nat Rev Genet. 2009;10(1):57–63.

23. Bonnefont CM, Toufeer M, Caubet C, Foulon E, Tasca C, Aurel MR, Bergonier D, Boullier S, Robert-Granie C, Foucras G, et al. Transcriptomic analysis of milk somatic cells in mastitis resistant and susceptible sheep upon challenge with Staphylococcus epidermidis and Staphylococcus aureus. BMC Genomics. 2011;12:208.

24. Vlassov AV, Magdaleno S, Setterquist R, Conrad R. Exosomes: current knowledge of their composition, biological functions, and diagnostic and therapeutic potentials. Biochim Biophys Acta. 2012;1820(7):940–8.

25. Robbins PD, Morelli AE. Regulation of immune responses by extracellular vesicles. Nat Rev Immunol. 2014;14(3):195–208.

26. Maddox JF, Mackay CR, Brandon MR. Surface antigens, SBU-T4 and SBU-T8, of sheep T lymphocyte subsets defined by monoclonal antibodies. Immunology. 1985;55(4):739–48.

27. Mackay CR, Beya MF, Matzinger P. Gamma/delta T cells express a unique surface molecule appearing late during thymic development. Eur J Immunol. 1989;19(8):1477–83.

28. Maddox JF, Mackay CR, Brandon MR. Ontogeny of ovine lymphocytes. II. An immunohistological study on the development of T lymphocytes in the sheep fetal spleen. Immunology. 1987;62(1):107–12.

29. Hall JG, Morris B. The origin of cells in the efferent lymph from a single lymph node. J Exp Med. 1965;121:901–10.

30. Hay JB, Hobbs BB. The flow of blood to lymph nodes and its relation to lymphocyte traffic and the immune response. J Exp Med. 1977;145:31–44.

31. Kimpton WG, Washington EA, Cahill RNP. Non-random migration of CD4+, CD8+ and gd+T19+ lymphocyte subsets following in vivo stimulation with antigen. Cell Immunol. 1990;130:236–43.

32. Washington EA, Kimpton WG, Cahill RNP. CD4+ lymphocytes are extracted from blood by peripheral lymph nodes at different rates from other T cell subsets and B cells. Eur J Immunol. 1988;18:2093–6.

33. Thompson GE. The Distribution of Blood-Flow in the Udder of the Sheep and Changes Brought About by Cold-Exposure and Lactation. J Physiol London. 1980;302:379–86.

34. Tatarczuch L, Bischof RJ, Philip CJ, Lee CS. Phagocytic capacity of leucocytes in sheep mammary secretions following weaning. J Anat. 2002;201(5):351–61.

35. Heath TJ, Kerlin RL. Lymph drainage from the mammary gland in sheep. J Anat. 1986;144:61–70.

15

Estimation of French cattle herd immunity against bluetongue serotype 8 at the time of its re-emergence in 2015

L. Bournez[1][*] ⓘ, L. Cavalerie[2], C. Sailleau[3], E. Bréard[3], G. Zanella[3], R. Servan de Almeida[4], A. Pedarrieu[4], E. Garin[5], I. Tourette[6], F. Dion[7], P. Hendrikx[1] and D. Calavas[8]

Abstract

Background: From 2006 to 2010, France experienced two bluetongue epidemics caused by serotype 1 (BTV-1) and 8 (BTV-8) which were controlled by mass vaccination campaigns. After five years without any detected cases, a sick ram was confirmed in August 2015 to be infected by a BTV-8 strain almost identical to that circulating during the previous outbreak. By then, part of the French cattle population was expected to be still protected, since bluetongue antibodies are known to last for many years after natural infection or vaccination. The objective of this study was to estimate the proportion of cattle in France still immune to BTV-8 at the time of its re-emergence in 2015.

Results: We used BTV group-specific cELISA results from 8525 cattle born before the vaccination ban in 2013 and 15,799 cattle born after the ban. Samples were collected from January to April 2016 to estimate seroprevalence per birth cohort. The overall seroprevalence in cattle at national and local levels was extrapolated from seroprevalence results per birth cohort and their respective proportion at each level. To indirectly assess pre-immune status of birth cohorts, we computed prevalence per birth cohort on infected farms in autumn 2015 using 1377 RT-PCR results. These revealed limited BTV circulation in 2015. Seroprevalence per birth cohort was likely to be connected to past exposure to natural infection and/or vaccination with higher seroprevalence levels in older animals. A seroprevalence of 95% was observed for animals born before 2008, of which > 90% were exposed to two compulsory vaccination campaigns in 2008-2010. None of the animals born before 2008 were found to be infected, unlike 19% of the young cattle which had never been vaccinated. This suggests that most ELISA-positive animals were pre-immune to BTV-8. We estimated that 18% (from 12% to 32% per *département*) of the French cattle population was probably pre-immune in 2015.

Conclusions: These results strongly suggest a persistence of antibodies for at least 5-6 years after natural infection or vaccination. The herd immunity of the French cattle population probably limited BTV circulation up to 2015, by which time more than 80% of cattle were naive.

Keywords: Bluetongue, Serotype 8, Seroprevalence, Herd immunity, Re-emergence, France, Cattle

Background

Bluetongue (BT) is a vector-borne viral disease of wild and domestic ruminants that can cause major losses in ruminant production especially in sheep. The BT virus (BTV) is transmitted by several species of biting midges of the genus *Culicoides*. To date, 27 BTV-serotypes have been identified [1]. In 2006, the emergence of BTV serotype 8 (BTV-8) in northern Europe initiated a widespread epidemic from 2006 to 2009 in central and western Europe [2]. In France, BTV-8 was introduced by the end of 2006 from Belgium and spread over most of the country, infecting at least 42,850 farms between 2006 and 2009 [3]. During the same period, France experienced another BT epidemic with BTV serotype 1 (BTV-1), first detected in November 2007 close to the Spanish border. BTV-1 mainly circulated in south-western France, although a few infected (i.e. RT-PCR positive) animals were later discovered throughout France,

---
* Correspondence: laure.bournez@anses.fr
[1]ANSES (French Agency for Food, Environmental and Occupational Health & Safety), Unité de coordination et d'appui à la surveillance, Direction des laboratoires, Maisons Alfort, France

without clear evidence of any local BTV circulation. Two years of mandatory vaccination followed by two voluntary campaigns were launched against both serotypes in throughout the French mainland from November 2008 to April 2010, and from November 2010 to April 2012 respectively. The cattle vaccination coverage against both serotypes was estimated to be 80% in 2008-2009, 90% in 2009-2010 and 25-30% in 2010-2011 [3, 4]. The vaccination coverage is unknown for 2011-2012 but assumed to be lower than in 2010-2011 due to both lack of interest and reluctance by farmers. The number of outbreaks drastically decreased in 2009 probably due to the high proportion of naturally infected (and thus immunised) and vaccinated animals. The last BTV-8 and BTV-1 outbreaks were detected respectively in December 2009 and June 2010, and mainland France officially recovered a BT-free status in December 2012. Vaccination was banned in mainland France on 31th May 2013.

On 11th September 2015, French authorities notified to the OIE a BTV-8 outbreak in the Allier *département* in central France. The virus was detected in a 5-year-old ram which showed clinical signs evocative of BT. The virus's genetic sequence was 99.9% similar to the virus circulating in 2006-2009 [5]. BTV-8 eradication through mandatory vaccination was not carried out in France in 2015. The limited amount of vaccines available in 2015-2016 was mainly used for animals leaving the restriction zone (i.e. within 150 km of outbreaks). From August 2015 to June 2016, 284 outbreaks were detected through pre-export tests (74%), active surveillance (21%) and clinical surveillance (5%), and were mainly located in the centre of France (the Allier and Puy-de-Dôme *départements*).

One of the main hypotheses for the resurgence of BTV-8 in France is that the virus had been circulating at low levels since 2009 and had remained undetected by the surveillance system [6–8]. Like other viral diseases, the intensity of BTV circulation may have increased along with the increasing proportion of naive animals due to ruminant population turn-over and/or loss of immunity. One can expect resurgence to occur when the proportion of naive animals towards BTV-8 is sufficiently high to lead to more intensive viral circulation and spread, but the value of this threshold is unknown. Cattle are much more attractive to *Culicoïdes* spp. than are sheep [9, 10] and hence more frequently infected by BT viruses [11]. They are considered as the primary reservoir and amplifying host for the virus [12, 13]. In France, they are much more numerous than sheep, with 19.2 and 7 million of head respectively (source: the French Livestock Institute Idele and GEB). By July 2015, 23% of the French cattle population was composed of animals born before 2010, which had therefore been present during the mandatory vaccination campaigns of 2008-2010 (source: National Identification Database

BDNI). Some of these animals might still be immune to BTV-8, but their proportion was unknown. Such data is not available for sheep.

The duration of BTV-8 immunity acquired after natural infection or vaccination and how it decreases over time depends on several factors. The protective immune status of animals with respect to BTV is generally assessed via their humoral immune response, even though cellular immune response might also be a determinant [14–17]. Although neutralising type-specific antibodies are generally preferred for estimating BT protective immune status, group-specific antibodies detected by ELISA can also be used to infer the immune status of animals against a serotype if this one serotype has been circulating or was targeted for vaccination in the area. Seroneutralising and ELISA results are relatively well correlated, although the proportion of ELISA positive results is generally higher [18–21]. Neutralising and group-specific antibodies against BTV-8 have been detected in cattle four years after natural infection and vaccination [18, 21–23]. However, different studies have observed large variation in the proportion of seropositive animals one year after vaccination ranging from 60% to 97% when evaluated by ELISA [18–20, 24–28]. Such a variation could be explained by a difference in the vaccination protocol (e.g. with or without a booster vaccination), the type of vaccine itself or the mean age of animals at vaccination [27]. In France, several inactivated commercial vaccines against BTV-8 and BTV-1 were successively used from 2008 to 2012 (Table 1). According to their birth date, animals received from one to several doses against serotype 1 or 8 (Table 2). All these factors may have influenced the proportion of cattle still immune in France in 2015, and made it difficult to infer this proportion without further investigation.

The objective of this paper was to estimate the proportion of cattle still immune to BTV-8 in France in 2015 at national and local levels. This was essential to better understand BT epidemiology and to empirically estimate the threshold levels of the proportion of naive animals in the cattle population required for BTV to re-emerge and spread at a detectable level. This was also important to better tailor surveillance measures and to provide advice on vaccination to French farmers. Considering the expected differences in exposure to viral infection over time and the successive vaccination protocols (Table 2), we estimated the proportion of seropositive cattle per birth cohort (see definition below) in winter 2015-2016 using a BTV group-specific competitive ELISA of 15,799 cattle and extrapolated it to the cattle population as a whole. In order to check the hypothesis that infection prevalence in autumn 2015 was lower in birth cohorts which were the more exposed to BTV-8 infection or vaccination, we estimated the proportion of cattle

Table 1 Vaccine products used from 2008 to 2013 in France against BTV-8 and BTV-1 in cattle

Year	BTV-8 vaccine	BTV-1 vaccine
2008-2009 (mandatory vaccination campaign)	Bovilis BTV-8® (Intervet, The Netherlands)	Zulvac® 1 Bovis (Fort Dodge Animal health, The Netherlands),Bluevac® 1 (CZ Veterinaria S.A., Spain)
2009-2010 (mandatory vaccination campaign)	BTVPUR® AlSap 8 (Merial, France)	Zulvac® 1 Bovis (Fort Dodge Animal health, The Netherlands), Bluevac® 1 (CZ Veterinaria S.A., Spain)
2010-2013 (voluntary vaccination campaign)	BTVPUR® AlSap 8, Bovilis BTV-8®, Zulvac® 8 Bovis (Fort Dodge Animal health, The Netherlands)	BTVPUR® AlSap 1, Bluevac® 1, Zulvac® 1 Bovis

positive to BTV-8 by RT-PCR per birth cohort on infected farms in autumn 2015. According to this hypothesis, we expected to find a lower number of infected animals in older birth cohorts.

Methods

Estimates of seroprevalence in cattle per birth cohort and the *département*'s BTV-8 status in 2015

Definition of birth cohorts and status of BTV-8 infected/non-infected areas

Given the differences in cattle exposure to BTV-8 and the different vaccination protocols implemented from 2007 to 2016, seroprevalence was expected to vary between animal birth dates (Table 2). We defined annual birth cohorts from 1st July to 30th June of the following year in order to take into account the period of BTV circulation (mainly from July to November, with a detection window by RT-PCR up to six months post-infection in cattle, i.e. up to May-June) and of vaccination campaigns (conducted during annual surveillance sampling for brucellosis and infectious bovine rhinotracheitis from October to April).

In order to estimate seroprevalence levels before and after the 2015 resurgence, seroprevalence per birth cohort was estimated separately for areas "infected" and "non-infected" by BTV-8 in 2015-2016. To define infected and non-infected areas, we chose the "*département*", the French administrative unit that is also the geographical area for BT management. A *département* was considered "infected" in 2015 when at least one

animal was found RT-PCR-positive by the surveillance system between 1st August 2015 and 30th June 2016 [29]. It is worth noting that a very high number of cattle from France (> 140,000) were tested by RT-PCR mainly for pre-export tests during this period. Blood samples were tested by a BTV group-specific RT-PCR and positive samples were then screened with a BTV-8 type-specific RT-PCR. These analyses were carried out by local veterinary laboratories. Results with a Ct value between 35 and 39 were confirmed by the National Reference Laboratory (ANSES, Maisons-Alfort). Given the long persistence of BT antibodies post-infection detected by ELISA, and the detection of viral genome up to six months post-infection by RT-PCR [30, 31], we used RT-PCR-positive results as a proof of an infection occurring in 2015-2016. The date of 30th June was defined as the end of the estimated period of the detection of the virus circulation in 2015-2016 considering several criteria: no vector activity between January to April-May [32], the possible detection of viral genome by RT-PCR up to 6 months post-infection [30, 31], an increase in the number of outbreaks associated with a decrease in Ct values of RT-PCR-positive results in July compared to previous months and first virus isolation in August 2016 [29].

Départements located in the main BTV-1 infected area in 2007-2008 (South-West France) were excluded to remove potential effects of past BTV-1 circulation on the antibody response of cattle to BTV-8, although a cross-immunity between these two serotypes is not expected [33]. In other *départements* we considered that BTV-1

Table 2 Exposure to BTV-8 and vaccination campaign characteristics per cattle birth cohort in France

Birth cohort	BTV-8 exposure	Vaccination campaign (estimated vaccination rate (%))			
	2007-2009	Mandatory 2008-2009 (~ 80%)	Mandatory 2009-2010 (~ 90%)	Voluntary 2010-2011 (~ 20-30%)	Voluntary 2010-2011 (unknown)
< July 2008	Yes	Yes	Yes	Yes	Yes
July 2008 - June 2009	Yes	Calves*	Yes	Yes	Yes
July 2009 - June 2010	No	No	Calves*	Yes	Yes
July 2010 - June 2011	No	No	No	Calves*	Yes
July 2011 - June 2012	No	No	No	No	Calves*
July 2012 - June 2013	No	No	No	No	No
> June 2013	No	No	No	No	No

*Cattle up to 12 months-old

circulation in 2007-2010 was too small to have a big influence on the response of BTV-8 antibodies. *Départements* with less than 20,000 cattle were also excluded (Fig. 1).

Sampling design

For cattle born after January 2013, we used data collected from the national serological survey organised by the French Ministry of Agriculture and carried out from December 2015 to March 2016 (see technical instruction DGAL/SDSPA/2016-35). Its objectives were to detect the presence of BTV in uninfected *départements* and to demonstrate *seasonally-free zones* of BT within the restriction zone. Animals born after January 2013 were targeted as they were considered to be naive and not vaccinated against BT; calves of less than 12 months old were excluded due to the potential persistence of colostral antibodies [34]. Between 167 and 680 cattle from 11 to 47 farms were sampled per *département,* according to the sampling scheme (Additional file 1).

For cattle born before January 2013, sampling was organised at the same time as the national survey of young animals in the winter of 2015-2016. The sampling scheme was adapted to practical and financial constraints inherent to surveillance programmes, especially in the context of re-emergence when animal health officers in the field are regularly asked to conduct surveys/investigations. More specifically, it was designed to limit the amount of work required of them and avoid delaying BT investigation in young animals. Therefore, the sampling scheme was not stratified per birth cohort. Ten farms were randomly selected per *département* and 15 animals not vaccinated in 2015 were randomly sampled per farm in order to reach 150 animals per *département.* This number was considered sufficient to have good estimates of the seroprevalence per birth cohort and per BTV-8 *département* status in 2015.

Serological analyses

BTV antibodies were detected in cattle serum by certified local veterinary laboratories using one of the authorised competitive ELISAs. Of the 37 laboratories involved in the surveillance system, 33 used ID Screen Bluetongue Competition (ID VET, France) and four used IDEXX Bluetongue Competition Ab (IDEXX, United Kingdom) (4% of the analyses). Analyses were performed and interpreted according to the manufacturers' instructions. For the ID-VET kit, samples with a competition percentage ≤ 35%,]35 -45%], > 45% were considered positive, doubtful or negative in that order. For the IDEXX kit, samples with a competition percentage ≤ 70%,]70 - 80%], and > 80% were considered positive, doubtful or negative in that order.

Seroprevalence estimates per birth cohort at national and regional levels

Animals that had moved from one *département* to another after July 2015 — considered the potential beginning of BTV circulation in 2015-2016 — were excluded from the analysis.

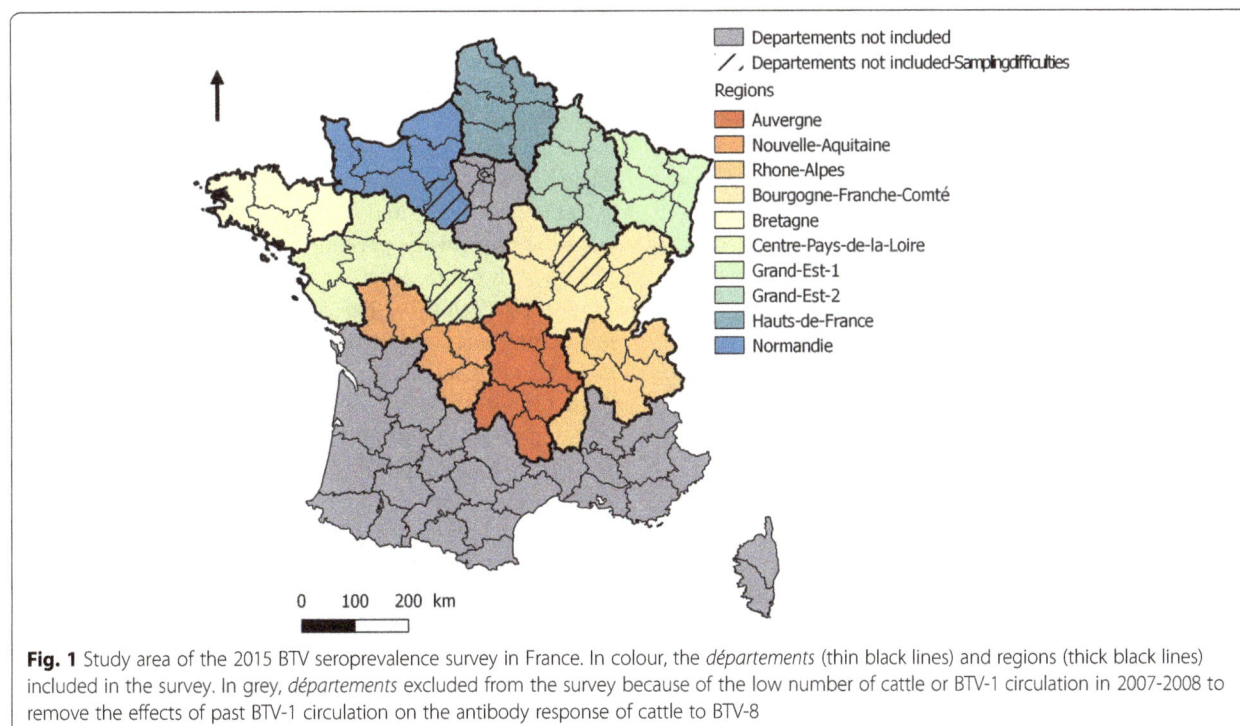

Fig. 1 Study area of the 2015 BTV seroprevalence survey in France. In colour, the *départements* (thin black lines) and regions (thick black lines) included in the survey. In grey, *départements* excluded from the survey because of the low number of cattle or BTV-1 circulation in 2007-2008 to remove the effects of past BTV-1 circulation on the antibody response of cattle to BTV-8

We calculated seroprevalence per birth cohort and *département* status (infected vs. non-infected in 2015-2016 as defined above) at national level. Confidence intervals of 95% (95% CI) were calculated using exact binomial law. Seroprevalences values of birth cohorts and *départment* status were compared with the Chi^2 test.

In order to assess if seroprevalence per birth cohort was spatially heterogeneous in 2015, we calculated seroprevalence and its 95% CI per birth cohort and region (an administrative unit including several *départements*; see Fig. 1). This analysis was carried out per region in order to reach a sufficient number of animals per birth cohort to enable birth period seroprevalence estimates within +/− 10%. The median and the range proportion of seroprevalence per birth cohort and per region, and the median range of the seroprevalence's 95% CI were then calculated.

Estimates of the seroprevalence of cattle population in France in 2015 at national, regional and *département* levels

As the distribution of birth cohorts is spatially hetergenous in France, we estimated the overall seroprevalence at three geographical levels: national, regional and *département* levels. To do so, we combined the estimate of the seroprevalence per birth cohort and the proportion of animals from each birth cohort in the cattle population of July 2015 at the defined geographical level.

We estimated the national seroprevalence in the cattle population given the estimate of seroprevalence per birth cohort as calculated above. We estimated this seroprevalence per *département* status: "BTV-infected in 2015" or "BTV-non-infected in 2015". We used the lower and upper limits of the 95% CI of the seroprevalence per each birth cohort to estimate the lower and upper bounds of the confidence interval for cattle seroprevalence. No serological data were available for calves less than 1 year old, but colostral antibodies are present only a few months after birth. We therefore considered them as naïve with a seroprevalence of 0%.

We estimated seroprevalence per region using the seroprevalence levels per birth cohort observed per region. The objective was to assess the spatial variation of seroprevalence due to the spatial variation of the cattle population and of the seroprevalence levels per birth cohort.

Finally, to estimate the seroprevalence of BT in cattle at *département* level, we used the national seroprevalence levels for each birth cohort estimated for the area considered free of BTV infection in order to exclude any effects of the 2015 BTV circulation.

Proportion of infected cattle per birth period on infected farms in autumn 2015 (RT-PCR)

In order to roughly estimate the likelihood of an animal being infected in 2015 according to its birth cohort, we calculated the proportion of RT-PCR positive animals per birth cohort on infected farms in autumn 2015. We used the results of two other surveys conducted in autumn 2015, organised by the Ministry of Agriculture.

We used RT-PCR results from 609 cattle on ten farms found to be infected during a survey conducted in September 2015. This survey targeted farms located within 2 km of the first BTV-8 outbreak detected in August 2015 in the Allier *département*. We also used results from 768 cattle on 26 farms found to be infected during a national cross-sectional survey conducted in September-October 2015 testing 30 cattle per farm on 1338 farms. All these infected farms were located in central France and were separated by a maximum distance of 220 km. Therefore, we considered that these animals were almost identically exposed to virus circulation in 2015 and were aggregated for the analysis of prevalence per birth period.

Data on birth date and cattle movements were extracted from the National Identification Database (BDNI). All analyses were performed using R software version 3.1.2 [35]. The protocol was designed, and the results analysed and interpreted within the "BT working group" of the French Platform for Epidemiological Surveillance in Animal Health ("ESA Platform").

Results

Estimates of seroprevalence in cattle per birth period and the *département*'s BTV-8 status in 2015

A total of 8525 animals born before 2013 and 15,799 animals born after 2013 from 55 *départements* were tested by ELISA. Because of difficulties related to sample collection, laboratory analyses or data recording, three *départements* were excluded from the analysis (Fig. 1). Among the sampled animals, 681 (2.8%) had left their *département* of birth since July 2015 and were therefore excluded from the analysis.

Nationally, we found large seroprevalence differences according to birth periods (Chi^2 = 16,950, df = 6, $p <$ 0.001), with higher levels among the birth cohorts before June 2009 (Table 3). No significant difference in seroprevalence was found between infected and non-infected *départements* in 2015 among animals born before July 2010 (Table 3). A difference of 2% to 6% was observed between infected and non-infected *départements* for animals born after July 2010. Similarly, a seroprevalence of 2-3% was also observed in animals born after July 2013 (animals that had never been vaccinated) and between July 2012 and June 2013 (animals that had probably never been vaccinated) in infected *départements*, while it was very low (0.5%) in non-infected *départements*.

Seroprevalence was high (> 75%) among cattle that were present during mandatory vaccination campaigns in 2008-2010 and the period of intense BTV circulation

Table 3 BTV-8 seroprevalence in French cattle in the winter of 2015-2016 per birth cohort and the *département*'s BTV-8 status

Birth cohort	BTV-8 non-infected *départements*						BTV-8 infected *départements*					
	No. analyses	ELISA positive		ELISA doubtful			No. analyses	ELISA positive		ELISA doubtful		
		No.	% (CI 95%)	No.	% (CI 95%)			No.	% (CI 95%)	No.	% (CI 95%)	
< July 2008	1310	1254	95.7 [94.5 - 96.8]	4	0.3 [0.1 - 0.8]		866	814	94.0 [92.2 - 95.5]	2	0.2 [0 - 0.8]	
July 2008 - June 2009	467	368	78.8 [74.8 - 82.4]	16	3.4 [2.0 - 5.5]		226	163	72.1 [65.8 - 77.9]	9	4.0 [1.8 - 7.4]	
July 2009 - June 2010	617	194	31.4 [27.8 - 35.3]	23	3.7 [2.4 - 5.5]		317	108	34.1 [28.9 - 39.6]	6	1.9 [0.7 - 4.1]	
July 2010 - June 2011	718	76	10.6 [8.4 - 13.1]	7	1.0 [0.4 - 2.0]		438	72	16.4 [13.1 - 20.2]	10	2.3 [1.1 - 4.1]	
July 2011 - June 2012	1134	37	3.3 [2.3 - 4.5]	5	0.4 [0.1 - 1.0]		710	57	8.0 [6.1 - 10.3]	13	1.8 [1.0 - 3.1]	
July 2012 - June 2013	4342	37	0.9 [0.6 - 1.2]	15	0.3 [0.2 - 0.6]		3179	75	2.3 [1.9 - 2.9]	2	0.1 [0 - 0.2]	
> June 2013*	5428	25	0.5 [0.4 - 0.7]	25	0.5 [0.3 - 0.7]		3891	102	2.6 [2.1 - 3.2]	5	0.1 [0 - 0.3]	

*Cattle > 12 months old

in 2007-2009. It was lower for cattle born between July 2008 and July 2009 (76.6%, 95% CI [73.3 – 79.7]) than those born before July 2008 (95.0%, 95% CI [94.0 – 95.9]) (Chi^2 = 11, df = 1, p < 0.001). Seroprevalence in animals born before 2008 did not vary greatly between regions (difference of 13%) with a minimum of 85.6% (Table 4). Conversely, a higher regional variability (difference of 25%) was observed for seroprevalence in cattle born between July 2008 and July 2009 which varied between 65.7% and 90.4%, but with a lower precision (median of the 95% CI range of 20.9%, Table 4).

As expected, seroprevalence was lower (< 40%) for cattle born in 2009-2010 and 2010-2011 (Table 3) given the lower proportion of cattle vaccinated during the two voluntary vaccination campaigns in 2010-2012 (20-30% in 2010-2011). These animals were only present during voluntary vaccination campaigns or were less than 12 months old during the previous mandatory vaccination campaigns and therefore not necessarily vaccinated at that time. Animals born in 2009-2010 had a seroprevalence of 32.3% (95% CI [29.3 – 35.4]) varying between 19.5 and 53.7% according to the region, whereas those born in 2010-2011 had a seroprevalence of 12.8% (95% CI [10.9 – 14.9]) varying between 1.1 and 21.7% according to the region (Tables 3 and 4).

The proportion of doubtful ELISA results varied between 0.2 and 3.6% according to birth period (Table 3). A higher proportion and a higher regional variability of doubtful results were observed for animals born between July 2008 and June 2010 (3.3% of doubtful results, which is significantly different from other classes Chi^2 = 233, df = 1, p-value< 0.001; 4-8% of regional variability; Tables 3 and 4) and those born between July 2010 and June 2012 (1.1% of doubtful results, which is significantly different from other classes Chi^2 = 37, df = 1, p-value< 0.001; 2% of regional variability; Tables 3 and 4). This proportion was higher in infected *départements* in 2015 than in non-infected ones for cattle born between July 2010 and June 2012 (2.5% vs. 0.6%, Chi^2 = 15.6, df = 1, p-value< 0.001, Table 3).

Estimates of seroprevalence in the French cattle population in 2015 at national, regional and *département* levels

In July 2015, 59% of cattle in mainland France were born after June 2012, including 29% of calves less than 12 months old. The proportion of animals born before July 2008, in 2008-2009, 2009-2010, 2010-2011 and 2011-2012 was respectively 12%, 5%, 6%, 8% and 10%. The structure of the cattle population varied per *département* but this was mainly due to the proportion of animals born before July 2008 (from 6.3% to 27%) and after June 2012 (from 44% to 67%). For other birth cohorts, smaller variations were reported (less than 5%).

We estimated a BT seroprevalence for the cattle population of 18.5% [CI 17.6 - 19.4] and 19.6% [CI 18.2 - 21.1] respectively in "non-infected" and "infected" *départements* in 2015. When using seroprevalence levels observed for each birth cohort per region, the proportion of seropositive animals per region in the cattle population of 2015 was estimated to range from 12.3% [10.8% – 14.6%] to 25.2% [22.3% – 28.1%] with a median of 15.7%. A higher seroprevalence (> 20%) was observed in the Auvergne and Aquitaine regions, where there was a higher proportion of beef farms and a higher proportion of older animals (> 20% of animals born before 2008). When using national seroprevalence figures per birth cohort, the seroprevalence per *département* was estimated to vary from 12.4% [11.6% – 13.2%] to 32.3% [31.3% – 33.2%] with a median of 18.1% at *département* level (Fig. 2). There was an increasing North-South gradient in the proportion of seropositive cattle per *département*. In the Allier and Puy-de-Dôme *départements* where BT outbreaks were mainly detected in 2015-2016, the seroprevalence was estimated to be 20.1 and 23.3% respectively.

Proportion of infected cattle per birth period on infected farms in autumn 2015

On infected farms in 2015, RT-PCR prevalence was the highest in animals born between July 2012 and June

Table 4 BTV-8 seroprevalence in cattle in the winter of 2015-2016 per birth cohort and region in France

Birth cohort	Proportion of positive results per region (%)				Proportion of doubtful results per region (%)			
	Median	Min - Max	Range	Median of the 95% CI range	Median	Min - Max	Range	Median of the 95% CI range
< July 2008	96.5	85.6 - 99.1	13.5	5.7	0.0	0.0 – 2.3	2.3	2.1
July 2008 - June 2009	75.4	65.7 – 90.4	24.7	20.9	3.6	1.2 – 5.4	4.2	10.5
July 2009 - June 2010	30.1	19.5 – 53.7	34.2	19.3	3.0	0.0 – 8.3	8.3	8.5
July 2010 - June 2011	11.9	1.1 – 21.7	20.6	13.4	1.6	0.0 – 3.6	3.6	4.5
July 2011 - June 2012	5.8	0.0 – 9.6	9.6	7.2	1.0	0.0 – 1.8	1.8	4.0
July 2012 - June 2013	1.0	0.2 – 4.6	2.6	1.7	0.0	0.0 - 1.4	1.4	0.6
> June 2013*	0.7	0.0 – 4.6	4.6	1.2	0.2	0.0 - 1.4	1.4	0.6

*Cattle > 12 months old

2013 (19%) and decreased in older birth cohorts down to 0% for animals born before 2008 (Table 5).

Discussion

Despite two massive mandatory vaccination campaigns in 2008-2010 where more than 90% of the cattle population was vaccinated, BTV-8 was again detected in France in 2015. We found an effect of birth cohorts on the seroprevalence related to their difference in exposure levels to BTV-8 or vaccination.

Our results suggest that BTV-8 circulation in 2015 was low and had little influence on the seroprevalence observed in cattle in the winter of 2015-2016. Seroprevalence in animals born after the vaccination ban in June 2013 can be used as an indicator of the level of recent viral circulation after this date. In *départements* where BTV circulation was evidenced in 2015-2016, 3% of those animals were seropositive. We observed similar seroprevalence levels for animals born between July 2012 and June 2013, suggesting poor vaccination coverage in 2012-2013. The difference in seroprevalence observed in the two birth cohorts of July 2010 to June 2012 between infected and non-infected *départements* (around 5%) was similar to the seroprevalence observed in young animals born after June 2012 in infected *départements* (around 3%). In 2015 or earlier, the virus may have been re-circulating in infected areas at a very low level with a maximum cumulative proportion of infected

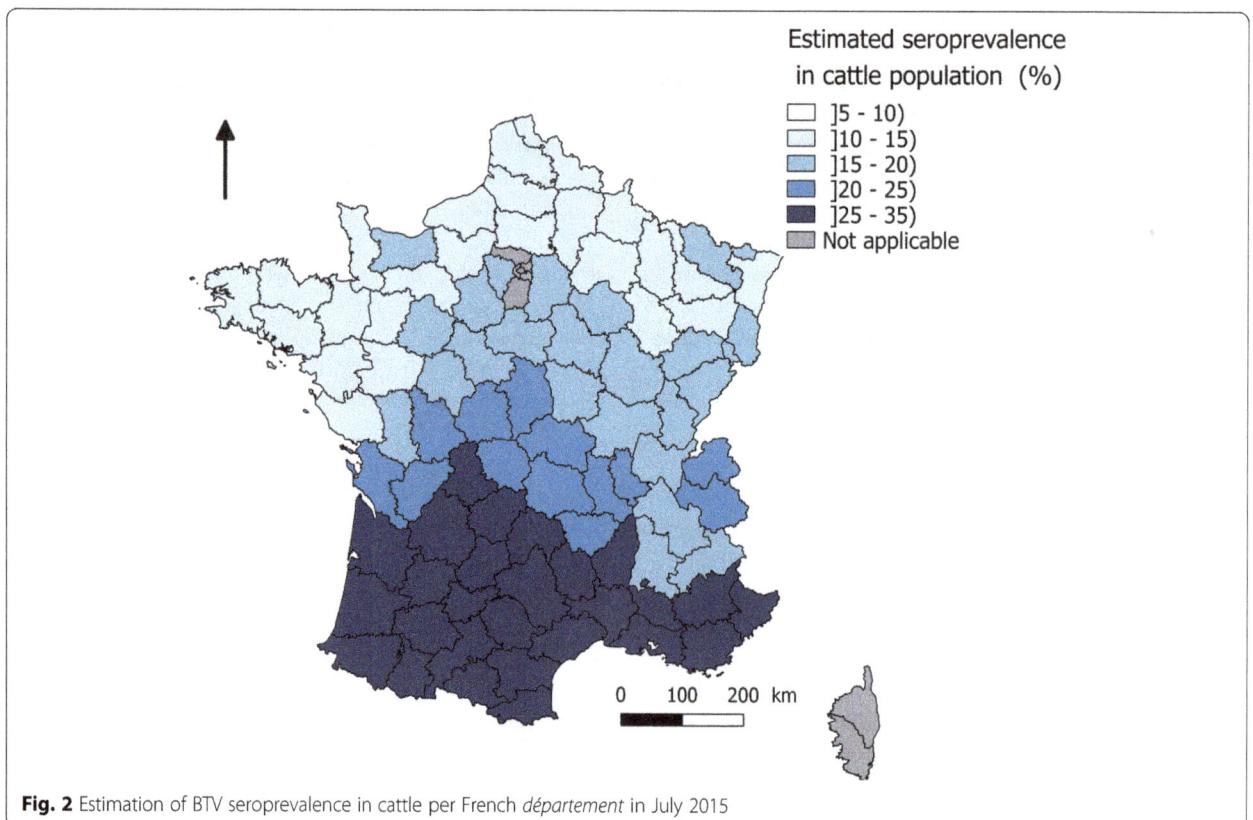

Fig. 2 Estimation of BTV seroprevalence in cattle per French *département* in July 2015

Table 5 RT-PCR results for cattle per birth cohort on farms found to be infected from September to October 2015 in France

Birth period	No. analyses	No. positives	Proportion of positive results (% [95 CI])
< July 2008	201	0	0 [0 - 1.8]
July 2008 - June 2009	66	3	4.5 [0.9 - 12.7]
July 2009 - June 2010	89	6	6.7 [2.5 - 14.1]
July 2010 - June 2011	113	10	8.8 [4.3 - 15.7]
July 2011 - June 2012	177	24	13.5 [8.9 - 19.5]
July 2012 - June 2013	206	39	18.9 [13.8 – 25.0]
July 2013-September 2014	180	24	13.3 [8.7 – 19.2]

animals of 5% in all *départements* infected in 2015-2016. It is worth noting that seroprevalence in young animals born after 2013 was higher (5-10%) in only two *départements* (Allier and Puy-de-Dôme in central France) considered as the epicentre of the 2015-2016 resurgence (data not shown). A very low BTV circulation in winter 2014-2015 (< 5% of infected animals among those born after the 2013 vaccination ban) has also been evidenced by a recent study conducted in seven *départements* including the Allier [8]. Therefore, although some *départements* in our study might be wrongly classified as non-infected due to undetected current or past BTV circulation, our study shows that viral circulation in 2015 did not significantly increase the proportion of seropositive animals. In these *départements*, the presence of BT antibodies is more likely to be due to past natural infection from 2007 to 2009 and/or vaccination from 2007 to 2012 and thus existed before the 2015 resurgence.

Seroprevalence was higher (more than 75%) for animals that were present during periods of intensive viral circulation from 2007 to 2009 or mandatory vaccination campaigns when 90% of cattle were vaccinated (i.e. animals born before July 2009) and lower for those only present during voluntary campaigns. This is consistent with previous reports of the long persistence of BTV-8 antibodies after natural infection or vaccination [18, 19, 21]. Our results suggest that high vaccination coverage in 2008-2010 allowed the seroprevalence level over mainland France to be homogenised, and most animals born before July 2008 and vaccinated during two consecutive years (probably > 90%) still carried antibodies six years later. Indeed, we found little variation in seroprevalence between regions (10–25%), similar to the regional variation estimated for vaccination coverage in 2009-2010 (20%). This spatial variation is lower than could be expected if the persistence of antibodies arises only from natural infection, given the proportion of notified outbreaks with a strong spatial variation in 2007-2009 (between 3.5 to 100% per region (MAAF unpublished data)).

Being younger than 12 months during the period of vaccination or the 2007/2009 epidemics appeared to influence 2015-2016 seroprevalence levels. Only 32% of animals born between July 2009 and June 2010 and thus being less than 12 months old during the mandatory vaccination campaign of 2009-2010 were seropositive, whereas a large majority was considered to have been vaccinated in 2009-2010 (> 90%). This proportion is closer to the estimate of vaccination coverage in 2010-2011 (around 30%). This "age effect" (i.e. cattle being younger than 12 months during the period of vaccination or intense BTV circulation) can be explained by a lower exposure to viral circulation and vaccination or by the absence of a long-term immune response after vaccination or natural infection for some calves born to immune dams (either infected or vaccinated) [34, 36]. Due to the persistence of colostral antibodies that interfered with the induction of the immune response after vaccination, Vitour et al. (2011) [34] predicted a vaccination success rate of only 50% for calves of 5-6 months old born to immune dams. As the requirement was to vaccinate all calves before they reached 6 months of age and most of them were probably not re-vaccinated later during the voluntary campaigns, it is likely that some of them did not develop an effective immune response. Similarly, among animals present during the two mandatory vaccination campaigns, animals born within the year of the first campaign from July 2008 to June 2009 presented a lower seroprevalence (77%) than those aged more than 12 months (95%). In addition to the "age effect" during the vaccination period, we also attributed this difference to a faster decline in antibodies among animals vaccinated only once (prime and boost injection against serotypes 1 and 8) compared to those vaccinated twice. The higher proportion of doubtful results for animals born in 2008-2009 and 2009-2010 compared to those born before 2008 might also suggest a lower serological response and a shorter persistence of antibodies in animals vaccinated only once. Although this decline might be slightly faster, it is worth pointing out that a high proportion of those animals still had antibodies five years later.

We used group-specific cELISA to infer the immune status of cattle towards BT although the level of antibodies detected is not type-specific and not directly correlated to protection. Given that vaccination was directed against both serotypes BTV-8 and BTV-1 throughout mainland France, ELISA-positive animals can have antibodies against either or both serotypes. On infected farms in autumn 2015, we found that older birth cohorts were infected less frequently than younger ones. As discussed above, we found in parallel that older birth cohorts were also those having the highest proportion of animals with pre-existing BT antibodies. None of the animals born before 2008 for which we observed in parallel high seroprevalence levels (> 95%) were found to be infected, for example. Conversely, 19% of naive animals born in 2012-2013 were infected in autumn 2015. These results suggest that most animals with pre-existing group-specific BT antibodies detected by ELISA were protected against BTV-8 infection in 2015. Moreover, the ID-VET ELISA kits used for 96% of the study's serological analyses offered both high sensitivity and specificity (> 99.8%) [37]. Therefore, we considered within the framework of this study that the undetected proportion of seropositive animals was probably low.

Given the respective proportion of animals per birth cohort in the cattle population, we estimated that 18-20% (resp. from 12 to 25%) of cattle were seropositive and hence probably immune in 2015 at the national level (resp. at the regional level). Our results indicated that the regional variation in cattle seroprevalence in 2015 was due more to regional variations in the cattle's age-population structure than in seroprevalence per birth cohort. These regional seroprevalence variations are mostly explained by the spatial variation in the proportion of animals born before 2008 (from 6.7 to 18.3% per region), highly seropositive, and in the proportion of naive young animals born after 2012 (from 56.7 to 64.5% per region). In considering only the spatial variation of the cattle population structure, we estimated that from 12 to 32% of the animals were approximately still seropositive and probably immune to BTV-8 in 2015 at *département* level before the resurgence. In central France, and more specifically in the Allier and Puy-de-Dôme *départements*, where most BT outbreaks were detected in 2015-2016, this proportion was estimated to be around 20% to 25%. Estimates of the proportion of immune cattle followed an increasing North-South gradient related to variations in cattle population structure and livestock farming types: most farms located in central and southern France are beef cattle farms with a lower turn-over (older and therefore immune cattle) compared to dairy cattle farms located in northern and western France. In our estimations, we considered all calves as naive whereas some of them — born to seropositive dams — may have colostral

antibodies and hence be immune from re-infection. This proportion was unknown, but by considering that only calves less than 7 months old and born to dams born before 2009 might have colostral antibodies to BT, we might have underestimated the overall seroprevalence in the cattle population by not more than 10% (BDNI data).

Conclusion

Our results indicate that at the time of t BTV-8 resurgence in 2015 in central France, the viral circulation in naive animal birth cohorts was limited (seroprevalence < 5-10%) and from 12% to 32% of the cattle population was probably immune. The long persistence of BT antibodies after natural infection or vaccination seems to have maintained a large proportion of immune animals in the cattle population which were protected from re-infection for many years. In 2015, the proportion of immune cattle seemed to be sufficiently high to limit BTV-8 circulation. There is no reliable information on the French sheep population to accurately estimate the levels of the residual sheep herd immunity in 2015. Considering their lower number in France and lower attractiveness to Culicoides, it has probably contributed less to slowing down BTV circulation than cattle herd immunity. However, further studies with dynamic mathematical modelling, for instance, might help to solve this issue. The immunity of the cattle population would continue to decrease along with the population turnover in non-infected areas. Based on our results and the respective proportion of each birth cohort in the cattle population of 2016-2017, we can hypothesise that 6% to 9% of the cattle population was still immune in those areas in 2016-2017. In the summer and autumn of 2016, the lower herd immunity and the higher number of infection sources may partly explain the higher BTV circulation observed in 2016 than in 2015 (1200 outbreaks between July and December 2016 vs. 284 outbreaks between August 2015 and June 2016) [29].

Abbreviations
BDNI: National Identification Database; BT: Bluetongue; BTV: Bluetongue virus; BTV-1: Bluetongue virus serotype 1; BTV-8: Bluetongue virus serotype 8; CI: Confidence interval; ELISA: enzyme-linked immunosorbent assay; RT-PCR: Real time polymerase chain reaction

Acknowledgements
The authors are very grateful to all the participants involved in BT surveillance in France (veterinarians, farmers, local and national farmers' associations, local laboratories, laboratory association, local and national veterinary services). They would like to thank more specifically all the members of the bluetongue surveillance working group of the ESA Platform who helped in protocol design, interpretation of results and communication, Mathilde Saussac for her help in extracting BDNI data.

Funding
This study was funded by the French Ministry of Agriculture within the framework of BT surveillance activities.

Authors' contributions
LB, LC, DC and PH designed the study. LB, LC, CS, EB, GZ, RSA, AP, IT, FD and EG were involved in protocol design, coordination of data collection and communication with stakeholders. LB performed the analysis. LB, DC and PH drafted the manuscript. All the authors read and approved the final manuscript.

Consent for publication
Not applicable.

Competing interests
The authors declare that they have no competing interests.

Author details
[1]ANSES (French Agency for Food, Environmental and Occupational Health & Safety), Unité de coordination et d'appui à la surveillance, Direction des laboratoires, Maisons Alfort, France. [2]Ministère en charge de l'Agriculture, Direction générale de l'Alimentation, Bureau de la santé animale, Paris, France. [3]ANSES, Laboratoire de santé animale, Université Paris-Est, Maisons-Alfort, France. [4]CIRAD, UMR ASTRE, Montpellier ; Inra, UMR ASTRE, Montpellier, France. [5]Coop de France, Paris, France. [6]GDS France, Paris, France. [7]Races de France, Paris, France. [8]ANSES, Laboratoire de Lyon, Unité Epidémiologie, Laboratoire de Lyon, Lyon, France.

References
1. Belbis G, Zientara S, Bréard E, Sailleau C, Caignard G, Vitour D, et al. Bluetongue virus: from BTV-1 to BTV-27: Advances in Virus Research. Elsevier; 2017. p. 161–97. https://doi.org/10.1016/bs.aivir.2017.08.003.
2. Wilson AJ, Mellor PS. Bluetongue in Europe: past, present and future. Philos Trans R Soc B Biol Sci. 2009;364:2669–81.
3. Languille J, Sailleau C, Bréard E, Zientara S. Bilan de la surveillance de la fièvre catarrhale ovine en France continentale en 2010: vers une maîtrise clinique de la maladie. Bull Epidemiol. 2010;46:24–5.
4. Languille J, Sailleau C, Bréard E, Desprat A, Viarouge C, Zientara S. Bilan de la surveillance et de la vaccination contre la fièvre catarrhale ovine en France continentale en 2011 : vers l'éradication de la maladie. Bull Epidemiol. 2011; 54:32–4.
5. Bréard E, Sailleau C, Quenault H, Lucas P, Viarouge C, Touzain F, et al. Complete genome sequence of bluetongue virus serotype 8, which reemerged in France in august 2015. Genome Announc. 2016;4:e00163–16.
6. Anses. Avis de l'Anses relatif à « l'évaluation du risque lié à la réapparition du sérotype 8 de la FCO en France continentale». Avis de l'Anses. 2015.
7. EFSA Panel on Animal Health and Welfare. Bluetongue: control, surveillance and safe movement of animals. EFSA J. 2017;15:e04698.
8. Courtejoie N, Durand B, Bournez L, Gorlier A, Bréard E, Sailleau C, et al. Circulation of bluetongue virus 8 in French cattle, before and after the re-emergence in 2015. Transbound Emerg Dis. 2017; https://doi.org/10.1111/tbed.12652.
9. Ayllón T, Nijhof AM, Weiher W, Bauer B, Allène X, Clausen P-H. Feeding behaviour of Culicoides spp. (Diptera: Ceratopogonidae) on cattle and sheep in northeast Germany. Parasit Vectors. 2014;7:34.
10. Bartsch S, Bauer B, Wiemann A, Clausen P-H, Steuber S. Feeding patterns of biting midges of the Culicoides obsoletus and Culicoides pulicaris groups on selected farms in Brandenburg. Germany Parasitol Res. 2009;105:373–80.
11. Durand B, Zanella G, Biteau-Coroller F, Locatelli C, Baurier F, Simon C, et al. Anatomy of bluetongue virus serotype 8 epizootic wave, France, 2007–2008. Emerg Infect Dis. 2010;16:1861–8.
12. Maclachlan NJ, Drew CP, Darpel KE, Worwa G. The pathology and pathogenesis of bluetongue. J Comp Pathol. 2009;141:1–16.
13. Caporale M, Di Gialleonorado L, Janowicz A, Wilkie G, Shaw A, Savini G, et al. Virus and host factors affecting the clinical outcome of bluetongue virus infection. J Virol. 2014;88:10399–411.
14. Jeggo MH, Wardley RC, Brownlie J. A study of the role of cell-mediated immunity in bluetongue virus infection in sheep, using cellular adoptive transfer techniques. Immunology. 1984;52:403.
15. Takamatsu H, Jeggo MH. Cultivation of bluetongue virus-specific ovine T cells and their cross-reactivity with different serotype viruses. Immunology. 1989;66:258.
16. Oura CAL, Wood JLN, Floyd T, Sanders AJ, Bin-Tarif A, Henstock M, et al. Colostral antibody protection and interference with immunity in lambs born from sheep vaccinated with an inactivated bluetongue serotype 8 vaccine. Vaccine. 2010;28:2749–53.
17. Savini G, Hamers C, Conte A, Migliaccio P, Bonfini B, Teodori L, et al. Assessment of efficacy of a bivalent BTV-2 and BTV-4 inactivated vaccine by vaccination and challenge in cattle. Vet Microbiol. 2009;133:1–8.
18. Batten CA, Edwards L, Oura CAL. Evaluation of the humoral immune responses in adult cattle and sheep, 4 and 2.5 years post-vaccination with a bluetongue serotype 8 inactivated vaccine. Vaccine. 2013;31:3783–5.
19. Oura CAL, Edwards L, Batten CA. Evaluation of the humoral immune response in adult dairy cattle three years after vaccination with a bluetongue serotype 8 inactivated vaccine. Vaccine. 2012;30:112–5.
20. Zanella G, Bréard E, Sailleau C, Zientara S, Viarouge C, Durand B. A one-year follow-up of antibody response in cattle and sheep after vaccination with serotype 8- and serotype 1-inactivated BT vaccines. Transbound Emerg Dis. 2014;61:473–6.
21. Eschbaumer M, Eschweiler J, Hoffmann B. Long-term persistence of neutralising antibodies against bluetongue virus serotype 8 in naturally infected cattle. Vaccine. 2012;30:7142–3.
22. Büchi M, Abril C, Vögtlin A, Schwermer H. Prevalence of antibodies against bluetongue virus serotype 8 in bulk-tank milk samples from dairy cattle herds located in risk areas for bluetongue virus transmission after a vaccination programme in Switzerland. Berl Munch Tierarztl Wochenschr. 2014;127:158–62.
23. Mughini-Gras L, Patregnani T, Nardelli S, Gagliazzo L, Comin A, Savini G, et al. Long-term persistence of vaccine-elicited bluetongue serotype 8 neutralizing antibodies in dairy cattle. Large Anim Rev. 2014;20:153–5.
24. Calistri P, Giovannini A, Savini G, Bonfanti L, Cordioli P, Lelli R, et al. Antibody response in cattle vaccinated against bluetongue serotype 8 in Italy: BTV8-inactivated vaccine in cattle. Transbound Emerg Dis. 2010;57:180–4.
25. Hultén C, Frössling J, Chenais E, Sternberg LS. Seroprevalence after vaccination of cattle and sheep against bluetongue virus (BTV) serotype 8 in Sweden: seroprevalence after BTV-8 vaccination in Sweden. Transbound Emerg Dis. 2013;60:438–47.
26. Hund A, Gollnick N, Sauter-Louis C, Neubauer-Juric A, Lahm H, Büttner M. A two year BTV-8 vaccination follow up: molecular diagnostics and assessment of humoral and cellular immune reactions. Vet Microbiol. 2012;154:247–56.
27. Vitale N, Radaelli MC, Chiavacci L, Paoletti M, Teodori L, Savini G. Factors affecting seroconversion rates in cattle vaccinated with two commercial inactivated BTV-8 vaccines under field conditions. Transbound Emerg Dis. 2016;63:175–83.
28. Wäckerlin R, Eschbaumer M, König P, Hoffmann B, Beer M. Evaluation of humoral response and protective efficacy of three inactivated vaccines against bluetongue virus serotype 8 one year after vaccination of sheep and cattle. Vaccine. 2010;28:4348–55.
29. Bournez L, Sailleau C, Bréard E, Servan de Almeida R, Pédarrieu A, Libeau G, et al. Bilan de la situation relative à la FCO de sérotype 8 en France continentale, au 31 décembre 2016. Centre de ressources de la Plateforme d'épidémiosurveillance en santé animale 2017. https://www.plateforme-esa.fr/article/bilan-de-la-situation-relative-la-fco-de-serotype-8-en-france-continentale-au-31-decembre.
30. Zanella G, Martinelle L, Guyot H, Mauroy A, De Clercq K, Saegerman C. Clinical pattern characterization of cattle naturally infected by BTV-8: clinical characterization of BTV-8 infected cattle. Transbound Emerg Dis. 2013;60:231–7.
31. Di Gialleonardo L, Migliaccio P, Teodori L, Savini G. The length of BTV-8 viraemia in cattle according to infection doses and diagnostic techniques. Res Vet Sci. 2011;91:316–20.
32. Garros C, Duhayon M, Lefrançois T, Fediaevsky A, Balenghien T. La surveillance entomologique des populations de Culicoides en France

pendant la période supposée d'inactivité vectorielle 2015 - 2016. Bull
Épidémiologique Santé Anim-Aliment. 2017; Numéro spécial maladies
animales reglementées Bilan 2015.

33. Maan S, Maan NS, Belaganahalli MN, Kumar A, Batra K, Rao PP, et al.
Genome sequence of bluetongue virus type 2 from India: evidence for
reassortment between outer capsid protein genes. Genome Announc. 2015;
3:e00045–15.

34. Vitour D, Guillotin J, Sailleau C, Viarouge C, Desprat A, Wolff F, et al.
Colostral antibody induced interference of inactivated bluetongue serotype-
8 vaccines in calves. Vet Res. 2011;42:18.

35. R Development Core team. R. A language and environment for statistical
computing. http://www.R-project.org. Vienna, Austria: R Foundation for
Statistical Computing; 2014.

36. Vangeel I, De Leeuw I, Méroc E, Vandenbussche F, Riocreux F, Hooyberghs
J, et al. Bluetongue sentinel surveillance program and cross-sectional
serological survey in cattle in Belgium in 2010–2011. Prev Vet Med. 2012;
106:235–43.

37. IDVet. ID screen bluetongue competition. Grabels: Manufacturer's manual; 2014.

16

Prevalence of tuberculous lesion in cattle slaughtered in Mubende district, Uganda

Daniel Pakasi Nalapa[1], Adrian Muwonge[2], Clovice Kankya[1*] and Francisco Olea-Popelka[3]

Abstract

Background: The aim of this study was to estimate the prevalence of gross pathology suggestive of bovine tuberculosis (TB-like lesions) and evaluate animal's characteristics associated with the risk of having bovine TB-like lesions among cattle slaughtered in Mubende district in the Uganda cattle corridor.

Method: We conducted a cross sectional study in which 1,576 slaughtered cattle in Mubende district municipal abattoir underwent post-mortem inspection between August 2013 and January 2014. The presence of bovine TB-like lesions in addition to the animal's sex, age, breed, and sub-county of origin prior to slaughter were recorded. Associations between the presence of bovine TB-like lesions and animal's age, sex, breed, and sub-county of origin prior to slaughter were initially analysed using a univariable approach with the chi-square test, and subsequently with a multivariable logistic regression model to assess the combined impact of these animal characteristics with the risk of having a bovine TB-like lesion. Additionally, and as a secondary objective, tissue samples were collected from all carcases that had a bovine TB-like lesion and were processed using standard *Mycobacterium* culture and identification methods. The culture and acid fast positive samples were tested using Capilia TB-neo® assay to identify *Mycobacterium tuberculosis* complex (MTC).

Results: Of 1,576 carcasses inspected, 9.7% (153/1,576) had bovine TB-like lesions from which *Mycobacterium* spp and Mycobacterium Tuberculosis Complex (MTC) were isolated in 13 (8.4%) and 12 (7.8%) respectively. Bovine TB-like lesions were more likely to be found in females (OR = 1.49, OR 95% CI: 1.06–2.13) and in older cattle (OR = 2.5, 95% CI: 1.64–3.7). When compared to Ankole cattle, Cross breed (OR = 6.5, OR 95% CI: 3.37–12.7) and Zebu cattle (OR = 2.57, 95% CI: 1.78–3.72) had higher odds of having bovine TB-like lesions. Animals from Kasanda (OR = 2.5, 95% CI: 1.52–4.17) were more likely to have bovine TB-like lesions than cattle from Kasambya.

Conclusions: The findings of study reveals that approximately one in ten slaughtered cattle presents with gross pathology suggestive of bovine TB in Mubende district in the Uganda cattle corridor district, however, we isolated MTC in only 8.4% of these bovine TB-like lesions. Therefore, there is a need to understand the cause of all the other bovine TB-like lesions in order to safe guard diagnostic integrity of meat inspection in Uganda.

Keywords: Tuberculous -like lesions, Uganda cattle corridor, Bovine tuberculosis

Background

Bovine tuberculosis (TB) is a chronic, infectious disease caused by *Mycobacterium bovis (M. bovis)* [1]. This disease has both public health and socioeconomic implications and in Uganda this disease is of concern particularly in rural communities [2, 3]. Apart from reduced productivity of infected cattle [4] and the possibility of carcass condemnation

at slaughter, consumption of infected uncooked meat and/ or unpasteurized milk (and milk products) increases the risk of zoonotic transmission of *M. bovis* to humans (zoonotic TB) in these communities [2, 3].

There is no official nationwide bovine TB control policy in Uganda, therefore, routine meat inspection at municipal abattoirs represent the first point of detection of this disease [2, 5]. It is noteworthy that not all animals are slaughtered at municipal abattoirs; therefore there is an unknown proportion of animals slaughtered that do not undergo meat inspection in Uganda. In countries

* Correspondence: clokankya@yahoo.com
[1]Department of Biosecurity Ecosystem and Veterinary Public Health, College of Veterinary Medicine, Animal Resources and Biosecurity, Makerere University, P.O. Box 7062, Kampala, Uganda

where a nationwide bovine TB control program is in place, detection of infected animals is synergised through annual TB skin testing and routine meat inspection [4, 6]. Slaughter detected TB-like lesions are confirmed at laboratory level through microscopic detection of acid fast bacilli before or after selective growth on media [4]. The cost of definitive diagnosis is reported to increases along the diagnostic cascade with molecular based diagnostic tools being the most expensive [7]. To reduce on this cost, rapid simple and inexpensive new generation assays for detecting members of *Mycobacterium tuberculosis* complex (MTC) including *Mycobacterium bovis, Mycobacterium bovis BCG, Mycobacterium canettii, Mycobacterium africanum, Mycobacterium pinnipedii, Mycobacterium microti, Mycobacterium caprae and Mycobacterium mungi.* have been developed, and one such assays is capilia TB-neo [7, 8]. This assay has been used in countries for this purpose and in Uganda it has been mostly used in the diagnosis of human TB [8, 9].

Under Uganda's field conditions, where abattoir surveillance is the only point where bovine TB can be identified, it is critical that the quality and integrity of meat inspection at this point is maintained by assuring the process is conducted appropriately and by regularly updating prevalence estimates which ensures optimal workload per *post mortem* inspector [10]. Furthermore, it is essential to understand disease drivers in order to strategically allocate scarce resources. Under such settings this process is usually hindered by the lack of optimal numbers of *M. bovis* caused bovine TB-like lesions in order to make meaningful risk assessment

studies. A recent study reported that one in fifty slaughtered pigs with bovine TB-like lesions were due to *M. bovis* in Mubende district. Such a finding is not only considered a spill-over event from cattle to pigs according to the epidemiology of bovine TB but it also inherently suggests a high prevalence of bovine TB in the cattle populations.

This study therefore aimed at estimating the prevalence of bovine TB-like lesions in cattle slaughtered at the Municipal abattoir in Mubende District in the Cattle Corridor in Uganda and to identify factors associated with the presence of bovine TB-like lesions among these cattle. Additionally, we aimed to isolate *Mycobacterium* species from these bovine TB-like lesions.

Methods
Study site
This cross-sectional study was conducted at the municipal abattoir of Mubende district in the Uganda cattle corridor (UCC) between August 2013 and January 2014. The UCC (Fig. 1) is a diagonal expanse of land spanning from the Northeast corner to the south Western corner of Uganda and holds ~ 45% of all the cattle in Uganda [11]. Mubende is located in the centre of the UCC with on average 610,000 and 35,000 human and cattle population respectively [12]. Most of the large herds of cattle are in pastoralist communities that tend to graze cattle communally.

Sampling and inclusion criteria
A preliminary survey of the study area was carried out in order to secure clearance from the District Veterinary

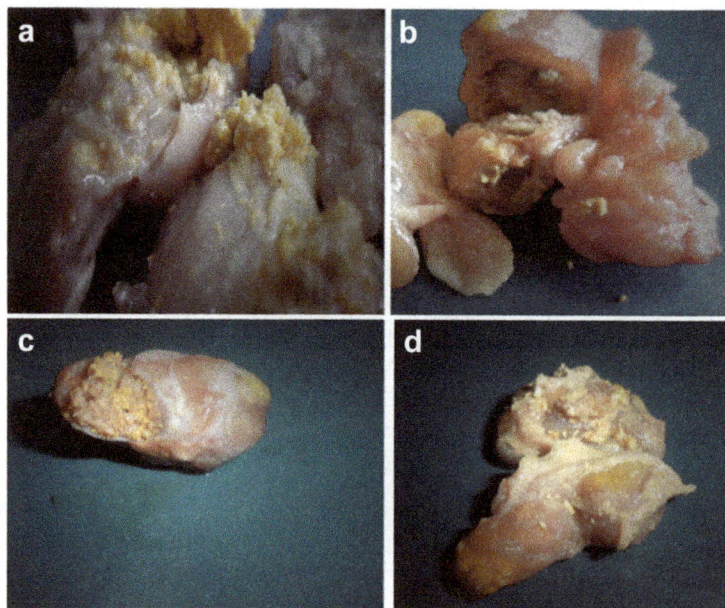

Fig. 1 shows a selection of observed BTB-like lesions found in our study. A deep yellow clumped gritty material in mesenteric lymph bodies, **b** a tubercle structure from a mediastinal lymph node, **c** & **d** caseous material from part of lymph nodes from the thoracic region

department and to understand the slaughter process in order to develop a sampling strategy that caused the least disturbance to the routine slaughter operations. The sampling sight was the Mubende municipal abattoir located in Mubende town council. All cattle slaughtered at this abattoir between August 2013 and January 2014 were inspected and tissue samples were obtained from all carcasses that presented with bovine TB-like lesions.

Ante mortem inspection and cattle characteristics

Physical examination of all the animals was conducted to obtain data on age (young and old), sex (male and female), and breed (Ankole-Sanga, East African Zebu, Friesian cross) before slaughter. Age estimation was done using the dental eruption and wear patterns [13]. Animals were classified as young cattle (less than 2.5 years old, and old (greater than 2.5 years old). Additional data on the sub-county of origin prior to slaughter of the animal and total number of animals slaughtered at the abattoir was retrieved from the abattoir records.

Post mortem inspection

All carcasses of cattle slaughtered during the study period were inspected for the presence of bovine TB-like (tuberculous) lesions. Examination of the carcasses was thoroughly conducted as sanction by the public health Act of Uganda and the World Organisation for Animal Health (OIE) [14]. Briefly, lymph nodes especially parotid, sub maxillary, mandibular, brachial, medial retropharyngeal, mediastinal, hepatic, mesenteric iliac, precrural, prescapular, supramammary, inguinal, and ischiatic lymph nodes, were palpated incised and inspected for the presence of bovine TB-like lesions. Organs including the livers, lungs, kidneys, spleens, intestines, mammary glands were also palpated and incised and inspected for bovine TB-like lesions. Lymph nodes that showed the following gross pathological characteristics were collected: increased size, colour changes, and granulation, calcification, and/or caseation texture. Bovine TB-like lesions were collected into a clean, sterile container to avoid contamination with other environmental mycobacteria. The samples were transported at 4 °C to the veterinary public health laboratory where they were stored at −20 °C until they were transferred to Mycobacteriology laboratory in the department of medical microbiology within the school of health sciences at Mulago for mycobacterial culturing.

Culturing, isolation and identification of *Mycobacterium Tuberculosis Complex* (MTC)

This selective growth was conducted in order to isolate *Mycobacterium spp* from bovine TB-like lesions. Briefly, fat and connective tissue was dissected away from the lymph node; then, 3 g of the tissue was cut into small pieces using a sterile scalpel blade. Tissue samples were crushed in a clean motor using a pestle, the minced tissue was then transferred into a clean, well labelled centrifuge tube. Ten (10) ml of 4% NaOH were added to the minced tissue for decontamination followed by a neutralisation Sterile with PBS (pH 6.8, 0.067 M) added up to the 50 ml mark and the tube was centrifuged at 3200 g for 15 min at 18 °C. All tissue was removed from the centrifuge tube and a phosphate buffered saline was added to 45 ml mark, followed by centrifugation at 3000xg for 15 min at 4° Celsius. The supernatant was carefully poured off leaving only the sediment in the centrifuge tubes. A half (0.5 ml) of the sediment was inoculated into mycobacteria growth indicator tube (MGIT), supplemented with PANTA/OADC and 0.2mls of Lowenstein-Jensen (LJ) supplemented with sodium pyruvate (to promote growth of *M. bovis*). The inoculated MGIT tubes were loaded into the instrument and incubated at 37° Celsius with 5-10% CO_2 for up 6 weeks. The media was monitored weekly for growth.

Acid-fast detection

MGIT tube with growth were examined for acid-fast bacilli by the Ziehl-Neelsen (ZN) by heat fixing the cells on a glass microscope slide and then flooded with carbolfuchsin stain as previously described [15]. The slide was gently and intermitently heated until it steamed for between 2 and 3 min, carbolfuchsin was then poured off and the slide washed thoroughly with water. This was followed by decolourisation with acid-alcohol for five minutes and again washing the slide with water thoroughly. Methylene blue counterstain was then flooded on the slide for one minute and the slide washed with water. Finally excess water was blotted off and the slide dried before viewing under light microscope (Olympus CX31) using a 10^3 magnification.

Capilia TB-Neo

The samples that were positive on the acid-fast test were then confirmed to be *Mycobacterium tuberculosis* complex (MTC) using capilia TB-Neo which detects the presence of MPB64 antigens specifically produced by MTC [8]. The kit includes a test plate with a carrier strip composed of a sample placing area, a reagent area including a colloidal gold labelled anti-MPB64 monoclonial antibody and a developing area that fixes the anti-MPB64 monoclonial antibody [7, 8]. Dissolved colony material of a sample (100 µl) was placed on the placing area of the test plate, the colloidal gold labelled anti-MPB64 antibody dissolves and forms an immune complex with MPB64 antigens in the sample. This immune complex migrates through the developing area by capillary action becoming captured by the anti-MPB64 antibody fixed in the developing area, and forms a red purple line of colloidal gold in the reading area. The red purple line visually displays the existence of MPB64 antigens in the sample. Regardless of the existence of the MPB64 antigens in the sample, excess colloidal gold-labelled anti-MPB64 antibody further migrates through the developing area, becoming captured by anti-mouse immunoglobulin antibodies fixed in the developing area, and red purple line in the reading. This means the colloidal gold-labelled anti-MPB64 antibody has migrated normally. Results are interpreted after 15 min.

Data management and analysis

Individual animal data collected from the abattoir was merged with laboratory results and entered into Microsoft excel 2007. Summary descriptive statistics including proportions and 95% confidence intervals were calculated. The chi-square test (univariable analysis), and multivariable logistic regression analyses were used to evaluate associations between cattle characteristics with the presence of bovine TB-like lesions at slaughter. Variables with a $p < 0.25$ in the univariate analysis were offered manually to the multivariable logistic regression model. The Hosmer-Lemeshow test was used to evaluate the overall model fit. Statistical significance was considered at $p < 0.05$.

Results
Animal characteristics

One thousand five hundred and 76 animals were inspected during the 6 months study period at the Mubende district Municipal abattoir. There were more male animals (62%) slaughtered than females, and a greater proportion (86%) of animals were young animals. Ankole-Sanga and East African Zebu accounted for over three quarters of the cattle slaughtered in Mubende district abattoir. Furthermore the highest proportion (35%) of animals slaughtered at this abattoir originated from Madudu sub-county while less than a quarter of the animals came from Mubende town council (Table 1).

Prevalence and risk factors associated with the presence of bovine TB-like lesions at slaughter

One hundred and 53 carcasses presented with bovine TB-like lesions. We observed that most of the bovine TB-like lesions were anatomically located in the thoracic region. These were mainly from the bronchial and mediastinal lymph nodes as shown in Fig. 1. Most of these lesions were caseocalcerous and grit-like particles embedded within the deep yellow to brownish coloured lesions. This grit-like texture was particularly felt on slicing through the lesion (Fig. 1).

Overall, 153 (9.7%, 95% CI 8.3–11.3) TB-like lesions were found among the 1,576 carcasses inspected

Table 1 Distribution of bovine TB-like lesions identified in cattle slaughtered at the Mubende Municipal abattoir, Uganda

Animal Characteristics	Inspected	Bovine TB-like Lesions	Risk (95% CI)	X^2	P-value
Total	1576	153	9.7 (8.3,11.3)	-	-
Sex					
Male	973	78	8.0 (6.4,9.9)	7.81	0.0052
Female	603	75	12.4 (9.9, 15.3)		
Breed					
Cross-breeds	53	18	34.0 (21.5, 48.3)	61.4	<0.0001
Zebu	619	83	13.4 (10.8, 16.3)		
Ankole	904	52	5.8 (4.3, 7.5)		
Age					
Young	1356	111	8.2 (6.8, 9.8)	24.5	<0.0001
Old	220	42	19.1 (14.1, 24.9)		
Origin					
Madudu	550	52	9.5 (7.1, 12.2)	14.71	0.002
Mubende T/C	210	17	8.1 (4.8, 12.6)		
Kasambya	395	25	6.3 (4.1, 9.2)		
Kasanda	421	59	14.0 (10.8, 17.7)		

* Univariable analysis using the Chi-square test
AUC=0.705, HL(df=8), *p* value= 0.1931

(Table 1). Females had a higher risk (12.4%, 95% CI 9.9–15.3) of presenting with a bovine TB-like lesion compared to males (8.0%, 95% CI 6.4–9.9) (Table 1). Older animals had a higher risk (19.1%, 95% CI: 14.1–24.9) of presenting with bovine TB-like lesions than younger animals (8.2, 95% CI: 6.8–9.8). Cross bred had the highest bovine TB-like lesion (34%, 95% CI: 21.5–48.3), while 13.4% (95% CI: 10.8–16.3) of Zebu cattle and 5.8% (95% CI: 4.3–7.5) of Ankole cattle presented with bovine TB-like lesions. Cattle originating in Kasanda had the highest risk (14%, 95% CI: 10.8–17.7) of presenting with a bovine TB-like lesion, followed by cattle from Madudu (9.5%, 95%CI: 7.5–12.2), Mubende (8.1%, 95% CI: 4.8–12.6), and Kasambya (6.3, 95% CI: 4.1–9.2) (Table 1). When adjusting for the combined effect of all these animal characteristics, results from the multivariable analysis (Table 2) indicate that bovine TB-like lesions were most likely to be observed in females (OR = 1.49, OR 95% CI: 1.06–2.13) than in males, and in older cattle (OR = 2.5, 95% CI: 1.64–3.7) when compared to young animals. When compared to Ankole cattle, Cross bred (OR = 6.5, OR 95% CI: 3.37–12.7) and Zebu cattle (OR = 2.57, 95% CI: 1.78–3.72) had higher odds of presenting with bovine TB-like lesions. Animal from Kasanda (OR = 2.5, 95% CI: 1.52–4.17) were more likely to disclose bovine TB-like lesions than cattle from Kasambya.

Table 2 Multivariable logistic regression results and adjusted odd ratios for animal sex, breed, age, and origin for the presence of bovine TB-like lesions among slaughtered cattle in Mubende district

Animal Characteristics	Adjusted Odd Ratio	Odds Ratio 95% CI	P-value
Sex			
Male	Ref	-	
Female	1.49	1.06–2.13	0.017
Breed			
Ankole	Ref	-	
Zebu	2.57	1.78–3.72	<0.001
Cross	6.5	2.7–12.57	<0.001
Age			
Young	Ref	-	-
Old	2.5	1.64-3.7	<0.001
Origin			
Kasambya	Ref	-	-
Madudu	1.66	1.00–2.76	0.05
Mubende TC	1.23	0.63-2.37	0.54
Kasanda	2.51	1.52-4.17	<0.001

Presence of *Mycobacterium tuberculosis* complex (MTC) in bovine TB- lesions at slaughter

Bacteria from the genus *Mycobacterium* was isolated from 13 bovine TB-like lesions (Table 3). Given this relatively small sample. Laboratory analysis shows that in 13

Table 3 Distribution of genus *Mycobacterium* among bovine TB-like lesions obtained at slaughter at the Mubende Municipal abattoir

Animal Characteristics	Number of bovine TB-like lesions	+ on ZN &/or Capilia (%)
Total	153	13 (8.5)
Sex		
Female	75	2 (2.66)
Male	78	11 (14.1)
Breed		
Zebu	83	6 (7)
Cross	18	1 (5.6)
Ankole	52	6 (12)
Age		
Old	42	2 (4.8)
Young	111	11 (9.9)
Origin		
Kasanda	59	4 (2.6)
Madudu	52	6 (3.9)
Kasambya	25	2 (1.3)
Mubende TC	17	1 (0.7)

(8.4%) of 153 bovine TB-like lesions belonged to the genus *Mycobacterium* based on the acid-fast test and MGIT culture. Twelve (7.8%) of the 13 were confirmed to be *Mycobacterium tuberculosis* complex (MTC) using the Capillia TB Neo. Two (1.3%) of the 12 are tentatively identified as *Mycobacterium bovis* based on their growth on MGIT, pyruvated LJ and positive result on Capillia TB Neo. We also recorded an 8.5 and 20% contamination rate on Pyruvated LJ and MGIT liquid culture system respectively (Table 4)

Discussion

The World Organisation for Animal Health (OIE) argues that the control of zoonotic TB in human population must start with controlling bovine TB in cattle populations [14]. In countries like Uganda insight into the magnitude of this health concern can only be achieved through applied utilisation of abattoir inspection data. This study therefore aimed at estimating the prevalence of bovine TB-like lesions at slaughter in Mubende district in the Uganda cattle corridor. Furthermore, we investigated factors associated with the presence of bovine TB-like lesions at slaughter in this area.

Approximately, one in ten slaughtered cattle had a bovine TB-like lesion identified during routine meat inspection. This observed prevalence of gross pathology suggestive of bovine TB is higher than previously reported in other districts like Kampala and Masaka [2, 5, 16]. This difference in prevalence between Kampala the capital of Uganda and this cattle corridor district is most likely a reflection of a "filtration process" where cattle are moved through rural cattle markets into the urban abattoirs (Kampala and Masaka), this process ensures that only cattle with a good body condition are sent to markets where they will fetch a higher price per unit kilogram of beef indirectly selecting for healthier animals [17]. This phenomenon has also been observed in Cameroon and Ethiopia [18, 19]. Although the bovine TB-like lesion prevalence here is comparable to Ethiopia [10, 20, 21], the currently observed prevalence is higher than the 7.3, 6.7 and 6.4% reported in Chad, Burkina Faso and Nigeria, respectively. This scenario in Mubende, could be contributing to the observed scenario with a spill-over into pigs, a phenomenon

Table 4 A comparison of diagnostic methods used to identify the causative agent of bovine TB-like lesions obtained from cattle slaughtered in Mubende Municipal abattoir between

Test outcome	TB-like lesions	Z&N	*Pyruvated LJ.	*MGIT	Capillia
Positive	153	13	2	13	12
Negative	0	140	138	108	1
Contaminated	N/A	N/A	13	32	NA
Total	153	153	153	153	13

associated with high transmissions pressure that has been previously reported in Mubende district [15].

Female cattle had a higher risk of having bovine TB-like lesions at slaughter in comparison to males, however, in a higher proportion of males genus *Mycobacterium* was isolated from bovine TB-like lesions. The results regarding MTC isolation should be viewed with caution due to the relatively small number of cultures obtained in this study. However, this finding is likely a reflection of cattle management practices especially with regards to the two sexes in pastoral settings. For example, a typical herd will have fewer bulls but kept them for a long time to maximize their breeding potential for the herd [22, 23], other males are kept as steers which are fattened and sold off for slaughter at an early age [17]. These different management practices are likely to increase the risk of exposure to *Mycobacterium* for males.

Older animals were more likely to have bovine TB-like lesions at slaughter which is in agreement with what has been previously reported in Cameroon and Ethiopia [24, 25]. The odds of cross-bred and Zebu animal having lesions at slaughter were 6.5 and 2.57 times those of an Ankole-Sanga breed, respectively. A similar association with breed has been reported in Cameroon [25]. However, there is a difference in risk status in these two studies, where being a mixed (Cross) breed was a protective factor in Cameroon while in our study in Mubende district cross bred cattle was found to have a higher risk of having bovine TB-like lesions when compared to Ankole cattle. In Cameroon up to 85% of the observed bovine TB-like lesions were due to genus *Mycobacterium* [25] while the opposite was found in this study, with only identifying MTC in 8.4% of bovine TB-like lesions. Animals with different geographical origins prior to slaughter had different risk of having bovine TB-like lesions, and this observation is similar to what had been reported in slaughtered pigs in Mubende [26]. Interestingly, bovine TB-like lesions in the study evaluating pigs in Mubende were mostly caused by non tuberculous mycobacteria [26].

Genus *Mycobacterium* was isolated from only 13 of the 153 lesions most of which belonged to the *Mycobacterium tuberculosis* complex (MTC). This represents 8.4% of MTC detected among all bovine TB-like lesions detected, and although this is lower than what has been reported in Ethiopia and Cameroon [19], these findings are in agreement with previous reports in Uganda [2, 5, 16]. These comparable prevalence estimates in different districts within the Uganda cattle corridor at different time points suggests that indeed the causative agents of bovine TB-like lesions are endemically present in the Uganda cattle corridor. However in-depth microbiological analysis reveals that over 19 per cent of the lesions are due to causes other-than *Mycobacterium*. There is therefore dire need to

understand what is causing these bovine TB-like lesions in order to safe guard the integrity of meat inspection as bovine TB detection tool.

From a laboratory diagnostic point of view, these results also reveal a substantial (20%) amount of contamination by other bacteria other than mycobacteria which triggered a much more stringent decontamination protocol on our part. This is reported to have an effect on the survival of mycobacteria in a sample especially those with a low bacillary load [27]. It is therefore possible that the decontamination process in this study reduced our ability to accurately identify/culture *Mycobacterium* spp from the obtained bovine TB-like lesion.

The presence of a large proportion of bovine TB-like lesions caused by other pathogens other than *Mycobacterium* has been reported in Europe [28], but in Africa, in Ghana and Chad 34 and 42% respectively of such lesions were reported to be caused by genus *Mycobacterium* [29, 30]. This has diagnostic implications as such gross pathology in which MTC is not identified, would be bovine TB 'false positives' which further compromises the diagnostic utility of the only routinely used public health first line detection of bovine TB. The recovery of members of MTC in isolates is particularly of public health concern as these are known to cause pulmonary and extra-pulmonary TB disease in human population [4]. This is expected to pose a considerable public health risk especially in such pastoral population where there is an overlap with HIV, human-TB, bovine TB and sociocultural practices that favour *M. bovis* transmission via direct contact with cattle and/or by consumption of animal products (mostly unpasteurized milk products).

Although, the main objective of this study was to investigate prevalence of bovine TB-like lesions (gross pathology) suggestive of bovine TB, we think it is essential to definitively establish if these bovine TB-like lesions are actually caused by *Mycobacterium* spp, and which species (and strains). For this specific study, insufficient funds and resources precluded the option of definitively identifying the *Mycobacterium* species found in these bovine TB-like lesions. Thus, we stored these isolates (alliquotes of DNA), and plan to analyse these samples further to species level. However, the presence of MTC at slaughter highlights the eminent risk of human exposure to bovine TB through foods of cattle origin in communities within the Uganda cattle corridor. This comes at a time when the need to identify zoonotic TB at-risk-communities is now more than ever critical in a bid to pre-empt the impact due to zoonotic TB as countries all over the world lay strategies for a TB free world in 2035 [31].

Conclusion

The findings in this study reveal that one in ten slaughtered cattle presents with gross pathology suggestive of bovine

TB (BTB-like lesions) in Mubende district in the Uganda cattle corridor districts, however we identified MTC in only 8.4% of these lesions. There is need therefore to understand the cause of all the other BTB-like lesions in order to safe guard diagnostic integrity of meat inspection.

Abbreviations

LJ: Lowenstein-Jensen; MGIT: Mycobacteria Growth Indicator Tube; MTC: Mycobacterium Tuberculosis Complex; OADCOleic: Acid, Albumin, Dextrose, And Catalase; OIE: World Organisation For Animal Health; PANTA: Polymyxin, Amphotericin B, Nalidixic Acid, Trimethoprim And Azlocillin; TB: Tuberculosis; UCC: Uganda Cattle Corridor; UNCST: Uganda National Council For Science And Technology; ZN: Ziehl-Neelsen

Acknowledgements

This work was done by the generous financial support from College of Veterinary Medicine and Biomedical Sciences Research Council, Colorado State University. The authors are grateful to the participant who contributed to all of the studies and the teams of investigators involved. We also would like to appreciate Gerald Mbowa and the district veterinary departments of the four districts for their corporation during sample collection.

Funding

Colorado State University, College of Veterinary Medicine and Biomedical Sciences, College Research Council (CRC) funding Program.

Authors' contributions

DPN contributed to the design, data collection, laboratory work, drafting and writing of the manuscript. CK contributed to conception, design, supervision and drafting of the manuscript. FO-P contributed to the acquisition of funds, data analysis and writing of the manuscript. AM contributed to conception, design, supervision and drafting of the manuscript. All authors have read and approved the final manuscript.

Competing interests

The authors declare that they have no competing interests.

Consent for publication

Not applicable.

Author details

[1]Department of Biosecurity Ecosystem and Veterinary Public Health, College of Veterinary Medicine, Animal Resources and Biosecurity, Makerere University, P.O. Box 7062, Kampala, Uganda. [2]Division of Genetics and Genomics, The Roslin Institute, University of Edinburgh, Easter bush Campus, EH259RG Edinburgh, UK. [3]College of Veterinary Medicine and Biomedical Sciences, Department of Clinical Sciences & Mycobacteria Research Laboratories, Colorado State University, Fort Collins, CO, USA.

References

1. Cosivi O, Meslin FX, Daborn CJ, Grange JM. Epidemiology of Mycobacterium bovis infection in animals and humans, with particular reference to Africa. Rev Sci Tech. 1995;14:733–46.

2. Oloya J, Kazwala R, Lund A, Opuda-Asibo J, Demelash B, Skjerve E, et al. Characterisation of mycobacteria isolated from slaughter cattle in pastoral regions of Uganda. BMC Microbiol. 2007;7:95.

3. Kankya C, Muwonge A, Olet S, Munyeme M, Biffa D, Opuda-Asibo J, et al. Factors associated with pastoral community knowledge and occurrence of mycobacterial infections in human-animal interface areas of Nakasongola and Mubende districts, Uganda. BMC Public Health. 2010;10:471.

4. Neill SD, Skuce RA, Pollock JM. Tuberculosis-new light from an old window. J Appl Microbiol. 2005;98:1261–9. Available from: http://onlinelibrary.wiley.com/doi/10.1111/j.1365-2672.2005.02599.x/epdf.

5. Asiimwe BB, Asiimwe J, Kallenius G, Ashaba FK, Ghebremichael S, Joloba M, et al. Molecular characterisation of Mycobacterium bovis isolates from cattle carcases at a city slaughterhouse in Uganda. Vet Rec. 2009;164:655–8.

6. Aagaard C, Govaerts M, Meikle V, Gutiérrez-Pabello J a, McNair J, Andersen P, et al. Detection of bovine tuberculosis in herds with different disease prevalence and influence of paratuberculosis infection on PPDB and ESAT-6/CFP10 specificity. Prev Vet Med [Internet]. Elsevier B.V.; 2010 [cited 2014 Nov 20];96:161–9. Available from: http://www.sciencedirect.com/science/article/pii/S016758771000173X.

7. Muyoyeta M, Mwanza WC, Kasese N, Cheeba-Lengwe M, Moyo M, Kaluba-Milimo D, et al. Sensitivity, specificity, and reproducibility of the Capilia TB-Neo Assay. J Clin Microbiol. 2013;51(12):4237–9.

8. Hillemann D, Rüsch-Gerdes S, Richter E. Application of the Capilia TB assay for culture confirmation of Mycobacterium tuberculosis complex isolates. Int J Tuberc Lung Dis. 2005;9:1409–11.

9. Muwonge A, Malama S, Johansen TB, Kankya C, Biffa D, Ssengooba W, et al. Molecular Epidemiology, Drug Susceptibility and Economic Aspects of Tuberculosis in Mubende District, Uganda. PLoS One. 2013;8(5):e64745.

10. Bekele M. Evaluation of Routine Meat Inspection Procedure to Detect Bovine Tuberculosis Suggestive Lesions in Jimma Municipal Abattoir, South West Ethiopia. Glob Vet. 2011;6:172–9.

11. WFO. Re-greening the Ugandan cattle corridor [Internet]. Water food Organ. 2012. [cited 2015 Jul 22]. Available from: http://waterandfood.org/2011/10/21/re-greening-the-ugandan-cattle-corridor/]

12. FAO. Cattle population [Internet]. Rome. 2005. Available from: http://www.fao.org/faostat/en/#search/Cattle.

13. Torell R, Bruce B, Kvasnicka B, Conley K. Methods of Determining Age of Cattle. 2003.

14. OIE. Bovine Tuberculosis. Paris: OIE; 2009.

15. Muwonge A, Johansen TB, Vigdis E, Godfroid J, Olea-Popelka F, Biffa D, et al. Mycobacterium bovis infections in slaughter pigs in Mubende district, Uganda: a public health concern. BMC Vet Res. 2012;8:168. Available from: http://www.biomedcentral.com/1746-6148/8/168.

16. Nasaka J. Occurence of Bovine Tuberculosis In Slaughtered Cattle. Msc thesis Environ Nat Sci Manag. 2014. pp. 1-86. Available from: http://citeseerx.ist.psu.edu/viewdoc/download?doi=10.1.1.904.8928&rep=rep1&type=pdf.

17. Karugia J, Wanjiku J, Nzuma J, Gbegbelegbe S, Macharia E, Massawe S, et al. The impact of non-tariff barriers on maize and beef trade in East Africa. East [Internet]. 2009;1–16. Available from: http://www.resakss.org/sites/default/files/pdfs/the-impact-of-non-tariff-barriers-on-maize-and-bee-42386.pdf.

18. Tschopp R, Schelling E, Hattendorf J, Aseffa A, Zinsstag J. Risk factors of bovine tuberculosis in cattle in rural livestock production systems of Ethiopia. Prev Vet Med. 2009;89:205–11. Available from: http://www.pubmedcentral.nih.gov/articlerender.fcgi?artid=2706391&tool=pmcentrez&rendertype=abstract.

19. Awah Ndukum J, Kudi AC, Bradley G, Ane-Anyangwe IN, Fon-Tebug S, Tchoumboue J. Prevalence of bovine tuberculosis in abattoirs of the littoral and Western highland regions of cameroon: a cause for public health concern. Vet Med Int. 2010;2010:495015. Available from: http://www.pubmedcentral.nih.gov/articlerender.fcgi?artid=2896641&tool=pmcentrez&rendertype=abstract.

20. Nega M, Mazengia H, Mekonen G. Prevalence and zoonotic implications of bovine tuberculosis in Northwest Ethiopia. Int J Med Med Sci. 2012;2:182–92. Available from: http://internationalscholarsjournals.org/download.php?id=226387237097549607.pdf&type=application/pdf&file=prevalence%20and%20zoonotic%20implications%20of%20bovine%20tuberculosis%20in%20northwest%20ethiopia%20.pdf.

21. Aylate A, Shah SN, Aleme H, Gizaw TT. Bovine tuberculosis: prevalence and diagnostic efficacy of routine meat inspection procedure in Woldiya municipality abattoir north Wollo zone, Ethiopia. Trop Anim Health Prod. 2013; 45:855–64. Available from: http://www.ncbi.nlm.nih.gov/pubmed/23080340.

22. Inangolet FO, Demelash B, Oloya J, Opuda-Asibo J, Skjerve E. A cross-sectional study of bovine tuberculosis in the transhumant and agro-pastoral cattle herds in the border areas of Katakwi and Moroto districts, Uganda. Trop Anim Health Prod. 2008;40:501–8.

23. Oloya J, Opuda-Asibo J, Djønne B, Muma JB, Matope G, Kazwala R, et al. Responses to tuberculin among Zebu cattle in the transhumance regions of Karamoja and Nakasongola district of Uganda. Trop Anim Health Prod. 2006; 38:275–83.

24. Ameni G, Aseffa A, Engers H, Young D, Gordon S, Hewinson G, et al. High prevalence and increased severity of pathology of bovine tuberculosis in

Holsteins compared to zebu breeds under field cattle husbandry in central Ethiopia. Clin Vaccine Immunol. 2007;14:1356–61. Available from: http://www.pubmedcentral.nih.gov/articlerender.fcgi?artid=2168113&tool=pmcentrez&rendertype=abstract.

25. Egbe N, Muwonge A, Ndip L, Kelly R, Sander M, Tanya V, Ngu Ngwa V, et al. Abattoir-based estimates of mycobacterial infections in Cameroon.Under-review. Sci Reports. 2016;6:24320.

26. Muwonge A, Kankya C, Johansen TB, Djønne B, Godfroid J, Biffa D, et al. Non-tuberculous mycobacteria isolated from slaughter pigs in Mubende district, Uganda. BMC Vet Res. 2012;8:52.

27. Chatterjee M, Bhattacharya S, Karak K, Dastidar S. Effects of different methods of decontamination for successful cultivation of Mycobacterium tuberculosis. Indian J Med Res. 2013;138:541–8.

28. Costello E, Quigley F, Flynn O, Gogarthy A, Mcguirk J, MURPHY A, et al. Laboratory examination of suspected tuberculous lesions detected on abattoir post mortem examination of cattle from non-reactor herds. Ir Vet J. 1998;51:248–50.

29. Atiadeve SK, Gyamfi OK, Mak-Mensah E, Galyuon IKA, Owusu D, Bonsu FA, et al. Slaughter surveillance for tuberculosis among cattle in three metropolitan abattoirs in Ghana. J Vet Med Anim Health. 2014;6:198–207. Available from: http://www.academicjournals.org/journal/JVMAH/article-full-text-pdf/AC9991745675.

30. Diguimbaye-djaibé C, Hilty M, Ngandolo R, Mahamat HH, Pfyffer E, Baggi F, et al. Tuberculous Lesions in Chadian Zebu Carcasses. Emerg Infect Dis. 2006;12:769–71.

31. Mario R. The end TB strategy. Geneva: World Health Organisation; 2014. p. 1–20. Available from: http://www.who.int/tb/End_TB_brochure.pdf?ua=1.

17

Oral administration of chestnut tannins to reduce the duration of neonatal calf diarrhea

F. Bonelli[1]* ⓘ, L. Turini[2], G. Sarri[1], A. Serra[2], A. Buccioni[3] and M. Mele[2]

Abstract

Background: Neonatal calf diarrhea is generally caused by infectious agents and is a very common disease in bovine practice, leading to substantial economic losses. Tannins are known for their astringent and anti-inflammatory properties in the gastro-enteric tract. The aim of this study was to evaluate the effect of the oral administration of chestnut tannins (*Castanea sativa* Mill.) in order to reduce the duration of calf neonatal diarrhea. Twenty-four Italian Friesian calves affected by neonatal diarrhea were included. The duration of the diarrheic episode (DDE) was recorded and the animals were divided into a control group (C), which received Effydral® in 2 l of warm water, and a tannin-treated group (T), which received Effydral® in 2 l of warm water plus 10 g of extract of chestnut tannins powder. A Mann-Whitney test was performed to verify differences for the DDE values between the two groups.

Results: The DDE was significantly higher in group C than in group T ($p = 0.02$), resulting in 10.1 ± 3.2 and 6.6 ± 3.8 days, respectively.

Conclusions: Phytotherapic treatments for various diseases have become more common both in human and in veterinary medicine, in order to reduce the presence of antibiotic molecules in the food chain and in the environment. Administration of tannins in calves with diarrhea seemed to shorten the DDE in T by almost 4 days compared to C, suggesting an effective astringent action of chestnut tannins in the calf, as already reported in humans. The use of chestnut tannins in calves could represent an effective, low-impact treatment for neonatal diarrhea.

Keywords: Calf, Neonatal diarrhea, Chestnut tannins, Phytotherapy

Background

Calf diarrhea is a very common disease in bovine practice, and is generally caused by infectious agents [1]. The most common pathogens involved are *Rotavirus*, *Coronavirus*, *Cryptosporidium parvum* and *Escherichia coli*, especially in animals less than one month old [1, 2]. Calf diarrhea causes substantial economic losses in the dairy industry, due treatment costs, a decreased growth rate in the calves, and a higher replacement rate due to culling or death [2].

In order to reduce the presence of antibiotic molecules in the food chain and in the environment, phytotherapic treatments are now commonly used both in human [3]

and in veterinary medicine. Various formulations of polyphenols, such as bacteriostatic, anti-clostridial and astringent, have been tested, especially in monogastric nutrition [4–7]. Some phytotherapic options have been evaluated for gastrointestinal diseases, such as gastric ulcers and gastric lesions, in rats, piglets, horses, and calves [8–10]. Pomegranate-residue supplements have been used in neonatal calves affected by *Cryptosporidium parvum*, resulting in a reduction in the fecal oocyst count, as well as in intensity and duration of diarrhea [11].

Tannins are a complex group of polyphenolic compounds which are present in several plants as secondary metabolites against pathogens [12]. In ruminants, in vitro and in vivo trials have demonstrated that tannins can improve animal performance and reduce the impact of gastrointestinal parasitism, nitrogen pollution, and

---

* Correspondence: francesca.bonelli@unipi.it
[1]Dipartimento di Scienze Veterinarie, via Livornese snc, Università di Pisa, via Livornese, 56122, San Piero a Grado, Pisa, Italy
Full list of author information is available at the end of the article

methane emission from rumen fermentation [13, 14]. Tannins may interfere with digestive processes by binding dietary protein, by modulating the activity of rumen micro-organisms, and by reducing the growth of the bacterial population [15–17].

In ruminants, tannins tend to decrease the rate of protein degradation in the rumen [18]. Tannins, in fact, inhibit the growth of proteolytic bacteria and form tannin–protein complexes in the rumen. This effect may significantly vary according to the chemical structure of the tannin and to the amount of tannin included in the diet. However, the reduction of protein degradation in the rumen increases the total amount of protein digested in the duodenum, because the tannin-protein complexes are dissociated in the abomasum, releasing protein. As a consequence, the use of moderate doses of tannins in the diet of ruminants usually improves animal performance [18, 19].

The astringent and anti-inflammatory effects of tannins on the gastro-enteric tract have been demonstrated in avian [5, 20] and swine species [9], and the effects of tannins on liver function have been evaluated in newborn calves [21].

The aim of the present study was to evaluate the effect of oral administration of chestnut tannins (*Castanea sativa*) in the treatment of calf neonatal diarrhea.

Methods

This in vivo blinded study was approved by the Institutional Animal Care and Use Committee (OBA, Pisa, prot. n. 33,479/2016 of 29.06.2016) and was carried out at the University of Pisa's dairy farm, where nearly 100 animals are maintained in free-stall conditions.

During the study period, the population consisted of 40 Italian Friesian calves aged between 1 and 60 days. All the calves underwent the same management condition. Briefly, immediately after birth they were weighed and then housed in a single straw-bedding pen (2.5 × 2 m) that leads contact between each other. Two L of good colostrum (≥50 g/L of Ig) from their own dam, or from the colostrum bank, were administered as soon as the calf could drink (30 min – 2 h). Another 2 L were administered within the next 4–8 h in order to achieve a good passive transfer immunity [22]. All the calves received a total of 4 L of colostrum, twice a day, until the third day of life. Then, they received 6 L of whole milk at 39 °C, twice a day until the third week of life.

All the feeding procedures were conducted by an expert operator and by using a nipple bucket. Even when there were cases of diarrhea, there were still no feeding restrictions or changes of feeding regime. From the third day of life, fresh and clean water were provided to each calf ad libitum. Free choice-hay was administered after the first week of life. The calves were weighed and moved into a collective pen after the third week of life.

The inclusion criteria were calves up to three weeks of life and the manifestation of diarrhea, defined as a fecal score ≥ 1 [1]. Briefly, the fecal score represents the evaluation of feces fluidity as reported in literature: score 0 means "normal", thus firm, but not hard and the original form is distorted slightly after dropping to the floor and settling; score 1 means "soft", thus does not hold its form, but piles and spreads slightly (like soft-serve ice cream); score 2 means "runny", thus spreads readily to about 6 mm depth. (i.e., pancake batter); score 3 means "watery", thus liquid consistency, splatters. (i.e., orange juice) [23]. Since the first day of diarrhea (T0), the fecal score (FS) was recorded daily by the same expert operator until the complete recovery from diarrhea, at the same time, the physical examination has been done. The duration of a diarrheic episode (DDE) was defined as the period (in days) between the first diarrhea outbreak (fecal score ≥ 1) and the normalizing of the FS (fecal score = 0). Once the diarrhea had started, a fecal sample was collected from each calf by manual restraint. A gloved, lubricated finger was passed gently through the anus in order to massage the rectal wall and to stimulate rectal evacuation. Fresh feces were then collected in two different sterile tubes: one aliquot was immediately tested with a rapid ELISA test (test strips for detection of *Rotavirus*, *Coronavirus*, *E. coli* F5 and *C. parvum* in bovine feces, Biox Diagnostics, Belgium), while the second aliquot was stored in a refrigerated bag and evaluated within one hour for gastrointestinal parasites, according to Izzo et al. (2015) [1].

At the inclusion time, calves were randomly assigned to a control group (C) or a tannin-treated group (T), both made up of 12 calves (6 males and 6 females in each of the two groups). All the calves enrolled in group C received Effydral® (Italy Zoetis Ltd.) (sodium chloride 2.34 g, potassium chloride 1.12 g, sodium bicarbonate 6.72 g, citric acid anhydrous 3.84 g, lactose monohydrate 32.44 g, glycine 2.25 g) in 2 L of warm water q24h. The calves assigned to group T received Effydral® (Italy Zoetis Ltd.) in 2 L of warm water plus 10 g of chestnut tannins as extract powder (750 g/kg of dry matter equivalent of tannic acid; Mauro Saviola Group srl, Radicofani, Siena, Italy) q24h. The chemical composition of the powder is described in Campo et al. (2016) [24]. The powder adopted in the present study was produced in a single batch and analyzed by the manufacturer at the beginning of the study. Both solutions were administered using a graduated calves bottle in order to ensure the intake of the entire quantity. The bottle was equipped with a flexible rubber nipple (10 cm of length) specific for calf feeding.

Calves in both groups received the solutions until the normalization of the FS. Before the administration of the treatments, all the calves were daily submitted to a complete physical examination and FS evaluation until the resolution of diarrhea. Physical examination included evaluation from a distance plus hands-on examination, focusing on: physical appearance, body weight, body condition, head, mouth, eyes, ears, neck and back, thorax, abdomen, umbilicus, musculoskeletal system, perianal region, body temperature, feces, urine, and external genitalia [25, 26]. Dehydration status was assessed with a score system [27]. Milk intake was recorded during the entire diarrheic episode.

At the conclusion of the study, the female calves were raised in the farm as rearing cows, while the male calves were sold for fattening as meat animals.

Data concerning the weight at birth and at the third week of life, the age of diarrhea onset and the T0 fecal scores (T0-FS) recorded for both groups were assessed for normal distribution by the Shapiro-Wilk normality test and then a Mann-Whitney test was applied in order to verify differences between the two groups at the inclusion time [28].

The average daily gain between birth date and third week of life was calculated for all the calves in both groups. Data concerning the average daily gain, the DDE and fecal scores recorded throughout the DDE were analyzed by a linear model including treatments and sex and their interaction as fixed factors. A Shapiro-Wilk normality test was performed to assess data distribution. A Mann-Whitney test was carried out to verify differences between the two groups regarding the average daily gain, the DDE and the fecal scoring [28]. Values with $P < 0.05$ were considered statistically significant.

Results

Twenty-four out of the 40 calves, 12 males and 12 females met the inclusion criteria, and were thus included in this study. The average weight at birth was 41 kg (minimum value of 34.5 kg and maximum value of 44 kg) and 37.5 kg (minimum value of 35 kg and maximum value of 48.5 kg) for groups C and T, respectively. The mean age at T0 was 7.9 ± 4.8 days for group C, and 7.2 ± 2.9 days for group T. The mean T0-FS were 1.6 ± 0.5 and 2.0 ± 0.7 for groups C and T, respectively. No significant differences were found in terms of the weight at birth, the age of diarrhea onset, and the T0-FS between the two groups ($P > 0.05$).

There were no alterations at the physical examination, and the hydration status was normal for all calves during the entire study period. No differences between groups were found for milk intake. Thus, no pharmacological treatments or fluid therapy were needed.

Seven out of twelve (7/12) calves belonging to group C resulted positive for *Cryptosporidium parvum*, 2/12 for *Rotavirus* and *Coronavirus,* and 3/12 negative. Nine/12 calves belonging to group T were positive to *Cryptosporidium parvum*, 2/12 for *Rotavirus* and *Coronavirus,* and 1/12 were negative to the rapid ELISA test performed. Fecal flotation did not show intestinal parasites.

The average weight at the third week of life was 53 kg (minimum value of 42 kg and maximum value of 64 kg) and 46.7 kg (minimum value of 46 kg and maximum value of 52.5 kg) for groups C and T, respectively. The average daily gain was 0.643 kg/day (minimum value of 0.143 kg/day and maximum value of 1.024 kg/day), and 0.442 kg/day (minimum value of 0.190 kg/day and maximum value of 0.524 kg/day), for groups C and T, respectively. Differences between the two groups were not significant.

The mean DDE in days was 10.3 ± 3.5 for group C, and 6.4 ± 3.9 for group T (Table 1). A significant difference between the two groups was found ($P = 0.01$). The mean fecal score recorded throughout the DDE for group C was 1.6 ± 0.5, while for group T the score was 1.4 ± 0.8, with a statistically significant difference between the two groups ($P = 0.03$). No statistically significant differences were found between male and female calves for the DDE and fecal score recorded throughout the DDE. No calves refused the tannins solution.

Discussion

Due to the high amount of antimicrobials used every year in the livestock industry and possible cross-resistance between human and animal pathogens [29, 30], for many years alternative treatments have been investigated [11].

Chestnut tannins have been proposed to modulate rumen fermentation and the degradability of food proteins in cattle [16, 17]. However, there is little information on the effects of chestnut tannins on diarrhea in neonatal calves, especially in relation to different pathogens.

The pathogen that was most isolated in our population of diarrheic calves was *Cryptosporidium parvum*, in line with those reported in the literature [1, 2, 11]. The mean age of diarrhea onset for all calves included was also in line with literature data [1, 2].

It is well known that natural plant tannins are able to control intestinal parasites in ruminants [15] and, more

Table 1 Duration of the diarrheic episode (DDE), fecal score recorded throughout the DDE of control group (C) and tannin-treated group (T) expressed as mean (X) ± standard deviation (SD)

	Group C (X ± DS)	Group T (X ± DS)	P value
DDE (days)	10.3 ± 3.5	6.4 ± 3.9	0.0184
Fecal score recorded throughout the DDE	1.6 ± 0.5	1.4 ± 0.8	0.618

in general, have antimicrobial properties [31]. The supplementation of chestnut extract in grazing heifers has been shown to lead to a significant increase in the average daily gain due to a decrease in fecal parasites infections, above all nematodes [32]. Several mechanisms of actions have been proposed for tannins, including enzyme inhibition and substrate or metal ions deprivation, and bacterial cell membrane integrity [31]. However, most studies have reported the effects of tannins (above all condensed tannins) on bacteria, fungi and nematode, whereas only few have investigated the effect of tannins on protozoa and, in particular, on *Cryptosporidium parvum* [11, 33].

The anti-protozoal activity of polyphenols has been reported for *Eimeria* [34–36]. A suggested mechanism is the ability of tannins to directly decrease the viability of the larval stage and disrupt egg hatching, as previously observed also for nematodes [15].

However, there have been conflicting results. Bhatta et al. (2009) found a decrease in protozoa when hydrolysable tannins from six different plant sources were applied [37]. Similar results were obtained by Tan et al. (2011), Benchaar et al. (2008) and Carulla et al. (2005) who reported that condensed tannins from *Leucaena*, quebracho and from *Acacia mearnsii* respectively, can decrease protozoal numbers [38–40]. In contrast, Vasta et al. (2010) found that quebracho tannins were able to increase protozoa in rumen liquor [41].

Differences in the concentration of tannins, plant sources, protozoal species and environment considered in such studies may explain the conflicting results obtained, because these parameters play an important role in the antiprotozoal activity of polyphenols.

To the best of our knowledge this is the first study that has investigated orally administrating tannins as a treatment for calf diarrhea. Tannins solution appeared to be pleasant for all the calves enrolled in the study and was easy to administer. The length of diarrhea was almost four days shorter in the group treated with tannins compared to the control group. Chestnut tannins seem to shorten the length of diarrhea in calves, as already reported in humans [19]. Chestnut polyphenols are ellagitannins and those from wood distillation are particularly rich in ellagic acid, such as pomegranate polyphenols [42].

Few studies have reported on the absorption and metabolism of ellagitannins in animal models. Studies of rat intestinal content demonstrated that at the caecum level ellagitannins are hydrolyzed to ellagic acid [43]. Other authors have detected free ellagic acid in human plasma after 1 h post ingestion of pomegranate and attributed this to the release after hydrolysis of ellagitannins according to an optimal physiological pH and gut microbiota activity. In addition, Cerdà et al. (2003;

2004; 2005) suggested a microbial involvement at the colon level in converting ellagic acid into urolithins, which are potent anti-inflammatory metabolites [44–46].

Although the length of diarrhea was significantly shorter in calves from group T, no significant difference was found in terms of average daily gain between control calves and calves that received chestnut tannins. Similarly, a study based on tannins from pomegranate reported no effects of tannins on average daily gain in the first 60 days of life in dairy calves [11].

Further studies are needed in order to investigate the possible effect of hydrolysable tannins on specific etiological pathogens, in particular on *Cryptosporidium parvum*, as already reported for the extract of pomegranate polyphenols [11]. Also, increasing the number of animals included might clarify the effect of chestnut tannins on the average daily gain in calves affected by diarrhea.

Conclusions

Chestnut tannins might represent a low-impact treatment of neonatal diarrhea in calves. However, the effects of chestnut tannins on the onset of diarrhea and on the weight gain of calves need further investigation due to the lack of data on the metabolism of ellagitannins in ruminants.

Abbreviations
C: control group; DDE: duration of a diarrheic episode; ELISA: Enzyme-Linked Immunosorbent Assay; FS: fecal score; T: tannin-treated group

Acknowledgements
The authors would like to thank Mauro Saviola Group srl, Radicofani, Siena, Italy for providing the chestnut tannin extract.

Funding
This research did not receive any specific grant from funding agencies in the public, commercial, or not-for-profit sectors.

Authors' contributions
FB, LT and MM designed and conducted the experiment, analyzed the resulting data and drafted the manuscript. FB, LT and GS conducted the experiment. AB and AS analyzed the resulting data and drafted the manuscript. All the authors read, revised, and approved the final manuscript.

Consent for publication
Not applicable.

Competing interests
The authors declare that they have no competing interests.

Author details

[1]Dipartimento di Scienze Veterinarie, via Livornese snc, Università di Pisa, via Livornese, 56122, San Piero a Grado, Pisa, Italy. [2]Centro di Ricerche Agro-ambientali "E. Avanzi" – Università di Pisa, via Vecchia di Marina, 6, 56122, San Piero a Grado, Pisa, Italy. [3]Dipartimento di Scienze delle Produzioni Agroalimentari e Ambientali, Università di Firenze, via delle Cascine, 5, 50100 Florence, Italy.

References

1. Izzo M, Gunn AA, House JK. Neonatal diarrhea. In: Smith BP, editor. Large Animal Internal Medicine. Elsevier: United States; 2015. p. 314–35.

2. Cho Y, Yoon KJ. An overview of calf diarrhea-infectious etiology, diagnosis, and intervention. J Vet Sci. 2014;15(1):1–17.

3. Tripathi NN, Mishra AK, Tripanthi S. Antibacterial potential of plant volatile oils: a review. Proc Nat Acad Sci India Sect B. 2011;81:23–68.

4. Laudato M, Capasso R. Useful plants for animal therapy. OA Alternative Medicine. 2013;1(1):1.

5. Tosi G, Massi P, Antongiovanni M, Buccioni A, Minieri S, Marenchino L, Mele M. Efficacy test of a hydrolysable tannin extract against necrotic enteritis in challenged broiler chickens. Italian J An Sci. 2013;12(3):386–9.

6. Mosele JI, Macià A, Motilva MJ. Metabolic and microbial modulation of the large intestine ecosystem by non-absorbed diet phenolic compounds: a review. Molecules. 2015;20(9):17429–68.

7. Brenes A, Viveros A, Chamorro S, Arija I. Use of polyphenol-rich grape by-products in monogastric nutrition A review. Anim Feed Sci Technol. 2016; 211:1–17.

8. Dekanski D, Risti S, Mitrovi DM. Antioxidant effect of dry olive (Olea europaea L.) leaf extract on ethanol-induced gastric lesions in rats. Med J Nutr Metab. 2009;2(3):205–11.

9. Ayrle H, Mevissen M, Kaske M, Nathues H, Gruetzner N, Melzig M, Walkenhorst M. Medicinal plants – prophylactic and therapeutic options for gastrointestinal and respiratory diseases in calves and piglets? A systematic review. BMC Vet Res. 2016;12:89.

10. Bonelli F, Busechian S, Meucci V, Caporrino G, Briganti A, Rueca F, Zappulla F, Ferini E, Ghiandai L, Sgorbini M. pHyloGASTRO in the treatment of equine gastric ulcer lesions. J Equine Vet Sci. 2016;46:69–72.

11. Weyl-Feinstein S, Markovics A, Eitam H, Orlov A, Yishay M, Agmon R, Miron J, Izhaki I, Shabtay A. Effect of pomegrate-residue supplement on Cryptosporidium parvum oocystis shedding in neonatal calves. J Dairy Sci. 2014;97:5800–5.

12. Frutos P, Hervas G, Giraldez FJ, Mantecon AR. Review. Tannins and ruminant nutrition. Span J Agric Res. 2004;2(2):191–202.

13. Animut G, Puchala R, Goetsch A, Patra A, Sahlu T, Varel V, Wells J. Methane emission by goats consuming different sources of condensed tannins. Anim Feed Sci Technol. 2008;144:228–41.

14. Getachew G, Pittroff W, Putnam DH, Dandekar A, Goyal S, Depeters EJ. The influence of addition of gallic acid, tannic acid, or quebracho tannins to alfalfa hay on in vitro rumen fermentation and microbial protein synthesis. Anim Feed Sci Technol. 2008;140:444–61.

15. Min BR, Hart SP. Tannins for suppression of internal parasites. J Anim Sci. 2003;81(E. Suppl. 2):E102–9.

16. Buccioni A, Pauselli M, Viti C, Minieri S, Pallara G, Roscini V, Rapaccini S, Marinucci MT, Lupi P, Conte G, Mele M. Milk fatty acid composition, rumen microbial population, and animal performances in response to diets rich in linoleic acid supplemented with chestnut or quebracho tannins in dairy ewes. J Dairy Sci. 2015;98(2):1145–56.

17. Buccioni A, Serra A, Minieri S, Mannelli F, Cappucci A, Benvenuti D, Rapaccini S, Conte G, Mele M. Milk production, composition, and milk fatty acid profile from grazing sheep fed diets supplemented with chestnut tannin extract and extruded linseed. Small Rum Res. 2015;130:200–7.

18. Patra AK, Saxena J. Exploitation of dietary tannins to improve rumen metabolism and ruminant nutrition. J Sci Food Agric. 2011;91(1):24–37.

19. Ashok PK, Upadhyaya K. Tannins are stringent. J Pharmacogn Phytochem. 2012;1(3):45–50.

20. Graziani R, Tosi G. In vitro anti-microbial activity of SILVA FEED ENC® tannin on bacterial strains of poultry origin. Worlds Poult Sci J. 2006;62:384–5.

21. Wieland M, Weber BK, Hafner-Marx A, Sauter-Louis C, Bauer J, Knubben-Schweizer G, Metzner M. A controlled trial on the effect of feeding dietary

22. chestnut extract and glycerol monolaurate on liver function in newborn calves. J Anim Physiol Anim Nutri. 2015;99:190–200.

22. Godden S. Colostrum management for dairy calves. Vet Clin North Am Food Anim Pract. 2008;24(1):19–39.

23. Larson LL, Owen FG, Albright JL, Appleman RD, Lamb RC, Muller LD. Guidelines toward more uniformity in measuring and reporting calf experimental data. J Dairy Sci. 1977;60(6):989–91.

24. Campo M, Pinelli P, Romani A. Hydrolyzable tannins from sweet chestnut fractions obtained by a sustainable and eco-friendly industrial process. Nat Prod Commun. 2016;11(3):409–15.

25. Bonelli F, Castagnetti C, Iacono E, Corazza M, Sgorbini M. Evaluation of some physical, haematological and clinical chemistry parameters in healthy newborn Italian Holstein calves. Am J Anim Vet Sci. 2015; 10(4):230–4.

26. House JK, Gunn AA, Chuck G, McGuirk SM. Initial management and clinical investigation of neonatal disease. In: Smith BP, editor. Large Animal Internal Medicine. Elsevier: United States; 2015. p. 294–301.

27. Constable PD, Gohar HM, Morin DE, Thurmon JC. Use of hypertonic saline-dextran solution to resuscitate hypovolemic calves with diarrhea. Am J Vet Res. 1996;57(1):97–104.

28. SAS. User's guide: statistics, version 8.0 edition. SAS Inst. Inc., Cary, NC, USA. 1999.

29. World Health Organization (WHO): Second joint FAO/OIE/WHO expert workshop on non-human antimicrobial usage and antimicrobial resistance: Management options. Oslo, WHO, 2004.

30. Marshall BM, Levy SB. Food animals and antimicrobials: impacts on human health. Clin Microbiol Rev. 2011;24:718–33.

31. Scalbert A. Antimicrobial properties of tannins. Phytochemistry. 1991;30(12): 3875–83.

32. Min BR, Pinchak WE, Hernandez K, Hernandez C, Hume ME, Valencia E, Fulford JD. Effects of plant tannin supplementation on animal responses and in vivo ruminal bacterial populations associated with bloat in heifers grazing wheat forage. The Professional Animal Scientist. 2012;28:464–72.

33. Al-Mathal EM, Alsalem AM. Pomegranate (Punica granatum) peel is effective in a murine model of experimental Cryptosporidium parvum. Exp Parasitol. 2012;131:350–7.

34. Landau S, Azaizeh H, Mukladaa H, Glasserc T, Ungara ED, Barama H, Abbasd N, Markovicsd A. Anthelmintic activity of Pistacia lentiscus foliage in two middle eastern breeds of goats differing in their propensity to consume tannin-rich browse. Vet Parasitol. 2010;173:280–6.

35. Markovics A, Cohen I, Muklada H, Glasser TA, Dvash L, Ungar ED, Azaizeh H, Landau SY. Consumption of Pistacia lentiscus foliage alleviates coccidiosis in young goats. Vet Parasitol. 2012;25:165–9.

36. Burke JM, Miller JE, Terrill TH, Orlik ST, Acharya M, Garza JJ, Mosjidis JA. Sericea lespdeza as an aid in the control of Emeria spp. in lambs. Vet Parasitol. 2013;193:39–46.

37. Bhatta R, Uyeno Y, Tajima K, Takenaka A, Yabumoto Y, Nonaka I, Enishi O, Kurihara M. Difference in the nature of tannins on in vitro ruminal methane and volatile fatty acid production and on methanogenic archaea and protozoal populations. J Dairy Sci. 2009;11:5512–22.

38. Carulla JE, Kreuzer M, Machmuller B, Hes HD. Supplementation of Acacia mearnsii tannins decreases methanogenesis and urinary nitrogen in forage-fed sheep. Austral J Agric Res. 2005;56:961–70.

39. Benchaar C, McAllister TA, Chouinard PY. Digestion, ruminal fermentation, ciliate protozoal populations, and milk production from dairy cows fed Cinnamaldehyde, Quebracho condensed tannin, or Yucca schidigera Saponin extracts. J Dairy Sci. 2008;12:4765–77.

40. Tana HY, Sieoa CC, Abdullaha N, Lianga JB, Huanga XD, Ho YW. Effects of condensed tannins from Leucaena on methane production, rumen fermentation and populations of methanogens and protozoa in vitro. Anim Feed Sci Technol. 2011;169:185–93.

41. Vasta V, Yanez-Ruiz DR, Mele M, Serra A, Luciano G, Lanza M, Biondi L, Priolo A. Bacterial and protozoal communities and fatty acid profile in the rumen of sheep fed a diet containing added tannins. Appl Envir Microbiol. 2010; 76(8):2549–55.

42. De Vasconcelos M, Bennett RN, Rosa EAS, Ferreira-Cardoso JV. Composition of european chestnut (Castanea sativa Mill.) and association with health effects:fresh and processed products. J Sci Food Agric. 2010;90:1559–78.

43. Daniel EM, Ratnayake S, Kinstle T, Stoner GD. The effects of pH and rat intestinal contents on the liberation of ellagic acid from purified and crude ellagitannins. J Nat Prod. 1991;54:946–52.

44. Cerdá B, Llorach R, Cerón JJ, Espín JC, Tomás-Barberán FA. Evaluation of the bioavailability and metabolism in the rat of punicalagin and antioxidant polyphenol from pomegranate juice. Eur J Nutr. 2003;42:18–28.
45. Cerdá B, Espín JC, Parra S, Martínez P, Tomás-Barberán FA. The potent in vitro antioxidant ellagitannins from pomegranate juice are metabolized into bioavailable but poor antioxidant hydroxy-6H-dibenzopyran-6-one derivatives by the colonic microflora in healthy humans. Eur J Nutr. 2004;43:205–20.
46. Cerda B, Tomas-Barberan FA, Espin JC. Metabolism of antioxidant and chemopreventive ellagitannins from strawberries, raspberries, walnuts, and oak-aged wine in humans: identification of biomarkers and individual variability. J Agric Food Chem. 2005;53:227–35.

18

Expanding behavior pattern sensitivity analysis with model selection and survival analysis

Casey L. Cazer[1]* ⓘ, Victoriya V. Volkova[2] and Yrjö T. Gröhn[2]

Abstract

Background: Sensitivity analysis is an essential step in mathematical modeling because it identifies parameters with a strong influence on model output, due to natural variation or uncertainty in the parameter values. Recently behavior pattern sensitivity analysis has been suggested as a method for sensitivity analyses on models with more than one mode of output behavior. The model output is classified by behavior mode and several behavior pattern measures, defined by the researcher, are calculated for each behavior mode. Significant associations between model inputs and outputs are identified by building linear regression models with the model parameters as independent variables and the behavior pattern measures as the dependent variables. We applied the behavior pattern sensitivity analysis to a mathematical model of tetracycline-resistant enteric bacteria in beef cattle administered chlortetracycline orally. The model included 29 parameters related to bacterial population dynamics, chlortetracycline pharmacokinetics and pharmacodynamics. The prevalence of enteric resistance during and after chlortetracycline administration was the model output. Cox proportional hazard models were used when linear regression assumptions were not met.

Results: We have expanded the behavior pattern sensitivity analysis procedure by incorporating model selection techniques to produce parsimonious linear regression models that efficiently prioritize input parameters. We also demonstrate how to address common violations of linear regression model assumptions. Finally, we explore the semi-parametric Cox proportional hazards model as an alternative to linear regression for situations with censored data. In the example mathematical model, the resistant bacteria exhibited three behaviors during the simulation period: (1) increasing, (2) decreasing, and (3) increasing during antimicrobial therapy and decreasing after therapy ceases. The behavior pattern sensitivity analysis identified bacterial population parameters as high importance in determining the trajectory of the resistant bacteria population.

Conclusions: Interventions aimed at the enteric bacterial population ecology, such as diet changes, may be effective at reducing the prevalence of tetracycline-resistant enteric bacteria in beef cattle. Behavior pattern sensitivity analysis is a useful and flexible tool for conducting a sensitivity analysis on models with varied output behavior, enabling prioritization of input parameters via regression model selection techniques. Cox proportional hazard models are an alternative to linear regression when behavior pattern measures are censored or linear regression assumptions cannot be met.

Keywords: Sensitivity analysis, Antimicrobial resistance, Antibiotic resistance, Beef cattle, Behavior pattern, Linear regression, Survival analysis

* Correspondence: clc248@cornell.edu
[1]Department of Population Medicine and Diagnostic Sciences, College of Veterinary Medicine, Cornell University, Ithaca, NY, USA
Full list of author information is available at the end of the article

Background

Mathematical models are commonly used to understand and study biological systems that are complex to replicate and describe in individual laboratory or field studies. Models facilitate testing hypotheses that may be difficult or unethical to test in vivo, identification of critical parameters within a system, and they guide hypothesis generation and experimental design. Different types of mathematical models have improved our understanding of important issues in veterinary medicine, including infectious disease management [1–4], drug pharmacokinetics [5–8], and antimicrobial resistance [9–12]. With the rise in computing power, models have evolved to be more complex, detailed, and data-intensive. Simple susceptible-infectious-recovered models [12, 13] have given way to meta-population models [1, 4, 9] and now agent-based models [2, 14, 15], in which individuals (animals, bacteria, etc.) are modeled with unique characteristics and behaviors [16]. However, each layer of complexity adds additional sources of uncertainty to the model outputs, particularly when parameter values are from varied sources or unavailable and when the modeler must make assumptions about model structure and parameter values.

A robust sensitivity analysis defends a mathematical model against the adverse effects of excessive uncertainty. In short, sensitivity analysis attempts to identify how input parameter natural variation and/or uncertainty affects model output [17, 18]. A modeler can use sensitivity analysis to achieve many goals, including elimination of uninfluential parameters (simplification), model structure and code validation, improved understanding of the modelled system, and prioritization of the most influential parameters [17, 18]. The first sensitivity analysis techniques developed compared changes in one input parameter at a time to changes in the model output and hence were termed 'local' or 'one-at-a-time' approaches. The local parameter influence can be evaluated directly via partial derivatives [18] or statistically with a correlation coefficient [19]. In contrast, 'global' sensitivity analysis considers the entire domain for all input parameters and assess how changes in each input affect the model output after accounting for the effects of the other inputs [18]. Correlation coefficients and regression models (regressing model output on inputs) are commonly used for global sensitivity analysis, including decomposition of output variance and Taguchi designs [17, 18, 20]. Regression approaches are employed in sensitivity analyses because the regression coefficients can be used to rank model input parameters in their effect on the outputs [18]; such regression models may be viewed as meta-models used for investigating statistical associations between the output and parameter values in mathematical models [17, 18]. These techniques rely upon the modeler defining a numerical model output of

interest and choosing a single time-point value of the output variable as the dependent variable (e.g., maximum or minimum value). Importantly, standard sensitivity analysis methods do not account for different output behaviors produced by one model and in some instances the behavior is of greater interest than a single time-point output value [20].

Behavior pattern sensitivity analysis has been proposed as an alternative and complement to standard sensitivity analysis when a model produces more than one mode of output behavior [20] because the behavior mode can confound associations between model parameters and outputs. We use the terms "behavior" and "behavior mode" to describe the model output pattern when the output is plotted over simulation time. Examples of behavior modes include, but are not limited to, oscillations, exponential decay, logarithmic growth, or sigmoid growth. Hekimoğlu et al. suggest a framework, similar to the first five steps of Fig. 1, for conducting and interpreting behavior pattern global sensitivity analysis on system dynamics models [20]. Behaviors are identified, characterized, and then regression models with standardized input parameter values are built to determine the association between input parameters and behavior pattern measures (Fig. 1) [20]. Such analysis can be useful for other model types that produce more than one behavior pattern or for situations where the output behavior matters more to the modeler than a single output value.

Here we apply behavior pattern sensitivity analysis for the first time to a pharmacokinetic-pharmacodynamic and bacterial ecology model of antimicrobial resistance in enteric bacteria in a beef steer during and after oral chlortetracycline administration. In 2011, the last time the U.S. national beef herd was surveyed about antimicrobial use, 71.7% of feedlots used chlortetracycline with 18.4% of all cattle receiving chlortetracycline in their feed during the feedlot period [21]. Most feedlots use chlortetracycline for disease prevention and control rather than disease treatment [21]. Although the U.S. Food and Drug Administration has recently prohibited in-feed use of medically important antimicrobials for growth promotion, such use for disease prevention, control and treatment is still permitted under the oversight of a veterinarian [22]. It is unclear whether this change will increase the use of chlortetracycline due to increased disease or decrease its use because of veterinarian oversight. Target pharmacologic models (empirical or physiologically based), with appropriate sensitivity analyses, can help create judicious antimicrobial-use policies by predicting antimicrobial concentrations in body compartments, such as the intestine, and evaluating alternative dosage regimens [7].

We build upon the original behavior pattern sensitivity analysis framework [20]. First, we incorporate principles

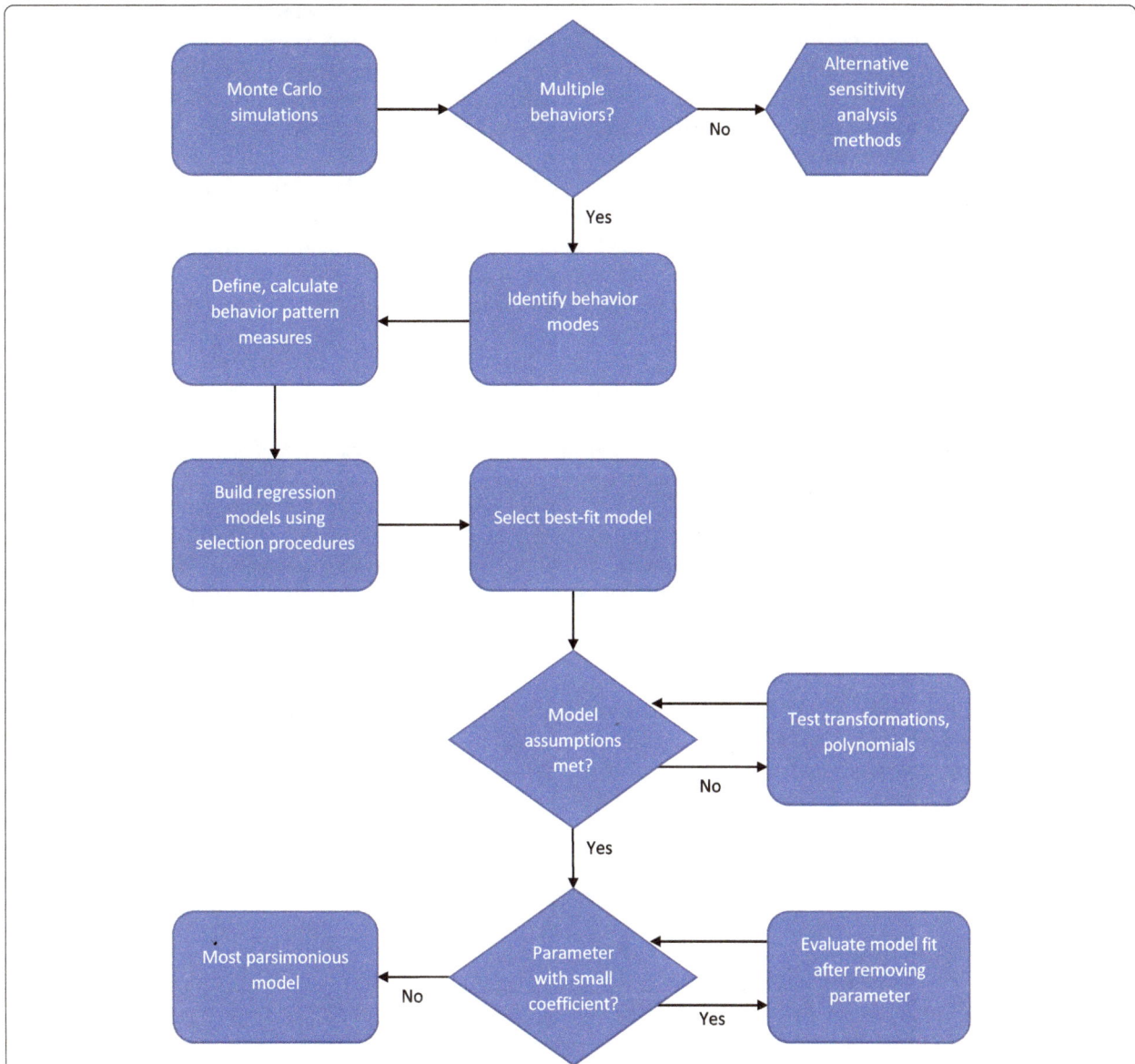

Fig. 1 Behavioral sensitivity analysis process. The outputs of Monte Carlo simulations of mathematical models are classified into behavior pattern modes and pattern measures are defined for each of the modes. Standardized input parameter values from Monte Carlo simulations are used to build regression models for each pattern measure from each of the behavior pattern modes. Smoothing spline curves are fit to the simulation outputs if necessary to eliminate noise and enable calculation of the behavior pattern measures. Variable selection and model fit evaluation methods are used to find each best-fit regression model. Validity of assumptions for the best-fit regression model is evaluated; dependent (simulation outputs) and independent (parameters) variable transformations or other appropriate approaches such as time-dependent coefficients are used to meet the regression model assumptions if necessary. To obtain a most parsimonious regression model, parameters with relatively small coefficients in the best-fit model are eliminated, starting with the smallest, if there is no substantial change in model fit or the other parameter coefficients. Validity of assumptions is re-evaluated for the most parsimonious regression model

and methods of model selection and model fit comparisons in order to improve parameter prioritization. By identifying which parameters contribute the most to model behavior variability, research efforts can be directed to produce data to reduce uncertainty in the values of those parameters. Next, we suggest techniques to address statistical assumption violations, including Cox proportional hazard models as an alternative to linear regression

when linear regression assumptions are not met. Finally, we recommend building parsimonious models by removing parameters with small coefficients from the regression models in order to improve interpretability without sacrificing model fit. Using the example model of resistant enteric bacteria, we demonstrate how behavior pattern sensitivity analysis can be used for parameter prioritization and improved understanding of the modelled system.

Results

The proportion of resistant enteric bacteria during and after chlortetracycline administration fits into one of three defined behaviors in all 1000 simulations: increasing, decreasing, or peaked. Sixty-seven simulations had increasing behavior of the proportion of resistant enteric *Escherichia coli* over time, 311 simulations had decreasing behavior, and 622 simulations had peaked behavior (an increase in proportion resistant during chlortetracycline administration followed by a decrease). We identified 3 behavior pattern measures that characterized these behaviors: equilibrium points, inflection points, and maximum points. These pattern measures were calculated in absolute terms (the proportion resistant at the time of occurrence) and in relative terms, in which the starting proportion resistant (Day 2) was subtracted from the proportion resistant at the time of occurrence. Specifically, increasing behavior was defined by absolute and relative equilibriums; decreasing behavior was defined by inflection points, and relative and maximum equilibriums; and peaked behavior was defined by absolute and relative maximum and equilibrium points. However, not every simulation achieved each behavior pattern measure during the simulated time period (90 days). The proportion resistant achieved equilibrium in only 36% of increasing behavior simulations by Day 90. Equilibrium was reached by 38% of decreasing behavior simulations and only 24% of peaked behavior simulations after chlortetracycline administration ended and before Day 90. Inflection points occurred

during chlortetracycline administration in 65% of decreasing behavior simulations. For each behavior pattern measure and behavior mode, only simulations that had the behavior pattern measure were included in the linear regression models (i.e. missing data was removed pairwise). A maximum proportion resistant during chlortetracycline administration could be calculated for all peaked behavior simulations. In 80% of such simulations, the maximum occurred at the last time step during chlortetracycline administration because the removal of chlortetracycline inevitably caused a decline in proportion resistant for this behavior mode.

The results from the final regression models for absolute and relative equilibrium levels of the proportion of resistant enteric bacteria are presented in Table 1. Five input parameters predominated in these models: p_r (proportion of resistance among bacteria flowing into the large intestine); $start_r$ (Day 0 proportion of resistant bacteria in the large intestine), λ_{in} (rate of bacteria flowing into the large intestine, proportional to the total bacteria population size), λ_{out} (rate of bacteria flowing out of the large intestine, proportional to the total bacteria population size), α (fitness cost of resistance for the intermediate and resistant bacteria). The parameter p_r consistently had the largest coefficient in the regression models, indicating that a one standard deviation change in p_r had the largest effect on the equilibrium proportion resistant in the large intestine. The parameter $start_r$ was significant in the relative equilibrium models only and it

Table 1 Linear regression models for proportion-resistant absolute and relative equilibrium levels of the three behavior modes

Behavior Mode	Behavior Pattern Measure	Most Parsimonious Model								Full Model		
		Standardized Input Parameters Coefficient (Standard Error)					Fit Statistics			Fit Statistics		
		p_r	$start_r$	λ_{in}	λ_{out}	α	AIC	BIC	Adj. R^2	AIC	BIC	Adj. R^2
Increasing	Equilibrium Level	0.101 (0.0006)		0.005 (0.0007)		−0.004 (0.0007)	−203	−197	.999	−363*	−335*	1*
Increasing	Relative Equilibrium Level	0.069 (0.002)	−0.068 (0.002)				−147	−142	.986	−239*	−211*	0.999*
Decreasing	Equilibrium Level	0.097 (0.0001)		0.001 (0.0002)	−0.001 (0.0001)	−0.001 (0.0001)	−1321	1305	0.999	1309	1231	0.999
Decreasing	Relative Equilibrium Level	0.065 (0.0004)	−0.063 (0.0004)	0.013 (0.0008)			−971	−957	0.997	−960	−882	0.997
Peaked	Equilibrium Level [A]	0.121 (0.0001)		0.003 (0.0002)	−0.001 (0.0001)	−0.002 (0.0001)	−2040	2020	0.999	2030	1936	0.999
Peaked	Relative Equilibrium Level [B]	0.081 (0.0005)	−0.081 (0.0005)	−0.007 (0.0007)			−1566	1549	0.994	−1562	1468	0.995

The example mathematical model was for the proportion of tetracycline-resistant enteric *Escherichia coli* in a beef steer during and after administration of oral chlortetracycline. A separate linear regression model was built for each behavior pattern measure of each behavior mode. The behavior pattern measure was the dependent variable in the linear regression models. Equilibrium was reached by 36% of increasing behavior simulations, 38% of decreasing behavior simulations and 24% of peaked behavior simulations after chlortetracycline administration ended and before the end of the simulation period. Simulations that did not reach equilibrium were excluded from these models. Coefficients and standard errors are listed for the standardized parameters that were included in each most parsimonious linear regression model. Full model refers to a linear regression model including all the parameters listed in Table 6 as independent variables. Akaike Information Criterion (AIC), Bayesian Information Criterion (BIC), and adjusted R^2 are given for the most parsimonious and the full model. *Full model excludes $\log_{10}(\beta_{ir})$, $\log_{10}(\beta_{sr})$, γ_s and MIC_i to prevent overfitting. [A]Peaked equilibrium level has 3 outliers removed (from reduced model and from full model). [B]Peaked relative equilibrium level has one outlier removed (from reduced model and from full model)

opposed the effect of p_r. In general, the reduced regression models had improved fit (smaller AIC and BIC) over the full models (full models included all 26 input parameters of the mathematical model as independent variables), except for the regression models for the increasing behavior mode. The final reduced regression models for all the equilibrium pattern measures and all three behavior modes had high explanatory power ($R^2 \geq 0.98$). Residual plots from the absolute and relative equilibrium level models are shown in Fig. 2: the models of absolute equilibrium level met the assumption of homoscedasticity of residuals, while the relative equilibrium level models for decreasing and peaked behavior violated that assumption.

A greater number of parameters were consequential in the final regression models for equilibrium time, compared to the equilibrium level models (Table 2). In addition, the parameters that were significant and consequential to model fit were different in each of the behavior modes, with the exception of p_r and $start_r$ which were in the equilibrium time models for all three behavior modes. The coefficients for these and other parameters had similar absolute values, indicating that the parameters had approximately equal contributions to explaining variation in equilibrium time in all three behavior modes. In contrast, in the equilibrium level models, p_r often had a coefficient that was 5 to 100 times larger than other parameter coefficients (Table 1). The equilibrium time models had lower explanatory power than the equilibrium level models ($R^2 < 0.74$) (Table 2 vs. Table 1).

Inflection points only occurred in the decreasing behavior mode and they occurred during chlortetracycline administration (between Day 2 and Day 30). Parameters related to the bacterial population dynamics and chlortetracycline pharmacokinetics-pharmacodynamics all made significant contributions to predicting the proportion of resistance at the inflection point (Table 3). In

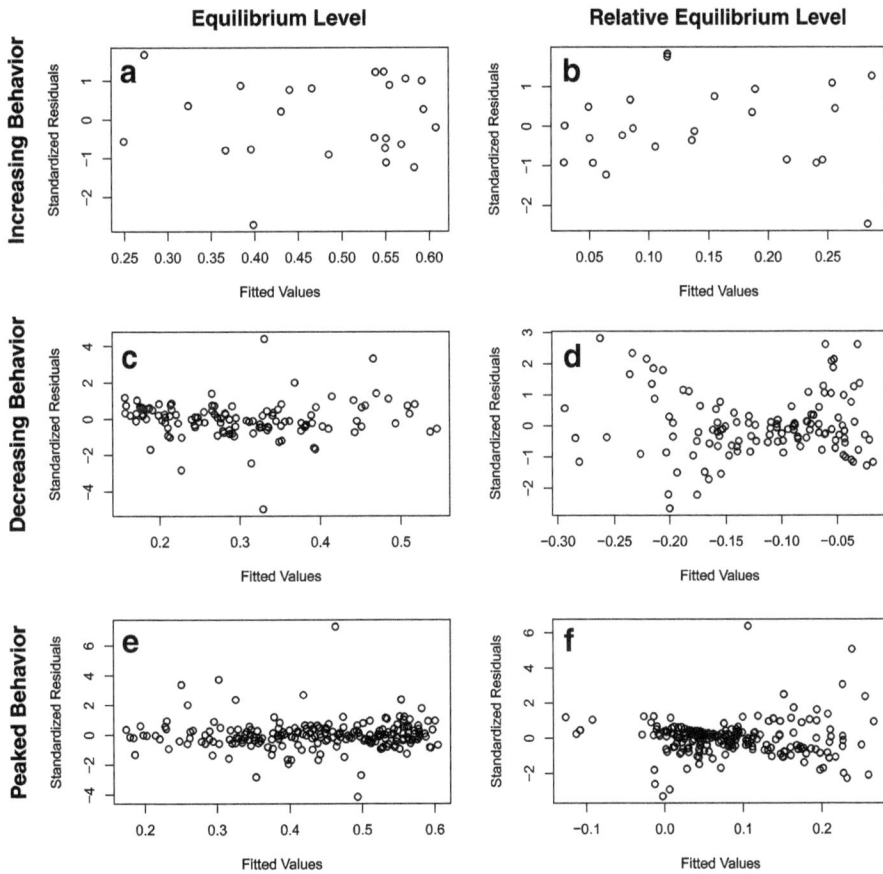

Fig. 2 Standardized residuals of the proportion-resistant equilibrium level linear regression models for the three behavior modes. The three behaviors of the resistant bacteria in the example mathematical model were: (**a**, **b**) increasing, (**c**, **d**) decreasing, and (**e**, **f**) increasing during antimicrobial therapy and decreasing after therapy ceases (peaked). Separate linear regression models were built for each of the absolute (**a**, **c**, **e**) and relative (**b**, **d**, **f**) equilibrium levels in each behavior mode and are described in Table 1. The relative equilibrium level (**b**, **d**, **f**) is the proportion of resistance at the equilibrium point minus the starting proportion of resistance. The fitted values of the equilibrium level outcome are shown on the x-axis and the standardized residual values are shown on y-axis

Table 2 Linear regression models for time to proportion-resistant equilibrium of the three behavior modes

Behavior Mode	Behavior Pattern Measure	Most Parsimonious Model — Standardized Input Parameters Coefficient (Standard Error)										Fit Statistics			Full Model Fit Statistics		
		p_r	start$_r$	λ_{in}	λ_{out}	δ	MIC_s	H_i	V_{LI}	η_{LI}	y_{LI}	AIC	BIC	Adj R^2	AIC	BIC	Adj R^2
Increasing	Equilibrium time	1868 (347)	-883 (313)		-1085 (352)			-1526 (352)	-884 (321)		-1141 (292)	427	437	0.733	436*	464*	-0.706*
Decreasing	Equilibrium Time	p_r -1527 (514); p_r^2 805 (408); p_r^3 552 (331); p_r^4 -410 (156)	1233 (207)	-3488 (396)		-750 (175)	-1107 (180)		-664 (179)			2141	2172	0.561	2149	2226	0.583
Peaked	Equilibrium Time	810 (175)	-667 (174)	-1807 (279)	660 (139)	-781 (123)	-1393 (154)	-842 (145)	-628 (146)			3812	3846	0.423	3824	3918	0.436

The example mathematical model was for the proportion of tetracycline-resistant enteric *Escherichia coli* in a beef steer during and after administration of oral chlortetracycline. A separate linear regression model was built for each behavior pattern measure of each behavior mode. The behavior pattern measure was the dependent variable in the linear regression models. Equilibrium was reached by 36% of increasing behavior simulations, 38% of decreasing behavior simulations and 24% of peaked behavior simulations after chlortetracycline administration ended and before the end of the simulation period. Simulations that did not reach equilibrium were excluded from these models. Coefficients and standard errors are listed for the standardized parameters that were included in each most parsimonious linear regression model. Full model refers to a linear regression model including all the parameters listed in Table 6 as independent variables. Akaike Information Criterion (AIC), Bayesian Information Criterion (BIC), and adjusted R^2 are given for the most parsimonious and the full model. *Excludes $\log_{10}(\beta_{lr})$, $\log_{10}(\beta_{sr})$, $\log_{10}(\beta_{sr})$, V_s and MIC_i to prevent overfitting

Table 3 Linear regression models for proportion-resistant inflection and maximum levels of decreasing and peaked behaviors, respectively

Behavior Mode	Behavior Pattern Measure	Most Parsimonious Model Standardized Input Parameters Coefficient (Standard Error)									Fit Statistics			Full Model Fit Statistics		
		p_r	$start_r$	λ_{in}	λ_{out}	$\log_{10}(\beta_{so})$	MIC_s	δ	V_{LI}	η_{LI}	AIC	BIC	Adj. R^2	AIC	BIC	Adj. R^2
Decreasing	Inflection Level	0.091 (0.002)		−0.031 (0.003)			MIC_s −0.007 (0.005) MIC_s^2 0.008 (0.003) MIC_s^3 −0.006 (0.002)	−0.02 (0.004)	−0.011 (0.002)	−0.011 (0.002)	−800	−767	0.885	−758	−664	0.869
Peaked	Max Level	0.086 (0.003)	0.022 (0.003)	−0.036 (0.003)	0.02 (0.002)	0.009 (0.002)	MIC_s −0.081 (0.003) MIC_s^2 0.032 (0.003)	−0.021 (0.002)	−0.022 (0.002)	−0.018 (0.002)	−1706	−1653	0.807	−1597	−1468	0.775
Peaked	Relative Max Level	0.06 (0.003)	−0.07 (0.003)	−0.043 (0.003)	0.02 (0.003)		−0.02 (0.003)	−0.068 (0.003)	−0.022 (0.003)	−0.021 (0.003)	−1685	−1641	0.734	−1709	−1585	0.752

The example mathematical model was for the proportion of tetracycline-resistant enteric *Escherichia coli* in a beef steer during and after administration of oral chlortetracycline. A separate linear regression model was built for each behavior pattern measure of each behavior mode. The behavior pattern measure was the dependent variable in the linear regression models. Inflection points occurred during chlortetracycline administration in 65% of decreasing behavior simulations. Simulations that did not have an inflection point were excluded from the inflection level model. A maximum proportion resistant during chlortetracycline administration could be calculated for all peaked behavior simulations. Coefficients and standard errors are listed for the standardized parameters that were included in each most parsimonious linear regression model. Full model refers to a linear regression model including all the parameters listed in Table 6 as independent variables. Akaike Information Criterion (AIC), Bayesian Information Criterion (BIC), and adjusted R^2 are given for the most parsimonious and the full model contrast, only parameters of the bacterial population dynamics were significantly associated with the proportion of resistance at equilibrium points. Consistent with the equilibrium level models, p_r had the largest coefficient in the inflection level model (Tables 1 and 3). The pharmacokinetic parameters δ (chlortetracycline abiotic degradation rate), V_{LI} (volume of large intestine), and η_{LI} (fraction of chlortetracycline adsorbed to digesta) all had small to moderate negative coefficients. Polynomial terms of MIC_s (susceptible *E. coli* MIC) were added to improve the linearity between MIC_s and inflection proportion of resistance (Fig. 3). The addition of polynomial terms resulted in a minor improvement in model fit compared to a model without polynomial terms ($R^2 = 0.885$, AIC $= -800$, BIC $= -767$ with polynomial terms compared to $R^2 = 0.864$, AIC $= -767$, BIC $= -741$ without). The final inflection level model explained a large amount of variation in the proportion of resistant *E. coli* at the inflection point ($R^2 = 0.885$).

The linear regression model for the inflection time in decreasing behavior simulations violated the assumption of normally distributed residuals but the residual distribution was significantly improved by modeling inflection time to the fourth power (Fig. 4). Only three parameters had a substantial impact on the inflection time: p_r, $start_r$, and MIC_s. The proportion of resistance in inflowing bacteria (p_r) and the starting proportion of resistance

($start_r$) had approximately equal but opposing effects (Table 4). A higher inflowing proportion of resistance shortened the time to the inflection point whereas a higher starting level of resistance increased the time to the inflection point. A higher chlortetracycline MIC for the susceptible bacteria also increased the time to the inflection point. The reduction from 26 to three parameters did not significantly affect the model fit, although the full and reduced models both had low explanatory power ($R^2 = 0.3$ for the reduced model).

Similar to the inflection level model for the decreasing behavior, the maximum level models for the peaked behavior incorporated parameters for the pharmacokinetics of chlortetracycline, *E. coli* population, and the pharmacodynamics (Table 3). In addition, the fit also improved from the full to the reduced model for absolute inflection level and the absolute maximum level. All the parameters from the absolute inflection level model were also included in the absolute and relative maximum level models and the coefficients had similar magnitude and direction (Table 3). The maximum level models also included effects for the starting proportion of resistance, outflow rate of bacteria, and one plasmid transfer term (absolute maximum level model only). The starting level of resistance had opposite effects on the absolute and relative maximum levels of resistance. An increase in the starting level of resistance had a positive association with

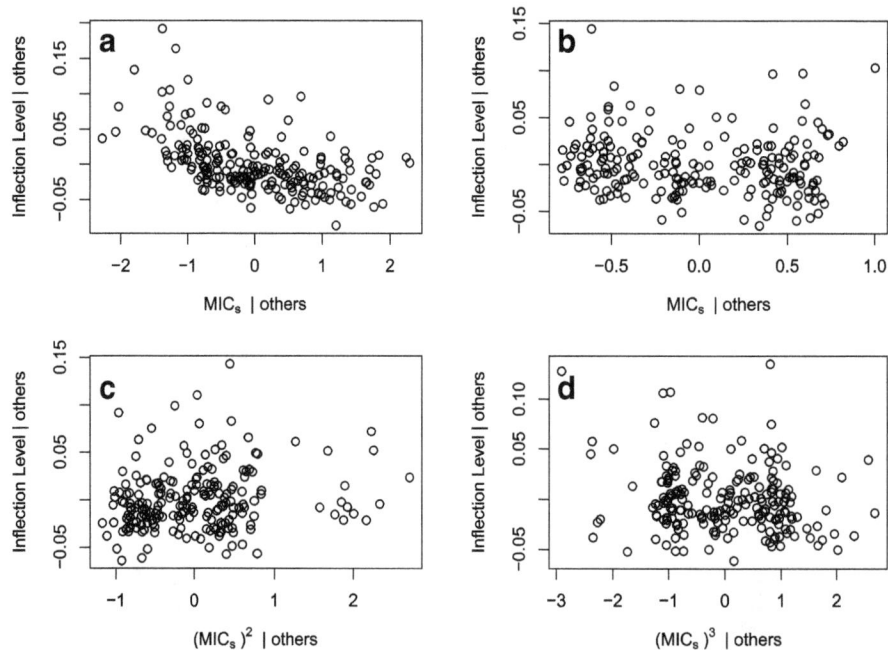

Fig. 3 Partial regression plots from the proportion-resistant inflection level linear regression models in decreasing behavior simulations. The example mathematical model was for the proportion of tetracycline-resistant enteric *Escherichia coli* in a beef steer during and after administration of oral chlortetracycline. **a** is a partial regression plot for a regression model that contains no polynomial terms and shows the effect of MIC_s on the inflection level of resistance after accounting for all other variables in the model. **b-d** are partial regression plots for a regression model that contains polynomial terms of MIC_s and show the effects of (**b**) MIC_s, (**c**) MIC_s^2, and (**d**) MIC_s^3 after accounting for all other variables in the polynomial model

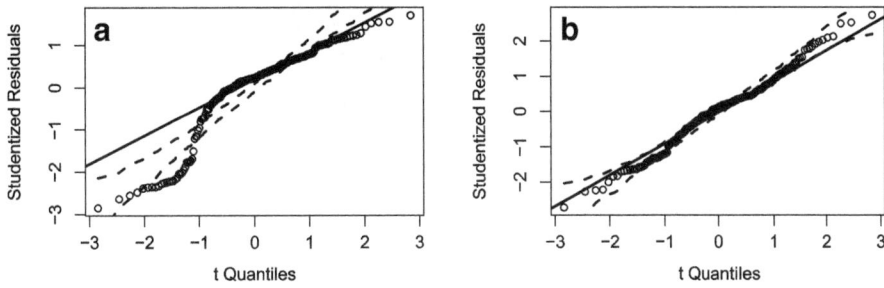

Fig. 4 Quantile-quantile plots for the decreasing behavior inflection time and inflection time to the fourth power. Residuals from a linear regression model with (**a**) inflection time as the output and (**b**) inflection time to the fourth power as the output are plotted on the y-axis. The theoretical quantiles of a normal distribution are plotted on the x-axis. The solid red line passes through the quartile-pairs and the dotted red lines encompass a 95% confidence interval for the theoretical normal distribution

the absolute maximum level of resistance but a negative association with the relative maximum level of resistance, reflecting the calculation of the relative level as the absolute minus the starting level.

The time of the maximum proportion resistant in peaked behavior simulations exhibited large violations of linear model assumptions (Fig. 5) and therefore should not be interpreted. This occurred because 80% of the simulations had a maximum proportion resistant at the last time-point of chlortetracycline administration, indicating that they had not reached an absolute maximum but instead had a local maximum due to the abrupt change in chlortetracycline input. Therefore, a Cox proportional hazard model was fit for the time of maximum resistance ('time to' the 'event' of maximum resistance). A non-censoring model was developed that considered all simulations to have reached a maximum, i.e. considering as the event a local or absolute maximum. In a second censored model, only reaching an absolute maximum was considered as the event; those simulations that had increasing resistance at the end of chlortetracycline administration and may have reached an absolute maximum at a later time-point were censored. The parameters retained in the best, most-parsimonious non-censored and censored models were the same, with moderate changes in parameter coefficients but no change in coefficient signs between the two models. However, several parameters in the best model (censored) violated the proportional hazard assumption: the proportion of resistant inflowing bacteria, the proportion of intermediate inflowing bacteria, the starting proportion of resistance, the inflow rate of bacteria, and the MIC of susceptible bacteria. The right-censored model was used to address the proportional hazards violation. A continuous time-dependent coefficient function could not be identified for these five parameters, therefore a step-function was used to create proportional hazards [23]. The range of maximum times was divided into equal thirds: Day 10 to Day 16.6, Day 16.7 to Day 23.3,

and Day 23.4 to Day 30. A right-censored Cox proportional model was then fit with this step function for the five non-proportional hazard parameters. The resulting model coefficients are presented in Table 5.

With the step function, all parameters met the assumption for proportional hazards. All three pharmacokinetic parameters (chlortetracycline abiotic degradation rate, large intestine volume, and adsorption of chlortetracycline to digesta) included in the maximum time Cox model had negative coefficients, indicating that an increase in those parameters is associated with a decrease in the hazard of reaching a maximum resistant proportion. For example, the hazard decreases by 29% when the degradation rate (δ) increases by one standard deviation. A decrease in hazard corresponds to a lower probability of reaching a maximum and hence a greater time to maximum. On the other hand, an increase in hazard corresponds to a greater probability of reaching maximum and a shorter time to maximum. An increase in the pharmacodynamic parameter MIC_s is also associated with a decrease in hazard but this association changes over the time strata. The MIC_s has a larger impact on hazard early in the chlortetracycline administration period, compared to later in the time period. This change over time was also true for p_r, although it had a more modest impact on the hazard of reaching maximum. An increase in p_i and $start_r$ both had a positive association with the hazard and a stronger effect at the beginning of the time period. Increases in both λ_{in} and λ_{out} also had positive associations with the hazard and the coefficient for λ_{in} changed over time.

Discussion

Behavior pattern sensitivity analysis can be useful for models that have multiple output behaviors because it includes separate behavior pattern measures and statistical analyses for each output behavior mode. In general, output behavior modes can be simple or complex, few or numerous [20]. Correctly classifying the behavior

Table 4 Linear regression models for proportion-resistant inflection and maximum times of decreasing and peaked behaviors, respectively

Behavior Mode	Behavior Pattern Measure	Most Parsimonious Model							Fit Statistics			Full Model		
		Standardized Input Parameters Coefficient (Standard Error)										Fit Statistics		
		p_r	p_i	$start_r$	$start_i$	λ_{out}	MIC_s	η_{LI}	AIC	BIC	Adj. R^2	AIC	BIC	Adj. R^2
Decreasing	Inflection Time[4]	−3.2e14 (4.5e13)		3.4e14 (4.6e13)			2.2e14 (4e13)		14,373	14,389	0.301	14,367	14,460	0.387
Peaked	Max Time	310.43 (31.71)	−226.84 (30.33)	−282.98 (31.6)	115.5 (30.3)	−102.45 (30.31)	270.5 (31.02)	61.23 (30.63)	10,014	10,054	0.348	9853	9982	0.512

The example mathematical model was for the proportion of tetracycline-resistant enteric *Escherichia coli* in a beef steer during and after administration of oral chlortetracycline. A separate linear regression model was built for each behavior pattern measure of each behavior mode. The behavior pattern measure was the dependent variable in the linear regression models. Inflection points occurred during chlortetracycline administration in 65% of decreasing behavior simulations. Simulations that did not have an inflection point were excluded from the inflection level model. A maximum proportion resistant during chlortetracycline administration could be calculated for all peaked behavior simulations. Coefficients and standard errors are listed for the standardized parameters that were included in each most parsimonious linear regression model. For the inflection time model, the dependent variable was inflection time raised to the fourth power. Full model refers to a linear regression model including all the parameters listed in Table 6 as independent variables. Akaike Information Criterion (AIC), Bayesian Information Criterion (BIC), and adjusted R^2 are given for the most parsimonious and the full model

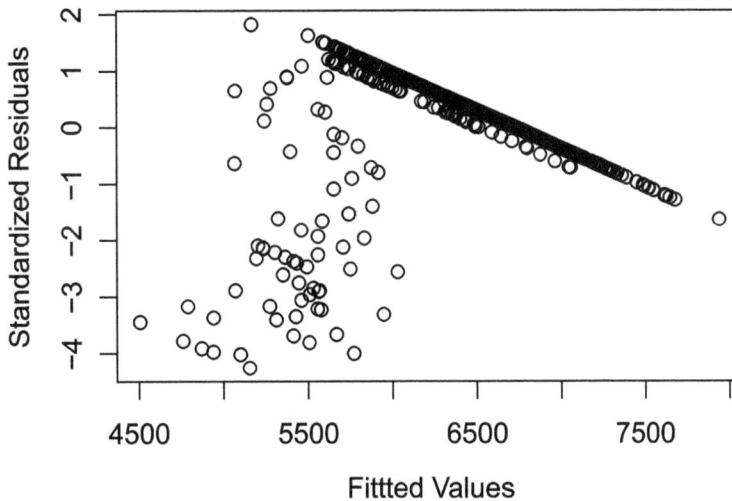

Fig. 5 Standardized residuals of the time of maximum proportion resistant regression model for the peaked behavior. The fitted values of the time of maximum outcome are shown on the x-axis and the standardized residual values are shown on y-axis

mode of each simulation is important to the validity of the analysis and misclassification could bias the regression models. At best, if the different behavior modes have similar associations with the same input parameters (as in Table 1), then there may be little to no discernable impact of behavior misclassification. However, if the different behavior modes are associated with different input parameters or have opposite associations with the same parameters (as in Table 2), then behavior misclassification can lead to bias towards the null and the effect of input parameters may be missed. Since the three behavior modes in the example mathematical model could be distinguished based on three timepoints (Day 2, Day 30, Day 90) and all simulations could be classified as one of the three behaviors, there was likely little to no misclassification in the example sensitivity analysis.

By applying model selection methods (e.g., stepwise variable selection) and model fit measures (R^2, BIC, AIC) to behavior pattern sensitivity analysis we have demonstrated how a large number of model input parameters can be efficiently prioritized. Appropriate parameter prioritization helps to focus research efforts on parameters that have the largest impact on the model output. Even though many parameters were statistically significant after the initial regression model selection, parameters that do not explain a large amount of variability in the outcome are likely not useful in manipulating real-world systems. By using standardized parameter values, the absolute values of the regression coefficients can be used to prioritize parameters for each behavior mode and behavior pattern measure [20]. Creating a parsimonious model by removing parameters with small coefficients (< 0.005 in the "level" models, < 240 in the

"time" models), without sacrificing model fit or affecting other parameter coefficients, facilitates interpretation of sensitivity analyses. This is similar to the motivation behind LASSO regression, which produces more interpretable linear models by constraining the sum of the coefficients [24].

Large associations between input parameters and behavior pattern measures can indicate that the parameters may have important roles in the biological system and are leverage points for manipulating the system [2, 20]. On the other hand, caution should be exercised because a large association could be due to inaccuracies in model structure or substantial uncertainty in the parameter estimate [17, 18]. In some modeling situations, sensitivity analyses are useful for identifying uninfluential parameters so they can be fixed at a best-estimate value to reduce model complexity [17]. Since we sought to build the smallest possible linear regression models without sacrificing model fit, we cannot directly evaluate which parameters had the smallest impact on the behavior pattern measures. When using regression modeling for behavior pattern sensitivity analysis, this goal is best achieved by examining the full regression models [20].

In our example, behavior pattern sensitivity analysis of a mathematical model of enteric antimicrobial resistance in beef cattle fed chlortetracycline, parsimonious regression model selection identified parameters that had a significant and substantial effect on the proportion of resistant bacteria during and after chlortetracycline administration. Some parameters (p_r, $start_r$, λ_{in}, λ_{out}, MIC_s, δ, V_{LI}, η_{LI}) were associated with several or all the resistant proportion behavior pattern measures and are therefore considered to have great importance in the model of

Table 5 Cox proportional hazard model for time of maximum proportion resistant of the peaked behavior mode

Standardized Input Parameter	Time Strata	Coefficient	Exponentiated Coefficient	Standard Error
p_r	1	−2.133	0.118	0.33
	2	−1.274	0.280	0.30
	3	−1.199	0.302	0.166
p_i	1	2.156	8.639	0.311
	2	1.748	5.745	0.305
	3	0.691	1.996	0.136
$start_r$	1	1.945	6.996	0.321
	2	1.249	3.485	0.291
	3	1.299	3.666	0.164
$start_i$		−0.856	0.425	0.110
λ_{in}	1	0.810	2.248	0.245
	2*	0.164	1.179	0.250
	3	1.038	2.822	0.143
λ_{out}		0.581	1.787	0.099
$\log_{10}(\beta_{sr})$		0.206	1.228	0.098
MIC_s	1	−8.913	0.0001	1.474
	2	−3.631	0.026	0.763
	3	−0.824	0.439	0.159
δ		−0.343	0.710	0.131
V_{LI}		−0.481	0.618	0.113
η_{LI}		−0.384	0.681	0.096

The example mathematical model was for the proportion of tetracycline-resistant enteric *Escherichia coli* in a beef steer during and after administration of oral chlortetracycline. The Cox model presented here uses right-censored data. In 80% of peaked behavior simulations the maximum proportion resistant occurred at the last time step of chlortetracycline administration and these simulations were considered to not have a maximum event occur in the Cox model (right-censored). Coefficients, exponentiated coefficients, and standard errors are listed for the standardized parameters that were included in the most parsimonious model. Time stratification, which allows the effect of a parameter to vary as a step-function between time strata, was used to meet the assumption of proportional hazards for p_r, p_i, $start_r$, λ_{in}, and MIC_s. The maximum times were divided into three time strata: (1) Day 10 to Day 16.6, (2) Day 16.7 to Day 23.3, and (3) Day 23.4 to Day 30. Coefficients for these five parameters were constant within a time stratum and different between time strata. Parameters that do not have time strata listed met the proportional hazard assumption without stratification and have just one coefficient, constant across time. For this model, Cox and Snell's R^2 was 0.248 (maximum possible 0.584). *$P > 0.05$

antimicrobial resistance. Most of these parameters were also previously identified via Spearman correlation as associated with the average proportion of enteric resistance during chlortetracycline administration, across all simulations and behaviors [11]. The proportion of inflowing resistance (p_r) consistently had a coefficient five to ten times larger than other parameters when predicting the level of enteric resistance (Tables 1, 3). Therefore, this variable could be useful as an intervention to alter the level of enteric resistance. Many

parameters have similarly large coefficients in the equilibrium time models (Table 2) and inflection time model (Table 4). Pharmacokinetic and pharmacodynamic parameters had a significant impact on the level of resistance during (Table 3) but not after (Table 1) chlortetracycline administration. Therefore, depending on the outcome and time of interest in the modelled system, it may or may not be worthwhile to expend effort to reduce the uncertainty in the pharmacokinetic and pharmacodynamic parameters. The behavior pattern sensitivity analysis also identified parameters (α, γ_{LI}, $\log_{10}(\beta sr)$, H_i) that were associated with only one or a few behavior pattern measures and were not identified as significant in the previous sensitivity analysis for the average resistant proportion during chlortetracycline administration [11]. These variables could be ranked as 'second-tier' importance during parameter prioritization.

In many practical modeling problems, it is important to have a detailed sensitivity analysis to translate model findings to real-world applications. For example, all three behavior modes had the proportion resistant at equilibrium affected by the same input parameters with similar coefficients (Table 1). Therefore, we can be confident that changes to those input parameters in the real-world will have consistent effects, despite individual variation in enteric resistance behavior. If the level of resistance after chlortetracycline treatment matters to beef producers and public health officials, then they can focus on changing the variables with the strongest, consistent associations with the equilibrium level of resistance. Diet changes could be used to reduce the proportion of inflowing resistant bacteria (p_r) and the rate of the inflow (λ_{in}) and thereby reduce the equilibrium level of resistant enteric bacteria. For example, added probiotics [25, 26] and silage-based diets [27] have been evaluated for such purposes. However, such an intervention may have complex effects on the level of enteric resistance during chlortetracycline treatment, when p_r has a positive association with resistance levels but λ_{in} has a negative association (Table 3). The effect of a diet change on the time required to reach the post-chlortetracycline equilibrium (effectively a resistance-based withdrawal period) may differ by the underlying behavior of enteric bacteria in individual animals (Table 2).

We used polynomial terms to address non-linear relationships between model input parameters and the behavior pattern measures, although polynomial terms can be difficult to interpret. For example, in the maximum proportion resistant linear regression model for the peaked behavior, the coefficient of MIC_s is negative and the coefficient of MIC_s^2 is positive. The combined effect is negative when MIC_s is between 0 and 2.5 standard deviations above its mean and positive when MIC_s is below its mean or greater than 2.5 standard deviations above its mean. We suggest plotting the polynomial equations to aid in interpretation; then the combined effect of the linear and polynomial terms at a given input parameter value can be compared relative to other input parameter coefficients in the linear regression model. Although using polynomials can correct violations of linear model assumptions, they may not substantially increase the model fit and therefore may not be worth the complexity of interpretation. We did not pursue interactions between input parameters because they can be difficult to interpret, particularly 3rd order and higher interactions. However, input parameters may interact in some model structures and therefore interactions may need to be included in regression models for the sensitivity analysis.

In their behavior pattern sensitivity analysis framework, Hekimoğlu et al. [20] do not address censored data, although censoring can occur in mathematical models when an event does not occur during the simulated time period. In our example model, some simulations did not reach an equilibrium point in the simulated time period, although all simulations tended towards an equilibrium and would presumably reach an equilibrium if the simulation time was extended. There are three potential solutions to this problem: (1) extend the simulation time until all simulations have the required event, (2) remove missing data pairwise, or (3) use statistical methods for censored data in the sensitivity analysis. Solution (1) may not reflect the reality of animal production systems. Animals are sent to slaughter at predetermined times and cannot necessarily be held at farms until their gut microbiota have equilibrized following removal of antimicrobial therapy. In addition, there are practical and legal limitations on how long antimicrobials can be fed to production animals [28], so the chlortetracycline administration time cannot be extended until all simulated animals reach an equilibrium or maximum during the drug administration. In most of linear regression models presented, we used strategy (2), but for the time to maximum in peaked behavior simulations we applied strategy (3) because the linear regression model severely violated the assumption of normally distributed residuals. We used survival analysis (Cox proportional hazard model) to account for 'censored' simulations—those that did not reach a maximum proportion resistant during chlortetracycline administration but rather had a recorded maximum at the last time-point of the administration. Although Cox proportional hazard models have fewer assumptions than linear regression models, the example model data violated the important assumption of proportional hazards, which was addressed by using a step-function (time dependent coefficients) within a right-censored model. The most parsimonious Cox model (Table 5) contained more input parameters than the most parsimonious linear regression model for the time to maximum (Table 4); the Cox model identified all the parameters in the linear regression model (Table 4) plus four additional parameters (Table 5), supporting its suitability for the purpose.

Conclusions

Behavior pattern sensitivity analysis is a flexible method that can be applied to models of bacterial antimicrobial resistance, including antimicrobial pharmacokinetic-pharmacodynamic and bacterial population dynamics models. It provides a detailed sensitivity analysis for each model output behavior and highlights similarities and differences in parameter importance among the behaviors. By using stepwise and best subsets model selection techniques, we have expanded the procedures for behavior pattern sensitivity analysis to efficiently identify the parameters that have the strongest association with each behavior pattern measure. We suggest techniques for addressing violations of linear regression models, including transformations of dependent and independent variables and alternative models (Cox proportional hazard models), thus expanding the techniques for behavior pattern sensitivity analysis. Finally, in the example model of enteric antimicrobial resistance in beef cattle administered an oral antimicrobial, we demonstrate that behavior pattern sensitivity analysis identifies important parameters that could be altered to reduce antimicrobial resistance.

Methods
Example mathematical model
The mathematical model we use as an example has previously been described in detail [11] and is represented schematically in Fig. 6. In short, the model combines a pharmacokinetic stock-flow model of chlortetracycline in the beef cattle gastrointestinal tract [5] with a population dynamics model of resistant and susceptible enteric *Escherichia coli* and a pharmacodynamic (sigmoid E_{max}) equation of the antimicrobial effect on the bacteria. For the application of behavior pattern sensitivity analysis we focused on one chlortetracycline indication and dosage: control of bacterial pneumonia in beef steers with chlortetracycline dosed at 350 mg/steer per day for 28 days. This dosage was chosen because disease prevention is the most common use of in-feed chlortetracycline in feedlot beef cattle [21]. The model reflected chlortetracycline flows between gastrointestinal, plasma, and tissue compartments and abiotic degradation. Within the large intestine, active chlortetracycline inhibited the growth of *E. coli* depending on their susceptibility; susceptible *E. coli* had slower growth in the presence of sub-inhibitory chlortetracycline concentrations than resistant *E. coli*. The model also reflected that resistance genes could be horizontally transferred between

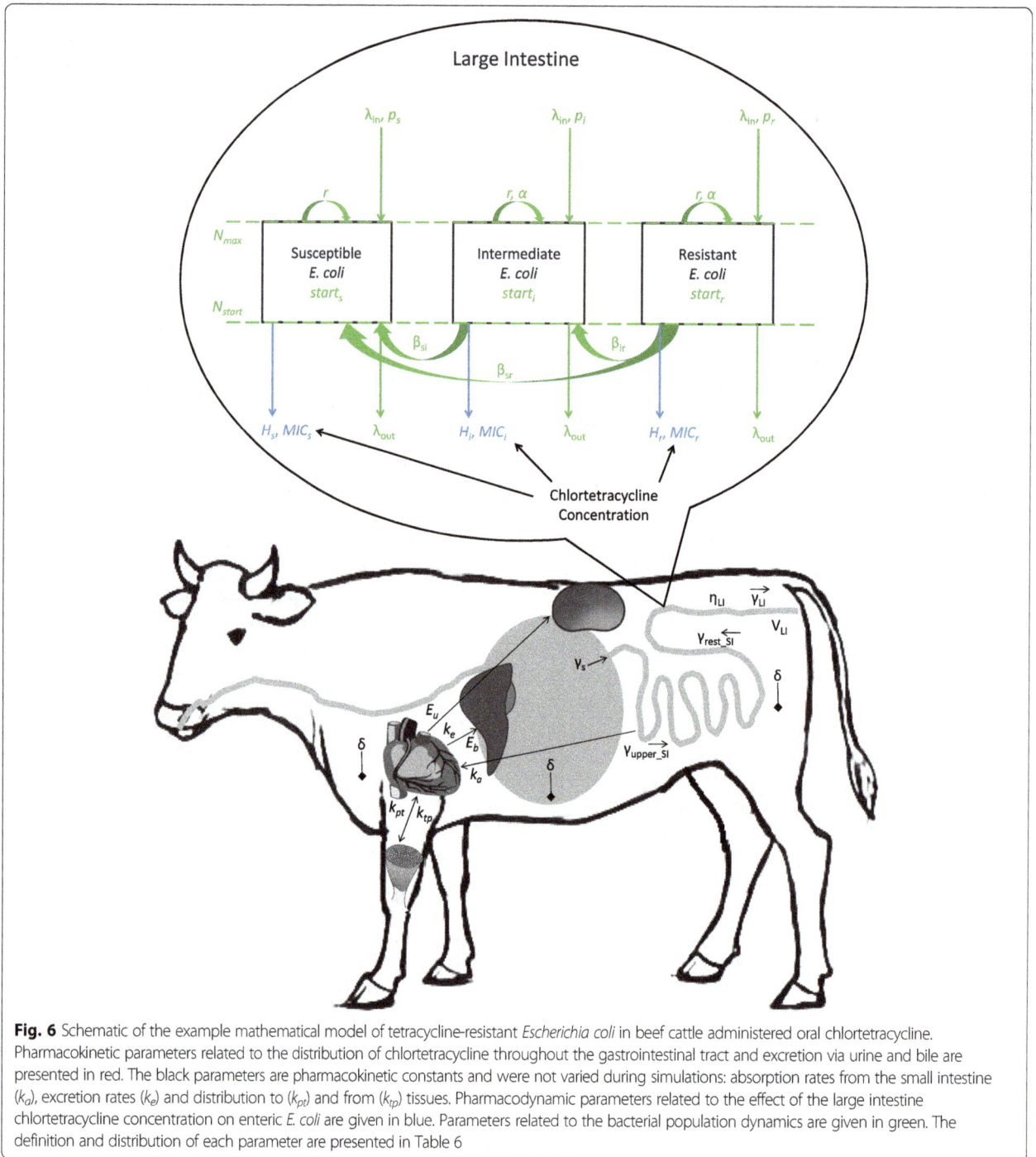

Fig. 6 Schematic of the example mathematical model of tetracycline-resistant *Escherichia coli* in beef cattle administered oral chlortetracycline. Pharmacokinetic parameters related to the distribution of chlortetracycline throughout the gastrointestinal tract and excretion via urine and bile are presented in red. The black parameters are pharmacokinetic constants and were not varied during simulations: absorption rates from the small intestine (k_a), excretion rates (k_e) and distribution to (k_{pt}) and from (k_{tp}) tissues. Pharmacodynamic parameters related to the effect of the large intestine chlortetracycline concentration on enteric *E. coli* are given in blue. Parameters related to the bacterial population dynamics are given in green. The definition and distribution of each parameter are presented in Table 6

resistant, intermediate, and susceptible *E. coli*, and the bacterial population could be replenished by ingested *E. coli*. Twenty-nine input parameters were assigned distributions based on published literature (Table 6), including parameters for chlortetracycline pharmacokinetics, chlortetracycline pharmacodynamics against *E. coli*, and *E. coli* ecology. Monte Carlo simulations ($n = 1000$) of the model were completed by randomly drawing values from the parameter distributions for each simulation [11];

the random values were recorded as a realization of each input parameter random variable. The model was simulated with a 0.1 h time-step for a total of 90 days: a two day burn-in period, 28 days of chlortetracycline administration, and an additional 60 days of follow-up after stopping chlortetracycline. The model output was the proportion of chlortetracycline-resistant *E. coli* out of total enteric *E. coli* over time. The model was built and simulated in MatLab® R2016b (MathWorks, Natick, MA, U.S.).

Table 6 Parameters, and their distributions, of the example mathematical model

Parameter	Distribution	Definition
CTC pharmacokinetics		
δ	Beta (0.54, 37.4)	CTC abiotic degradation rate
γ_s	Uniform (0.0535, 0.0895)	CTC flow rate from stomachs to small intestine
γ_{upper_si}	Uniform (0.250, 0.416)	CTC flow rate through the upper 1/3 small intestine
γ_{rest_si}	Uniform (0.100, 0.166)	CTC flow rate through the lower 2/3 small intestine
γ_{LI}	Uniform (0.100, 0.166)	CTC flow rate through large intestine
E_b	Uniform (0.39, 0.64)	Fraction CTC eliminated via bile
η_{LI}	Uniform (0.69, 0.89)	Fraction CTC adsorbed to digesta in the large intestine
V_{LI}	Uniform (6, 22)	Large intestine contents volume
CTC pharmacodynamics		
H_s	Uniform (1.62, 2.23)	Hill coefficient for susceptible bacteria
H_i	Uniform (5.71, 9.53)	Hill coefficient for intermediate bacteria
H_r	Uniform (6.42, 10)	Hill coefficient for resistant bacteria
MIC_s	Uniform (0, 4)	Anaerobic MIC for susceptible bacteria
MIC_i	Uniform (2.7, 16)	Anaerobic MIC for intermediate bacteria
MIC_r	Uniform (14.7, 128)	Anaerobic MIC for resistant bacteria
Bacterial population dynamics in the large intestine		
r	Uniform (0.05, 0.5)	Bacterial growth rate in the large intestine
a	Uniform (0, 0.03)	Fitness cost for intermediate and resistant bacteria
$\log_{10}(N_{max})$	Weibull (14.03, 20.32) − 7.59	Large intestine carrying capacity for the bacteria
N_{start}	Uniform (0.1, 0.9) * N_{max}	Starting bacterial population size
$\log_{10}(\beta_{jk})$	Gamma (94.17,0.16) − 22.57	transposon transfer rate between /transposon transfer rate between $E.$ $coli$ subpopulations
λ_{in}	Uniform (0.001, 0.01)	Bacterial in-flow rate to the large intestine
λ_{out}	Uniform (0.01, 0.02)	Bacterial out-flow rate from the large intestine
p_i	Uniform (0.02, 0.15)	Proportion intermediate in in-flowing bacteria
p_r	Uniform (0.16, 0.61)	Proportion resistant in in-flowing bacteria
p_s	1- p_i - p_r	Proportion susceptible in in-flowing bacteria
$start_j$	Same as p_j	Starting proportions of resistant ($start_r$), intermediate ($start_i$), and susceptible ($start_s$) bacteria in the large intestine

The model was for the proportion of tetracycline-resistant enteric *Escherichia coli* in a beef steer during and after administration of oral chlortetracycline, and has been previously published (see text). The parameter symbols, definitions, and distributions are given here. Chlortetracycline (CTC)

Behavior pattern sensitivity analysis framework

The general process for behavior patterns sensitivity analysis is detailed in Fig. 1 and described below. We followed the framework laid out by Hekimoğlu et al. [20] for a behavior pattern sensitivity analysis on system dynamics models:

1. Run Monte Carlo simulations with predetermined parameter distributions
2. Identify and separate different output behavior modes
3. Define and compute output behavior pattern measures for every behavior mode

4. Perform regression analyses with output behavior pattern measures as dependent variables and standardized input parameter values as independent variables

Behavior mode identification and separation

After running 1000 Monte Carlo simulations, we examined the model output behaviors by plotting the proportion of tetracycline-resistant enteric bacteria over time. We visually examined all 1000 simulations by plotting the outputs of 100 simulations at a time and noting behavior trends. Three distinct behaviors were identified and confirmed by examining the proportion of resistance over time from a subset of simulations (Fig. 7): (1) an increase in the proportion of resistant *E. coli* across the entire 90 days, (2) a decrease in the proportion resistant across the entire 90 days, and (3) an increase in the proportion resistant during chlortetracycline administration followed by a decrease after stopping chlortetracycline. These behavior modes were formally defined by comparing the proportion resistant at the start of chlortetracycline administration (Day 2), at the end of administration (Day 30), and at the end of the simulation period (Day 90). Increasing behavior was defined as the proportion resistant at Day 2 < Day 30 < Day 90. Decreasing behavior was defined as the proportion resistant at Day 2 > Day 30 > Day 90. The third behavior was termed "peaked" and was defined as the proportion resistant at Day 2 < Day 30 > Day 90.

Behavior pattern measures

Behavior pattern measures become the dependent variables in the regression analyses, with a separate regression model for each behavior pattern measure from each behavior mode. Hekimoğlu et al. suggest that the modeler should identify a set of behavior pattern measures that completely characterize each behavior mode. Suggested pattern measures included the output levels and timing of the equilibriums, inflection points, tipping points (peaks), as well as the oscillation periods and oscillation amplitude slopes [20]. The three behaviors observed in our model were characterized by the output equilibrium level, time

to equilibrium, the output inflection point level, inflection point time, the output maximum level, and time of maximum. Maximums were found for the peaked behavior simulations by using the *max* MatLab function during the period of chlortetracycline administration (Day 2 to Day 30). We searched for equilibriums after ending chlortetracycline (> Day 30) and for inflection points during chlortetracycline administration (Day 2 to Day 30). Equilibriums and inflection points can be calculated based on the first and second derivatives of a curve, respectively. However, the proportion of resistant *E. coli* contained too much noise to calculate informative derivatives or gradients because chlortetracycline was fed in 12 h intervals, which created small oscillations in the *E. coli* proportions. Therefore, we first fit a smoothing spline curve with the least-squares method (using the *spaps* MatLab function) to each simulation's output with a tolerance of 0.01, such that the spline was within 0.01 of the simulated proportion resistant at each time-point. The spline fit was verified visually by plotting the spline and the model output (proportion resistant over time) for a subset of the 1000 simulations. Nineteen of the 1000 simulations could not be fit with a spline and therefore equilibriums and inflection points could not be found for those simulations. An equilibrium point was defined as the first time-point where the absolute value of the spline's first derivative was $< 1 \times 10^{-9}$ and remained $< 1 \times 10^{-9}$ ten time-steps (1 h total) later. This equilibrium cut-off value had to be sufficiently small to correctly identify equilibriums in simulations that had relatively small changes in proportion

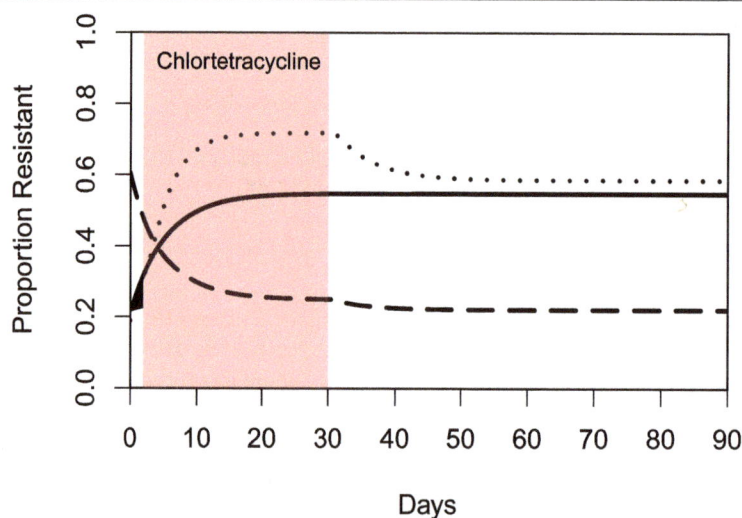

Fig. 7 Examples of the behaviors of tetracycline-resistant enteric *Escherichia coli* in beef cattle administered oral chlortetracycline. The day of the simulation is shown on the x-axis and the proportion of tetracycline-resistant enteric *Escherichia coli* is shown on the y-axis. The red shaded box is the period of chlortetracycline administration from Day 2 to Day 30. The solid line is an example of the increasing behavior, with the proportion of resistance at Day 2 < Day 30 < Day 90. The dashed line is an example of the decreasing behavior, with the proportion of resistance at Day 2 > Day 30 > Day 90. The dotted line is an example of the peaked behavior, with the proportion of resistance at Day 2 < Day 30 > Day 90

resistant over time (hence consistently small derivatives), but still demonstrated clear behavioral patterns. For example, one simulation with increasing behavior may have a 20 percentage point change in proportion resistant from Day 2 to Day 90 while another simulation may have only a 5 percentage point change from Day 2 to Day 90. The equilibrium cut-off value was chosen by examining a subset of simulations visually and assessing the derivative value at the approximate visual equilibrium. The inflection point in a simulation was found by searching for the first pair of consecutive time points where the spline's second derivative changed from positive to negative or vice versa. The first time-point of the pair was taken as the time of the inflection point. The starting proportion resistant (Day 2) was subtracted from the maximum levels and equilibrium levels in order to investigate relative maximum levels and relative equilibrium levels.

Linear regression model building and selection

Three of the parameters (N_{start}, p_s, $start_s$) were collinear with other parameters (Pearson correlation coefficient > 0.9) and therefore excluded from the regression models. N_{start} was a function of N_{max} so N_{start} was excluded and N_{max} was included in the regression models. The incoming proportion of antimicrobial-susceptible $E.\ coli$ (p_s) was a function of the incoming proportions of intermediate and resistant bacteria, since all three proportions must sum to 1. Similarly, the starting proportion susceptible ($start_s$) was a function of the starting proportions of intermediate and resistant. Therefore p_s and $start_s$ were also excluded from the regression models. Within each behavior mode, the input parameter values (x_{ik}) were standardized for each of the remaining 26 parameters, k, to make them dimensionless and facilitate regression coefficient comparisons [20, 29], as follows:

$$\tilde{x}_{ik} = \frac{x_{ik} - \bar{x}_k}{\sigma_{x_k}}$$

$\tilde{x}_{ik}$ = standardized value of parameter k for simulation i
x_{ik} = non-standardized value of parameter k for simulation i
$\bar{x}_k$ = mean of parameter k across all simulations
σ_{x_k} = standard deviation of parameter k across all simulations

Linear regression models were built using R 3.4.3 software [30] with the RStudio 1.0.136 (RStudio, Inc., Boston, MA, U.S.) user interface. Each pattern measure for each individual behavior mode was modeled separately as the dependent variable, resulting in 14 regression models. All input parameters (except for the three excluded for collinearity) were eligible to be independent variables. Thus the full regression models contained 26 input parameters as independent variables. Reduced models were built using two model selection techniques to obtain a best-fit, most parsimonious model. First, the models were fit with the 'forward,' 'backward,' and 'both' (stepwise) variable selection procedures (step, package stats) using the AIC (Akaike Information Criteria) as the fit criteria during the selection

process. The reduced models suggested by the three variable selection routines were compared. Second, the best model subsets of up to 10 variables were identified using regsubsets (package leaps), which compares models of the same size with several selection criteria [31]. The models suggested by the variable selection routines and the best subsets were compared using BIC (Schwarz's Bayesian Information Criteria), AIC, and adjusted R^2. Of the models with similarly small BIC and AIC, and large adjusted R^2, the most parsimonious model was selected as the best-fit model and was tested for model assumption validity.

Linear regression models must meet six assumptions in order to make valid predictions:

1. Observations are independent
2. Residuals follow a Normal distribution with a mean of zero
3. Linear relationship between dependent and independent variables
4. Homoscedasticity of residuals
5. Minimal or no multicollinearity of independent variables
6. Outlier observations do not drive the parameter estimates and predictions

We assessed whether each of the 14 best-fit regression models met these assumptions by examining the QQ plots, partial regression plots, standardized residuals versus fitted dependent-variable values, standardized residual histograms, Cook's D plots, and calculating variance inflation factors. Violations of assumptions were addressed by transforming independent and dependent variables and excluding outlier output values when possible. Transformations included using polynomial terms for both independent and dependent variables in order to meet the assumption (3) of a linear relationship between dependent and independent variables. Input parameters were not re-standardized after transformations. In cases of large assumption violations, alternative regression models (e.g. Cox proportional hazard models) were considered for the behavior pattern measures.

We then further simplified each of the 14 best-fit models when possible by eliminating independent variables with small coefficients (< 0.005 for "level" models and < 240 for "time" models) as long as the model fit was not significantly altered (< 10% increase in BIC, < 2 percentage point decrease in adjusted R^2, and no substantial change in residual plots) and coefficients of other parameters were generally unchanged (< 20% change). If removing a parameter did alter the model fit or other parameter coefficients, it was returned to the model. This process resulted in a best-fit, most-parsimonious model, which was again confirmed to meet linear regression assumptions.

Cox proportional hazard model building and selection

Survival analysis was considered as an alternative model to linear regression for the 'time to' pattern measures (e.g. time to maximum proportion resistant). Survival analysis handles censored data, which can occur if a behavior pattern measure (i.e. 'event') does not occur before the end of the simulated time period. Behavior pattern measures can be considered right-censored if the 'event' could occur if the simulated process time or overall simulated time was extended. For example, some peaked-behavior simulations did not reach an absolute maximum proportion resistant during chlortetracycline administration (i.e. the proportion was still increasing when chlortetracycline administration stopped). Although these simulations did reach a local maximum at the end of chlortetracycline administration, if chlortetracycline administration was simulated for a longer time, those simulations may have reached an absolute maximum. Hence this pattern measure can be considered right-censored; specifically, it has end-of-study censoring. Cox proportional hazard models [32] were built (*coxph*, package *survival*) for the 'time to' pattern measures with the occurrence of a behavior pattern measure (e.g. maximum, equilibrium) as the 'event' and the time of the occurrence as the 'event time.' Both right-censored (absolute maximum is considered the event, which was not reached in some of the simulations that are then considered censored) and non-censored (local maximum is considered the event, which is reached in all the simulations and in some of those it is reached at the last time of chlortetracycline administration) Cox models were built. The standardized input parameters of the mathematical model were independent predictors. Stepwise selection based on the AIC was used to select the best-fit, most parsimonious right-censored and non-censored models. The best-fit model between the right-censored or non-censored was chosen based on the AIC and BIC, and for that model the assumption of proportional hazards was validated by testing for no interaction between Schoenfeld residuals and time (*cox.zph*, package *survival*) [33, 34]. Violations of the proportional hazard assumption were addressed by using a time-dependent coefficient for the violating parameter (i.e. specifying an interaction between time and the parameter that allows the coefficient to change continuously over time) or by stratifying the simulated time period into strata and making the parameter coefficient a step-function of the time strata (i.e. the coefficient is a constant value within each time stratum) [34, 35].

Abbreviations
AIC: Akaike Information Criteria; BIC: Schwarz's Bayesian Information Criteria; CTC: Chlortetracycline; MIC: Minimum inhibitory concentration

Acknowledgements
Not applicable.

Funding
This project was supported by Agriculture and Food Research Initiative Competitive Grant no. 2016–68003-24607 from the USDA National Institute of Food and Agriculture. CC was supported by the Office of the Director, National Institutes of Health of the National Institutions of Health under award number T32OD011000. VV was supported by Kansas Bioscience Authority via their funding for the Institute of Computational Comparative Medicine at Kansas State University. The content is solely the responsibility of the authors and any opinions, findings, conclusions or recommendations expressed in this publication do not necessarily represent the official views of the National Center for Research Resources, the National Institutes of Health, Kansas Bioscience Authority, or the U.S. Department of Agriculture.

Authors' contributions
CC developed the example model, performed the behavior pattern sensitivity analysis, and drafted this manuscript. VV developed the example model, suggested methods for the sensitivity analysis, evaluated the behavior pattern sensitivity analysis results, and revised this manuscript. YG contributed to the example model development, suggested methods for the sensitivity analysis, evaluated the behavior pattern sensitivity analysis results, and revised this manuscript. All authors read and approved the final manuscript.

Consent for publication
Not applicable.

Competing interests
The authors declare that they have no competing interests.

Author details
[1]Department of Population Medicine and Diagnostic Sciences, College of Veterinary Medicine, Cornell University, Ithaca, NY, USA. [2]Department of Diagnostic Medicine/Pathobiology, College of Veterinary Medicine, Kansas State University, Manhattan, KS, USA.

References
1. Ayscue P, Lanzas C, Ivanek R, Grohn YT. Modeling on-farm Escherichia coli O157:H7 population dynamics. Foodborne Pathog Dis. 2009;6:461–70.
2. Al-Mamun MA, Smith RL, Schukken YH, Grohn YT. Modeling of Mycobacterium avium subsp. paratuberculosis dynamics in a dairy herd: an individual based approach. J Theor Biol. 2016;408:105–17.
3. Keeling MJ, Woolhouse ME, May RM, Davies G, Grenfell BT. Modelling vaccination strategies against foot-and-mouth disease. Nature. 2003;421: 136–42.
4. Russell CA, Real LA, Smith DL. Spatial control of rabies on heterogeneous landscapes. PLoS One. 2006;1:e27.
5. Cazer CL, Volkova VV, Grohn YT. Use of pharmacokinetic modeling to assess antimicrobial pressure on enteric bacteria of beef cattle fed chlortetracycline for growth promotion, disease control, or treatment. Foodborne Pathog Dis. 2014;11:403–11.
6. Wu H, Baynes RE, Leavens T, Tell LA, Riviere JE. Use of population pharmacokinetic modeling and Monte Carlo simulation to capture individual animal variability in the prediction of flunixin withdrawal times in cattle. J Vet Pharmacol Ther. 2013;36:248–57.
7. Lin Z, Gehring R, Mochel JP, Lave T, Riviere JE. Mathematical modeling and simulation in animal health - part II: principles, methods, applications, and value of physiologically based pharmacokinetic modeling in veterinary medicine and food safety assessment. J Vet Pharmacol Ther. 2016;39:421–38.
8. Riviere JE, Gabrielsson J, Fink M, Mochel J. Mathematical modeling and simulation in animal health. Part I: moving beyond pharmacokinetics. J Vet Pharmacol Ther. 2016;39:213–23.
9. Graesboll K, Nielsen SS, Toft N, Christiansen LE. How fitness reduced, antimicrobial resistant bacteria survive and spread: a multiple pig-multiple bacterial strain model. PLoS One. 2014;9:e100458.
10. Ahmad A, Graesboll K, Christiansen LE, Toft N, Matthews L, Nielsen SS. Pharmacokinetic-pharmacodynamic model to evaluate intramuscular tetracycline treatment protocols to prevent antimicrobial resistance in pigs. Antimicrob Agents Chemother. 2015;59:1634–42.

11. Cazer CL, Ducrot L, Volkova VV, Gröhn YT. Monte Carlo simulations suggest current chlortetracycline drug-residue based withdrawal periods would not control antimicrobial resistance dissemination from feedlot to slaughterhouse. Front Microbiol. 2017;8:1753.

12. Abatih EN, Alban L, Ersboll AK, Lo Fo Wong DM. Impact of antimicrobial usage on the transmission dynamics of antimicrobial resistant bacteria among pigs. J Theor Biol. 2009;256:561–73.

13. Dietz K. Epidemics and Rumours: a survey. J R Stat Soc Ser A (General). 1967;130:505.

14. Huang L, Huang Y, Wang Q, Xiao N, Yi D, Yu W, Qiu D. An agent-based model for control strategies of Echinococcus granulosus. Vet Parasitol. 2011; 179:84–91.

15. Robins J, Bogen S, Francis A, Westhoek A, Kanarek A, Lenhart S, Eda S. Agent-based model for Johne's disease dynamics in a dairy herd. Vet Res. 2015;46:68.

16. Lanzas C, Chen S. Complex system modelling for veterinary epidemiology. Prev Vet Med. 2015;118:207–14.

17. Iooss B, Lemaître P. A review on global sensitivity analysis methods. In: Uncertainty Management in Simulation-Optimization of Complex Systems. Boston: Springer; 2015. p. 101–22.

18. Saltelli A, Tarantola S, Campolongo F, Ratto M. Sensitivity analysis in practice: a guide to assessing scientific models. Chichester, England: John Wiley & Sons, Ltd; 2004.

19. Ford A, Flynn H. Statistical screening of system dynamics models. Syst Dyn Rev. 2005;21:273–303.

20. Hekimoğlu M, Barlas Y, Luna-Reyes L. Sensitivity analysis for models with multiple behavior modes: a method based on behavior pattern measures. Syst Dyn Rev. 2016;32:332–62.

21. USDA. Feedlot 2011 Part IV: health and health management on US feedlots with a capacity of 1,000 or more head. USDA-APHIS-VS-CEAH-NAHMS. Fort Collins, CO, 2013: #638.0913.

22. US Food and Drug Administration. Guidance for Industry #209: The judicious use of medically important antimicrobial drugs in food-producing animals. Rockville, MD: Administration UFaD; 2012.

23. Therneau T, Crowson C, Atkinson E. Using time dependent covariates and time dependent coefficients in the cox model. In: Survival vignettes. The comprehensive R archive network; 2017.

24. Tibshirani R. Regression shrinkage and selection via the lasso. J R Stat Soc Ser B Methodol. 1996;58(1):267–88.

25. Sharma C, Rokana N, Chandra M, Singh BP, Gulhane RD, Gill JPS, Ray P, Puniya AK, Panwar H. Antimicrobial resistance: its surveillance, impact, and alternative management strategies in dairy animals. Front Vet Sci. 2017;4:237.

26. McAllister TA, Beauchemin KA, Alazzeh AY, Baah J, Teather RM, Stanford K. Review: the use of direct fed microbials to mitigate pathogens and enhance production in cattle. Can J Anim Sci. 2011;91:193–211.

27. Alexander TW, Yanke LJ, Topp E, Olson ME, Read RR, Morck DW, McAllister TA. Effect of subtherapeutic administration of antibiotics on the prevalence of antibiotic-resistant Escherichia coli in feedlot cattle. Appl Environ Microbiol. 2008;74:4405–16.

28. Zoetis: Aureomycin 50 Granular A [package insert]. In: Chlortetracycline Type A Medicated Article, vol. 10012901. Kalamazoo: Zoetis; 2017.

29. Kleijnen JPC. Sensitivity analysis and optimization of system dynamics models: regression analysis and statistical design of experiments. Syst Dyn Rev. 1995;11:275–88.

30. R Core Team. R: a language and environment for statistical computing. Vienna: R Foundation for Statistical Computing; 2017.

31. Thomas Lumley based on Fortran code by Alan Miller: leaps: Regression Subset Selection. 2017.

32. Cox DR. Regression models and life-tables. J R Stat Soc Ser B Methodol. 1972;34:187–220.

33. Therneau TM. A package for survival analysis in S; 2015.

34. Hess KR. Graphical methods for assessing violations of the proportional hazards assumption in cox regression. Stat Med. 1995;14:1707–23.

35. Zhang Z, Reinikainen J, Adeleke KA, Pieterse ME, Groothuis-Oudshoorn CGM. Time-varying covariates and coefficients in cox regression models. Ann Transl Med. 2018;6:121.

19

Assessment of a rtPCR for the detection of virulent and benign *Dichelobacter nodosus*, the causative agent of ovine footrot, in Australia

Nickala Best[1], Lucas Zanandrez[2], Jacek Gwozdz[3], Eckard Klien[4], Nicky Buller[4], Robert Suter[5], Grant Rawlin[3] and Travis Beddoe[1]* (iD)

Abstract

Background: Ovine footrot is a highly contagious bacterial disease of sheep, costing the Australian sheep industry millions of dollars annually. *Dichelobacter nodosus*, the causative agent of footrot, is a gram-negative anaerobe classed into virulent and benign strains as determined by thermostability of their respective protesases. Current methods for detection of *D. nodosus* are difficult and time-consuming, however new molecular techniques capable of rapidly detecting and typing *D. nodosus* have been reported.

Results: A competitive real-time PCR (rtPCR) method, based on the ability to detect a 2 nucleotide difference in the *aprV2* (virulent) and *aprB2* (benign) extracellular protease gene has been tested on Australian samples for determining detection rates, along with clinically relevant cut-off values and performance in comparison to the traditional culturing methods. The rtPCR assay was found to have a specificity of 98.3% for virulent and 98.7% for benign detection from samples collected. Sheep with clinical signs of footrot showed a detection rate for virulent strains of 81.1% and for benign strains of 18.9%. A cut-off value of a Ct of 35 was found to be the most appropriate for use in Victoria for detection of sheep carrying virulent *D. nodosus*.

Conclusions: In summary, the rtPCR assay is significantly more capable of detecting *D. nodosus* than culturing, while there is no significant difference seen in virotyping between the two methods.

Keywords: Australia, Benign, Footrot, Real-time polymerase chain reaction, Sensitivity, Sheep, Specificity, Victoria, Virulent

Background

Ovine footrot is a highly contagious bacterial disease of sheep, causing lesions in the hoof and lameness [1]. The primary aetiologic agent of ovine footrot is *Dichelobacter nodosus*, a gram-negative anaerobe [2]. Many strains of *D. nodosus* exist consisting of multiple serogroups that are classified in Australia into virulent or benign based on extracellular protease activity. Infections with benign strains may appear as inflammation of the interdigital skin (interdigital dermatitis), while infections with virulent strains

may vary from interdigital dermatitis to severe lesions with extensive necrosis and separation of the horn from the soft tissue [3, 4]. Footrot lesions are graded using a simple scoring system ranging from 0 (clinically healthy) to 5 (severe underrunning of the hard horn of the hoof) [3, 4]. The severity of lesions produced by virulent strains is reliant on environmental conditions, with temperatures of above 10 °C and consistent rainfall required for the full expression of virulence factors [5]. Subsequently, when environmental conditions are not optimal, infection with virulent strains may not be apparent clinically or it may present itself as mild infection imitating benign footrot [6]. The virulence potential of the *D. nodosus* strains may be determined by measuring the

* Correspondence: t.beddoe@latrobe.edu.au
[1]Department of Animal, Plant and Soil Science, Centre for AgriBioscience (AgriBio), La Trobe University, Bundoora, Melbourne, VIC, Australia

thermostability of serine proteases of isolates using the gelatin gel (gelatinase) test [7]. However, culture-based assays have been reported to have modest diagnostic power [8]. Furthermore, culture-based tests are labour-intensive and requires several weeks for the results to become available [9]. It has been reported that the acidic protease 2 (AprV2) plays a key role in virulence of *D. nodosus* [10]. Virulent strains have the *aprV2* gene encoding a thermostable protease. Benign strains have the gene *aprB2* encoding a thermolabile protease. The *aprV2* and *aprB2* alleles vary by a two-base pair substitution. This difference has been exploited in the probe design of a real-time polymerase chain reaction (rtPCR) assay [11]. Using this rtPCR assay, the presence of *D. nodosus* and its virulence can be determined within 1 day. The test is also capable of detecting both benign and virulent in the same clinical sample. There are limited data on the diagnostic performance of the rtPCR in sheep in Europe [12]. The objective of this paper was to evaluate the rtPCR assay with clinical samples collected from sheep in Victoria, Australia, with confirmed or suspected virulent footrot and sheep considered to be free of infection with *D. nodosus*.

Methods
Sample collection
The clinical samples were obtained retrospective from Victoria Government Veterinary Diagnostic unit collection. The samples were submitted by Victorian District Veterinary Officers, Animal Health Officers and private veterinary practitioners from the interdigital skin of lame sheep for routine diagnostic testing. Flocks were selected for this study if individual foot scores corresponding to sample labelling from individual sheep were provided, as well as the clinical history of the flock as determined by the submitting persons. Three hundred eighteen sheep from 10 flocks (#1 to 10) considered free of footrot, 170 sheep from 13 flocks (#12 to 24) with confirmed and/or suspected virulent footrot and 27 sheep from a closed flock with the history of virulent footrot that was deemed successfully eradicated. Prior to sampling, each sheep was examined and foot lesions scored and recorded. The interdigital skin of the foot with the highest score was sampled using two sterile swabs. One swab was placed into Stuarts Transport Media for culture and the second swab was placed into 800 µL of phosphate buffered saline (PBS) (8.1 mM Na_2HPO_4, 137 mM NaCl, 1.4 mM KH_2PO_4 and 2.6 mM KCl) with 20 mM ethylenediaminetetraacetic acid (EDTA), pH 8.0), for rtPCR. The origin, breed, sex, age, foot score of the sheep sampled and their flock history are presented in Tables 1 and 2. Samples were kept at 4 °C after collection and sent the following day to AgriBio, 5 Ring Rd., La Trobe,

Bundoora, Victoria, 3083. Samples were submitted for disease investigations pertaining to footrot.

Culturing of *D. nodosus*
Swabs collected from sheep from flocks 1 to 14 were plated at AgriBio one to 2 days after collection onto 4% (*w/v*) agar with 3% (w/v) ground hoof media (Footrot Reference Laboratory, Department of Agriculture, Perth, Australia) and anaerobically incubated at 37 °C for 7 days. Plates were examined for *D. nodosus* growth and suspect colonies subcultured and gram stained as described elsewhere [9]. *D. nodosus* isolate A198 (*aprV2*) (AC: 6466), virulent control, and isolate C305 (*aprB2*) (AC: 6465), benign control, were obtained from the Footrot Reference Laboratory and grown concurrently with sample plates. Isolates morphologically consistent with *D. nodosus* were sent on ice to the Footrot Reference Laboratory for gelatinase testing. Swabs collected from sheep from flocks 15 to 24 were sent on ice by overnight courier to the Footrot Reference Laboratory, Perth, Western Australia for culture and gelatinase testing.

DNA extraction
DNA was extracted from swabs using a commercial extraction kit (MagMAX™ – 96 Viral RNA isolation kit, Ambion, Austin, USA) and purification system (Kingfisher-96 magnetic particle handling system, Thermo Fisher Scientific, Finland). Swabs from two positive culture controls were used as positive extraction controls. Swabs collected from sheep from flocks 1 to 14 were subject to DNA extraction on two separate occasions (runs).

aprV2/B2 rtPCR
Primers, probes and cycling conditions as described by Stauble et al., 2014 were used [11]. A commercial rtPCR kit (AgPath-ID™ One Step RT-PCR Kit, Ambion, Austin, USA) was used as master mix according to manufacturer's instructions, with final concentrations of 300 nM primers, 100 nM DnAprTM-vMGB, 250 nM DnAprTM-bMGB and 5 µL of template DNA. Primers and probes were synthesised commercially (Primers and probes, Applied Biosystems, California, USA) Reactions were carried out and analysed (7500 Fast Real-Time PCR System, Life Technologies) with a set threshold of 0.05. Two DNA extracts derived from sheep from flocks 1 to 14 were assayed by the rtPCR in two separate runs. Singular DNA extracts derived from sheep from flocks 15 to 24 were assayed by the rtPCR in one run.

Positive extraction controls from live cultures of isolate A198 and C305, and purified and isolated genomic DNA from the same culture isolates were used as rtPCR controls in each run. The rtPCR run was considered valid when results obtained in rtPCR controls were

Table 1 Descriptive characteristics of 11 Australian sheep flocks considered free of virulent footrot were sampled between June '15 and August '15 for evaluation of the specificity of an rtPCR for detection of virulent (*aprV2*) and benign (*aprB2*) protease genes of *D. nodosus*

Flock ID	No. animals sampled	Sampling date	Flock origin (shire or city)	Breed	Age	Sex	Comments/Flock history
1	18	02.06.2015	City of Broken Hill, NSW	Merino	Lambs	Mixed	Abbatoir line.
2	18	02.06.2015	Blayney Shire, NSW	Mixed	Mixed	Mixed	Abbatoir line.
3	19	02.06.2015	Shire ofArarat, Vic	Mixed	Rams	Male	Abbatoir line.
4	18	02.06.2015	Shire of Ararat, Vic	Merino	Mixed	Female	Abbatoir line.
5	19	02.06.2015	Shire of Ararat, Vic	Merino	Lambs	Mixed	Abbatoir line.
6	18	02.06.2015	City of Wagga Wagga, NSW	Merino	Ewes	Female	Abbatoir line.
7	18	02.06.2015	Southern Grampians Shire, Vic	Crossbreed	Ewes	Female	Abbatoir line.
8	55	16.06.2015	Yarra Ranges Shire, Vic	Coopworth crosses	Ewes	Female	Farm has had previous intermittent lameness, footrot has not been confirmed as the cause.
9	81	01.07.2015	Strathbogie Shire, Vic	Merino	Wethers	Male	Farm has no history of footrot, but lameness occasionally observed. Sheep footbathed late 2014.
10	54	17.07.2015	Wellington Shire, Vic	Merino	Wethers	Male	Well managed merino stud, no history of footrot. Wether had strayed into adjoining properties and had been cought and shorn 2 days prior to sampling.
11[a]	27	05.08.2015	East Gippsland Shire, Vic	Merino	Ewes	Female	Virulent footrot first introduced in the 80's, treated by footbathing. Second footrot introduced in 1995; eradicated by footbathing, antibiotic regime and culling. A closed flock currently. No footbathing for ≥ 10 years.

[a] Flock 11 has been excluded from specificity and sensitivity calculations because of its history

concordant. Results are reported as cycling threshold (Ct) values, the point at which the sample signal exceeds the threshold of 0.05. Samples producing a probe-specific fluorescent signal were defined as being positive. The effect of two cut-off values; Ct < 40 and Ct < 35, on the rtPCR detection rate and specificity, was assessed.

Data analysis

Considering the culture/gelatinase method lacks adequate diagnostic accuracy and both the virulent and benign strains of *D. nodosus* may produce subclinical or mild, clinically unapparent infection, detection rates of the *aprB2* rtPCR and *aprV2* rtPCR were calculated using data obtained from 135 sheep with foot lesions and 35 healthy sheep from the 13 flocks with confirmed or suspected virulent footrot. Because of the lack of the gold diagnostic standard, the specificities of the *aprB2* rtPCR and *aprV2* rtPCR were calculated using data derived only from the 297 healthy sheep from the 10 flocks considered to be free of footrot. Twenty-one sheep with foot lesions scored 1 and 2 were excluded from the specificity calculations. Both the detection rate of clinically infected animals and specificity of the rtPCR were calculated for cut-offs of Ct of 35 and 40 values respectively, using data obtained in the first rtPCR run. Descriptive statistics

were performed with Microsoft Excel 2010, Microsoft Corporation, Redmond, WA. Further analyses including Cohen's kappa statistic (agreement between rtPCR runs, agreement between gelatinase gel test and rtPCR virulence designations), using the Altman scheme where ≤0 = worse than chance alone, < 0.20 = poor, 0.21–0.40 = fair, 0.41–0.60 = moderate, 0.61–0.80 = good, and 0.81–0.99 = very good, 1.00 = perfect and McNemar's test for comparisons of foot lesions vs. rtPCR, culture vs. rPCR and rtPCR run vs. rtPCR run were also performed with GraphPad Prism version 4.01 for Windows, GraphPad Software, La Jolla, CA.

Results

The rtPCR results in relation to foot scores and culture/gelatinase test results obtained in 10 flocks considered free of footrot, 13 flocks with confirmed and/or suspected virulent footrot and a closed flock (#11) with the history of eradicated virulent footrot are presented in Table 3.

For all data, Ct's of 40 and 35 were investigated as suitable cut-off values to interpret positive results. From data collected through the clinically healthy trial, the use of a Ct of 40 showed poor discrimination between healthy and clinically affected sheep in both *aprV2* and *aprB2* results ($p > 0.005$, both, Table 4). When using a Ct

Table 2 Descriptive characteristics of 11 Australian sheep flocks considered having virulent footrot were sampled between October '14 and July '15 for evaluation of the specificity of an rtPCR for detection of virulent (*aprV2*) and benign (*aprB2*) protease genes of *D. nodosus*

Flock ID	No. animals sampled	Sampling date	Flock origin (shire or city)	Breed	Age	Sex	Comments/Flock history
12	26	27.07.2015	Wangaratta Rural City, Vic	Dorper	Rams	Male	Farm has a history of virulent footrot. Footbathing and foot pairing done frequently. Minimal lameness and lesions currently present. Sheep reviously footbathed in May 2015.
13	10	10.07.2015	East Gippsland Shire, Vic	Merino Cross	Mixed	Mixed	Footrot introduced December 2014 by purchased rams. Owners observed lameness of about 1 in 150 animals in March/April 2015.
14	36	23.06.2015	Mitchell Shire, Vic	Merino	Rams	Male	Farm suspected of virulent footrot, samples taken on a confirmatory visit.
15	10	13.10.2014	Strathbogie Shire, Vic	Merino	NR	Female	History of footrot and lameness in flock
16	10	13.10.2014	Greater Shepparton City, Vic	Merino cross	2.5 y	Female	A mob of South African Merino X Merino yearling ewes purchased in Nov-Dec 2013. Sheep had been on agistment. A few sheep lame when they arrived; footbathed 2–4 weeks before sampling.
17	10	14.10.2014	Shire of Newstead, Vic	Merino	Mixed	Female	Property has a footrot history - previously treated sucessfully with Footrite®. This season a recurrance of lameness, some mobs reached a 20%.
18	9	20.10.2014	Indigo Shire, Vic	Dorper	Adult	Female	Lameness in more than one foot. Lesions suggestive of benign footrot.
19	10	03.11.2014	Shire of Glenelg, Vic	NR	3.5 years	Female	Footrot appeared in June; source not determined, appears to be clinically aggressive; high prevalence of score 4/5 (20%+). These sheep last footbathed ~ 3 weeks before sampling.
20	10	04.11.2014	Southern Grampians Shire, Vic	NR	2 years	Female	Footrot has probably been present for a long time. Controlled by regular footbathing. Last month have had 30% of average long term rainfall for this time of the year. These sheep last footbathed ~ February 2014
21	10	17.11.2014	Bass Coast Shire, Vic	Droper	Mixed	Mixed	Many lambs, ewes and some rams reported lame with lesions very suggestive of footrot. Treated with footbath (formalin) and antibiotics. The lesions look in a process of healing but are still obvious (inflammation limited to interdigital space).
22	10	18.11.2014	Mitchell Shire, Vic	Merino	NR	NR	Virulent footrot for several years.
23	10	24.11.2014	Southern Grampians Shire, Vic	NR	Adult	Female	Footrot appeared in June (source not determined) with high prevalence of score 4/5 (20%+). Last 3 months have had 30% of average long term rainfall for this time of the year. These sheep last footbathed ~ 3 weeks earlier.
24	9	01.12.2014	Colac Ottway Shire, Vic	Crossbred	Adult	Female	10 of 15 seep examined had feet lesions scored ≥ 2

of 35 and under to indicate positive rtPCR results, detection of footrot using rtPCR differed significantly from using clinical signs alone ($p = 0.0014$ for *aprB2*, and $p = 0.019$ for *aprV2*). The agreement between repeated runs at Ct of 40 for virulent results was good (kappa = 0.731), and perfect when using a Ct of 35 (kappa = 1). A similar result is seen for the benign rtPCR results, with a reasonable agreement for a Ct of 40 between repeats (kappa = 0.302), and a highly significant agreement when using a Ct of 35 (kappa = 1) (Table 4). In the data from clinically affected animals, there was a significant difference between numbers of positive results obtained by the *aprV2* rtPCR and *aprB2* rtPCR at a cut-off of 40 and

35 respectively, in the clinically affected ($n = 135$) and healthy ($n = 35$) sheep. Using a Ct of 35 resulted in a better agreement between replicates of the same sample (Table 5). Results using a Ct of 35 only will be reported further due to the increased agreement between repeats.

Using 297 animals from historically healthy farms scored 0, specificity was calculated to be 98.3% for *aprV2* detection, and 98.7% for *aprB2* detection when using a Ct of 35 as the cut-off. From these flocks, 318 animals were tested in total, with 21 scored 1–5 (Table 3). One sheep of the 318 from the flocks considered free of footrot yielded growth of *D. nodosus* of undetermined gelatinase profile. This sheep (Flock 2) also produced strong reactions

Table 3 Clinical scores and results of *D. nodosus* culture, gelatinase test and rtPCR (aprB2/aprV2 positive, two cut offs, duplicate runs) in 318 sheep from flocks considered to be free of footrot (clinically healthy), 170 sheep from flocks considered having virulent footrot (clinically affected) and 27 sheep from a closed flock (#11) that apparently eradicated virulent footrot more than 10 years ago. Flocks were sampled from October 2014 to August 2015

	Flock ID	No. sheep tested	Results							Culture[c]	Gelatinase test[d]			aprB2 rtPCR[d]				aprV2 rtPCR[d]			
			Clinical Score[b]											Run 1		Run 2		Run 1		Run 2	
			0	1	2	3	4	5	Mean		Benign	Virulent	Undetermined	Cut-off		Cut-off		Cut-off		Cut-off	
														40	35	40	35	40	35	40	35
Clinically healthy	1	18	18	–	–	–	–	–	0	–	–	–	–	–	–	–	–	–	–	–	–
	2	18	18	–	–	–	–	–	0	1	–	–	1	–	–	–	–	2	2	2	2
	3	19	17	–	1	–	1	–	0.1	–	–	–	–	–	–	1	–	–	–	–	–
	4	18	11	1	6	–	–	–	0.6	–	–	–	–	–	–	4	–	–	–	–	–
	5	19	19	–	–	–	–	–	0	–	–	–	–	–	–	2	–	–	–	–	–
	6	18	11	3	1	1	–	2	0.8	–	–	–	–	–	–	–	–	5	4	6	4
	7	18	16	2	–	–	–	–	0.1	–	–	–	–	–	–	–	–	–	–	–	–
	8	55	55	–	–	–	–	–	0	–	–	–	–	5	1	1	1	3	2	4	2
	9	81	81	–	–	–	–	–	0	2	2	–	–	5	3	4	3	2	–	5	–
	10	54	51	3	–	–	–	–	0.1	–	–	–	–	8	–	–	–	2	1	2	1
Subtotal		318	297	9	8	1	1	2		3	2	0	1	18	4	#	4	14	9	19	9
	11[a]	27	11	16	–	–	–	–	0.6	–	–	–	–	16	10	20	13	–	–	–	–
Clinically affected	12	26	14	12	–	–	–	–	0.5	7	–	5	2	–	–	–	–	26	25	26	26
	13	10	1	3	–	5	1	–	2.2	4	2	2	–	5	3	3	3	9	6	9	6
	14	36	20	8	3	–	–	5	1.1	2	–	2	–	6	5	6	5	17	14	21	13
	15	10	–	–	1	–	2	7	4.5	2	–	2	–	–	–	ND	ND	10	9	ND	ND
	16	10	–	–	4	3	2	1	3	5	–	5	–	–	–	ND	ND	10	10	ND	ND
	17	10	–	–	4	2	4	–	1	10	4	6	–	4	4	ND	ND	6	6	ND	ND
	18	9	–	9	–	–	–	–	4.9	3	–	3	–	–	–	ND	ND	9	9	ND	ND
	19	10	–	–	–	–	1	9	2.7	3	–	3	–	–	–	ND	ND	9	9	ND	ND
	20	10	–	–	4	5	1	–	1.1	3	2	1	–	10	10	ND	ND	10	10	ND	ND
	21	10	–	9	1	–	–	–	3.5	–	–	–	–	2	2	ND	ND	2	1	ND	ND
	22	10	–	–	1	5	3	1	3.3	6	1	5	–	–	–	ND	ND	10	10	10	10
	23	10	–	–	–	–	10	–	4	–	–	–	–	–	–	ND	ND	10	10	ND	ND
	24	9	–	–	5	2	2	–	2.6	–	–	–	–	–	–	ND	ND	9	8	ND	ND
Subtotal		170	35	41	23	22	26	23		45	9	34	2	27	24	9	8	137	127	66	55

[a] Flock 11 has been excluded from specificity and sensitivity calculations because of its history
[b] Results expressed as a number of sheep in which the highest rated foot was in the particular score
[c] Results expressed as a number of sheep from which an organism morphologically consistent with *D. nodosus* was isolated
[d] Results expressed as a number of sheep that tested positive

(Ct ≤ 31) in the *aprV2* rtPCR in both runs. Two other sheep (Flock 9) tested positive for the benign strain of *D. nodosus* by culture/gelatinase test and also by the *aprB2* rtPCR in both runs (Ct ≤ 33.74). One of these two sheep gave also a weak reaction (Ct 38.39) in the *aprV2* rtPCR in one run. All three sheep that yielded growth of *D. nodosus* and tested positive by the *aprV2* rtPCR and/or *aprB2* rtPCR were clinically healthy (Table 3).

From the 170 animals clinically affected or suspected, 135 were scored 1 or above and considered clinically affected, while 35 were scored 0. The rtPCR produced positive virulent results in 112 of the 135 clinically affected sheep, giving a detection rate of 83% when using a Ct of 35, or 81.1% overall when using all 170 animals (Table 3).

In comparison, 45 animals from the 170 in the affected group had *D. nodosus* isolates successfully obtained and 43 had the gelatinase test performed (Table 6). There was a significant difference between the two methods capabilities to detect virulent *D. nodosus* (p < 0.0001). Comparing the gelatin gel designation of virulence to

Table 4 Results of duplicate runs at cut offs Ct 40 and 35 for *aprV2* and *aprB2* rtPCR in 297 clinically healthy sheep and 21 sheep with foot lesions from 10 flocks considered free of footrot

Foot lesion	Cut off 40				Cut off 35			
	Run 1 +ve	Run 1 -ve	Run 2 +ve	Run 2 -ve	Run 1 +ve	Run 1 -ve	Run 2 +ve	Run 2 -ve
aprB2								
Positive	0	21	2	19	0	21	0	21
Negative	18	279	10	287	4	293	4	293
Specificity		93.90%		96.60%		98.70%		98.70%
McNemars two tailed *p* value		0.7488		0.1374		0.0014		0.0014
aprB2 cut off agreement								
Run 2 +ve	5	7			4	0		
Run 2 -ve	13	293			0	314		
Kappa	0.302	"fair"			1	"perfect"		
aprV2								
Positive	4	17	4	17	4	17	4	17
Negative	10	287	15	282	5	292	5	292
Specificity		96.60%		94.90%		98.30%		98.30%
McNemars two tailed *p* value		0.2482		0.8597		0.019		0.019
aprV2 cut off agreement								
Run 2 +ve	12	7			9	0		
Run 2 -ve	2	297			0	309		
Kappa	0.713	"good"			1	"perfect"		

Specificity is shown along with *p*-value for McNemar's test for independence between lesion score and rtPCR result, and kappa statistic for agreement between rtPCR runs

Table 5 Results from two cut offs of Ct 40 and 35 from the *aprB2* rtPCR and *aprV2* rtPCR obtained from two runs, in 72 sheep randomly sampled from 3 flocks considered having virulent footrot

Foot lesion	Cut off 40				Cut off 35			
	Run 1 +ve	Run 1 -ve	Run 2 +ve	Run 2 -ve	Run 1 +ve	Run 1 -ve	Run 2 +ve	Run 2 -ve
aprB2								
Positive	9	28	8	29	7	30	7	30
Negative	2	33	1	34	1	34	1	34
Overall % positive		15.30%		12.50%		11.10%		11.10%
McNemars two tailed *p* value		< 0.0001		< 0.0001		< 0.0001		< 0.0001
aprB2 cut off agreement								
Run 2 +ve	8	1			8	0		
Run 2 -ve	3	60			0	64		
Kappa	0.768	"good"			1	"perfect"		
aprV2								
Positive	33	4	33	4	30	7	29	8
Negative	19	16	23	12	15	20	16	19
Overall % positive		72.20%		77.80%		62.50%		62.50%
McNemars two tailed *p* value		0.0035		0.0005		0.1356		0.153
aprV2 cut off agreement								
Run 2 +ve	51	5			44	1		
Run 2 -ve	1	15			1	26		
Kappa	0.779	"good"			0.941	"very good"		

The *p*-value for McNemar's test for independence between lesion score and rtPCR result is shown along with kappa statistic for agreement between rtPCR runs

Table 6 Gelatinase gel test and rtPCR results for 45 individual samples within the clinically affected data set that successfully had *D. nodosus* isolated, where S is thermostable (virulent) and U is thermolabile (benign)

Flock ID	Score[a]	Gelatinase gel test	rtPCR
12	1	S	*aprV2*
	1	S	*aprV2*
	0	S	*aprV2*
	0	NA	*aprV2*
	0	S	*aprV2*
	0	NA	*aprV2*
	0	S	*aprV2*
13	1	U	*aprV2* and *aprB2*
	1	S	*aprV2* and *aprB2*
	3B	U	*aprV2*
	3C	S	*aprV2*
14	5	S	*aprV2*
	1	S	*aprV2* and *aprB2*
15	5	S	*aprV2*
	5	S	*aprV2*
16	3	S	*aprV2*
	2	S	*aprV2*
	2	S	*aprV2*
	3	S	*aprV2*
	4	S	*aprV2*
17	4	S	*aprV2*
	2	S	*aprV2*
	2	U	*aprB2*
	4	U	*aprB2*
	4	U	*aprB2*
	3	U	*aprB2*
	3	S	*aprV2*
	2	S	*aprV2*
	4	S	*aprV2*
	2	S	*aprV2*
18	1	S	*aprV2*
	1	S	*aprV2*
	1	S	*aprV2*
19	5	S	*aprV2*
	5	S	*aprV2*
	5	S	*aprV2*
20	3A	S	*aprV2* and *aprB2*
	2	U	*aprV2* and *aprB2*
	3B	S	*aprV2* and *aprB2*
22	3A	S	*aprV2*
	3A	U	*aprV2*
	4	S	*aprV2*

Table 6 Gelatinase gel test and rtPCR results for 45 individual samples within the clinically affected data set that successfully had *D. nodosus* isolated, where S is thermostable (virulent) and U is thermolabile (benign) (*Continued*)

Flock ID	Score[a]	Gelatinase gel test	rtPCR
	3A	S	*aprV2*
	3	S	*aprV2*
	4	S	*aprV2*

[a]Foot scores are according to Stewart et al., 1983 [4]. An additional file describes the foot scoring in more detail (see Additional file 1)

the rtPCR results, there is no significant difference between the two tests when using 37 of the isolates (McNemars Test, $p = 0.479$). Samples that tested positive for *aprV2* and *aprB2* via rtPCR and had *D. nodosus* successfully isolated were excluded from the above calculation as no sample had more than one isolate obtained. None of the 135 clinically affected sheep and 35 clinically healthy sheep tested negative by the *aprV2* rtPCR but positive for the virulent strain of *D. nodosus* by the culture/gelatinase test. The agreement between results produced by the *aprV2* rtPCR and that obtained by the culture/gelatinase test ranged from fair (Kappa = 0.2–0.222) to poor (kappa = 0.082–0.158) (Table 7). In total, when the affected flock samples were cultured, an overall detection rate of 25% was obtained (Table 7).

There was a significant difference ($p \leq 0.0015$) between numbers of animals tested positive for the virulent strain of *D. nodosus* by the *aprV2* rtPCR and culture/gelatinase test among the 35 clinically healthy sheep. In this group, the *aprV2* rtPCR gave positive results in 15 (42.9%) and 16 (45.7%) of clinically healthy animals at a Ct of 35 cut-off value in run 1 and run 2, respectively, whereas the gelatinase test gave positive results in 3 of the healthy animals (Table 7).

The same method was applied to *aprB2* rtPCR positive samples, resulting in a detection rate of 18.9%, with 23 of 135 clinically affected animals (17%) positive for *aprB2* via rtPCR. Using culturing and the gelatin gel test, 9 (6.7%) of the 135 clinically affected animals tested positive for benign *D. nodosus*. Again, there was a significant difference between the two methods capabilities to detect benign *D. nodosus* ($p = 0.0022$). Two of the 135 clinically affected sheep tested positive for the benign strain of *D. nodosus* by the culture/gelatinase test but negative by the *aprB2* rtPCR. The agreement between results produced by the *aprB2* rtPCR and that obtained by the culture/gelatinase test ranged from fair (Kappa = 0.336–0.378) to poor (kappa = 0–0.163) (Table 8).

Among the 35 clinically healthy sheep, there was no significant difference ($p = 1$) between numbers of animal's positive for the benign strain of *D. nodosus* by the *aprB2* rtPCR and culture/gelatinase test. In this group, the *aprB2* rtPCR at a Ct of 35 gave a positive reaction in

Table 7 Comparisons of identification of *D. nodosus* by the *aprV2* rtPCR and culture/gelatinase gel test from subsets of presence/absence or both of foot lesions from the clinically affected flocks, consisting of 135 clinically affected sheep and 35 clinically healthy sheep

Flocks	Sheep clinical status (foot lesions)	No. sheep tested	aprV2 PCR Run/Cut-off	Culture Gelatinase Thermostable (Virulent) vs aprV2 PCR						Percentage positive results	
				Concordant results		Culture +ve PCR -ve	Culture -ve PCR +ve	p value	Kappa	Culture	aprV2 PCR
				+ve	-ve						
12 to 24	+ve	135	Run 1 / 40	31	17	0	87	< 0.0001	0.082	23.0%	87.4%
12 to 24	+ve	135	Run 1 / 35	31	20	0	84	< 0.0001	0.1	23.0%	85.2%
12 to 14	-ve	35	Run 1 / 40	3	16	0	16	0.0002	0.146	8.6%	54.3%
12 to 14	-ve	35	Run 2 / 40	3	12	0	20	< 0.0001	0.093	8.6%	65.7%
12 to 14	-ve	35	Run 1 / 35	3	20	0	12	0.0015	0.222	8.6%	42.9%
12 to 14	-ve	35	Run 2 / 35	3	19	0	13	0.0009	0.2	8.6%	45.7%
12 to 14	-ve & +ve	72	Run 1 / 40	9	20	0	43	< 0.0001	0.104	12.50%	72.20%
12 to 14	-ve & +ve	72	Run 2 / 40	9	16	0	47	< 0.0001	0.078	12.50%	77.80%
12 to 14	-ve & +ve	72	Run 1 / 35	9	27	0	36	< 0.0001	0.158	12.50%	62.50%
12 to 14	-ve & +ve	72	Run 2 / 35	9	27	0	36	< 0.0001	0.158	12.50%	62.50%
12 to 24	-ve & +ve	170	Run 1 / 40	34	33	0	103	< 0.0001	0.114	20.0%	80.6%
12 to 24	-ve & +ve	170	Run 1 / 35	34	40	0	96	< 0.0001	0.143	20.0%	76.5%

The *p*-value for McNemar's test for independence between culture gelatinase thermostable (Virulent) and *aprV2* PCR result is shown. Detection rates of both culture and rtPCR are presented as percentage positive results

Table 8 Comparisons of identification of *D. nodosus* by the *aprB2* rtPCR and culture/gelatinase gel test from subsets of presence/absence or both of foot lesions from the clinically affected flocks, consisting of 135 clinically affected sheep and 35 clinically healthy sheep

Flocks	Sheep clinical status (foot lesions)	No. sheep tested	aprB2 PCR Run and/or Cut-off	Culture Gelatinase Thermolabile (Benign) vs aprB2 PCR						Percentage positive results	
				Concordant results		Culture +ve PCR -ve	PCR +ve Culture -ve	p value	Kappa	Culture	aprB2 PCR
				+ve	-ve						
12 to 24	+ve	135	Run 1 / 40	7	108	2	18	0.0008	0.348	6.7%	18.5%
12 to 24	+ve	135	Run 1 / 35	7	110	2	16	0.0022	0.378	6.7%	17.0%
12 to 14	-ve	35	Run 1 / 40	0	33	0	2	0.4795	0	0.0%	5.7%
12 to 14	-ve	35	Run 2 / 40	0	34	0	1	1	0	0.0%	2.9%
12 to 14	-ve	35	Run 1 / 35	0	34	0	1	1	0	0.0%	2.9%
12 to 14	-ve	35	Run 2 / 35	0	34	0	1	1	0	0.0%	2.9%
12 to 14	-ve & +ve	72	Run 1 / 40	1	60	1	10	0.0159	0.112	2.8%	15.3%
12 to 14	-ve & +ve	72	Run 2 / 40	1	62	1	8	0.0455	0.143	2.8%	12.5%
12 to 14	-ve & +ve	72	Run 1 / 35	1	63	1	7	0.0771	0.163	2.8%	11.1%
12 to 14	-ve & +ve	72	Run 2 / 35	1	63	1	7	0.0771	0.163	2.8%	11.1%
12 to 24	-ve & +ve	170	Run 1 / 40	7	141	2	20	0.0003	0.336	5.3%	15.9%
12 to 24	-ve & +ve	170	Run 1 / 35	7	141	2	17	0.0013	0.376	5.3%	14.1%

The *p*-value for McNemar's test for independence between culture gelatinase thermostable (benign) and *aprB2* PCR result is shown. Detection rates of both culture and rtPCR are presented as percentage positive results

1 (2.9%) animal. None of the 35 clinically healthy sheep tested positive for the benign strain of *D. nodosus* by the culture/gelatinase test.

Discussion

New molecular techniques capable of rapidly detecting and typing *D. nodosus* are required for improved diagnostics in Australia. A rtPCR method, capable of detecting and discrimination virulent *D. nodosus* strains, has been developed under European conditions. This rtPCR was assessed on Australian samples for detection rates, along with clinically relevant cut-off values. Ct values of 40 and 35 for positive identification cut off were investigated with regards to repeatability between runs, with signals above Ct 35 showing more discrepancies than those below 35. This is common in rtPCR assays, often with results past the Ct of 35 commonly seen as outliers [13]. In addition to higher specificities and more significant results at a Ct of 35 for positive samples (*aprV2*, $p = 0.019$, *aprB2*, $p = 0.002$, McNemar's two-tailed p-value), no animal scored 1 or above returned a positive result above a Ct of 35, supporting the use of a cut-off set at 35 in diagnosing animals with some form of the disease. Interestingly, 6 animals that are clinically negative were *aprV2* positive after lowering the cut-off. This could suggest that these animals be monitored for clinical signs of footrot when favourable environmental conditions occur. They may potentially be asymptomatic carriers depending on circumstance, with the ability to re-infect the flock [14, 15].

The diagnostic power of rtPCR was assessed using the current clinical scoring system to judge if an animal was diseased, or free from disease, as this is the currently accepted method of diagnosis in Victoria. The detection of *D. nodosus* on 3 animals within the healthy population, by both rtPCR and culturing, suggests the presence of infection but not the disease [15]. As the population used for the detection of *D. nodosus* on clinically affected animals was deliberately chosen for clinical virulence, a low detection rate for benign *D. nodosus* was anticipated. From the animals that had the *aprB2* gene detected in the rtPCR, 81% were in association with a co-infection, where *aprV2* was also detected. As the virulent form of footrot is the clinical disease of interest, no additional investigation into a purely benign population was conducted. The reported overall detection rate for *aprV2* at a Ct of 35 is conservative, with an increase being seen when using a Ct of 40 (89.7%), and also when only using animals scored 3–5, which are traditionally considered virulent [9]. However, as the population used had confirmed or suspected of virulent footrot, the full range of scores were used to indicate the presence of disease. As *D. nodosus* is found in the sheep hoof environment and its presence does not always result in disease [16], it was expected that a number of the clinically

negative sheep, throughout the whole data set, would return a positive result due to the nature of the sheep hoof and its environment. Further research monitoring the development of disease in association with environment on these animals may provide insights into the usefulness of *D. nodosus* detection prior to lesion formation, and therefore have practical management applications.

Time taken to receive a result using the rtPCR was typically within 1 day of sample collection or receipt. This is in comparison to the average of 2 to 4 weeks taken for a result when using the culturing method, while the rtPCR also provided better detection of *D. nodosus* from the samples collected. The advantage this provides would allow for a timelier confirmation of the presence of *D. nodosus*, confirming the clinical symptom is in association with the presence of bacteria. The gelatinase test relies on the phenotypic expression of proteases and the associated thermostability of those produced, and the culturing of *D. nodosus* is difficult and requires specialist skills and media. There are also inherent disadvantages to using this method, including the chance of missing strains of *D. nodosus* in the sample, and in the instance of this study, no facility was available in Victoria for the virotyping of isolates, so transport was required. This increases the likelihood of damage to the bacteria in transport and may affect the expression of proteases, the amount made or the ability to reliably have thermostability measured [17]. This may have been the cause of the disagreements from one sample in the clinically healthy data, and two from the clinically affected data. Despite this, there was no significant difference between the two tests when it came to identifying the virulence of the isolates. Instances of results where the rtPCR has detected both virulent and benign protease from a sample, yet culturing missed one or the other strain, are shown in 6 cases from the clinically affected data.

There are many challenges with footrot and the assessment of new diagnostic testing methods due to complex interplay between *D. nodosus*, environment and the host which may result in clinical signs of disease [18]. The method of assessment here reflects the way implementation and sample collection would occur in the field, and so the analysis is appropriate for Victoria's methods of disease investigation. The most prominent difficulty is that the development of lesions is required for a visual diagnosis, to which rtPCR detection rate has been evaluated against. Factors like sheep breed, management, weather, and timing of inspection may all contribute to the lack of lesion, yet *D. nodosus* may still be present and found by the rtPCR. This would contribute to reduced specificity due to being rtPCR positive for *D. nodosus*, yet lesion negative, or lesions that are healing and not indicative of the infecting strain [19, 20].

Conclusions

The improved speed and detection of virulent *D. nodosus* by this rtPCR assay could lead to a change of animal husbandry practices if the focus were to shift from clinical disease to the detection of virulent *D. nodosus*, indicating infection [15]. The ability to pool samples for this type of rtPCR has also been demonstrated [21, 22], an advantage in time and cost over culturing. Testing of this nature is also capable of detecting and quantifying the bacteria associated with the clinical disease, providing the basis to measure the success of various management practices for both treatment and prevention of footrot [23].

Abbreviations
EDTA: Ethylenediaminetetraacetic acid; rtPCR: Competitive real-time PCR

Acknowledgements
We would like to thank the district veterinary officers, animal health officers, and the farmers for their time, effort and sheep. Additional thanks to the DEDJTR molecular and bacteriology diagnostic teams, specimen reception and the WA Animal Health Laboratories. We thank the three anonymous reviewers whose comments/suggestions helped improve and clarify this manuscript.

Funding
The Victorian Sheep and Goat Compensation fund.

Authors' contributions
NB, LZ was responsible for the experimental work, data analysis, and writing of the manuscript. JG contributed to the writing of the manuscript and reviewed the drafts. EK and NB provided knowledge, training and culturing of *D. nodosus*. GR and TB coordinated the project, contributed to the experimental design, and reviewed the drafts. All authors read and approved the final version of the manuscript.

Consent for publication
Not applicable.

Competing interests
The authors declare that they have no competing interests.

Author details
[1]Department of Animal, Plant and Soil Science, Centre for AgriBioscience (AgriBio), La Trobe University, Bundoora, Melbourne, VIC, Australia. [2]Federal University of Minas Gerais, Belo Horizonte, Minas Gerais, Brazil. [3]Department of Economic Development, Jobs, Transport and Resources Centre for AgriBioscience (AgriBio), Victorian Government, Bundoora, Melbourne, VIC, Australia. [4]DAFWA Diagnostics and Laboratory Services, Biosecurity and Regulation, Department of Agriculture and Food, South Perth, Western Australia. [5]Agriculture Services and Biosecurity Operations, Department of Economic Development, Jobs, Transport and Resources, Attwood, Victoria, Australia.

References
1. Egerton JR, Roberts DS, Parsonson IM. The aetiology and pathogenesis of ovine foot-rot. I A histological study of the bacterial invasion. J Comp Pathol. 1969;79(2):207–15.
2. Bennett G, Hickford J, Sedcole R, Zhou H. Dichelobacter nodosus, fusobacterium necrophorum and the epidemiology of footrot. Anaerobe. 2009;15(4):173–6.
3. Egerton JR, Roberts DS. Vaccination against ovine foot-rot. J Comp Pathol. 1971;81(2):179–85.
4. Stewart DJ, Clark BL, Peterson JE, Griffiths DA, Smith EF, O'Donnell IJ. Effect of pilus dose and type of Freund's adjuvant on the antibody and protective responses of vaccinated sheep to Bacteroides nodosus. Res Vet Sci. 1983; 35(2):130–7.
5. Graham NP, Egerton JR. Pathogenesis of ovine foot-rot: the role of some environmental factors. Aust Vet J. 1968;44(5):235–40.
6. Liu D, Roycroft C, Samuel J, Webber J. A retrospective study of clinical and laboratory characteristics of ovine footrot. Vet Microbiol. 1994;42(4):373–81.
7. Liu D, Roycroft C, Samuel J, Webber J. Relationship between clinical manifestations of footrot and specific DNA products of Dichelobacter nodosus amplified through PCR. Res Vet Sci. 1995;59(2):102–5.
8. Liu D, Webber J. A polymerase chain reaction assay for improved determination of virulence of Dichelobacter nodosus, the specific causative pathogen for ovine footrot. Vet Microbiol. 1995;43(2–3):197–207.
9. Buller NB, Eamens G. Ovine Footrot - Australian and New Zealand Standard Diagnositic Procedure. Australian Goverment; 2014. http://www.agriculture. gov.au/SiteCollectionDocuments/animal/ahl/ANZSDP-Ovine-footrot.pdf.
10. Riffkin MC, Wang LF, Kortt AA, Stewart DJ. A single amino-acid change between the antigenically different extracellular serine proteases V2 and B2 from Dichelobacter nodosus. Gene. 1995;167(1–2):279–83.
11. Stauble A, Steiner A, Frey J, Kuhnert P. Simultaneous detection and discrimination of virulent and benign Dichelobacter nodosus in sheep of flocks affected by foot rot and in clinically healthy flocks by competitive real-time PCR. J Clin Microbiol. 2014;52(4):1228–31.
12. Stauble A, Steiner A, Normand L, Kuhnert P, Frey J. Molecular genetic analysis of Dichelobacter nodosus proteases AprV2/B2, AprV5/B5 and BprV/ B in clinical material from European sheep flocks. Vet Microbiol. 2014;168(1): 177–84.
13. Goni R, Garcia P, Foissac S. The qPCR data statistical analysis. Integromics SL, Integromics White Paper; 2009. p. 1–9.
14. Depiazzi LJ, Roberts WD, Hawkins CD, Palmer MA, Pitman DR, McQuade NC, Jelinek PD, Devereaux DJ, Rippon RJ. Severity and persistence of footrot in merino sheep experimentally infected with a protease thermostable strain of Dichelobacter nodosus at five sites. Aust Vet J. 1998;76(1):32–8.
15. Locher I, Greber D, Holdener K, Luchinger R, Haerdi-Landerer C, Schuepbach-Regula G, Frey J, Steiner A. Longitudinal Dichelobacter nodosus status in 9 sheep flocks free from clinical footrot. Small Rumin Res. 2015;132:128–132.
16. Green LE, George TR. Assessment of current knowledge of footrot in sheep with particular reference to Dichelobacter nodosus and implications for elimination or control strategies for sheep in Great Britain. Vet J. 2008;175(2):173–80.
17. Liu D, Yong WK. Improved laboratory diagnosis of ovine footrot: an update. Vet J. 1997;153(1):99–105.
18. McPherson AS, Dhungyel OP, Whittington RJ. Evaluation of genotypic and phenotypic protease virulence tests for Dichelobacter nodosus infection in sheep. J Clin Microbiol. 2017;55(5):1313–26.
19. Witcomb LA, Green LE, Calvo-Bado LA, Russell CL, Smith EM, Grogono-Thomas R, Wellington EM. First study of pathogen load and localisation of ovine footrot using fluorescence in situ hybridisation (FISH). Vet Microbiol. 2015;176(3–4):321–7.
20. Witcomb LA, Green LE, Kaler J, Ul-Hassan A, Calvo-Bado LA, Medley GF, Grogono-Thomas R, Wellington EM. A longitudinal study of the role of Dichelobacter nodosus and fusobacterium necrophorum load in initiation and severity of footrot in sheep. Prev Vet Med. 2014;115(1–2):48–55.
21. Frosth S, Konig U, Nyman AK, Aspan A. Sample pooling for real-time PCR detection and virulence determination of the footrot pathogen Dichelobacter nodosus. Vet Res Commun. 2017;41(3):189–93.
22. Greber D, Locher I, Kuhnert P, Butty MA, Holdener K, Frey J, Schupbach-Regula G, Steiner A. Pooling of interdigital swab samples for PCR detection of virulent Dichelobacter nodosus. J Vet Diagn Investig. 2018;30(2):205–10.
23. Greber D, Bearth G, Luchinger R, Schuepbach-Regula G, Steiner A. Elimination of virulent strains (aprV2) of Dichelobacter nodosus from feet of 28 Swiss sheep flocks: a proof of concept study. Vet J. 2016;216:25–32.

20

Interrelationship between tetracycline resistance determinants, phylogenetic group affiliation and carriage of class 1 integrons in commensal *Escherichia coli* isolates from cattle farms

Kuastros Mekonnen Belaynehe, Seung Won Shin and Han Sang Yoo[*] (iD)

Abstract

Background: Carriage of antibiotic-resistant foodborne pathogens by food production animals is one of many contributors to treatment failure in health care settings, and it necessitates an integrated approach to investigate the carriage of resistant pathogens harboring integrons in food-producing animals.

Methods: *Escherichia coli* isolates with reduced susceptibility to tetracycline antibiotics ($n = 92$) were tested for associations between carriage of class1 integrons, phylogenetic group affiliation and tetracycline resistance determinants using the MIC method, PFGE analysis, PCR and sequencing.

Results: Phylogroups B1 and A were the most common (58.7 and 19.6%, respectively), followed by groups D (20.7%) and B2 (1.1%). All isolates carried at least one of the *tet* genes examined. In addition, 88 (95.7%) of all tetracycline-resistant isolates carried *tet(A)* or *tet(B)*, while 47 (51.1%) and 41 (44.6%) harbored only *tet(A)* or *tet(B)*, respectively. Likewise, isolates harboring these genes had a higher chance ($P < 0.05$) of carrying class 1 integrons. Of the tested isolates, 38 (41.3%) carried the *intI1* gene. Classical integrons with complete genes (*sul1* and *qacEΔ1*) at the 3′-CS were recognized in 27 isolates. PCR screening and subsequent sequencing demonstrated that 84.2% (32/38) of the *intI1*-positive isolates harbored resistance gene cassettes. Overall, seven gene cassettes were identified, either solely or combined with another gene cassette. The most common gene was *aadA1* (10 isolates), followed by a combination of *aadA1-dfrA1* (seven isolates), *aadA1-dfrA12* (six isolates) and *aadA1-aadA2-dfrA12* (three isolates). Genetic typing using PFGE showed minimum clonal relatedness with 28 different clusters and 12–25 discernible DNA fragments.

Conclusions: This study brings new insight into the relationships between the presence of integrons, phylogenetic group association and characteristics of tetracycline antibiotic resistance determinants in commensal *E. coli* strains.

Keywords: *E. coli*, Class 1 integrons, Tetracycline resistance, Phylogenetic group, PFGE

* Correspondence: yoohs@snu.ac.kr
Department of Infectious Diseases, College of Veterinary Medicine, Seoul
National University, Seoul 08826, Republic of Korea

Background

The spread and emergence of resistance to antimicrobial drugs among bacteria has been observed over the past several decades, and this constraint has been a constant impediment to effective infectious disease therapy for as long as antibiotics have been used [1]. In many cases, multidrug resistance was determined to be associated with transmissible plasmids, and the importance of integrons in the acquisition of resistance genes constitute the major vector of multidrug resistance in Gram-negative and, to a lesser extent, in Gram-positive bacteria [2–4].

Over the last few years, rigorous exploration of the diversity of integrons in natural environments has indicated that they are more than just a curious feature of antibiotic-resistant pathogens but that they play a more general and crucial role in the genomic evolution and adaptation of bacteria [5]. To date, five mobile integron classes have been described and characterized based on variations in the *intI* sequences. However, class 1 integrons are ubiquitous and the most frequently encountered among clinical and commensal isolates; therefore, they have been the focus of numerous studies [1, 2, 6].

The basic structure of class 1 integrons includes two conserved segments (CSs) that are usually separated by a variable region that includes mobile cassettes containing antibiotic resistance genes. The 5′-CS carries an integrase class 1 (*intI1*) gene encoding an integrase enzyme and a recombination site (*attI1*), whereas *qacEΔ1* and *sul1*, which confer resistance to quaternary ammonium compounds and sulfonamides, respectively, are localized at the 3′-CS [1, 6–8]. The site-specific recombination system between *attI* and *attC* has enabled a diverse array of resistance determinants to be drawn by individual class 1 integrons [1, 2, 9].

The coding regions of the gene cassettes have no promoters; however, most cassettes encode various antimicrobial resistance genes, with more than 130 distinguishable resistance genes having been found to date [6]. The majority of class 1 integrons harbors an aminoglycoside adenyltransferase gene (*aadA*) and a dihydrofolate reductase gene (*dfr*), which confer resistance to streptomycin and spectinomycin, and trimethoprim, respectively [10, 11].

Tetracycline has been used in human and veterinary medicine and as a growth promoter in animal husbandry. The major mechanisms of tetracycline resistance in *Enterobacteriaceae* are mediated through one of several mechanisms; namely, efflux pump activity, ribosomal protection, and enzymatic inactivation. The predominant genes that confer tetracycline resistance via efflux pump activity are *tet(A)*, *tet(B)*, *tet(C)*, *tet(D)*, *tet(E)*, and *tet(G)*. Indiscriminate application of tetracyclines in food-producing animals enhances multidrug resistance due to antibiotic selective pressure induced by the presence of high environmental concentrations of the antibiotics. This selective pressure ultimately leads to increased prevalence of tetracycline resistance via the *tet* genes and promotes the dissemination of mobile genetic elements in bacteria [12, 13].

E. coli isolates belong to four major phylogenetic groups, A, B1, B2, and D, and strains from the B2 phylogenetic group happen to be the least resistant to antimicrobials. Moreover, there is a tendency towards lower integron carriage among phylogroup B2 [14]. Nevertheless, due to certain factors, such as the level of resistance to antimicrobials, the site of the infection and geographical location, there is variation in the prevalence of different phylogenetic groups [15]. Isolates in the phylogenetic groups B1 and D tend to harbor class 1 integrons, and a previous report also showed that *intI*-positive B2 strains were the least prevalent [16]. There are various observations on the interrelationship of different phylogroups and integron carriage for environmental, human and animal isolates raising the hypothesis that the two phenomena are connected and indicating that various genetic elements are involved in strains with different phenotypes. Characterization of this association will help to better understand the infection process and will reflect the possible different survival strategies of *E. coli* phylogroups under different circumstances.

Antimicrobial resistant bacteria derived from animals seriously compromises public health by causing food-borne infections and raises a food safety issue globally. The effects of such bacteria are not only limited to food safety but also pose occupational hazards for animal handlers, meat inspectors and veterinarians. In particular, carriage of antibiotic-resistant foodborne pathogens by food-production animals is one of many contributors to treatment failure in health care settings, and it establishes the need for a detailed and thorough investigation of the carriage of such antibiotic-resistant pathogens harboring integrons in food-producing animals [12]. Furthermore, integrons are not only limited to pathogenic organisms but have also been isolated from bacteria recovered from environmental samples and healthy animals [17]. Similarly, the lack of sufficient and current information describing the association between antibiotic resistance and phylogenic groups with respect to integron carriage in commensal *E. coli* isolates of cattle from Korea necessitates further research. Therefore, in the present study, we investigated the role of integrons and their associated diverse gene cassettes in mediating antimicrobial resistance in commensal *E. coli* isolates recovered from cattle. Moreover, we examined the relationship between class 1 integron carriage with respect to phylogroups and patterns of tetracycline resistance.

Methods

Selection of bacterial strains for the study

In total, 247 commensal *E. coli* isolates obtained by our research group between 2014 and 2015 from fecal samples from 405 tested animals at four healthy beef cattle farms located in four different cities in South Korea (Pyeongchang, Anyang, Yangpyeong, and Cheonan) were used in the present study [18]. The beef farms consisted of different age groups of cattle, such as weaned calves, bulls and steers. Since the farms are intensive, cattle were kept in confinement in a conventional housing system. Fecal samples were freshly collected from the rectum of each cattle and a single bacterial isolate was recovered per animal. All bacterial strains were routinely cultured in tryptic soy broth (TSB) (Oxoid, Basingstoke, UK) for 18 h at 37 °C. Among the 247 isolates, 92 *E. coli* isolates demonstrating resistance or decreased susceptibility by microbroth dilution assays to any of the tetracycline antibiotics referred to below were selected for further investigation.

Susceptibility testing

Phenotypic characterization for all isolates was performed using the disc diffusion method, and the following antibiotic discs were analyzed in this study: tetracycline (TE, 30 μg), streptomycin (S, 10 μg), chloramphenicol (C, 30 μg), ampicillin (AMP, 10 μg), amoxicillin-clavulanic acid (AMC, 30 μg), ciprofloxacin (CIP, 5 μg), nalidixic acid (NA, 30 μg) and trimethoprim-sulfamethoxazole (SXT, 25 μg) (Sigma-Aldrich, St. Louis, MO, USA). The MICs for oxytetracycline, doxycycline, tetracycline, minocycline and tigecycline were determined using cation-adjusted Mueller-Hinton broth (Oxoid, Basingstoke, UK). All susceptibility testing was performed according to the procedures and interpretive criteria specified by the Clinical Laboratory Standards Institute (CLSI), and *E. coli* ATCC 25922 was used as a quality control strain [19].

Phylogenetic group determination

E. coli phylogenetic groups (A, B1, B2 and D) were investigated by amplifying two genes and a DNA fragment using multiplex PCR as previously described [20].

Analysis of antimicrobial resistance genes

PCR amplification to investigate the tetracycline resistance-encoding genes was conducted for all isolates. The following genes, encoding the tetracycline efflux mechanism, were investigated as previously described: *tet(A)*, *tet(B)*, *tet(C)*, *tet(D)*, *tet(E)* and *tet(G)* [12, 21–24]. Furthermore, genes conferring resistance to sulfonamide antibiotics (*sul1*, *sul2* and *sul3*) and genes conferring chloramphenicol/florfenicol resistance (*cat1*, *cmlA* and *floR*) were also analyzed. PCR amplification of the resistance genes was conducted using the primers presented in Table 1 [12, 21–30].

Detection and characterization of class 1 integrons and their gene cassettes

Total DNA was extracted by boiling a suspension of overnight-cultured bacterial cells [grown on tryptic soy agar plates (TSA) at 37 °C for 10 min] in 200 μl of sterile RNase/DNase-free distilled water. All *E. coli* isolates were PCR screened for the presence of *IntI1* gene-encoding class 1 integrons. Further testing was performed on the integron positive isolates for the presence of gene cassettes in the variable region and the *sul1* and *qacEΔ1* genes in the 3′-CS. All primers and PCR conditions are presented in Table 1. Gel purification of all PCR products was conducted using PCR quick-spin PCR product purification kits (iNtRON Biotechnology, USA), after which the samples were sequenced (Macrogen Co., Seoul, Korea). Following sequencing, the gene cassettes within the variable regions of the class 1 integrons were determined by using BLAST (Basic Local Alignment Search Tool) searches of the NCBI database (National Center for Biotechnology Information).

Clonal relationships among integron positive strains

Determination of the genetic relationship between the integron positive isolates was accomplished by pulsed-field gel electrophoresis (PFGE) analysis according to the protocols and criteria previously established by the Centers for Disease Control and Prevention (CDC) using *XbaI* as the restriction enzyme. Briefly, following 18 to 20 h growth on TSA at 37 °C, genomic DNA was digested with 50 U *XbaI* (TaKaRa, Japan) for 2 h at 37 °C, then the DNA fragments were subsequently separated on a 1.0% SeaKem Gold agarose gel (Lonza, USA) in 0.5× Tris-borate-EDTA (TBE) buffer using a CHEFMapper gel apparatus (Bio-Rad Laboratories, California, USA). The conditions for electrophoresis were as follows: pulse time, 2–30s at 14 °C; run time, 18 h; voltage, 6 V/cm. Analysis of the image was performed by using the Bionumerics software (Applied Maths, Belgium).

Statistical analysis

All experiment data are stored in Excel 2010, and the susceptibility testing was analyzed using IBM SPSS/Statistics, version 24. The association between the *tet* genes and the presence of class 1 integron gene was analyzed by Fisher's exact test or Pearson's χ^2 test, contingent on cell frequencies. The median MICs for the respective

Table 1 Primers used for the PCR detection of resistance genes

Primer name	Target gene	Nucleotide sequence	Annealing temperature (°C)	Amplicon size (bp)	Reference
TetA-F	*tet(A)*	GGCGGTCTTCTTCATCATGC	55	502	[21]
TetA-R		CGGCAGGCAGAGCAAGTAGA			
TetB-F	*tet(B)*	CATTAATAGGCGCATCGCTG	55	930	[21]
TetB-R		TGAAGGTCATCGATAGCAGG			
TetC-F	*tet(C)*	GCTGTAGGCATAGGCTTGGT	55	888	[21]
TetC-R		GCCGGAAGCGAGAAGAATCA			
TetD-F	*tet(D)*	GAGCGTACCGCCTGGTTC	55	780	[12]
TetD-R		TCTGATCAGCAGACAGATTGC			
TetE-F	*tet(E)*	AAACCACATCCTCCATACGC	55	278	[22]
TetE-R		AAATAGGCCACAACCGTCAG			
TetG-F	*tet(G)*	GCTCGGTGGTATCTCTGCTC	55	468	[23]
TetG-R		AGCAACAGAATCGGGAACAC			
Sul1-F	*sul1*	CGGCGTGGGCTACCTGAACG	57	433	[24]
Sul1-R		GCCGATCGCGTGAAGTTCCG			
Sul2-F	*sul2*	CGGCATCGTCAACATAACCT	57	721	[21]
Sul2-R		TGTGCGGATGAAGTCAGCTC			
Sul3-F	*sul3*	CAACGGAAGTGGGCGTTGTGGA	57	244	[25]
Sul3-R		GCTGCACCAATTCGCTGAACG			
Cat-F	*Cat*	GGT GAG CTG GTG ATA TGG	55	209	[26]
Cat-R		GGG ATT GGC TGA GAC GA			
Flor -F	*flor*	CAC GTT GAG CCT CTA TAT	55	868	[27]
Flor -R		ATG CAG AAG TAG AAC GCG			
CmlA -F	*cmlA*	TGT CAT TTA CGG CAT ACT CG	55	455	[27]
CmlA -F		ATC AGG CAT CCC ATT CCC AT			
Var1-F	*var1*	GGCATCCAAGCAGCAAG	55	Variable	[28]
Var1-R		AAGCAGACTTGACCTGA			
qacEΔ1 F	*qacEΔ1*	ATCGCAATAGTTGGCGAAGT	60	225	[29]
qacEΔ1 R		CAAGCTTTTGCCCATGAAGC			
Intl1-F	*intl1*	GGGTCAAGGATCTGGATTTCG	60	483	[30]
Intl1-R		ACATGCGTGTAAATCATCGTCG			

tetracycline antibiotics between the isolates with and without *Intl1* was analyzed using the Mann-Whitney test. A $P < 0.05$ was considered to indicate statistical significance.

Results

Antimicrobial resistance phenotypes

The resistance percentages to the tested antibiotics were as follows: streptomycin, 84 (91.3%); nalidixic acid, 36 (39.1%); ampicillin, 35 (38%); chloramphenicol, 28 (30.4%); trimethoprim-sulfamethoxazole, 24 (26.1%); ciprofloxacin, 11 (12%) and amoxicillin-clavulanic, 2 (2.2%) (Fig. 1). The MIC range for the 92 tetracycline resistant isolates was > 256 µg/ml to 16 µg/ml, and their MIC_{50} and MIC_{90} values were 128 and 256 µg/ml, respectively. Oxytetracycline resistance was identified in all isolates

(MIC range > 256 µg/ml–32 µg/ml), of which 49 isolates were highly resistant (MIC ≥256 µg/ml). Moreover, 80 strains (87%) were resistant to doxycycline, 41 (44.6%) to minocycline and none to tigecycline. Significantly higher median oxytetracycline MICs were observed for isolates with class 1 integrons than for isolates without class 1 integrons ($P < 0.006$; Table 2); however, there were no significant statistical differences in the median MIC values for tetracycline, doxycycline, minocycline and tigecycline between class 1 integron-positive and integron-negative strains.

E. coli phylogenetic groups

Of the 92 isolates, phylogenetic groups B1 and D were the most common (54 isolates; 58.7% and 19 isolates; 20.7%, respectively), followed by group A, which was

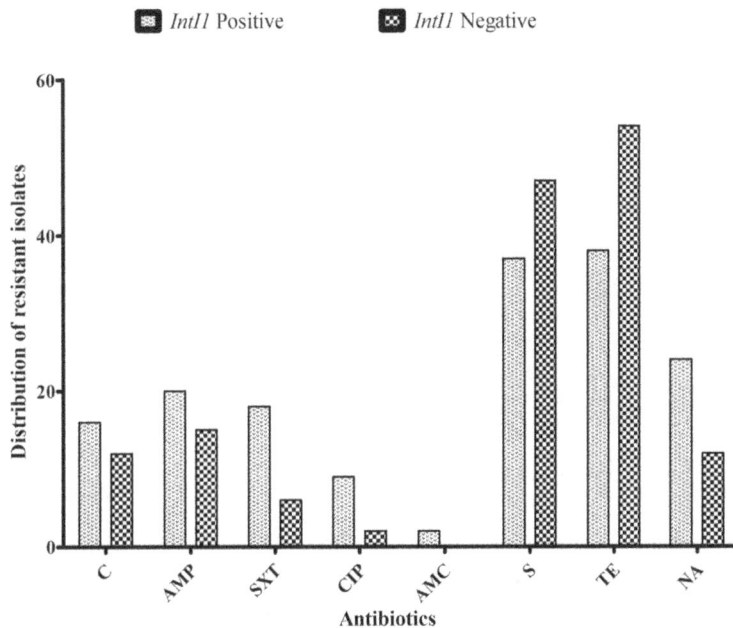

Fig. 1 Different classes of antibiotics with respect to their presence or absence in integrons

assigned to 18 isolates (19.6%). Group B2 was rare, occurring in only 1 isolate. We also compared integron-positive and integron-negative isolates across the phylogenetic groups, and phylogenetic group D (16 isolates) was the most prevalent among the *intI1*-positive isolates, whereas B1 (40 isolates) was most prevalent among the *intI1*-negative isolates. Our results showed an association between the presences of class 1 integrons and affiliation with phylogenetic groups D and B1 (*P* < 0.01). The frequencies of integron-negative and integron-positive strains for in A and B2 phylogenetic groups were similar, with no statistically significant differences (Fig. 2).

Characterization of antimicrobial resistance genes

All isolates carried at least one of the *tet* genes examined. Overall, 88 (95.7%) of the tetracycline-resistant isolates carried *tet(A)* or *tet(B)*, with 47 (51.1%) and 41

(44.6%) isolates harboring only *tet(A)* or *tet(B)*, respectively. The *tet(C)* and *tet(G)* genes were found in only five (5.4%) and six (6.5%) isolates, respectively, and the *tet(C)* gene was found in isolates that were not harboring integrons. Moreover, eight isolates harbored two *tet* genes, while none of the isolates carried the *tet(D)* or *tet(E)* genes. The distribution of tetracycline resistance genes among the integron-positive and -negative isolates is shown in Table 3. *E. coli* isolates carrying class 1 integrons were more likely to harbor the *tet(A)* gene (*P* < 0.01). In addition, the following determinants for chloramphenicol/florfenicol resistance were identified: *cat1* (47.4%), *floR* (50%), and *cmlA* (18.4%). Among the 92 *E. coli* isolates investigated, 28 isolates harboring integrons and eight isolates without integrons (*n* = 36; 39.1%) had the *sul1* gene. Moreover, the *sul2* and *sul3* genes were identified in 16 (17.4%) and seven (7.6%) isolates, respectively.

Table 2 Susceptibility to various tetracycline antibiotics stratified by the presence or absence of class 1 integrons

Antimicrobial agents	IntI1 present					IntI1 Absent					P value
	MIC (μg/ml)				Number (%) of resistant isolates	MIC (μg/ml)				Number (%) of resistant isolates	
	Range		MIC$_{50}$	MIC$_{90}$		Range		MIC$_{50}$	MIC$_{90}$		
Tetracycline	64	> 256	128	256	38 (41.3)	16	> 256	128	256	54 (58.7)	0.244
Doxycycline	4	128	16	64	36 (39.1)	4	64	32	64	44 (47.8)	0.975
Oxytetracycline	64	> 256	> 256	> 256	38 (41.3)	32	> 256	256	> 256	54 (58.7)	*P<0.006*
Minocycline	4	64	8	32	14 (15.2)	2	64	16	32	28 (30.4)	0.267
Tigecycline	0.25	2	1	1	–	0.25	8	0.5	1	1 (1.1)	0.054

Numbers indicated in italics indicate significance difference (*P*< 0.05)

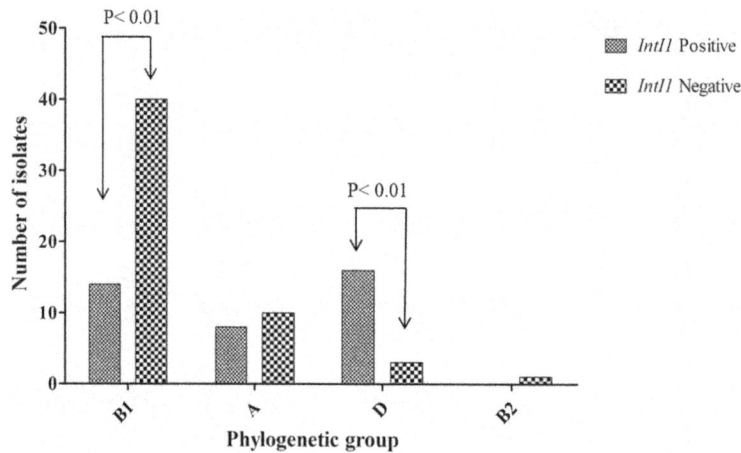

Fig. 2 Association between phylogenetic group and isolates carrying integrons

Detection of the *intI1* gene and characterization of gene cassettes in the *E. coli* strains

Integrase gene-encoding class 1 integrons were detected by PCR in 38 (41.3%) isolates. Resistance to quaternary ammonium compounds and sulfonamides conferred by the *qacEΔ1* and *sul1* genes in the 3′-CS, respectively, was identified in 36 integron-positive isolates. Among these, 27 contained the entire 3′-CS (*qacEΔ1-sul1*) structure, whereas nonclassical integrons lacking the 3′-CS were found in only two of the 38 *intI1*-positive *E. coli* isolates. Of the 38 isolates, one had only *sul1* in the 3′-CS and eight possessed only *qacEΔ1* in the 3′-CS. The class 1 integron variable regions were amplified in 32 (84.2%) of the 38 *intI1*-positive isolates, and their genetic contents were ascertained via PCR amplification of the integron variable regions and subsequent full sequence analysis. Different lengths of PCR products ranging from ∼ 1–2.5 kb were observed for strains having variable regions. Of these, the predominant cassette amplicons carried by the isolates were 1 kb in 18 strains, 1.5 kb in 10 strains and 2.5 kb in four strains (Table 4).

Overall, seven gene cassettes and eight distinct profiles of gene cassette arrays, namely, *aadA1* (10 isolates),

aadA2 (two isolates), *dfrA12* (two isolates), *aadA1-dfrA1* (seven isolates), *aadA1-dfrA12* (six isolates), *aadA5-dfrA17* (one isolate), *aadA1-aadA2-dfrA12* (three isolates) and *aadA1-aadA5-dfrA5* (one isolate), were described. The 2.5 kb amplicon consists of *aadA1-aadA2-dfrA12* and *aadA1-aadA5-dfrA5* (Table 4).

PFGE analysis of isolates containing class 1 integrons

The genetic relatedness among the multidrug resistant *E. coli* isolates carrying integrons was established based on their *Xba*I-digested chromosomal DNA fragments, and the most commonly detected genotypes are depicted in Table 4 and Fig. 3. Several profiles were observed, with 12–25 discernible DNA fragments from 38 isolates when analyzed by the Dice coefficient method. When an 80% cut-off band pattern similarity was used, 28 different PFGE clusters were observed, whereas 26 clusters were detected when a 70% cut-off band pattern similarity was applied. Strong relationships (> 90% similarity) were encountered in six clusters constituting 12 isolates sharing the same antibiotics resistance spectrum and resistance gene pattern. For instance, isolates EC174 and EC175 had more than 97% band pattern similarity, as well as the same antibiotic resistance pattern (TE-S-NA), integron gene cassette arrays (*aadA1-dfrA12*), and resistance genes [*tet(A), sul1* and *floR*].

Discussion

Our study investigated the association of class 1 integron carriage, phylogenetic group affiliation and different tetracycline antibiotics resistance patterns in commensal *E. coli* strains isolated from cattle farms in Korea. All 92 *E. coli* isolates were significantly resistant to tetracycline and oxytetracycline. These findings indicate a widespread application of these antibiotics either for therapeutic purposes or as a supplement for promoting growth, and this continuous exposure to tetracyclines

Table 3 Association between integron-positive and integron-negative *E. coli* isolates and the frequencies of *tet* genes

tet genes	Class 1 integron presence		*P* value
	intI1 positive	*intI1* negative	
tet(A)	26	21	*P < 0.01*
tet(B)	12	30	*0.023*
tet(C)	–	5	0.054
tet(G)	1	5	0.205
tet(A) + tet (B)	–	1	0.399
tet(A) + tet (C)	–	1	0.399
tet(B) + tet (G)	1	5	0.205

Numbers indicated in italics indicate significance difference (*P*< 0.05)

Table 4 Characterization of *E. coli* isolates harboring class 1 integrons and description of their associated gene cassettes

Isolates No.	3′CS	Cassette amplicons (bp)	Other resistance gene pattern	Integron gene cassette arrays	PFGE pattern	Resistance pattern
EC151	*qacEΔ1-sul1*	1500	*tetA, sul1, sul2, cat1*	*aadA1-dfrA1*	B	TE-S-C-SXT
EC139	*qacEΔ1-sul1*	1500	*tetB, tetG, sul1, sul2, cat1, floR*	*aadA1-dfrA1*	A	TE-S-C-AMP-SXT-CIP-NA
EC143	–	1500	*tetA, sul3, cmlA, cat1*	*aadA1-dfrA12*	U	TE-S-AMP-SXT-CIP-NA
EC147	*qacEΔ1-sul1*	1000	*tetA, sul1*	*aadA2*	H	TE-S-NA
EC152	*qacEΔ1-sul1*	1000	*tetA, sul1*	*aadA1*	H	TE-S-AMP-SXT-NA
EC153	*qacEΔ1-sul1*	2500	*tetA, sul1*	*aadA1-aadA2-dfrA12*	H	TE-S-NA
EC155	qacEΔ1-sul1	–	*tetB, sul1, sul2, cat1*	–	S	TE-S-C-SXT-CIP-NA
EC156	*qacEΔ1-sul1*	1000	*tetA, sul1, floR*	*aadA1*	H	TE-S-AMP-NA
EC157	*qacEΔ1-sul1*	1500	*tetB, sul1, sul2, cat1*	*aadA5-dfrA17*	G	TE-S-AMP-SXT-CIP-NA
EC159	*qacEΔ1-sul1*	1500	*tetB, sul1, sul2, cat1*	*dfrA12*	R	TE-S-C-AMP-SXT-CIP-NA
EC160	*qacEΔ1-sul1*	1500	*tetB, sul1, sul2, cat1*	*dfrA12*	R	TE-S-C-AMP-SXT-CIP-NA
EC161	–	–	*tetB, sul3, cmlA*	–	Q	TE-S-C
EC162	*qacEΔ1-sul1*	1000	*tetA, sul1, floR*	*aadA1*	N	TE-S-NA
EC163	*qacEΔ1-sul1*	1000	*tetA, sul1, floR*	*aadA1-dfrA1*	L	TE-S-NA
EC164	*qacEΔ1-sul1*	1000	*tetA, sul1, floR*	*aadA1*	M	TE-S-SXT-NA
EC165	*qacEΔ1-sul1*	1000	*tetB, sul1, sul2, cat1, floR*	*aadA1*	T	TE-S-C-AMP-SXT-CIP-NA
EC166	*qacEΔ1-sul1*	–	*tetB, sul1, sul2, cat1*	–	G	TE-S-C-AMPCIP-NA
EC167	*qacEΔ1-sul1*	2500	*tetA, sul1, floR*	*aadA1-aadA2-dfrA12*	K	TE-S-NA
EC172	*qacEΔ1-sul1*	1000	*tetA, sul1, floR*	*aadA1*	J	TE-S-NA
EC173	*qacEΔ1-sul1*	1000	*tetA, sul1, floR*	*aadA1*	J	TE-S-NA
EC174	*qacEΔ1-sul1*	1000	*tetA, sul1, floR*	*aadA1-dfrA12*	I	TE-S-NA
EC175	*qacEΔ1-sul1*	1000	*tetA, sul1, floR*	*aadA1-dfrA12*	I	TE-S-NA
EC176	*qacEΔ1*	1000	*tetB, sul2, sul3, cmlA, floR*	*aadA1*	W	TE-S-C
EC177	*qacEΔ1-sul1*	1000	*tetA, sul1, floR*	*aadA1*	Y	TE-S-AMP
EC178	*qacEΔ1-sul1*	2500	*tetA, sul1, floR*	*aadA1-aadA5-dfrA5*	Y	TE-S-AMP-NA
EC179	*qacEΔ1*	2500	*tetB, sul2, sul3, cml1, floR*	*aad1-aadA2-dfrA12*	O	TE-S-C
EC180	*qacEΔ1*	1000	*tetB, sul2, sul3, cml1, floR*	*aadA2*	P	TE-S-C
EC181	*qacEΔ1-sul1*	1500	*tetB, sul1, sul2, cat1*	*aadA1-dfrA1*	D	TE-S-C-AMP-SXT-CIP-AMC-NA
EC185	*qacEΔ1-sul1*	1500	*tetA, sul1, sul2*	*aadA1-dfrA1*	Z	TE-S-AMP-SXT-AMC-NA

Table 4 Characterization of *E. coli* isolates harboring class 1 integrons and description of their associated gene cassettes *(Continued)*

Isolates No.	3'CS	Cassette amplicons (bp)	Other resistance gene pattern	Integron gene cassette arrays	PFGE pattern	Resistance pattern
EC191	qacEΔ1	1000	tetA, sul3, cml1, cat1, floR	aadA1-dfrA12	A2	TE-S-C-SXT
EC194	qacEΔ1-sul1	1500	tetA, sul1, cat1	aadA1-dfrA1	F	TE-S-AMP-SXT
EC198	qacEΔ1-sul1	1500	tetA, sul1, cat1	aadA1-dfrA1	F	TE-S-AMP-SXT
EC209	qacEΔ1	1000	tetA, sul3, cml1, cat1, floR	aadA1-dfrA12	X	TE-S-C-SXT
EC230	qacEΔ1	–	tetA, sul2, cat1	–	E	TE-S-C-AMP-SXT-NA
EC231	qacEΔ1	–	tetA, sul2, cat1	–	E	TE-S-AMP
EC254	qacEΔ1-sul1	1000	tetA, sul1, cat1	aadA1-dfrA12	V	TE-S-*AMP*
EC258	sul1	–	tetA, sul1	–	A1	TE-AMP-SXT
EC262	qacEΔ1	1000	tetA, sul2, cat1, floR	aadA1	C	TE-S-C-AMP

has led to a higher percentage of tetracycline-resistant *E. coli* isolates. The findings of our work were comparable with other observations where a high prevalence of resistance to antimicrobials commonly used with livestock, such as tetracycline and streptomycin were observed in commensal *E. coli* isolated from food-producing animals in South Korea. For instance, Lim et al. [31] observed tetracycline (30.5%) as the most frequently observed resistance in *E. coli* isolates of cattle origin, and Kang et al. [32] showed that *E. coli* isolates recovered from swine with diarrhea were highly resistant to streptomycin (99.0%) and tetracycline (97.1%); furthermore, the work of Shin et al. [33] also demonstrated that the most prevalent resistance phenotype observed was streptomycin (63.1%), followed by tetracycline (54.5%). Tetracycline antibiotics have long been the

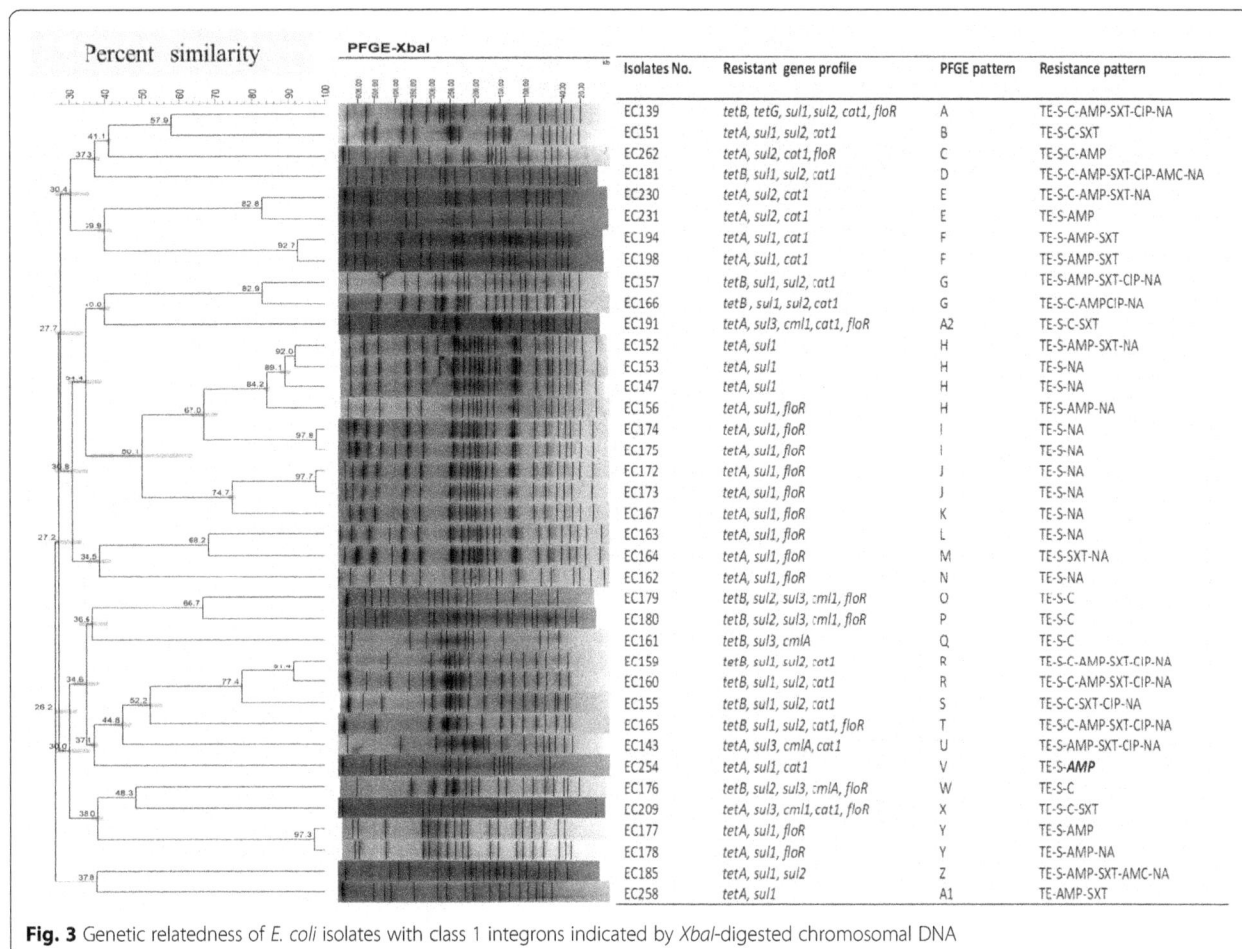

Fig. 3 Genetic relatedness of *E. coli* isolates with class 1 integrons indicated by *Xbal*-digested chromosomal DNA

single most commonly used class of antimicrobial in livestock, accounting for around 50% of the total amount of antimicrobial consumption both in the USA [34] and Korea [35], and it is, therefore, not surprising to observe tetracycline resistance as the most frequently antimicrobial resistance class in *E. coli* isolates. As indicated by the animal and plant quarantine agency of Korea (APQA) [36], although gradually decreasing since 2003, tetracyclines still comprise the predominant antibiotics sold for veterinary use.

The investigated isolates were recovered from clinically healthy animals; accordingly, B1 (58.7%), which is commonly associated with nonpathogenic commensal strains, was the common phylogroup classified. Accordingly, only a single isolate was classified into phylogenetic group B2, which is normally linked with pathogenicity [37]. Moreover, no isolates categorized as B2 carried the *intI1* gene, which is similar to the results of a previous study that demonstrated that the B2 phylogroup has a lower tendency to harbor integrons than other phylogroups [16]. In the present study, significant differences in the numbers of isolates with and without integrons were observed (*P* < 0.01) between phylogroups D and B1. These agree with those of a previous study that demonstrated that strains associated with phylogroups A and B1 tend to carry integrons more often than those associated with B2 and D [10, 38]. In contrast, affiliation with a specific phylogenetic group was not linked to the presence of integrons in *E. coli* strains recovered from river water [39]. This variability is likely because of ecological differences among the sites from which the *E. coli* isolates were recovered that may influence their ability to harbor integron genes.

In the present study, the *tet(A)* gene was the predominant resistance determinant, followed by the *tet(B)* gene. There is general agreement regarding the widespread importance of the link between the *tet(A)* and *tet(B)* genes and resistance to tetracycline antibiotics in *Enterobacteriaceae* as reported by multiple investigators [12, 40, 41]. In this study, isolates having more than one *tet* gene were also observed in 8.7% of the strains, which is a common phenomenon in *E. coli* isolates of cattle origin. Previous studies have shown similar results, in which 3.5% [42], 5.4% [43] and 22.2% [44] of isolates had two *tet* genes, with only a slight difference in the total number of isolates used between the studies. The acquisition of more than one *tet* gene by a given strain is attributed to powerful selection pressures due to the high level of tetracycline in the environment rather than to a special selective advantage conferred by the *tet* genes [12].

There are varying accounts of which *tet* gene is most frequently reported in different countries. For example, Karami et al. [13] reported that *tet(B)* was the most frequently observed (51%) among commensal *E. coli* strains

from Sweden, while Shin et al. [41] and Dessie et al. [45] reported that *tet(A)* accounted for 46.5% and 63.2% of all *tet* genes detected in Korea, respectively. A significantly higher frequency of the *tet(A)* gene (*P* < 0.01) was also observed in isolates with integrons, demonstrating an association between *tet(A)* carriage and presence of class 1 integrons. This observation has previously been reported by others, who found that *intI1* and *tet(A)* coexisted on the same large transferable plasmid or other genetic elements in *E. coli*, validating an established association between tetracycline resistance genes and class 1 integrons [46, 47]. *sul1* was identified in 39.1% of isolates; since it is commonly linked to integrons and transposons as a component of the 3′-CS, previous studies have similarly reported it among bacteria of the family *Enterobacteriaceae* [48].

In the present study, 41.3% of *E. coli* isolates harbored *intI1* gene-encoding class 1 integrons. A comparable result regarding the prevalence of class 1 integrons was previously reported in Korea and other countries; for instance, 40% of the *E. coli* isolates carried class 1 integrons in Lithuania [49], 49.8% in Italy [50], and 27% in the United States [51], as well as 44% of the commensal *E. coli* isolates from poultry in Korea [52]. Non-classical integrons lacking the normal 3′-CS were detected in only two class 1 integron-positive isolates. Similar observations were made for *intI1*-positive *E. coli* isolates that originated from food, animals, and healthy humans [53]. Moreover, 32 (84.2%) of the 38 *intI1*-positive isolates had variable regions containing gene cassettes. Overall, our analysis showed that the *aad* and *dfr* families comprise the majority of class 1 integron gene cassettes, similar to the results reported for *E. coli* originating from beef cattle [38]. In the present study, *aadA1-dfrA1* was the most commonly detected combination, which is in agreement with previous reports on isolates recovered from clinical and healthy animals, humans and food samples [27, 49, 52, 54]. Furthermore, 27 (71.1%) of the cassette arrays contained the *aadA1* gene, either alone or in combination with other gene cassette arrays that encode aminoglycoside adenyltransferases, which confer resistance to streptomycin/spectinomycin [39]. When we made a comparison between these environmental isolates and clinical isolates from the same region, commensal *E. coli* isolates from animals mostly carried a single gene cassette, whereas clinical *E. coli* isolates from humans had multigene cassettes [52]. In addition, we found between one and three gene cassettes in a single isolate, which is a distinguishing feature of class 1 integrons in which no more than 6 gene cassettes are carried in the variable region [4].

Based on the results of the *Xba*I-PFGE, the *E. coli* isolates carrying class 1 integrons could be categorized into 28 and 26 different PFGE cluster groups when 80% and

70% cut-off band pattern similarities were applied, respectively. In this study, *E. coli* isolates carrying integrons showed a high degree of polymorphism. This diverse clonal relationship resulted from the horizontal transfer of resistance genes between different strains, rather than a dissemination of a single clonal strain, as previously described [55].

Conclusions

Due to their carriage of resistant genes and class 1 integrons, commensal *E. coli* isolates have a significant implication in public health through their ability to disseminate antibiotic resistant genes via contamination of the food chain. A positive association was observed between isolates harboring the *intI1* and *tet(A)* genes, confirming that isolates containing the *tet(A)* gene are more likely to carry class 1 integrons. Likewise, affiliation with phylogroup D was positively associated with the presence of class 1 integrons. Further detailed investigation of the class 1 integron genetic content should be conducted to provide a more complete understanding of the molecular mechanisms responsible for multidrug resistance in *E. coli* strains. Moreover, the interrelation of integron related resistance genes with other factors should be studied by integrating environmental and veterinary factors and factors associated with the food chain. Accordingly, the resulting advances could have a profound effect on clinical practice, infection control measures and treatment options, both in veterinary and human medicine.

Abbreviations

APQA: Animal and plant quarantine agency of Korea; CS: Conserved segment; KFDA: Korean Food and Drug Administration; MIC: Minimum inhibitory concentration; PCR: Polymerase chain reaction; PFGE: Pulsed-field gel electrophoresis; USFDA: United States Food and Drug Administration

Acknowledgments

We are grateful to Marie Salomé GABRIAC from IUT Nancy Brabois-Université de Lorraine for her technical support and assistance during laboratory work.

Consent to participate

Not applicable

Funding

This study was supported by the "Cooperative research program for agriculture science and technology development (project No. PJ00897001)", Rural Development Administration and Research Institute of Veterinary Science and BK21 PLUS program, Seoul National University, Republic of Korea.

Authors' contributions

Conceived the study and participated in its design: KMB SWS HSY. Performed isolation and identification of isolates: KMB. Performed the experiments: KMB. Analyzed the data: KMB HSY. Wrote the paper: KMB. All authors critically reviewed the draft and approved the final manuscript.

Authors' information

Kuastros Mekonnen Belaynehe: DVM, MSC, PhD research student.
Seung Won Shin: DVM, PhD.
Han Sang Yoo: DVM, PhD (Professor of Veterinary Microbiology and director of the Institute of Green Bio Science and Technology, Seoul National University).

Consent for publication

Not applicable.

Competing interests

The authors declare that they have no competing interests.

References

1. Mazel D. Integrons: agents of bacterial evolution. Nat Rev Microbiol. 2006; 4(8):608–20.
2. Cambray G, Guerout AM, Mazel D. Integrons. Annu Rev Genet. 2010;44:141–66.
3. Ponce-Rivas E, Muñoz-Márquez ME, Khan AA. Identification and molecular characterization of class 1 integrons in multiresistant *Escherichia coli* isolates from poultry litter. Appl Environ Microbiol. 2012;78(15):5444–7.
4. Escudero JA, Loot C, Nivina A, Mazel D. The integron: adaptation on demand. In: Mobile III DNA, editor. MicrobiolSpectrum. USA: American Society of Microbiology; 2015. p. 139–61.
5. Gillings MR. Integrons: past, present, and future. Microbiol Mol Biol Rev. 2014;78(2):257–77.
6. Deng Y, Bao X, Ji L, Chen L, Liu J, Miao J, Chen D, Bian H, Li Y, Yu G. Resistance integrons: class 1, 2 and 3 integrons. Ann Clin Microbiol Antimicrob. 2015;14(1):45.
7. Recchia GD, Hall RM. Gene cassettes: a new class of mobile element. Microbiology. 1995;141(12):3015–27.
8. Carattoli A. Importance of integrons in the diffusion of resistance. Vet Res. 2001;32(3–4):243–59.
9. Stokes HW, Nesbø CL, Holley M, Bahl MI, Gillings MR, Boucher Y. Class 1 integrons potentially predating the association with Tn402-like transposition genes are present in a sediment microbial community. J Bacteriol. 2006; 188(16):5722–30.
10. Singh T, Das S, Ramachandran V, Wani S, Shah D, Maroof KA, Sharma A. Distribution of Integrons and phylogenetic groups among Enteropathogenic *Escherichia coli* isolates from children <5 years of age in Delhi, India. Front Microbiol. 2017;8:561.
11. Mazel D, Dychinco B, Webb VA, Davies J. Antibiotic resistance in the ECOR collection: integrons and identification of a novel *aad* gene. Antimicrob Agents Chemother. 2000;44(6):1568–74.
12. Koo HJ, Woo GJ. Distribution and transferability of tetracycline resistance determinants in *Escherichia coli* isolated from meat and meat products. Int J Food Microbiol. 2011;145(2):407–13.
13. Karami N, Nowrouzian F, Adlerberth I, Wold AE. Tetracycline resistance in *Escherichia coli* and persistence in the infantile colonic microbiota. Antimicrob Agents Chemother. 2006;50(1):156–61.
14. Johnson JR, Kuskowski MA, Owens K, Gajewski A, Winokur PL. Phylogenetic origin and virulence genotype in relation to resistance to fluoroquinolones and/or extended-spectrum cephalosporins and cephamycins among *Escherichia coli* isolates from animals and humans. J Infect Dis. 2003;188(5): 759–68.
15. Bukh AS, Schønheyder HC, Emmersen JM, Søgaard M, Bastholm S, Roslev P. *Escherichia coli* phylogenetic groups are associated with site of infection and level of antibiotic resistance in community-acquired bacteraemia: a 10 year population-based study in Denmark. J Antimicrob Chemother. 2009; 64(1):163–8.
16. Skurnik D, Le Menac'h A, Zurakowski D, Mazel D, Courvalin P, Denamur E, Andremont A, Ruimy R. Integron-associated antibiotic resistance and phylogenetic grouping of *Escherichia coli* isolates from healthy subjects free of recent antibiotic exposure. Antimicrob Agents Chemother. 2005; 49(7):3062–5.
17. Petersen A, Guardabassi L, Dalsgaard A, Olsen JE. Class I integrons containing a *dhfrI* trimethoprim resistance gene cassette in aquatic *Acinetobacter* spp. FEMS Microbiol Lett. 2000;182(1):73–6.

18. Belaynehe KM, Shin SW, Park HT, Yoo HS. Occurrence of aminoglycoside modifying enzymes (AMEs) among isolates of *Escherichia coli* exhibiting high levels of aminoglycoside resistance isolated from Korean cattle farms. FEMS Microbiol Lett. 2017;364:1–9.

19. Clinical and Laboratory Standards Institute (CLSI). Performance standards for antimicrobial susceptibility testing; twenty-third informational supplement. Wayne: Clinical and Laboratory Standards Institute; 2013. M100-S23.

20. Clermont O, Bonacorsi S, Bingen E. Rapid and simple determination of the *Escherichia coli* phylogenetic group. Appl Environ Microbiol. 2000;66(10): 4555–8.

21. Lanz R, Kuhnert P, Boerlin P. Antimicrobial resistance and resistance gene determinants in clinical *Escherichia coli* from different animal species in Switzerland. Vet Microbiol. 2003;91(1):73–84.

22. Levy SB, McMurry LM, Barbosa TM, Burdett V, Courvalin P, Hillen W, Roberts MC, Rood JI, Taylor DE. Nomenclature for new tetracycline resistance determinants. Antimicrob Agents Chemother. 1999;43(6):1523–4.

23. Zhao J, Aoki T. Nucleotide sequence analysis of the class G tetracycline resistance determinant from *Vibrio anguillarum*. Microbiol Immunol. 1992; 36(10):1051–60.

24. Kerrn M, Klemmensen T, Frimodt-Møller N, Espersen F. Susceptibility of Danish *Escherichia coli* strains isolated from urinary tract infections and bacteraemia, and distribution of *sul* genes conferring sulphonamide resistance. J Antimicrob Chemother. 2002;50(4):513–6.

25. Kozak GK, Pearl DL, Parkman J, Reid-Smith RJ, Deckert A, Boerlin P. Distribution of sulfonamide resistance genes in *Escherichia coli* and *Salmonella* isolates from swine and chickens at abattoirs in Ontario and Quebec, Canada. Appl Environ Microbiol. 2009;75(18):5999–6001.

26. Orman BE, Pineiro SA, Arduino S, Galas M, Melano R, Caffer MI, Sordelli DO, Centrón D. Evolution of multiresistance in nontyphoid *Salmonella* serovars from 1984 to 1998 in Argentina. Antimicrob Agents Chemother. 2002;46(12): 3963–70.

27. Sáenz Y, Briñas L, Domínguez E, Ruiz J, Zarazaga M, Vila J, Torres C. Mechanisms of resistance in multiple-antibiotic-resistant *Escherichia coli* strains of human, animal, and food origins. Antimicrob Agents Chemother. 2004;48(10):3996–4001.

28. Levesque C, Piche L, Larose C, Roy PH. PCR mapping of integrons reveals several novel combinations of resistance genes. Antimicrob Agents Chemother. 1995;39(1):185–91.

29. Wan MT, Chou CC. Class 1 integrons and the antiseptic resistance gene (*qacEΔ1*) in municipal and swine slaughterhouse wastewater treatment plants and wastewater—associated methicillin-resistant *Staphylococcus aureus*. Int J Environ Res Publ Health. 2015;12(16):6249–60.

30. Vali L, Hamouda A, Hoyle DV, Pearce MC, Whitaker LH, Jenkins C, Knight HI, Smith AW, Amyes SG. Antibiotic resistance and molecular epidemiology of *Escherichia coli* O26, O103 and O145 shed by two cohorts of Scottish beef cattle. J Antimicrob Chemother. 2007;59(3):403–10.

31. Lim SK, Lee HS, Nam HM, Cho YS, Kim JM, Song SW, Park YH, Jung SC. Antimicrobial resistance observed in *Escherichia coli* strains isolated from fecal samples of cattle and pigs in Korea during 2003–2004. Int J Food Microbiol. 2007;116(2):283–6.

32. Kang SG, Lee DY, Shin SJ, Ahn JM, Yoo HS. Changes in patterns of antimicrobial susceptibility and class 1 integron carriage among *Escherichia coli* isolates. J Vet Sci. 2005;6(3):201.

33. Shin SW, Byun JW, Jung M, Shin MK, Yoo HS. Antimicrobial resistance, virulence genes and PFGE-profiling of *Escherichia coli* isolates from south Korean cattle farms. J Microbiol. 2014;52(9):785–93.

34. United States Food and Drug Administration (USFDA). Summary Report on Antimicrobials Sold or Distributed for Use in Food-producing Animals US Food and Drug Administration; 2015. p. 2016.

35. Korean Food and Drug Administration (KFDA). Establishment of control system of antibiotics for livestocks. Seoul: Korean Food and Drug Administration; 2004. p. 19–43.

36. Animal and Plant Quarantine Agency of Korea (APQA). Antimicrobial use in livestock and monitoring report of antimicrobial resistance in animal. South Korea: Ministry of Agriculture, Food and Rural Affairs; 2012. p. 1–98.

37. Carlos C, Pires MM, Stoppe NC, Hachich EM, Sato MI, Gomes TA, Amaral LA, Ottoboni LM. *Escherichia coli* phylogenetic group determination and its application in the identification of the major animal source of fecal contamination. BMC Microbiol. 2010;10(1):161.

38. Wu R, Alexander T, Li J, Munns K, Sharma R, McAllister T. Prevalence and diversity of class 1 integrons and resistance genes in antimicrobial-resistant *Escherichia coli* originating from beef cattle administered subtherapeutic antimicrobials. J Appl Microbiol. 2011;111(2):511–23.

39. Koczura R, Mokracka J, Barczak A, Krysiak N, Kaznowski A. Association between the presence of class 1 integrons, virulence genes, and phylogenetic groups of *Escherichia coli* isolates from river water. Microb Ecol. 2013;65(1):84–90.

40. Thaker M, Spanogiannopoulos P, Wright GD. The tetracycline resistome. Cell Mol Life Sci. 2010;67(3):419–31.

41. Shin SW, Shin MK, Jung M, Belaynehe KM, Yoo HS. Prevalence of antimicrobial resistance and transfer of tetracycline resistance genes in *Escherichia coli* isolated from beef cattle. Appl Environ Microbiol. 2015;81: 5560–6.

42. Marshall B, Tachibana C, Levy S. Frequency of tetracycline resistance determinant classes among lactose-fermenting coliforms. Antimicrob Agents Chemother. 1983;24(6):835–40.

43. Sengeløv G, Halling-Sørensen B, Aarestrup FM. Susceptibility of *Escherichia coli* and *Enterococcus faecium* isolated from pigs and broiler chickens to tetracycline degradation products and distribution of tetracycline resistance determinants in *E. coli* from food animals. Vet Microbiol. 2003;95(1):91–101.

44. Bryan A, Shapir N, Sadowsky MJ. Frequency and distribution of tetracycline resistance genes in genetically diverse, nonselected, and nonclinical *Escherichia coli* strains isolated from diverse human and animal sources. Appl Environ Microbiol. 2004;70(4):2503–7.

45. Dessie HK, Bae DH, Lee YJ. Characterization of integrons and their cassettes in *Escherichia coli* and *Salmonella* isolates from poultry in Korea. Poult Sci. 2013;92(11):3036–43.

46. Sunde M, Norström M. The prevalence of, associations between and conjugal transfer of antibiotic resistance genes in *Escherichia coli* isolated from Norwegian meat and meat products. J Antimicrob Chemother. 2006; 58(4):741–7.

47. Boerlin P, Travis R, Gyles CL, Reid-Smith R, Lim NJH, Nicholson V, McEwen SA, Friendship R, Archambault M. Antimicrobial resistance and virulence genes of *Escherichia coli* isolates from swine in Ontario. Appl Environ Microbiol. 2005;71(11):6753–61.

48. Vinué L, Sáenz Y, Somalo S, Escudero E, Moreno MÁ, Ruiz-Larrea F, Torres C. Prevalence and diversity of integrons and associated resistance genes in faecal *Escherichia coli* isolates of healthy humans in Spain. J Antimicrob Chemother. 2008;62(5):934–7.

49. Povilonis J, Šeputienė V, Ružauskas M, Šiugždinienė R, Virgailis M, Pavilonis A, Sužiedėlienė E. Transferable class 1 and 2 integrons in *Escherichia coli* and *Salmonella enterica* isolates of human and animal origin in Lithuania. Foodborne Pathog Dis. 2010;7(10):1185–92.

50. Cavicchio L, Dotto G, Giacomelli M, Giovanardi D, Grilli G, Franciosini MP, Trocino A, Piccirillo A. Class 1 and class 2 integrons in avian pathogenic *Escherichia coli* from poultry in Italy. Poult Sci. 2015;94(6):1202–8.

51. Shaheen BW, Oyarzabal OA, Boothe DM. The role of class 1 and 2 integrons in mediating antimicrobial resistance among canine and feline clinical *E. coli* isolates from the US. Vet Microbiol. 2010;144(3):363–70.

52. Kang HY, Jeong YS, Oh JY, Tae SH, Choi CH, Moon DC, Lee WK, Lee YC, Seol SY, Cho DT. Characterization of antimicrobial resistance and class 1 integrons found in *Escherichia coli* isolates from humans and animals in Korea. J Antimicrob Chemother. 2005;55(5):639–44.

53. Sáenz Y, Vinué L, Ruiz E, Somalo S, Martínez S, Rojo-Bezares B, Zarazaga M, Torres C. Class 1 integrons lacking *qacEΔ1* and *sul1* genes in *Escherichia coli* isolates of food, animal and human origins. Vet Microbiol. 2010; 144(3):493–7.

54. Cocchi S, Grasselli E, Gutacker M, Benagli C, Convert M, Piffaretti JC. Distribution and characterization of integrons in *Escherichia coli* strains of animal and human origin. FEMS Immunol Med Microbiol. 2007;50(1):126–32.

55. Sawant AA, Hegde NV, Straley BA, Donaldson SC, Love BC, Knabel SJ, Jayarao BM. Antimicrobial-resistant enteric bacteria from dairy cattle. Appl Environ Microbiol. 2007;73(1):156–63.

21

Profiling of rumen fermentation, microbial population and digestibility in goats fed with dietary oils containing different fatty acids

I. Nur Atikah[1], A. R. Alimon[2], H. Yaakub[2], N. Abdullah[1], M. F. Jahromi[1], M. Ivan[1] and A. A. Samsudin[2]* (iD)

Abstract

Background: The effects of the dietary oils with differing fatty acid profiles on rumen fermentation, microbial population, and digestibility in goats were investigated. In Experiment I, rumen microbial population and fermentation profiles were evaluated on 16 fistulated male goats that were randomly assigned to four treatment groups: i) control (CNT), ii) olive oil (OL), iii) palm olein oil (PO), and iv) sunflower oil (SF). In Experiment II, another group of 16 male goats was randomly assigned to the same dietary treatments for digestibility determination.

Results: Rumen ammonia concentration was higher in CNT group compared to treatment groups receiving dietary oils. The total VFA and acetate concentration were higher in SF and OL groups, which showed that they were significantly affected by the dietary treatments. There were no differences in total microbial population. However, fibre degrading bacteria populations were affected by the interaction between treatment and day of sampling. Significant differences were observed in apparent digestibility of crude protein and ether extract of treatment groups containing dietary oils compared to the control group.

Conclusions: This study demonstrated that supplementation of different dietary oils containing different fatty acid profiles improved rumen fermentation by reducing ammonia concentration and increasing total VFA concentration, altering fibre degrading bacteria population, and improving apparent digestibility of crude protein and ether extract.

Keywords: Dietary oil, Digestibility, Goat, Rumen fermentation, Rumen microbial population

Background

Ruminant acquires energy from plant materials through the activity of microbial fermentation and plant degradation mainly by groups of cellulolytic bacteria. The dynamics of major cellulolytic bacterial population found in the rumen, in particular *Fibrobacter succinogens, Ruminococcus albus,* and *Ruminococcus flavefaciens,* have been widely studied in response to dietary shift [1] or between species [2] using molecular approaches of quantitative real-time polymerase chain reaction (qPCR). The importance of cellulolytic bacteria in ruminant nutrition is due to the fact that this particular group of bacteria plays a critical role not only in

utilizing feeds that are not suitable for monogastric animals, but also in facilitating animals to survive on poor quality fibrous forages [3]. During the fermentation process, energy is released in the form of adenosine triphosphate (ATP), which is used to fuel different activities of rumen microorganism. This energy can be improved in ruminant by supplementing the animals with dietary fat, an approach that has been commonly practiced.

Other studies have shown different effects of vegetable oil supplementations in rumen fermentation and microbial population using cattle. For example, [4] reported that supplementation of linseed oil to dairy cow did not affect ruminal pH, ammonia, and total volatile FA concentrations. Similarly, [5] reported that flaxseed supplementation to calves has no effects on rumen fermentation parameters. However, fish oil supplementation in steer ruminal

* Correspondence: anjas@upm.edu.my
[2]Department of Animal Science, Faculty of Agriculture, Universiti Putra Malaysia, UPM, 43400 Serdang, Selangor, Malaysia

fluid had lower ruminal acetate and butyrate but greater propionate concentration, as reported by [6]. Another study from [7] reported that feeding oilseeds from flaxseed to cows had no effects on pH and concentrations of NH_3-N and total volatile fatty acids while acetate: propionate ratio was decreased. In the same study, they also described that oilseeds decreased protozoa and increased total cellulolytic bacteria population in rumen fluid. Similarly, [8] showed that supplementation with palm oil in cows reduced protozoa population, but it did not affect cellulolytic bacteria population in cows. Other vegetable oils such as coconut oil [9] and soybean oil [10] are used as energy sources and have the potential to manipulate the microbial ecosystem of the rumen to enhance fibrous feed digestibility, reduce methane emission, and reduce nitrogen excretion by ruminants [11].

There is not much information available that emphasizes the effect of diets supplemented with olive oil that contains oleic acid (C18:1), sunflower oil that contains linoleic acid (C18:2), and palm olein oil that consists of linoleic acid (C18:2) and palmitic acid (C16:0) on rumen fermentation, rumen microbial populations, and digestibility in goats. Therefore, the aim of this study was to investigate rumen microbial population, fermentation profile and nutrient digestibility for local goats fed diets supplemented with sunflower oil (SF), olive oil (OL), or palm olein oil (PO).

Results

Rumen pH and volatile fatty acids
The results of ruminal pH and VFA concentration are presented in Table 1. The mean of ruminal pH ranged between 6.26 (PO) and 6.80 (OL) and was affected by day of sampling ($P < 0.01$). Different treatment diets had no significant effect on ruminal pH. However, OL fed group tended to have a slightly higher rumen pH value than that of the other groups.

The total VFA concentration (mmol) was significantly higher ($P < 0.05$) in OL and SF compared to the CNT and PO groups (Table 1). The OL and SF groups had higher ($P < 0.05$) concentration of acetic acid compared to the other groups. Diets supplemented with PO did not show significant differences in acetate when compared to the control group. Supplementation of OL showed an increased level of isobutyric acid concentration when compared to other treatment groups. Propionate, butyrate, valerate, and isovalerate concentrations and acetic acid to propionic acid ratio (A/P) were not affected by the different types of oil supplementation. However, the significant effect of sampling day can be observed in the concentrations of the total VFA, acetate, butyrate, and valerate.

Ammonia
The mean values of ruminal ammonia are presented in Table 1. Addition of dietary oils significantly decreased the concentration of ruminal ammonia. The concentration (mg/l) of ammonia-N in the rumen fluid was significantly affected by diet ($P < 0.05$), day of sampling ($P < 0.01$), and diet × day of sampling interaction ($P < 0.05$). Higher amount of ammonia concentration was observed in CNT (42.6 mg/l) compared to other groups (36.4–37.9 mg/l).

Microbial population
The effects of dietary oils on rumen microbial population are presented in Table 2. Higher numbers of total bacteria could be observed in the treatment groups compared to CNT although not statistically significant ($P > 0.05$). No significant difference was observed in the *F. succinogenes*, *R. albus*, and *R. flavefaciens* populations although the highest level of log10 copy no./g was recorded in SF

Table 1 Rumen fermentation parameters (mean ± SE) of goats fed diet supplemented with different types of oils

Parameters	Treatment				P-value		
	CNT	OL	PO	SF	Tr	Day	Tr × Day
pH	6.29 ± 0.11	6.80 ± 0.11	6.26 ± 0.06	6.33 ± 0.07	NS	**	NS
Ammonia (mg/l)	42.6 ± 1.65[a]	37.9 ± 1.36[b]	36.4 ± 1.08[b]	36.9 ± 1.24[b]	**	*	*
Total VFA (mmol)	89.42 ± 8.59[b]	95.58 ± 7.81[a]	91.11 ± 6.44[b]	95.79 ± 3.15[a]	*	*	NS
Acetate (%)	59.22 ± 5.83 [b]	63.70 ± 5.46[a]	56.79 3.56[b]	63.72 ± 2.01[a]	*	*	*
Propionate (%)	18.21 ± 2.11	19.62 ± 1.47	22.15 ± 2.81	20.22 ± 0.95	NS	NS	NS
Butyrate (%)	8.19 ± 0.89	8.83 ± 0.99	8.13 ± 0.96	8.37 ± 0.71	NS	*	NS
Isobutyrate (%)	0.76 ± 0.13[b]	0.88 ± 0.11 [a]	0.73 ± 0.04[b]	0.68 ± 0.04[b]	*	NS	NS
Valerate (%)	2.18 ± 0.48	2.42 ± 0.13	2.14 ± 0.24	1.72 ± 0.28	NS	*	*
Isovalerate (%)	0.86 ± 0.25	1.13 ± 0.13	1.17 ± 0.14	1.08 ± 0.16	NS	NS	NS
Acetate/Propionate	3.25 ± 0.18	3.72 ± 0.22	2.56 ± 0.30	3.15 ± 0.19	NS	NS	NS

CNT Control diet, *OL* Olive oil diet, *PO* Palm olein diet, *SF* Sunflower oil diet, *Tr* Treatment
*Significant level at $P < 0.05$; ** Significant level at $P < 0.01$
[a, b]Means in the same row with different superscripts are statistically different ($P < 0.05$)

Table 2 Effects of supplementation with different types of oils on microbial population (mean ± SE) in the rumen of goats

Parameter	Treatment				P-value		
	CNT	OL	PO	SF	Treatment	Day	Treatment×Day
Total Microbes (Log10 copy No/g)	9.56 ± 0.21	10.20 ± 0.15	10.01 ± 0.13	9.83 ± 0.16	NS	NS	NS
Fibrobacter succinogenes (Log10 copy No/g)	4.20 ± 0.18	3.67 ± 0.18	4.20 ± 0.14	4.21 ± 0.14	NS	*	*
Rumonococcus albus (Log10 copy No/g)	7.81 ± 0.21	7.65 ± 0.18	7.95 ± 0.22	8.07 ± 0.19	NS	*	*
Ruminococcus flavefaciens (Log10 copy No/g)	5.02 ± 0.19	5.06 ± 0.26	5.32 ± 0.20	5.36 ± 0.10	NS	*	*
Methanogenic archea (Log10 copy No/g)	3.60 ± 0.12[b]	4.23 ± 0.22[a]	3.91 ± 0.14[ab]	4.39 ± 0.21[a]	*	*	*
Protozoa (Log10 copy No/g)	3.30 ± 0.22	2.63 ± 0.22	3.30 ± 0.13	3.22 ± 0.15	NS	*	NS

CNT Control diet, *OL* Olive oil diet, *PO* Palm olein diet, *SF* Sunflower oil diet, *Tr* Treatment
*Significant level at $P < 0.05$
[a, b]Means ± std. error in the same row with different superscripts are statistically different ($P < 0.05$)

groups. Significant differences ($P < 0.05$) were observed for methanogenic archea population, and the numbers were influenced by treatment, day of sampling, and the interaction of treatment × day. Population of protozoa demonstrated that there were no significant differences among all the treatment groups although the CNT and PO groups had higher values of log10 copy no./g and were significantly affected by day of sampling ($P < 0.05$). As shown in Table 2, significant differences are seen in the methanogens population while the protozoa population is not significantly affected by the treatment diets.

Apparent digestibility

The results of apparent digestibility study are presented in Table 3. The DM, OM, NDF, and ADF also followed a similar pattern with the treatment group ($P > 0.05$). The CP apparent digestibility was significantly improved ($P < 0.05$) in all treatment groups containing oil, in which OL had the highest CP apparent digestibility (85.04%). The apparent digestibility of EE also followed a similar pattern where the three treatments of OL, PO, and SF had a higher EE digestibility ($P < 0.05$) compared to CNT, with SF groups having the highest apparent digestibility percentage (91.13%). However, no significant difference was observed in fibre digestibility although the PO and SF groups numerically tended to have higher ADF digestibility.

Discussion

Rumen pH and volatile fatty acids

Ruminal pH values were within normal range, and the increment has minimal effects on rumen cellulolytic processes of fibre and protein digestion (6.0–7.0) [12]. This result suggests that the microbial population of rumen is able to adapt to the diet given, regardless of the additions and differences in composition of dietary oil supplemented [13]. Adequate roughage supply in the diet reduced the negative effect of dietary oil on rumen fermentation because the fibre fraction creates a supporting environment for rumen microbes to hydrolyze the dietary oils [6, 14, 15]. The findings of the present study are consistent with those of [8] who reported that supplementation of palm oil did not give negative effects on ruminal pH in dairy cows. In addition, other studies using different types of dietary oils in other ruminants have also reported similar observations [16–18].

The increased level of the total VFA concentration in OL and SF groups in the present study indicates the efficiency of nutrient digestion. It confirms the fact that the notable effect of supplementing dietary oils on rumen fermentation depends on the type and level of fatty acids [10]. A similar result was reported by [19], where supplementation of C_{18} fatty acid increased the total VFA concentration although less influence was seen in the different types of fatty acid supplemented. Nevertheless,

Table 3 Apparent digestibility (% DM) of nutrient (mean ± SE) in goats fed diet supplemented with different types of oils

Apparent digestibility (%)	Treatment				P-value
	CNT	OL	PO	SF	
Dry matter	75.81 ± 2.77	73.49 ± 1.98	75.44 ± 1.09	75.37 ± 1.57	NS
Organic matter	80.05 ± 1.78	80.84 ± 1.57	80.17 ± 1.61	78.81 ± 2.34	NS
Crude protein	77.65 ± 0.81[b]	85.04 ± 1.19[a]	82.20 ± 2.27[a]	82.43 ± 1.44[a]	*
Ether extract	59.89 ± 10.39[b]	88.11 ± 1.46[a]	87.92 ± 1.93[a]	91.13 ± 2.76[a]	*
NDF	75.19 ± 4.31	71.59 ± 2.84	71.82 ± 3.47	71.28 ± 4.31	NS
ADF	47.42 ± 1.99[b]	54.75 ± 5.27[a]	57.54 ± 4.49[a]	55.01 ± 4.41[a]	*

CNT Control diet, *OL* Olive oil diet, *PO* Palm olein diet, *SF* Sunflower oil diet
[a, b]Means in the same row with different superscripts are statistically different ($P < 0.05$)

different responses were observed in the work done by [20] that showed the reduction in the concentration of VFA supplemented with fatty acids.

A significant increase ($P < 0.05$) in the acetate level in the OL and SF can also suggest a modification of the ruminal microbial population [19, 21]. However, the validation on this reason needs to be done in future studies. This result is thought to be due to the modification of the ruminal microbial ecosystem, as occurs with 18-carbon polyunsaturated FA. A decrease in cellulolytic and methanogenic bacteria is observed with most fat sources inclusion. For branched fatty acid concentration, there was a significant increment in the molar proportion of isobutyrate in the OL group. The finding is consistent with a study done by [20] who reported that branched fatty acid concentration was increased with the supplementation of fatty acid.

Ammonia

Fat supplementation in ruminant diets has been shown to consistently depress rumen ammonia concentration [22, 23]. In the current study, the ammonia level recorded in the oil supplemented groups was within the normal range as reported by [24]. The optimum ammonia level that favors the ruminal microbial activity in animals fed with lignocellulosic materials was between 16.5 and 37.9 mg/l. There were also significant differences observed by the interaction of treatments with day of sampling, suggesting that ammonia level in rumen might be associated with the shift of the microbial population of rumen by time, due to the addition of dietary oils. A report by [25] suggested that the increase in the ammonia level was due to the reduction in protozoal predation toward rumen bacteria thus, reducing the recycling of bacteria protein in the rumen. A similar result by [26] also reported that ammonia concentration tended to increase when the longer chain of unsaturated fatty acid was present in the diet. [27] reported that rumen ammonia concentration reduction corresponded with lowered ammonia flow to the duodenum and was similar to other studies [28, 29] in sheep as well as in cattle [18, 22]. However, contradictory results with regard to ammonia level in previous studies on ruminant have been noted with supplementations of linoleic acid by [19, 30] and supplementations of sunflower oil in cattle by [31].

Microbial population

Supplementation of vegetables oils did not alter the fibre degrading bacteria and total microbial populations in the present study, indicating that these microorganisms are not sensitive to dietary oils supplementation [32]. Another possible reason was mentioned by [33] who reported that the negative effects toward ruminal fibrolytic bacteria were neglected in the case of high grain-fed

diet. In their study, neither NDF nor fibrolytic bacteria population has shown significant responses toward fatty acid supplementation. Vegetable oil supplementations have shown inconsistent results of rumen microbial population in other studies. [6] observed a decrease of *F. succinogenes* population but not the *R. albus,* and *R. flavefaciens* populations in steers. Furthermore, [10] reported a decrease of *F. succinogenes* and *R. flavefaciens* population. In addition, [34] observed an increase of *F. succinogenes, R. albus,* and *R. flavefaciens* populations in goats. [32] on the other hand, observed no effects on *F. succinogenes* population and a decrease of *R. albus* and *R. flavefaciens* populations in goats. The significant differences observed by sampling day and interaction of treatments with sampling day on fibre degrading bacteria, methanogens, and protozoa populations in the present study suggest that the ruminal microbial populations shifted by time due to the addition of dietary oils.

The population of protozoa in rumen often correlates to the population of methanogens. It has been reported that a reduction in protozoa reduced the methanogens population since methanogens live in association with protozoa, linked by hydrogen transfer within the interspecies [35]. However, the association of those methanogens is only 0.1–0.2% of the total population, whereby the others that exist freely in the rumen might not be affected by the supplementation to the same extent. Due to that, the reduction of methanogens does not always follow the population pattern similar to protozoa. A decrease in rumen protozoa population was observed with the supplementation of blended canola and palm oil by [34] and supplementation of linseed oil and coconut oil by [36]. A similar observation was also reported by [10] who observed that dietary soybean oil reduced the population of methanogens in lambs.

Apparent digestibility

An increase in CP digestibility in the treatment groups suggests that oil supplementation can act as a source of energy for rumen microbes to convert feed protein into microbial protein, which is more digestible. Besides, the increased CP digestibility may be due to the reduction in the microbial degradation by protozoa, which in turn increased the level of protein available in the lower sections of the gastrointestinal tract [37].

The higher apparent digestibility of EE in the treatment groups reported in this study is in agreement with a study using lambs by [38]. Diet rich in dietary fats tends to have a higher hydrolysis percentage in the rumen compared to the conventional diet [39]. Lipases that are involved in rumen lipid hydrolysis have been shown to be more active in diets with high fibre and protein contents [40]. A previous study by [15] showed that fatty acid had higher digestibility with increasing number of double bonds. In this study, although there were no significant changes, the SF group showed higher values in EE digestibility. It may be due to the presence of linoleic acid (C18:2) in sunflower oil compared to the presence of oleic acid (C18:1) in both olive oil and palm olein.

Supplementation of dietary oils tended to coat the particle of fibre, thus preventing them from the attack of rumen microbes [41]. In the present study, although there was a decreased pattern of NDF digestibility, there were no significant differences observed. This may suggest that the level of oil inclusion is not enough to deliver the effect. This result is also supported by [42] who indicated that fibre digestion would be limited when fat content in ruminant diet is higher than 70 g/kg DM intake, a level which is higher than the level used in the present study. Another reason is that NDF digestibility follows ruminal protozoa population, as supported in a study by [43]. Similar results by [44] did not find variations in NDF digestibility whereas ADF digestibility was higher in in vitro study. The increase in nutrient digestibility observed in the present study might have been caused by the increase in ruminal retention time as suggested by [45, 46].

Conclusion

Supplementation with olive oil, palm olein oil, and sunflower oil improved and developed better ruminal microorganism population to a certain extent in goats. In terms of digestibility, oil supplementation improved both protein and fat digestibility. For future research, it is recommended that studies on the effects of C_{18} fatty acids on the metabolic activity of microbial population in rumen, particularly methanogens, be deeply clarified, including determination of the long-term effects of fatty acids on in vivo rumen fermentation, methanogenesis, and animal performance. With regard to rumen physiology, understanding the association, symbiotic relationship, and cross feeding among microorganism is important in predicting the response of microorganisms when given a new diet.

Methods
Animals and diet

The experiment was carried out at Department of Animal Science, Faculty of Agriculture, Universiti Putra Malaysia following the guidelines approved by the Institutional Animal Care and Use Committee (IACUC) (Approval No. R064/2016) of the Universiti Putra Malaysia. Sixteen mature local Katjang-crossed male goats aged between 20 and 24 months with an average weight of 28.32 ± 1.85 kg and fitted with rumen cannula were used for rumen fermentation profile and microbial population study (Experiment I). The animals were properly maintained by treated against endo and

ectoparasites prior to the commencement of the experimental procedure. In addition, all of the animals were under supervisory of veterinarian for assessing their health. In another study (Experiment II), 16 Katjang-crossed male goats aged between 10 and 12 months with an average weight of 23.17 ± 0.94 kg were used to determine the digestibility of nutrients in the diets. The diets were formulated to have approximately equal amount of crude protein (CP) and energy content [47]. All of the animals used in these two studies were purchased from a commercial farm, De Kebun Enterprise in Selangor.

The animals were randomly assigned into four groups to receive four different dietary treatments: i) basal diet (CNT or control), ii) basal diet + olive oil (OL), iii) basal diet + palm olein oil (PO), and iv) basal diet + sunflower oil (SF). The oil content was supplemented at the rate of 6% of the total feed ingredients. The oil content was supplemented at the rate of 6% of the total feed ingredients. The ingredients, including the oils, were purchased from commercial sources. Dry matter (DM), organic matter (OM), crude protein (CP), neutral detergent fibre (NDF), and acid detergent fibre (ADF) of the experimental diets were analyzed according to [48, 49]. The ingredients and chemical composition of the diets are presented in Table 4.

Animal housing and management

The experiments were conducted in Serdang, Selangor, Malaysia (3° 2′ 0″ North, 101° 43′ 0″ East). The cannulated animals were kept individually in separate pens while in the digestibility study, the animals were kept in an individual metabolic crate and had free access to water.

Experimental procedure and sampling schedule

Experiment I was conducted for 28 days of adjustment period followed by 30 days of treatment period. During the adjustment period, all the animals were fed with basal diet that acted as the control diet. The diet was offered ad libitum to the goats at 09:00 daily. After the adjustment period, the animals were randomly assigned according to a completely randomized design into four groups, with each group consisted of four goats and received one of the four dietary treatments. The random selection was done in Microsoft Excel using tag number of animals. Rumen content was collected on days 27 and 28 of the adjustment period, which is considered as the initial sampling day. The data were pooled and recorded as day 0 of the experimental period. Rumen samples were collected from different parts of the rumen 2 h after morning feeding through the cannula. Similarly, rumen samples were also collected on day 2, 4, 6, 12, 18, 24, and 30 of treatment period.

Table 4 Ingredients and chemical composition of treatment diets

	Treatments			
	CNT	OL	PO	SF
Ingredient (as fed)				
Rice straw	30.8	25.8	25.8	25.8
Barley grain	35.0	35.0	35.0	35.0
Soybean meal	30.0	30.0	30.0	30.0
Molases	2.0	1.0	1.0	1.0
Vitamin mineral-mix	0.5	0.5	0.5	0.5
Limestone	1.3	1.3	1.3	1.3
Sodium Sulphate	0.4	0.4	0.4	0.4
Olive oil	–	6.0	–	–
Palm Oil	–	–	6.0	–
Sunflower oil	–	–	–	6.0
Chemical analysis (DM %)				
DM	76.17	76.02	78.27	78.73
OM	93.60	93.34	94.61	94.96
CP	15.76	15.48	15.9	16.00
EE	1.86	4.56	4.70	4.74
NDF	63.53	58.76	58.54	51.27
ADF	17.04	18.26	20.66	21.41
Fatty acid (g/100 g total fatty acid DM)				
C-16:0	8.09	11.72	44.50	4.92
C-18:0	2.70	1.40	3.73	6.28
C-18:1, n 9	27.88	73.20	41.45	28.47
C-18:2, n 6	55.76	12.9	9.62	59.36
C-18:3, n 3	5.57	0.80	0.30	0.97

CNT Control diet, OL Olive oil diet, PO Palm olein diet, SF Sunflower oil diet, DM Dry matter, OM Organic matter, CP Crude protein, EE Ether extract, NDF Neutral detergent fiber, ADF Acid detergent fiber, C-16:0 Palmitic acid, C-18:0 Stearic acid, C-18:1, n 9 Oleic acid, C-18:2, n 6 Linoleic acid, C-18:3, n 3 Linolenic acid

Determination of rumen microbial population was done on the rumen content collected on day 0, 12, and 30 using real-time PCR. Briefly, about 500 ml of rumen fluid was collected. Rumen pH was immediately measured and then divided into two portions. The first portion of rumen fluid collected was squeezed through four layers of cheesecloth to eliminate larger solid feed particles. Immediately, two drops of sulphuric acid were added to stop further fermentation. The samples were then kept at 20 °C until they were further processed for VFA and ammonia determination. The second portion of the rumen fluid was immediately kept in ice and stored at – 20 °C until further analyses of microbial population study using a qPCR procedure.

For Experiment II, all animals were placed in an individual metabolic crate throughout the 19 days of the experiment (14 days of adjustment period to experimental

diets and 5 days of sampling). The goats were divided into four groups of four goats and fed with the respective diets. They had free access to clean water. The respective diets were offered ad libitum to the goats at 09:00 daily until day 14. The feed intake of each animal on days 11–14 was recorded. On days 15–19, the animals were fed with 90% of the recorded intake. Fecal samples were collected daily from day 15 until day 19, and approximately 10% of the total collections were kept frozen at – 20 °C until further chemical analyses of nutrients. At the end of this study, all goats were fasted for 12 h with free access to drinking water, transported to the abattoir, allowed to rest, and then weighed before slaughter. The goats were slaughtered in accordance with the procedures outlined in MS1500:2009 (Department of Standards Malaysia, 2009) which allows animal to be slaughtered, without being stunned, with a razor sharp knife. In this study the slaughter was performed by a certified and highly experienced technician with a sharp knife. The goats were to be used for another study to determine the effects of different oils on carcass and meat quality which involves a food tasting study (not reported in the current study) [50].

Chemical analyses of feed and fecal samples

Feed and fecal samples were analyzed for DM, OM, CP, and EE using the procedure by [48], whereas NDF and ADF were determined using the procedure by [49].

Rumen pH and volatile fatty acid determination

The pH of the rumen content was measured immediately after the collection of the rumen fluid using a portable pH meter (Eco Testr pH 1, Eutech Instruments). For VFA determination, rumen filtrated samples were thawed at 4 °C prior to analysis following the procedure described by [51] with some modifications. One ml 3:1 v/v solution of 24% metaphosphoric acid and 5% formic acid was added into 5 ml of the rumen filtrate. The mixture was left to stand for 30 min before being centrifuged at 12,000 x g for 20 min. Then 0.5 ml of supernatant was collected and kept in 2-ml vials, and 0.5 ml internal standard (4-methyl-n-valeric acid) was pipetted into the vials. The samples were analyzed using gas chromatography, equipped with Flame Ionization Detector (FID) and capillary column (DB-FFAP, 122–3232).

Ammonia determination

Rumen filtrated samples were centrifuged at 12,000 x g for 20 min. 5 ml of supernatant was collected and kept for further determination of the ammonia content using a protocol described by [52]. A standard solution was prepared using 1.908 g of ammonium chloride dissolved in 500 ml distilled water to give 1000 mg/l ammonia- nitrogen (ammonia-N). A standard 0.2, 0.5, 1.0, and

2.0 ppm solution were prepared by dissolving 0.02, 0.05, 0.10, and 0.20 ml of the stock solution with 100 ml distilled water, respectively. 5 ml of water (blank) or standard was added in an Erlenmeyer flask, and 0.2 ml of the phenol solution was added and swirled. In sequence, 0.2 ml of nitroprusside and 0.5 ml of oxidizing solution were added. The flask was then swirled, stopped and allowed to stand for 1 h at room temperature. The absorbance was then determined at 640 nm. Regression equation was determined from blank and standard samples before ammonia-N was estimated in the samples.

DNA extraction and quantification using qPCR

Total genomic DNA from rumen content samples on day 0, day 12, and day 30 was extracted using QIAamp DNA Stool Mini Kit (Qiagen Inc., Valencia, CA, USA). The guideline on the protocol was provided by the manufacturer. The extracted DNA was stored at – 20 °C until subsequent procedures. Real-time PCR was used to determine the population of total bacteria, *F. succinogens*, *R. albus*, *R. flavefaciens*, methanogens, and protozoa. Species-specific PCR primers used to amplify partial 16S rDNA regions were chosen from literatures as presented in Table 5. Real-time PCR amplification and detection were performed using CFX 96 system (Bio-Rad, Hercules, CA, USA). The amplification reaction was conducted in a final volume of 25 μl containing 12.5 μl Maxima SYBR Green qPCR Master Mix, 1 μl species-specific PCR forward primer, 1 μl species-specific PCR reverse primer, 8.5 μl RNAse-free distilled water, and 2 μl of DNA elution. The PCR conditions of all species were as follows: an initial denaturation 95 °C for 15 s, followed by 39 cycles of denaturing of 15 s at 95 °C, 30 s at annealing temperature, and 20 s at 72 °C for an extension. The standards used in this study were prepared according to the protocol demonstrated by [53]. Briefly, DNA was extracted from a pure culture of microorganisms of interest to produce a high concentration of the target DNA using normal PCR.

Later, the products of the PCR were purified using MEGAquick-spin™ (Intron Biotechnology, Inc.) The concentrations of the products were then measured using a Nanodrop ND-1000 spectrophotometer. An online formula [54] was used to calculate the number of copies of a template DNA per ml of elution buffer. Finally, standard curves were constructed using a serial dilution of plasmid DNA of each microbial group.

Determination of apparent digestibility

Apparent digestibility of each nutrient was calculated by measuring the feed intake and feces excreted. The feed and fecal samples were analyzed for nutrient of interest using similar procedures in the previous section of chemical analysis of feed and fecal samples. The differences between the amounts of nutrient consumed and excreted in the fecal samples are the amount of nutrient digested and absorbed:

$$\text{Apparent digestibility } (\%)$$
$$= \frac{\textit{amount of nutrient consumed} - \textit{amount of nutrient excreted in faeces}}{\textit{amount of nutrient consumed}} \times 100$$

Statistical analysis

The sample size calculation for this study was established using the Resource Equation Approach.

The data for rumen fermentation and rumen microbial population were statistically analyzed using repeated measures of general linear model (GLM) procedure of [19]. It was used to analyze the parameters as affected by dietary treatments, days of sampling, and treatment × day of sampling interaction in the model. The Duncan multiple range test was used to further compare means at $P < 0.05$. The parameters for digestibility were analyzed using a one-way analysis of variance (ANOVA) using the GLM procedure of [55]. Mean differences were determined using the Duncan multiple range test at $P < 0.05$.

Table 5 The PCR primer used for quantification of rumen microorganism

Microbes	Primer		Amplicon (base pairs)	Ref
	Forward	Reverse		
General bacteria	5′-CGGCAACGAGCGCAACCC-3′	5′-CCATTGTAGCACGTGTGTAGCC-3′	130	[56, 57]
Fibrobacter succinogenes	5′-GTTCGGAATTACTGGGCGTAAA-3′	5′-CGCCTGCCCCTGAACTATC-3′	121	[56, 57]
Ruminococcus albus	5′-CCC TAA AAG CAG TCT TAG TTC G-3′	5′-CCT CCT TGC GGT TAG AAC A-3′	175	[58]
Ruminococcus flavefaciens	5′-CGAACGGAGATAATTTGAGTTTACTTAGG-3′	5′-CGGTCTCTGTATGTTATGAGGTATTACC-3′	132	[56, 57]
Methanogenic archea	5′-TTCGGTGGATCDCARAGRGC-3′	5′-GBARGTCGWAWCCGTAGAATCC-3′	140	[19]
Protozoa	5′-GCTTTCGWTGGTAGTGTATT-3′	5′-CTTGCCCTCYAATCGTWCT-3′	223	[19]

Ref References

Abbreviations
CNT: Control; OL: Olive oil; PO: Palm oil; SF: Sunflower oil; VFA: Volatile fatty acids

Acknowledgements
We would like to thank the staff members of the Institute Tropical Agriculture Laboratory, Animal Science Department Laboratory and Goat Unit Ladang 2, Universiti Putra Malaysia for their assistance.

Funding
This research was funded by RUGS (9199720) awarded by Universiti Putra Malaysia. Nur Atikah I. was a recipient of MyBrain15 Scholarship from the Malaysia Ministry of Higher Education.

Authors' contributions
Conceived and designed the experiment: MI, ARA, AAS. Provide supervision for animal health assessment during experiment: AAS, HY. Performed the experiment: NAI. Analyzed the data: NAI, MI, MFJ. Data interpretation and scientific discussion: NAI, ARA, AAS, MFJ, HY, NA. Contributed reagents/materials: AAS, ARA, HY, NA. Writing the manuscript: NAI, AAS. All authors have read and approved the final manuscript.

Consent for publication
Not applicable.

Competing interests
The authors declare that they have no competing interests.

Author details
[1]Institute of Tropical Agriculture, Universiti Putra Malaysia, UPM, 43400 Serdang, Selangor, Malaysia. [2]Department of Animal Science, Faculty of Agriculture, Universiti Putra Malaysia, UPM, 43400 Serdang, Selangor, Malaysia.

References
1. Mosoni P, Chaucheyras-Durand F, Béra-Maillet C, Forano E. Quantification by real-time PCR of cellulolytic bacteria in the rumen of sheep after supplementation of a forage diet with readily fermentable carbohydrates: effect of a yeast additive. J Appl Microbiol. 2007;103(6):2676–85.
2. Chanthakhoun V, Wanapat M, Kongmun P, Cherdthong A. Comparison of ruminal fermentation characteristics and microbial population in swamp buffalo and cattle. Livest Sci. 2012;143(2):172–6.
3. Ørskov ER. Protein nutrition in ruminants. (London) Ltd: Academic Press Inc; 1982.
4. Benchaar C, Romero-Pérez GA, Chouinard PY, Hassanat F, Eugene M, Petit HV, Côrtes C. Supplementation of increasing amounts of linseed oil to dairy cows fed total mixed rations: effects on digestion, ruminal fermentation characteristics, protozoal populations, and milk fatty acid composition. J Dairy Sci. 2012;95(8):4578–90.
5. Kadkhoday A, Riasi A, Alikhani M, Dehghan-Banadaky M, Kowsar R. Effects of fat sources and dietary C18: 2 to C18: 3 fatty acids ratio on growth performance, ruminal fermentation and some blood components of Holstein calves. Livest Sci. 2017;204:71–7.
6. Liu SJ, Bu DP, Wang JQ, Liu L, Liang S, Wei HY, Zhou LY, Li D, Loor JJ. Effect of incremental levels of fish oil supplementation on specific bacterial populations in bovine ruminal fluid. J Anim Physiol Anim Nutr. 2012;96(1):9–16.
7. Ivan M, Petit HV, Chiquette J, Wright AD. Rumen fermentation and microbial population in lactating dairy cows receiving diets containing oilseeds rich in C-18 fatty acids. Brit J Nutr. 2013;109(7):1211–8.
8. Anantasook N, Wanapat M, Cherdthong A, Gunun P. Effect of plants containing secondary compounds with palm oil on feed intake, digestibility, microbial protein synthesis and microbial population in dairy cows. Asian-Australas J Anim Sci. 2013;26(6):820–6.
9. Kongmun P, Wanapat M, Pakdee P, Navanukraw C, Yu Z. Manipulation of rumen fermentation and ecology of swamp buffalo by coconut oil and garlic powder supplementation. Livest Sci. 2011;135(1):84–92.
10. Mao HL, Wang JK, Zhou YY, Liu JX. Effects of addition of tea saponins and soybean oil on methane production, fermentation and microbial population in the rumen of growing lambs. Livest Sci. 2010;129(1):56–62.
11. Patra AK, Kamra DN, Agarwal N. Effect of plant extracts on in vitro methanogenesis, enzyme activities and fermentation of feed in rumen liquor of buffalo. Anim Feed Sci Technol. 2006;128(3):276–91.
12. Wales WJ, Kolver ES, Thorne PL, Egan AR. Diurnal variation in ruminal pH on the digestibility of highly digestible perennial ryegrass during continuous culture fermentation. J Dairy Sci. 2004;87(6):1864–71.
13. Ørskov E. Recent advances in understanding of microbial transformation in ruminants. Livest Prod Sci. 1994;39:53–60.
14. Messana JD, Berchielli TT, Arcuri PB, Reis RA, Canesin RC, Ribeiro AF, Fiorentini G, Fernandes JJ. Rumen fermentation and rumen microbes in Nellore steers receiving diets with different lipid contents. Rev Bras Zootecn. 2013;42(3):204–12.
15. Jenkins TC. Lipid metabolism in the rumen. J Dairy Sci. 1993;76(12):3851–63.
16. Adeyemi KD, Sazili AQ, Ebrahimi M, Samsudin AA, Alimon AR, Karim R, Karsani SA, Sabow AB. Effects of blend of canola oil and palm oil on nutrient intake and digestibility, growth performance, rumen fermentation and fatty acids in goats. Anim Sci J. 2016;87(9):1137–47.
17. Chelikani PK, Bell JA, Kennelly JJ. Effects of feeding or abomasal infusion of canola oil in Holstein cows 1. Nutrient digestion and milk composition. J Dairy Res. 2004;71:279–87.
18. Zinn RA. Influence of level and source of dietary fat on its comparative feeding value in finishing diets for feedlot steers: metabolism. J Anim Sci. 1989;67:1038–104.
19. Zhang CM, Guo YQ, Yuan ZP, Wu YM, Wang JK, Liu JX, Zhu WY. Effect of octadeca carbon fatty acids on microbial fermentation, methanogenesis and microbial flora in vitro. Anim Feed Sci Technol. 2008;146(3):259–69.
20. Harvatine KJ, Allen MS. Effects of fatty acid supplements on ruminal and total tract nutrient digestion in lactating dairy cows. J Dairy Sci. 2006;89(3):1092–103.
21. Maia MR, Chaudhary LC, Figueres L, Wallace RJ. Metabolism of polyunsaturated fatty acids and their toxicity to the microflora of the rumen. Antonie Van Leeuwenhoek. 2007;91(4):303–14.
22. Tamminga S, Van Vuuren AM, Khattab HM, van Gils LG. Further studies on the effect of fat supplementation of concentrates fed to lactating dairy cows, 3: effect on rumen fermentation and site of digestion of dietary components. Neth J Agric Sci. 1983;31:249–58.
23. Hall KL, Goetsch AL, Landis KM, Forster LA Jr, Brake AC. Effects of a fat and ground maize supplement on feed intake and digestion by cattle consuming bermudagrass hay (Cynodon dactylon). Anim Feed Sci Technol. 1990;30(3–4):275–88.
24. Ørskov ER, MacLeod NA. The determination of the minimal nitrogen excretion in steers and dairy cows and its physiological and practical implications. Br J Nutr. 1982;47(3):625–36.
25. Wallace RJ, Broderick GA, Brammall ML. Microbial protein and peptide metabolism in rumen fluid from faunated and ciliate-free sheep. Br J Nutr. 1987;58(1):87–93.
26. Scollan ND, Dhanoa MS, Choi NJ, Maeng WJ, Enser M, Wood JD. Biohydrogenation and digestion of long chain fatty acids in steers fed on different sources of lipid. J Agric Sci. 2001;136(3):345–55.
27. Tesfa AT. Effects of rape-seed oil supplementation on digestion, microbial protein synthesis and duodenal microbial amino acid composition in ruminants. Anim Feed Sci Technol. 1993;41(4):313–28.
28. Kowalczyk J, Orskov ER, Robinson JJ, Stewart CS. Effect of fat supplementation on voluntary food intake and rumen metabolism in sheep. Brit. J. Nutr. 1977;37:251–7.
29. Ikwuegbu OA, Sutton JD. The effect of varying amount of linseed oil supplementation on rumen metabolism in sheep. Brit J Nutr. 1982;48:365–75.
30. Szumacher-Strabel M, Cieślak A, Nowakowska A. Effect of oils rich in linoleic acid on in vitro rumen fermentation parameters of sheep, goats and dairy cows. J Anim Feed Sci. 2009;18(3):440–52.
31. Shingfield KJ, Chilliard Y, Toivonen V, Kairenius P, Givens DI. Trans fatty acids and bioactive lipids in ruminant milk. In: Bosze Z, editor. Bioactive

Components of Milk. New York: Springer; 2008. p. 3–65.

32. Ebrahimi M, Rajion MA, Adeyemi KD, Jafari S, Jahromi MF, Oskoueian E, Meng GY, Ghaffari MH. Dietary n-6: n-3 fatty acid ratios alter rumen fermentation parameters and microbial populations in goats. J Agr Food Chem. 2017;65(4):737–44.

33. Yabuuchi Y, Matsushita Y, Otsuka H, Fukamachi K, Kobayashi Y. Effects of supplemental lauric acid-rich oils in high-grain diet on *in vitro* rumen fermentation. Anim Sci J. 2006;77(3):300–7.

34. Adeyemi KD, Ahmed MA, Jotham S, Roslan NA, Jahromi MF, Samsudin AA, Sazili AQ. Rumen microbial community and nitrogen metabolism in goats fed blend of palm oil and canola oil. Ital J Anim Sci. 2016;15(4):666–72.

35. Finlay BJ, Esteban G, Clarke KJ, Williams AG, Embley TM, Hirt RP. Some rumen ciliates have endosymbiotic methanogens. FEMS Microbiol Lett. 1994;117(2):157–61.

36. Sutton JD, Knight R, McAllan AB, Smith RH. Digestion and synthesis in the rumen of sheep given diets supplemented with free and protected oils. Br J Nutr. 1983 May;49(3):419–32.

37. Ruiz R, Albrecht GL, Tedeschi LO, Jarvis G, Russell JB, Fox DG. Effect of monensin on the performance and nitrogen utilization of lactating dairy cows consuming fresh forage. J Dairy Sci. 2001;84(7):1717–27.

38. Manso T, Castro T, Mantecón AR, Jimeno V. Effects of palm oil and calcium soaps of palm oil fatty acids in fattening diets on digestibility, performance and chemical body composition of lambs. Anim Feed Sci Technol. 2006; 127(3):175–86.

39. Bauchart D, Legay-Carmier F, Doreau M. Ruminal hydrolysis of dietary triglycerides in dairy cows fed lipid-supplemented diets. Reprod Nutr Dev. 1990;30(Suppl2):187s–187s.

40. Gerson T, John A, King ASD. The effects of dietary starch and fibre on the *in vitro* rates of lipolysis and hydrogenation by sheep rumen digesta. J Agric Sci. 1985;105(1):27–30.

41. Oldick BS, Firkins JL. Effects of degree of fat saturation on fibre digestion and microbial protein synthesis when diets are fed twelve times daily. J Anim Sci. 2000;78(9):2412–20.

42. Palmquist DL. The role of dietary fats in efficiency of ruminants. J Nutr. 1994; 124(suppl_8):1377S–82S.

43. Bhatt RS, Soren NM, Tripathi MK, Karim SA. Effects of different levels of coconut oil supplementation on performance, digestibility, rumen fermentation and carcass traits of Malpura lambs. Anim Feed Sci Technol. 2011;164(1–2):29–37.

44. Hoover WH, Miller TK, Stokes SR, Thayne WV. Effects of fish meals on rumen bacterial fermentation in continuous Culture1. J Dairy Sci. 1989; 72(11):2991–8.

45. Choi BR, Palmquist DL, Allen MS. Cholecystokinin mediates depression of feed intake in dairy cattle fed high fat diets☆ 1. Domest Anim Endocrinol. 2000;19(3):159–75.

46. Piantoni P, Lock AL, Allen MS. Palmitic acid increased yields of milk and milk fat and nutrient digestibility across production level of lactating cows. J Dairy Sci. 2013;96(11):7143–54.

47. NRC (National Research Council). Nutrient requirements of small ruminant. 6th ed. Washington: National Academy Press; 2007.

48. AOAC (Association of Official Analytical Chemists). Official methods of analysis. 15th ed. Washington: Association of Analytical Chemists; 1990.

49. Van Soest PV, Robertson JB, Lewis BA. Methods for dietary fibre, neutral detergent fibre, and nonstarch polysaccharides in relation to animal nutrition. J Dairy Sci. 1991;74(10):3583–97.

50. Malaysian Standard M.S. 1500. Halal food – production, preparation, handling and storage – general guidelines (second revision). Kuala Lumpur: Department of Standards Malaysia; 2009.

51. Cottyn BG, Boucque CV. Rapid method for the gas-chromatographic determination of volatile fatty acids in rumen fluid. J of Agricult and Food Chem. 1968;16(1):105–7.

52. Parsons TR, Maita Y, Lalli CM. A manual of chemical and biological methods for seawater analysis. Oxford: Pergamon Press; 1984.

53. Navidshad B, Liang JB, Jahromi MF. Correlation coefficients between different method of expressing bacterial quantification using real time PCR. Int J Mol Sci. 2012;13(2):2119–32.

54. Staroscik A. Calculator for determining the number of copies of a template. URI Genomics & Sequencing Center. 2004. https://cels.uri.edu/gsc/cndna.html. Accessed 10 Jan 2014.

55. SAS Institute Inc. SAS User's Guide, Version 9.1. 2nd ed. Cary: SAS Institute Inc; 2003.

56. Samsudin AA, Wright AD, Al Jassim R. The effect of fibre source on the numbers of some fibre-degrading bacteria of Arabian camel's (*Camelus dromedaries*) foregut origin. Trop Anim Health and Prod. 2014;46(7):1161–6.

57. Denman SE, McSweeney CS. Development of a real-time PCR assay for monitoring anaerobic fungal and cellulolytic bacterial populations within the rumen. FEMS Microbiol Ecol. 2006;58(3):572–82.

58. Koike S, Kobayashi Y. Development and use of competitive PCR assays for the rumen cellulolytic bacteria: *Fibrobacter succinogenes, Ruminococcus albus and Ruminococcus flavefaciens*. FEMS Microbiol Lett. 2001;204(2):361–6.

22

Esophageal groove dysfunction: a cause of ruminal bloat in newborn calves

Tamirat Kaba[1*], Berhanu Abera[2] and Temesgen Kassa[2]

Abstract

Background: Esophageal groove dysfunction is one of the major causes of ruminal bloat. This condition is fatal in new born calves if it is not treated early. In healthy, suckling calves, milk should bypass the forestomach (rumen and reticulum) and enter into the abomasum where enzymatic digestion of milk proteins takes place. However, failure of the esophageal groove allows milk to enter into the forestomach, which results in the production of excess gases by microbial fermentation. Consequently, this increase in abdominal distention particularly on the left side in ruminants is an imminent manifestation of excess gases in the foresomach.

Case presentation: A 10-day-old crossbred male calf presented with a distended left abdomen and manifesting dyspnea at a dairy farm. The calf was weak, reluctant to move, and had visibly congested mucus membranes. Regarding the calf's feeding, milk was the only thing ingested and the calf had not started on dry feeds (hay, concentrates, and roughages). According to the herdsman of the farm, the calf had a mild-to-moderate form of bloat and 3–5 h after milk feeding the bloat would disappear spontaneously. During bloat, an increase in pulse rate, respiratory rate (tachypnea), and shallow breathing was noted. Physical examination revealed severe distention of the left side of the abdomen, and on percussion, accumulation of gases mixed with fluid in the left abdomen was detected. An attempt was made to release gases from forestomach by introducing a stomach tube with oral antibiotics; however, the case was not resolved. The calf suffered from frequent recurrence of bloat after every milk feed, and in response to the refractory outcome to conventional treatment, a rumenostomy was indicated and a better treatment response was achieved. In addition, IV fluid and other supportive therapy were provided while milk was withheld. However, considering the fact that milk is a natural feed that should not be taken away from every calf at this age, we had to encourage calf to consume milk as it would not result in bloat as far as rumen fistula is being created. Furthermore, encouraging calves to consume starter feed (fresh grasses and hay) earlier than usual recommended period whilst decreasing milk intake would hasten the rumen function.

Conclusions: Cases like this are successfully managed by a rumenostomy when conventional options fail.

Keywords: Esophageal groove, Ruminal bloat, Rumenostomy

Background

Calf mortality has been an important problem in the dairy industry for more than a hundred years, and the causes of death are multifactorial, from environmental and infectious agents, to host phenotype. Although the knowledge about neonatal diseases in calves has increased in recent years, the mortality rate is still rather high [1]. One of the major reasons for high mortality rate among neonatal calves is ruminal bloat. Ruminal bloat occurs when gas produced during fermentation builds up in the rumen and is unable to escape. It is usually a secondary problem in newborn calves. Ruminal bloat can become life threatening within a few hours and often requires medical attention [2]. The most common cause of ruminal bloat in calves that solely consume milk, is failure of esophageal groove closure [3]. For the first two weeks after birth, a calf is monogastric, a simple stomached animal, using only the abomasum to digest the milk or milk replacer. When the calf suckles milk, milk bypasses the rumen and reticulum to enter into the abomasum, where digestion and absorption takes place. Milk entering into the rumen and reticulum

* Correspondence: kabatamirat@yahoo.com
[1]College of Veterinary Medicine, Haramaya University, P.O.Box-138, Dire Dawa, Ethiopia
Full list of author information is available at the end of the article

is both wasteful and dangerous to the newborn calf; hence the importance of the esophageal groove in diverting milk from the esophagus into the abomasum [4].

The physiology of esophageal groove closure was studied by many scholars. For instance, in 1826, Tiedeman and Gmelin were the first workers to report that milk passed directly to the abomasum in young lambs and calves [5]. Since then, a couple of physiologists had investigated what triggers the closure of the esophageal groove and have come up with very diverging opinions. Colin [6] concluded that the "esophageal groove closed when boli was swallowed during rumination and that this was the main route for passage of solid matter from the rumen to the omasum and abomasum". Schalk and his colleague [7], and a non-peer reviewed compilation Costello, [8] and Wise [9] suggested that if calves suckled milk from a rubber nipple it usually passed into the abomasum, while it often passed to the rumen if it was drunk from a bucket. They concluded that the closure of the esophageal groove is trigged when the calf directly suckled the milk from a dam. Larry [10] stated that the esophageal groove closure depends upon the liquid ingested which stimulates the nerve receptors in the mouth. Other studies suggest that esophageal groove closure and dilatation of the omaso-abomasal canal is initiated by the stimulation of the vagus nerve through contact with sensory receptors in the oral cavity and pharyngeal area [11]. Gradually (after a few weeks of weaning), this response fades so that the groove is no longer functional. Dysfunction of the esophageal groove results in leakage of fluid into the forestomach. Spillage from the esophageal groove may result from either a complete failure of groove closure or sequential opening and closure during drinking. According to Gentile [12] pathological conditions (diarrhea, phlebitis of jugular vein, cough, otitis and anorexia), irregular feeding (irregular feeding times, forceful feeding, bucket feeding of milk, abnormal milk temperature) and stress factors (long distance transportation) are some of the causes of esophageal groove dysfunction. However, many studies [3, 8, 11] indicate that esophageal groove dysfunction is unusual in calves that suckle directly from the dam. This report presents a single clinical case of ruminal bloat associated with a putative esophageal groove dysfunction in a 10-day-old calf. We believe that the therapeutic intervention made in the field was a better management approach in respect to the area's lack of facilities to conduct a laboratory investigation.

Case presentation

A 10-day-old male crossbred (Frisian x local indigenous) calf presented with a severely distended abdomen (Fig. 1). Due to the distention the paralumbar fossa, especially on the left, was not visible. The calf was reluctant to suckle

Fig. 1 A Photo showing ruminal bloat in 10 days old crossbred calf

from the dam, unable to walk, exhibited rapid and shallow breathing, and had visibly congested mucus membranes. Percussion of the left abdomen revealed a drum-like gaseous sound. On auscultation of the left abdomen, a dull fluid sound was detected. The anamnesis indicated that the calf had been dribbling urine continuously, unable to defecate, or had irregularly voided very little, hard, and pasty feces. General physical examination revealed no esophageal obstruction, but the calf was weak and with an abnormal gait. The calf was suckling its dam twice in a 12 h interval (at morning and evening) and had not started feeding the hay/roughage/concentrate or the calf starter at the moment. The physiological parameters of the calf were as follows: Rectal temperature = 39.8 degree Centigrade (°C), Pulse =175 beats/minute, Respiration =60 breaths/minute.

Differential diagnosis

Abomasal bloat and choke.

Treatment approach

In order to release trapped gases and check the patency of the esophagus, a flexible stomach tube coated with mineral oil was inserted into the esophagus, and advanced down into the rumen. A fermented watery-like fluid accompanied by some clots of milk and gases was released from the rumen through the stomach tube. Procaine penicillin (Pen Aqueous; Zoetis Canada), 10 ml (ml), 10,000 international unit per milliliter (Iu/ml of solution) mixed with 0.25Liter (L) of mineral oil was administered orally for 3 days, while milk was withheld to reduce the microbial burden and coalescence of gas. An isotonic solution containing 0.9% Sodium Chloride (Jiangsu HFQ Bio-Technology Co., Ltd), 8.4% Sodium Bicarbonate (Vet One, Nova-Tech, Grand Island, USA) and 5% Dextrose in water 1000 ml injection (Addis Pharmaceutical factory) was administered intravenously (IV) at a rate of 100 ml/kilogram (kg) over 3–5 h for 2 days. Before administration of IV fluid, the calf was sedated using Xylazine hydrochloride, 20 mg/ml (xylazine® immunological LTD, Hyderabad, India) intramuscularly (IM). This was administered during every fluid therapy, and the calf was tied up with rope in a lateral recumbent position. The hair around the jugular groove of the neck was clipped and the area was cleaned and disinfected using diluted 70% Ethanol (Addis Pharmaceutical factory). The superficial jugular vein was catheterized using 20 Gage, 0.8 in. butterfly catheter (Unolok, Hindustan syringe, Medical device LTD Faridabad, India) and secured with adhesive tape around the neck.

Response to treatment

After 3 days of treatment, bloat reoccurred. Treatment was initiated a second time by giving antibiotic pen strep (Pen & Strep®, 100 ml, York Vet, USA): 5 ml, (IM), every 24 h (q24hrs) for 2 days while the calf had been fastening. Additional supportive therapy of 40% glucose (100 ml/kg/day IV), isotonic saline solution (10 ml/kg/hrs IV) and a multivitamin (Multivitamin injection 100 ml, Norbrook Laboratories Limited, Ireland), was administered 10 ml IM once at a time (Stat.) during the time that the milk was withheld. After 2 days of treatment, the calf was allowed to suckle milk from the dam; however, the calf exhibited bloat again 5 h after milk consumption.

Rumenostomy

Ruminal fistulation (rumenostomy) was conducted to prevent recurrence according to a procedure described by Turner and Mcilwraith [13]. Before the surgical procedure milk was withheld from the calf overnight while IV fluids and glucose were administered at the dose rate explained above. The left paralumbar fossa was prepared by shaving the hair and washing skin aseptically using 7.5% povidone-iodine surgical scrub (Povidone-iodine cleansing solution, Wockhard LTD, Mumbai, India) while the calf was standing. A circular area of 6 cm (cm) in diameter just below the transverse process of the lumbar vertebrae was marked and infiltrated with local anesthetic, 2% Lidocaine (Zoetis Canada, Kirkland,Quebec), at the concentration of 20 mg per milliliter (mg/ml). Approximately a 2 cm diameter circular incision was made to remove the skin. After skin removal the abdominal muscles were dissected bluntly to expose the rumen. The rumen was grasped using sponge forceps and pulled to the exterior. The rumen wall was then tacked to the edge of the skin by four horizontal mattress sutures at "quarter hour" positions (12, 3, 6 and 9 o'clock). These sutures acted as stay sutures using a non-absorbable suture (Sofsilk™ 6–0 Black,

Medtronic, USA). The rumen wall was incised carefully at one half centimeter from the wound margin/apposing skin. As the contents of the rumen came out during the procedure, we observed a high amount of milk that had entered into the rumen (Fig. 2).

Post Rumenostomy management

The calf was separated from the herd for 10 days to maintain close observation. Since rumenostomy is considered a clean contaminated surgery, we had to give parenteral antibiotic, penstrep, (Penstrep-400, Metaalweg, 85,804 CG Venray, Netherlands) 5 ml for 4 days, q24hrs IM to reduce the risks of peritonitis. A dexamethasone injection at 2 mg/ml (Sparhawk laboratories Inc., Lemexa, KS66215, USA) was given every eight hours (q8hrs) IM, and also served as an anti-inflammatory agent. Moreover, Deltamethrin 1% (w/v) pour-on ready-for-use formulation (Appropriate Applications Ltd., USA) at a dose rate of 10 ml per 100 kg body weight was used to prevent insect infestation and miyiasis. The surgical wound was examined and monitored every day until closure for any complications such as wound dehiscence or rumen attachment to the skin. Rumen contents leaking out onto the flank area and outer surgical site were cleaned by using antiseptic solution (Chlorhexidine) and clean towels. The rumen was repeatedly flushed through the fistula with 0.5–1 l of warm tap water adjusted to the calf's body temperature. This flushing helped to prevent desiccation and was used for buffering purpose. The calf was allowed to suckle milk from its dam twice a day during the follow-up period. After 10 days post-operation, the calf was provided with some hay and fresh grasses to stimulate

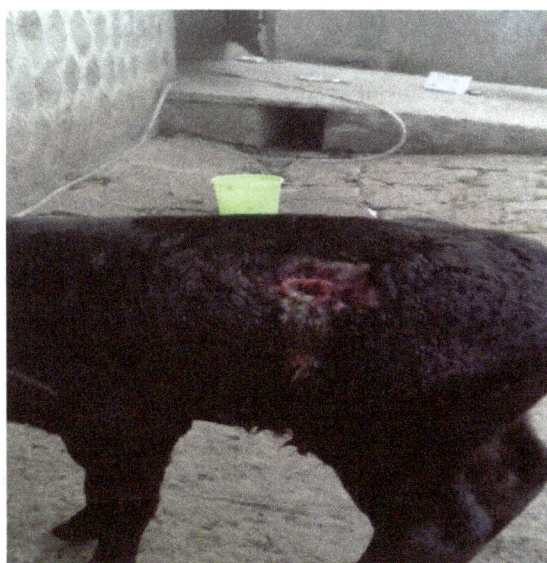

Fig. 2 A photo showing a calf with fistulated rumen (taken right after surgery)

rumen function. Bloat resolved by the time the calf started solid feeds and the wound was closed surgically just after a week of feeding grasses and hay.

Response to Rumenostomy

The calf was followed for 6 months after the procedure. Shortly after wound closure, the amount of milk that the calf was getting was reduced to encourage the intake of hay and grasses. During this time, bloat did not occur as it had been observed prior to surgery. Although ruminal contents spilled onto the flank post-operatively, this did not appear to upset the calf, and its general condition improved gradually. We recommended that the owner reintroduce the calf with the existing herd 6 months after the surgery, and advised the owner to inform us of any observable complications. We promised the owner that we would visit the calf at one year; however, the owner had sold the calf at 9 months of age to a beef farmer in another area of the country.

Discussion and conclusion

The occurrence of bloat in calves that have started on a hay/grass/concentrate or a calf starter is not a new phenomenon, however, in newborn calves that are only suckling milk, it is unusual [14]. According to some studies, [3, 12] ruminal bloat is a secondary consequence to esophageal groove dysfunction in calves at this age. Esophageal groove dysfunction is the major cause of ruminal bloat in newborn calves that are directly suckling its dam, or could be the result of overfeeding concentrate feed [15]. The later cause of ruminal bloat doesn't seem to be appearing a factor for this case because the calf wasn't turned onto concentrate feed at the moment, and was suckling only the milk from its dam. We were confronted with a paradox justification because in some studies [3, 4, 8, 12] esophageal groove dysfunction in calves directly suckling their dam is not common, but failure of the esophageal groove can occur when calves drink cold milk, are tube fed, or fed from a bucket [16]. In normally functioning esophageal grooves, milk should bypass the rumen and reticulum [17]. The presence of milk in the rumen was confirmed during surgery, when the ruminal contents were observed and a high quantity of fermented milk and milk clots were noted. Abomasal bloat can be a differential diagnosis for this case, however, there were certain things that set ruminal bloat apart from abomasal bloat. For instance, in ruminal bloat abdominal distention is higher on the left side [3], which was seen in this case. In addition, during ruminal bloat rumen contents can easily be released out with the help of a stomach tube whereas, in abdominal bloat it is difficult to introduce a stomach tube into the abomasum and flush its contents out while the animal is in a standing position [2]. When we manipulated the stomach tube

while the calf was standing, we were confident that the content was coming out from the forestomach. In the case of choke, bloat must accompany drooling of saliva [3] and recurrences should not have occurred after checking the patency of esophagus using the stomach tube.

The largest problem that we have failed to demonstrate was the underlying cause or factor for esophageal groove failure. Gentile [12] reported that pathological conditions (diarrhea, otitis, phlebitis, vagus nerve problem, etc.), inadequate feeding technique (irregular feeding time, bucket feeding, very cold milk feeding etc) and stress are some of the causes of esophageal groove dysfunction. In our investigation, the calf wasn't exhibiting diarrhea or any other gross pathological conditions except abdominal distention. Furthermore, we also investigated the feeding technique of the calf and realized that suckling was the only feeding technique and it was regular, twice everyday (12 h interval of milk feed per day).

Bloat in older animals is associated with grazing legumes in legume-dominant pastures, feeding high-grain diets, and impaired eructation processes [18]. Despite the primary cause of bloat being multifactorial, it's clear that the esophageal groove is not functional in those animals as it regresses when they start solid feeds [19]. Hence, "esophageal groove dysfunction" cannot be an ideal term to use to describe bloat in older animals. Apart from other treatment protocols, several scholars [20–22] suggest that rumenostomy is a therapeutic option for animals with recurrent or non-resolving bloat in young or older animals. Amanda et al. 2015 [23] mentioned that of 42 rumenostomy treated cases, 20 cases were indicated for bloat. According to the authors, half of the calves were followed for long periods in the herd and they had better health conditions until they were culled. While the primary associated factor of esophageal groove failure is unclear, the presence of milk clot and fermented fluid in high amounts in the rumen at an early age suggests a malfunction in the normal physiology of the esophageal groove.

To the best of our knowledge, case like this has never been reported so far in naturally suckling calves. As treatment intervention, withholding milk whilst giving IV fluid would give temporary relief. However, considering the fact that milk is a natural feed that should not be taken away from every calf at this age, we rather encourage calves to consume milk as it would not results in bloat as far as rumen fistula is being created. Furthermore, encouraging calves to consume starter feed (fresh grasses and hay) earlier than usual recommended period whilst decreasing milk intake would hasten the rumen function in those calves. Therefore we concluded that esophageal groove dysfunction should be suspected when severe and recurrent bloat occurs in calves that consume only milk by suckling. Nevertheless, since we did not investigate the underlying cause, detailed study on the primary causes of esophageal groove dysfunction in young calves should be encouraged. We also found that rumenostomy is a better management option over conservative approaches in similar clinical cases. Despite rumenostomy considered a better option, it degrades the appearance and the value of the animal, and we suggest additional studies on alternative treatment methods.

Abbreviations
Im: Intramuscular; Iu: International unit; Iv: Intravenous; q24hrs: every 24 hours; q8hrs: every 8 hours; Stat.: Once at a time; *W/V*: weight per volume

Acknowledgements
We would like to thank Slagel Clare, C. for devoting her time in editing language and technical aspects.

Authors' contributions
T Kassa has recorded all the information of case history, diagnosis of the case and treatment intervention, BA involved post operative management and follow up of the case; T Kaba compiled data for write up of the manuscript. All authors read and approved the final manuscript.

Consent for publication
The owner gave informed consent for publication.

Competing interests
The authors declare that they have no competing interests.

Author details
[1]College of Veterinary Medicine, Haramaya University, P.O.Box-138, Dire Dawa, Ethiopia. [2]Ethiopian Institutes of Agricultural Research, Holleta Agricultural Research Center, P.O. Box: 2003, Holleta, Ethiopia.

References
1. Assefa K, Mamo G, Goshu G, Regassa F. A retrospective study on calf mortality in Wolaita Soddo Jersey cattle breeding and multiplication center, Wolaita Soddo, Southern Ethiopia. J Biol Agric Healthc. 2014;4:63–7.
2. Stock R, Rasby RJ, Rice D. Bloat Prevention and Treatment. Historical Materials from University of Nebraska-Lincoln Extension. 1974; Paper 251.
3. Andrews AH, Blowey RW, Boyd H, Eddy RG. Bovine medicine: diseases and husbandry of cattle. In: Blowey RW (eds). Digestive disorders of calves. Wiley-Blackwell: United Kingdom; 2004. p. 231–5.
4. Costello R. Bloat in young calves and other pre-ruminant livestock, a subsidiary of merrick animal nutrition, 2012, available at http://www.merricks.com. Accessed 18 Apr 2016.
5. Orskov ER. Reflex closure of the esophageal groove and its potential application in ruminant nutrition. S Afri J Anim Sci. 1972;2:169–76.
6. Colin G. Traite de Physiologie Comparee des Animaux. 3rd ed. Paris: Vol. I; 1886.
7. Schalk AF, Amadon RS. Physiology of ruminant stomach. North Dakot Agricultural Experiment Station Bulletin 1928, No. 216.
8. Costello R. Bottles vs. pails: are there differences between calf feeding methods? A subsidiary of merrick animal nutrition, 2010, available at http://www.merricks.com. Accessed 10 Apr 2016.
9. Wise GH, Petersen WE, Gullickson TW. Inadequacy of a whole milk ration for dairy calves as manifested in changes of blood composition and in other physiological disorders. J Dairy Sci. 1939;22:559–72.
10. Larry RE. Review of veterinary physiology. 5th ed. USA: Teton, New media, South Hwy; 2008.
11. Comjine RS, Titchen DA. Reflex contraction of the esophageal groove in young ruminants. J Physiol. 1951;115:210–6.
12. Gentile A. Ruminal acidosis in milk-fed calves. Large Animal Vet Rounds. 2004;4:9.
13. Turner AS, Mcilwraith CW. Technique in large animal surgery. 2nd ed. London: Lippincott Williams and Wilkins; 1989.
14. Orskov ER, Benzie D, Kay RNB. The effects of feeding procedure on closure of esophageal groove in young sheep. Br J Nutr. 1970;24:785–95.
15. Wenham G, Robinson JJ. Colostrums by stomach tube. Vet Rec. 1979;104:99.
16. Lateur-Rowet HJM, Breukink HJ. The failure of the esophageal groove reflex when fluids are given with an esophageal feeder to newborn and young calves. Vet Quart. 1983;5:68–74.
17. Jim Q. Rumen acidosis and rumen drinking in milk-fed calves. Calf notes, 2005. Available at http://www.calfnotes.com. Accessed 3 Mar 2016.
18. Meyer NF, Bryant TC. Diagnosis and management of rumen acidosis and bloat in feedlots. Vet Clin North Am Food Anim Pract. 2017;33:481–98.
19. Braun U, Brammertz C. Ultrasonographic examination of the oesophageal groove reflex in young calves under various feeding conditions. Schweiz Arch Tierheilkd. 2015;157:457–63.
20. Haskell SRR. Rumen fistula surgery for the private practitioner. Retrieved from the University of Minnesota Digital Conservancy. 2002. http://hdl.handle.net/11299/108760. Accessed 12 Mar 2017.
21. Chigerwe M, Tyler WJ, Dawes ME, Nagy DW, Schultz LG, Luby CD, et al. Enteral feeding of 3 mature cows by rumenostomy. J Vet Intern Med. 2005;19:779–81.
22. Callan RJ, Applegate T. Temporary rumenostomy for the treatment of forestomach diseases and enteral nutrition. Vet Clin North Am Food Anim Pract. 2017;33:525–37.
23. Amanda KH, Andrew JN, Marjolaine R, Rebecca LP, Matt DM, David EA. Indications for and factors relating to outcome after rumenotomy or rumenostomy in cattle: 95 cases (1999-2011). J Am Vet Med Assoc. 2017; 247:659–64.

23

Diversity of indigenous sheep of an isolated population

Caroline Marçal Gomes David[1][*] (ID), Celia Raquel Quirino[1], Wilder Hernando Ortiz Vega[1], Aylton Bartholazzi Junior[1], Aparecida de Fátima Madella-Oliveira[2] and Ricardo Lopes Dias Costa[3]

Abstract

Background: Because of the influence of genetics on animal production and the risk of losing genetic diversity of naturally adapted breeds, this study evaluated the genetic diversity of sheep of the Morada Nova breed belonging to an animal science institute in Brazil. The herd in question is one of the country's most representative of the breed. Samples of DNA extracted from the plasma of 61 animals were used for later analysis of the genotypes using microsatellite molecular markers.

Results: The polymorphic information content was 0.66, the observed heterozygosity was 0.65 and the fixation index was 0.048. According to the results, there is moderate genetic diversity in the studied population, suggesting the implantation of breeding programs aimed at conservation of the observed genetic diversity.

Conclusion: The results obtained in this study will be of great importance to decisions on herd structure, besides contributing to other work to be carried out at the research center.

Keywords: Genetic improvement, Molecular biology, *Ovis aries*, Polymorphism

Background

Morada Nova is one of the main hair sheep breeds in Brazil. They are small animals that stand out among naturalized sheep breeds sheep due to the good production of meat and hides. Because of traits such as high prolificacy [1], parasite resistance [2] and adaptability to edaphoclimatic conditions, this breed is considered useful for crossbreeding, in particular in industrial breeding programs with breeds having higher zootechnical performance. Morada Nova herds are mostly found Brazil's northeastern semi-arid region, where they are one of the main sources of income of smallholders [3].

Currently, small herds of Morada Nova sheep can also be found in the Southeast and Midwest of Brazil, populations are distributed mainly in research centers, most frequently in São Paulo state. The Instituto de Zootecnia in the municipality of Nova Odessa has one of the main herds representative of the breed, assisting and monitoring producers them in addition to developing research in areas like nutrition, health, reproduction and breeding, among others. Since genetics exerts an influence on production indexes, either positive or negative, so the best use of this potential depends on selection, which in turn requires a diversified number of alleles that can favor larger numbers of combinations.

In addition to support for genetic improvement programs, genetic diversity is a determining factor for the conservation of genetic resources and the evolutionary persistence of species. Naturally adapted breeds may be a source of genetic material able to improve resistance of other breeds to unfavorable conditions [4]. The ability of a population to respond adaptively to environmental changes depends on its level of variability or genetic diversity [5, 6].

Genetic diversity of indigenous breeds is a major concern considering the need to preserve what may be irreplaceable richness regarding new productive demands [7]. A species without enough genetic diversity is thought to be unable to cope with changing environments or evolving competitors and parasites [8]. Conservation should be based on thorough knowledge of the genetic resources of the specific breed [9]. Therefore, it is important to genetically characterize indigenous

* Correspondence: carolinedavid.mg@gmail.com
[1]Universidade Estadual do Norte Fluminense, Centro de Ciências e Tecnologias Agropecuária Laboratório de Reprodução e Melhoramento Genético Animal, 2000, Alberto Lamego Ave, Campos dos Goytacazes, Rio de Janeiro 28016-811, Brazil

breeds [9], for which the application of molecular genetics has many important advantages [5].

The aim of obtaining higher animal production rates is impaired by loss of diversity and the introduction of exotic breeds in crosses carried out indiscriminately, modifying the population structure and compromising a breed's existence. Genetic depression due to inbreeding is also an extremely important factor, also impairing herd performance [10].

To maintain diversity, studies need to be conducted to learn the genetic structure of populations. Microsatellite markers are used to estimate parameters of genetic diversity and herd structure in genetic resource conservation programs due to their dominance and high sensitivity [11].

Considering this, the objective of this work was to evaluate the genetic diversity of sheep of the Morada Nova breed belonging to the Instituto de Zootecnia in Nova Odessa, São Paulo. The results obtained in this study will serve as an aid to future research and conservation of the breed.

Results

Of the 15 microsatellite markers tested, just INRA63 did not amplify loci and 14 were polymorphic, for a total of 100 detected alleles. We did not observe the presence of null alleles, that is, all samples amplified the 14 microsatellite loci tested. The mean number of alleles (Na) was 7.14 and the mean number of effective alleles (Ne) was 3.68. The locus ILSTS08 presented the lowest polymorphism with two alleles and Ne of 1.44, while locus OARFCB304 was the most polymorphic, with Na 11 and Ne of 7.728.

According to the polymorphic information content (PIC) test, all markers presented PIC above 0.5, considered highly informative, except for the OARAE129 marker, which presented a low PIC of 0.19.

Of the 14 microsatellites tested, seven were in Hardy Weinberg equilibrium with significant p-values: $p < 0.001$ (ETH152, INRA06, OARFCB304); $p < 0.01$ (MAF65, OARAE129); and $p < 0.05$ (CSRD24, ILSTS87). Another seven were not in equilibrium, with $p > 0.05$ (ILSTS08, INRA05, INRA172, MCM42, MCM527, OARCP49 and OARFCB20).

In the evaluation of allelic richness (number of different alleles for the same genome region), the mean value of all markers in this population was 4.51, showing the long-term potential of adaptation and persistence of the population. When observed individually by marker, the lowest values of allelic richness were for the loci ILST08 (1.95), OAR129 (2.168) and MCMX42 (3.49), the highest value was in the OAR304 locus (7.06), and the other markers obtained medium values ranging from 4.15 to 5.30.

The mean observed heterozygosity (Ho) in the population was 0.65 while the expected (He) was 0.66. The diversity estimates applied for each locus studied are presented in Table 1.

The fixation index (F) reflects the probability that two alleles within the same individual are identical in offspring. Six of the 14 markers tested had negative indices, that is, they were not fixed in the population. For the other markers, two presented moderate indices (ETH152, ILSTS08) and one a high index (OARAE129). The overall mean considering all the microsatellite markers was low (F = 0.048).

The 14 microsatellite loci analyzed showed 91 possible combinations, a total of 16 combinations of markers, the genotypes in one locus are not independent of the genotypes in the other locus. Of the 16 combinations in linkage disequilibrium, only one combination was syntenic (Table 2).

The statistical cluster analysis by ascending hierarchical ordering, grouped the 61 animals into two clusters (Fig. 1).

Based on the Evanno method for identification of the ideal K in the analysis of population structure (Fig. 2), the individuals were grouped into two populations (K = 2) from the 14 microsatellite markers used in the analysis performed with the Structure software, based on the delineation of clusters of individuals in relation to their genotypes (Fig. 3).

Table 1 Mean number of alleles (Na), effective number of alleles (Ne), observed (Ho) heterozygosity and expected (He), Fixation index (F), allele richness (AR)

Locus	Na	Ne	Ho	He	F	AR
CSRD247	7	3.935	0.721	0.746	0.033	5.04
ETH152	6	4.309	0.656	0.768	0.146	5.19
ILSTS08	2	1.441	0.246	0.306	0.196	1.96
ILSTS87	8	4.731	0.820	0.789	−0.039	5.13
INRA05	10	4.678	0.787	0.786	−0.001	5.30
INRA06	7	4.064	0.705	0.754	0.065	5.04
INRA172	9	2.988	0.705	0.665	−0.060	4.96
MASF65	8	3.832	0.705	0.739	0.046	4.51
MCM42	7	2.612	0.672	0.617	−0.089	3.49
MCM527	7	2.967	0.672	0.663	−0.014	4.15
OARAE129	3	1.265	0.131	0.210	0.375	2.17
OARCP49	8	3.585	0.656	0.721	0.091	4.68
OARFCB20	7	3.510	0.787	0.715	−0.100	4.53
OARFCB304	11	7.728	0.852	0.871	0.021	7.06
Mean	7.14	3.68	0.65	0.66	0.048	4.51
SE	0.63	0.42	0.05	0.05	0.034	1.20

Table 2 The linkage disequilibrium between pairs of microsatellites markers

Locus#1	Locus#2	P-Value	SE	LD	Synteny
ETH152	MAF65	0.0000	0,0000	+	N
INRA06	MCM42	0.0000	0,0000	+	N
INRA172	MCM527	0.0028	0,0021	+	N
INRA172	MAF65	0.0029	0,0021	+	N
OARCP49	OARFCB20	0.0037	0,0025	+	N
MCM527	OARFCB20	0.0079	0,0045	+	N
MAF65	MCM527	0.0110	0,0054	+	N
INRA06	MAF65	0.0116	0,0059	+	N
INRA172	OARCP49	0.0131	0,0082	+	N
INRA05	MCM42	0.0162	0,0059	+	N
MCM42	OARFCB30	0.0167	0,0077	+	N
ETH152	OARAE129	0.0189	0,0059	+	S
INRA05	OARFCB30	0.0203	0,0104	+	N
ETH152	ILSTS87	0.0246	0,0097	+	N
CSRD247	OARFCB30	0.0345	0,0157	+	N
MAF65	OARFCB20	0.0371	0,0152	+	N

P-Value significance = 0,05. *SE* standard error, *N* Non-syntenic, *S* Syntenic

Discussion

Microsatellite markers are commonly used in paternity tests and are considered an excellent alternative 22 for breed profiling [11, 12], genetic diversity, breeding and conservation programs, [13, 14]. The use of these markers is considered to be reliable because they have characteristics such as high polymorphism, good specificity, high mutation rate and wide genome distribution, besides being easy to analyze [14].

The inter simple sequence repeat (ISSR) technique also presented highly polymorphic loci and has been used previously in genetic diversity studies [9]. These markers are useful to find markers associated with major and minor genes controlling important traits. Several studies have been conducted on associations of ISSR markers with important characteristics, including sheep production traits [7, 15].

Currently, breeding values in genomic selection are generally predicted based on SNP markers (single nucleotide polymorphism). These have also been used in analyses of genetic diversity [16]. There are more than 54,000 validated SNPs for sheep breeds of major economic interest, including Brazilian breeds such as Morada Nova and Santa Inês (Illumina, San Diego, CA). According to Fischer [17], the potential of SNPs to describe the genetic structure of the population depends heavily on the density of the chips used. However, the high cost to purchase these chips restricts research in countries that have low scientific investment, thus justifying preference for the use of microsatellite markers [18]. In studies using a low number of SNPs and microsatellites, the microsatellites presented similar or better results than SNPs [19, 20].

In this study, the microsatellite markers used reflected the genetic diversity evaluated in the population. The 14 markers were highly informative, with PIC above 0.50. PIC values above 0.50 are considered highly, values between 0.50 and 0.25 are moderately informative and values below 0.25 are considered slightly informative [20]. This high observed value is also an indicative parameter for the classification and selection of the microsatellite used, since these presented efficiencies in the detection of polymorphisms. The locus used in this study are commonly used in paternity test and genetic diversity in sheep and are recommended by the International Society for Animal Genetics (ISAG). The linkage disequilibrium observed between the locus is probably due to the inheritance of common alleles of common lineages. According to Flint-Garcia [21], the mating system of the species (selfing versus outcrossing),

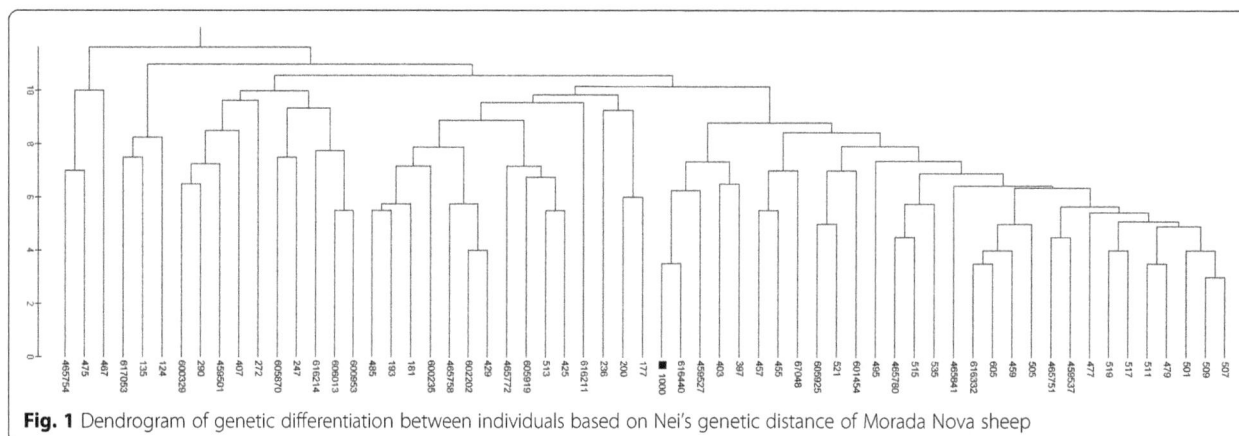

Fig. 1 Dendrogram of genetic differentiation between individuals based on Nei's genetic distance of Morada Nova sheep

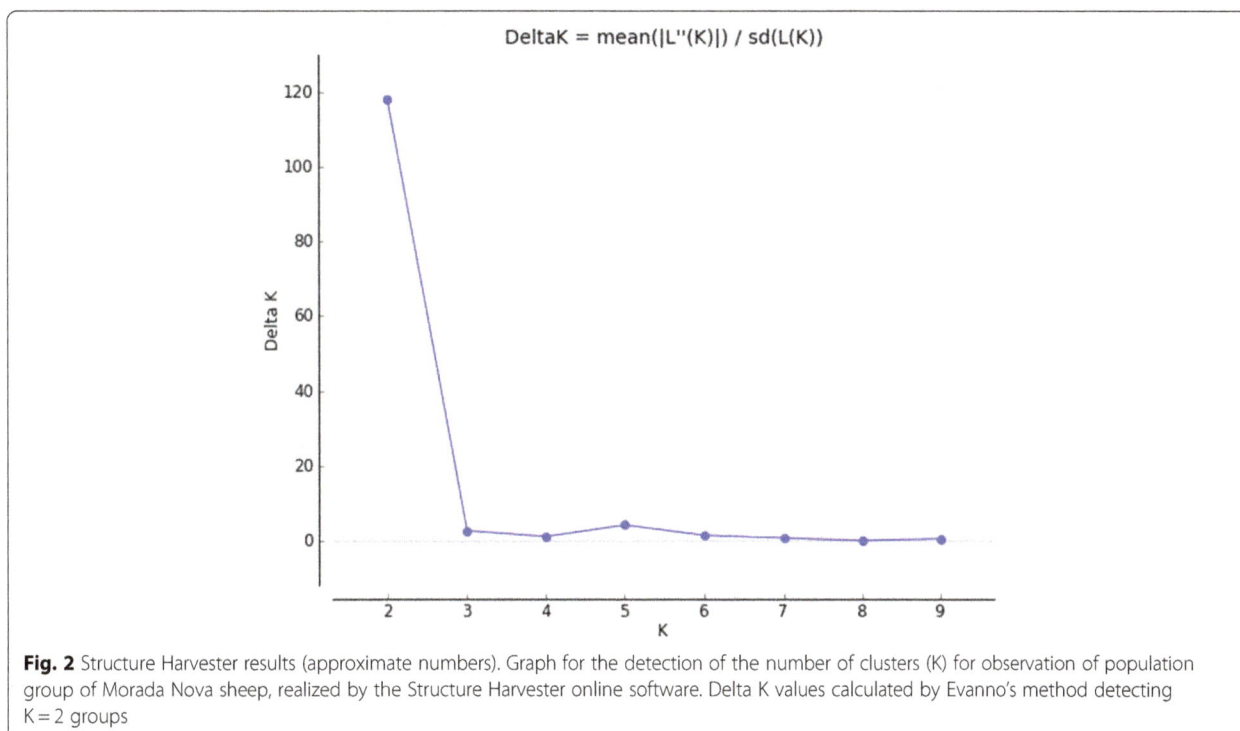

Fig. 2 Structure Harvester results (approximate numbers). Graph for the detection of the number of clusters (K) for observation of population group of Morada Nova sheep, realized by the Structure Harvester online software. Delta K values calculated by Evanno's method detecting K = 2 groups

and phenomena such as population structure can strongly influence patterns of LD.

The number of alleles within a population as one of the parameters of genetic diversity and is a major factor in animal production, since a greater number of alleles within a population allows for greater possibilities of recombination. It is common to note that in commercial or geographically isolated herds, selection pressure exerts forces contrary to diversity, leading to genetic drift. However, the herd studied, although new and composed of small sheep, had a high Na value compared to other studies conducted in different regions analyzing the same breed.

We found 100 alleles, with average number of alleles per locus of (Na) 7.14. The highest number of allele markers were in OARFCB304, with 11 alleles, and INRA05 with 10 alleles, while least number were in ILST08 and OARAE129, with 2 and 3 alleles respectively. This result indicates that even though the population studied is isolated, the Na was higher than that described in other studies evaluating Morada Nova herds. In Ceará, 89 alleles (3.28 alleles per locus)

were observed using 17 microsatellite markers in a study performed in 13 different farms [22]. Ferreira [23] used 15 microsatellites, of which, 5 were the same as those studied by us (MAF65, MCM527, INRA172, INRA05, OARF304) in the same breed in a single population, and described Na of 81 alleles and average of alleles 4.62 per locus. Understanding of the relation between the effective size and the real size of a population is of fundamental importance for planning conservation strategies. The mean number of effective alleles (Ne) was 3.68, a medium value considering the value of Na (7.14). The loci with lowest Ne values were also the lowest in the observed heterozygosity. The means were higher than those observed in five distinct populations of Kermani sheep, which presented means of Na 2.94 ± 0.23 and Ne $2.31 + 0.23$ [9]. Even though the mentioned study was carried in different populations, the number of animals was low, which showed a loss of diversity of indigenous animals. The authors emphasized that a better understanding of the potential of native species is necessary to support long-term genetic improvement.

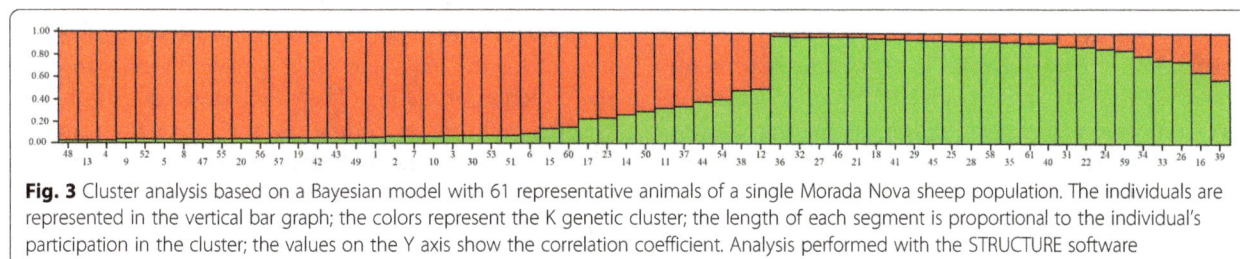

Fig. 3 Cluster analysis based on a Bayesian model with 61 representative animals of a single Morada Nova sheep population. The individuals are represented in the vertical bar graph; the colors represent the K genetic cluster; the length of each segment is proportional to the individual's participation in the cluster; the values on the Y axis show the correlation coefficient. Analysis performed with the STRUCTURE software

Our study revealed the presence of low allelic richness. The decrease of allelic richness may lead to a reduction in the population's potential for adaptation to future environmental changes, since this diversity is the raw material for evolution through natural selection [17].

The value of 4.51 identified for allelic richness in this study suggests unequal distribution of allelic frequencies with the presence of low-frequency alleles, such as ILST08 (1.96) and OAR129 (2.17). The average of allelic richness observed by GebreMichael [24] was 6.79 for animals of the same breed from different regions. The author observed phenotypic variability among populations and within populations, and attributed this variability as characteristic of traditional populations that were not subjected to a strong selection. However, that author only compared the allelic richness with values reported for other domestic sheep populations, and these values are similar to those observed in our study. On the other hand, the AR values observed by [24] in different breeds were reported to be above 8.0 for six of the seven breeds studied.

Like the AR values, the heterozygosity indices also estimate genetic variation within a population and are one of the most widely used parameters of genetic diversity [25]. Heterozygosity expresses great miscegenation and lower allele fixation. A higher proportion of heterozygous individuals than expected, according to the rate of segregation of a population, is desirable to maintain genetic diversity. The estimates of observed heterozygosity (Ho) and expected heterozygosity (He) in this study (0.65 and 0.67, respectively) represented moderate genetic diversity.

Values of Ho (0.53) and He (0.59) were attributed to the low diversity of the Morada Nova breed [22]. In that study, the authors used animals from eight farms in three different states. That result indicates that the isolated herd used in our study is representative of the breed, with moderate heterozygosity considering the number of animals studied.

Estimates of diversity in populations of sheep from Colombia presented mostly low heterozygosity, with values of Ho below the values of He, with means of 0.68 and 0.77 respectively [26]. In that study, the authors used 513 animals from 13 breeds from 56 farms, and values indicative of diversity were lower than those found in our study. The desired genetic diversity is Ho greater than He. Like the values observed in five Kermani sheep populations in Iran, which presented He of 0.56 ± 0.06 and Ho of 0.97 ± 0.12 [9]. The authors reported enormous biodiversity among domestic animals in developing countries and warned of the loss of this diversity because of the introduction of different breeds and the importance in conserving naturalized breeds.

One of the main factors that causes loss of heterozygosity is the fixation of alleles of commercial interest and mating directed to obtain higher production rates, common in industrial production. This heterozygosity loss increases with rising number of alleles in homozygosis, resulting in allele fixation. According to the FAO (1998), the estimated value through the fixation index (F) should not exceed 1%. In this study, the estimated value of F was 4.8%. The ETH152, ILSTS08 and OARAE markers presented values of 14.6, 19.6 and 37.5, respectively. These large values indicate high homozygous for these loci and possible loss of alleles. Five markers presented values below 6% and another five negative values, with these also being the loci with highest Ho. This fact can be attribuited to the lack of gene flow and rigorous mating control. However, the low number of Morada Nova sheep in this geographic region makes it hard to introduce new genes to increase the diversity.

Likewise, in a study of diversity study of Morada Nova herds located in five Brazilian states including São Paulo, the authors found F values between 4 and 12% and reported consanguinity between the herds [27]. Inbreeding among herds of different regions observed in studies shows that the introduction of animals from different herds alone may not be sufficient to increase the breed's diversity, so intensified conservation work is important.

The Bayesian group factor analysis performed by the Structure software determined the structure of the sheep population studied and presented a population structure clearly divided into two representative groups, with some individuals presenting a mixture of the two gene groups, possibly due to the miscegenation of the main cluster present. This structure is possibly due to the origin of the herd, which was formed of individuals coming from different states.

The ability of the Bayesian algorithm to detect the most probable number of clusters that explain the distribution of the genetic groups of the population studied was based on a model that correlates the allelic frequencies within the population (admixture model). This contributes to the detection of subpopulations or genetic subgroups in a single population when individuals may be particularly related and presents the same detection ability as the independent allele frequency model, such as the absence of high levels of relationship between the individuals or populations evaluated [28].

The structure of the population presented here is important for the maintenance of genetic diversity through the direction of mating. The dendrogram allows observing the animals according to their genetic distance, which facilitates the identification of the best genetic combinations to avoid endogamous mating.

Diversity is conserved by the maintenance of individuals with few genetic relationships. Knowledge about these relationships helps in the management herds with better use of existing genotypes in breeding programs, besides being fundamental in programs for conservation of genetic resources.

Conclusion

The herd of Morada Nova of the Instituto de Zootecnia presented moderate genetic diversity. Strategies to maintain or increase genetic diversity should be implemented considering the need to preserve what may be irreplaceable richness and the importance of the herd in the conservation of the breed regarding new production demands, mainly as a maternal lineage in industrial mating.

Methods
Locale and population studied
The study was carried out on 61 animals of the Morada Nova breed from the Sheep Unit of the Center for Research and Development of Diversified Animal Husbandry, of the Instituto de Zootecnia, part of the Paulista Agribusiness Technology Agency of the Department of Agriculture, located in Nova Odessa,SP (22 ° 42 'S and 47 ° 18' W).

DNA extraction and genotyping
DNA was extracted from blood plasma using the commercial NucleoSpin Blood kit (Macherey-Nagel, Düren, Germany) following the manufacturer's instructions, with subsequent quantification of DNA concentrations measured with a spectrophotometer (NanoDrop™ 2000-Thermo Science). The extraction and purification steps were performed at the Animal Genetic Improvement Laboratory of Norte Fluminense State University in Rio de Janeiro.

Genotyping was performed using a capillary sequencer (MegaBACE 1000 DNA Analysis System - GE Healthcare). We used 15 microsatellite markers recommended for identification of kinship and paternity in sheep by the International Society for Animal Genetics (ISAG): CSRD24, ETH152, ILSTS08, ILSTS87, INRA05, INRA06, INRA172, MASF65, MCM42, MCM527, OARAE129, OARCP49, OARFCB20, OARFCB304 AND INRA 63.

Statistical analysis
The GenAlEx software version 6.502 was used to evaluate population genetics, to estimate: number of alleles (Na), effective number of alleles (Ne), observed heterozygosity (Ho), expected heterozygosity (He), fixation index (F); and polymorphic information content analysis (PIC). Linkage disequilibrium was estimated for all marker pairs using GenePop software version 1.2 [29].

The distribution of genetic diversity was studied using the genetic distance analysis of Nei. For the Hardy-Weinberg equilibrium (HWE), Fisher's exact test was

applied. The population allele richness (A_R) measure of the 14 markers was estimated using the HP-RARE 1.0 software. Using Nei's genetic distance matrix with standardized data the dendrogram was constructed from the unweighted pair group method with arithmetic mean using the software MEGA-5 (Molecular Evolutionary Genetics Analysis).

The population structure was analyzed using the software STRUCTURE 2.3.4, which employs functions based on Bayesian clustering algorithms, considering a mixed ancestry model (admixture model) and correlating the allelic frequency in the population. The burn-in period was 200,000 rounds followed by 500,000 Monte Carlo Markov Chain interactions. To observe numbers of possible clusters in the population, the independent "K" test was performed from 1 to 10 clusters with 20 replicates, thus verifying the consistency of the results. The ideal "K" value was evaluated after the analysis of the result file by the Evanno method using the web-software STRUCTURE Harvester version 0.6.94.2012 software. After identification of the number of subpopulations, a test was performed to select the optimal "K" to generate the illustrative population plot.

Abbreviations
Ar: Allele richness; F: Fixation index; He: expected heterozygosity; Ho: observed heterozygosity; HWE: Hardy-Weinberg equilibrium; ISAG: International Society for Animal Genetics; ISSR: Inter simple sequence repeat; LD: Linkage disequilibrium; MEGA-5: Molecular Evolutionary Genetics Analysis; Na: Number of alleles; Ne: effective number of alleles; PIC: Polymorphic information content

Funding
This study was financed in part by the Coordenação de Aperfeiçoamento de Pessoal de Nível Superior - Brasil (CAPES) - Finance Code 001.
National Council for Technological and Scientific Development (CNPq) for the research grant.

Authors' contributions
CMGD and CRQ conceived and designed the experiments. CRQ directed the study. CMGD, ABJ and WHOV analyzed the results and wrote the manuscript. ABJ, CMGD, AFMO and RLDC participated in the data collection and revised the manuscript. All authors read and approved the manuscript.

Consent for publication
Not applicable.

Competing interests
The authors declare they have no competing interests.

Author details
[1]Universidade Estadual do Norte Fluminense, Centro de Ciências e Tecnologias Agropecuária Laboratório de Reprodução e Melhoramento Genético Animal, 2000, Alberto Lamego Ave, Campos dos Goytacazes, Rio de Janeiro 28016-811, Brazil. [2]Instituto Federal de Educação, Ciência e Tecnologia do Espírito Santo, 482 Rod Br, Km 47, s/n - Rive, Alegre, Espírito Santo 29520-000, Brazil. [3]Instituto de Zootecnia de Nova Odessa, 56, Heitor Penteado St., Nova Odessa, São Paulo 13460-000, Brazil.

References

1. Lacerda TS, Caetano AR, Facó O, de Faria DA, McManus CM, Lôbo RN, et al. Single marker assisted selection in Brazilian Morada Nova hair sheep community-based breeding program. Small Rumin. Res. [Internet]. Elsevier B.V.; 2016;139:15–9. Available from: https://doi.org/10.1016/j.smallrumres.2016.04.009
2. Issakowicz J, Issakowicz ACKS, Bueno MS, da Costa RLD, Katiki LM, Geraldo AT, et al. Parasitic infection, reproductive and productive performance from Santa Inês and Morada Nova ewes. Small Rumin. Res. [Internet]. Elsevier B.V.; 2016;136:96–103. Available from: http://www.sciencedirect.com/science/article/pii/S0921448816300165
3. Facó O, Paiva SR, Alves LRN, Lôbo RNB, Villela LCV. Raça Morada Nova: Origem, características e perspectiva. Sobral: Embrapa Caprinos; 2008. p. 43.
4. Notter DR. The importance of genetic diversity in livestock populations of the future. J Anim Sci. 1999;77:61–9.
5. Mousavizadeh A, Abadi MM, Torabi A, Nassiry MR, Ghiasi H, Koshkoieh AE. Genetic polymorphism at the growth hormone locus in Iranian Talli goats by polymerase chain reaction-single strand conformation polymorphism (PCR-SSCP). Iran J Biotechnol. 2009;7:51–3.
6. Mohammadabadi MR, Sattayimokhtari R. Estimation of (Co) Variance Components of Ewe Productivity Traits in Kermani Sheep. Slovak J Anim Sci. 2013;2013:45–51.
7. Mohammadabadi M. Inter-simple sequence repeat loci associations with predicted breeding values of body weight in kermani sheep. Genet Third Millenn. 2016;14:4383–90.
8. Khodabakhshzadeh R, Mohammadabadi MR, Esmailizadeh AK, Moradi Shahrebabak H, Bordbar F, Ansari Namin S. Identification of point mutations in exon 2 of GDF9 gene in Kermani sheep. Pol J Vet Sci. 2016;19:281–9.
9. Mohammadabadi M, Esfandyarpoor E, Mousapour A. Using inter simple sequence Repear multi-loci markers for studying genetic diversity in Kermani sheep. J Reseach Dev. 2017;5:1–4.
10. Paiva SR, Facó O, Faria DA, Lacerda T, Barretto GB, Carneiro PLS, et al. Molecular and pedigree analysis applied to conservation of animal genetic resources: the case of Brazilian Somali hair sheep. Trop Anim Health Prod. 2011;43:1449–57.
11. Sheriff O, Alemayehu K. Genetic diversity studies using microsatellite markers and their contribution in supporting sustainable sheep breeding programs: A review. Cogent Food Agric. [Internet]. Cogent; 2018;4. Available from: https://www.cogentoa.com/article/10.1080/23311932.2018.1459062
12. Crispim B do A, Matos MC, Seno LDO, Grisolia AB. Molecular markers for genetic diversity and phylogeny research of Brazilian sheep breeds. African J Biotechnol. 2012;11:15617–25.
13. Phyu PP, Pichler R, Soe O, Aung PP, Than M, Shamsuddin M, et al. Genetic diversity, population structure and phylogeography of Myanmar goats. Small Rumin. Res. [Internet]. Elsevier B.V.; 2016;148:33–42. Available from: https://doi.org/10.1016/j.smallrumres.2016.12.028
14. Ligda C, Altarayrah J, Georgoudis A. Genetic analysis of Greek sheep breeds using microsatellite markers for setting conservation priorities. Small Rumin Res. 2009;83:42–8.
15. Zamani P, Akhondi M, Mohammadabadi M. Associations of Inter-Simple Sequence Repeat loci with predicted breeding values of body weight in sheep. Small Rumin. Res. [Internet]. Elsevier; 2015 [cited 2018 Sep 10];132:123–7. Available from: https://www.sciencedirect.com/science/article/pii/S0921448815300894
16. Kijas JW, Lenstra JA, Hayes B, Boitard S, Neto LRP, San M, et al. Genome-wide analysis of the world's sheep breeds reveals high levels of historic mixture and strong recent selection. PLoS Biol. 2012;10:e1001258.
17. Fischer MC, Rellstab C, Leuzinger M, Roumet M, Gugerli F, Shimizu KK, et al. Estimating genomic diversity and population differentiation – an empirical comparison of microsatellite and SNP variation in Arabidopsis halleri. BMC Genomics. 2017;18:69 Available from: http://bmcgenomics.biomedcentral.com/articles/10.1186/s12864-016-3459-7.
18. Väli Ü, Einarsson A, Waits L, Ellegren H. To what extent do microsatellite markers reflect genome-wide genetic diversity in natural populations? Mol Ecol. 2008;17:3808–17.
19. Singh N, Choudhury DR, Singh AK, Kumar S, Srinivasan K, Tyagi RK, et al. Comparison of SSR and SNP markers in estimation of genetic diversity and population structure of Indian rice varieties. PLoS One. 2013;8:1–14.
20. Botstein D, White RL, Skolnick M, Davis RW. Construction of a genetic linkage map in man using restriction fragment length polymorphisms. Am J Hum Genet. 1980;32:314–31.
21. Flint-Garcia SA, Thornsberry JM, Buckler ES. Structure of linkage disequilibrium in plants. Annu Rev Plant Biol. 2003;54:357–74 Available from: http://www.annualreviews.org/doi/10.1146/annurev.arplant.54.031902.134907.
22. Silva RCB da, Pimenta Filho EC, Ribeiro MN, Silva EC da, Facó O, Paiva SR. Diversidade genética de ovinos Morada Nova no estado do Ceará, Brasil Regina. 48a Reun. Anu. da Soc. Bras. Zootec. Belém – PA: In: Reunião Anual da Sociedade Brasileira de Zootecnia; 2011. p. 1–3.
23. Ferreira JSB, Paiva SR, Silva EC, McManus CM, Caetano AR, Façanha DAE, et al. Genetic diversity and population structure of different varieties of Morada Nova hair sheep from Brazil. Genet Mol Res. 2014;13:2480–90.
24. GebreMichael SG. Sheep resources of Ethiopia: genetic diversity and breeding strategy; 2008.
25. Toro MA, Caballero A. Characterization and conservation of genetic diversity in subdivided populations. Philos Trans R Soc. 2005;360:1367–78.
26. Ocampo R, Cardona H, Martí-nez R. Genetic diversity of Colombian sheep by microsatellite markers. Chil J Agric Res. 2016;76:40–7 Available from: http://www.scielo.cl/scielo.php?script=sci_arttext&pid=S0718-58392016000100006&lng=en&nrm=iso&tlng=en.
27. Arandas, J. K. G., Ribeiro, M. N., Pimenta Filho, E. C., Sliva, R. C. B., Facó, O., Esteves SN. Estrutura Populacional de Ovinos da Raça Morada Nova. IX Simpósio Bras. Melhor. Anim. João Pessoa, PB: Embrapa Caprinos e Ovinos -Artigos em Anais e Congressos In: SIMPÓSIO BRASILEIRO DE MELHORAMENTO ANIMAL, 9; 2012. p. 3.
28. Porras-Hurtado L, Ruiz Y, Santos C, Phillips C, Carracedo Á, Lareu MV. An overview of STRUCTURE: applications, parameter settings, and supporting software. Front Genet. 2013;4:1–13.
29. Rousset F, Raymond M. GENEPOP (version 1.2): population genetics software for exact tests and ecumenicism. J Hered. 1995;86:248–9.

24

A study to identify the practices of the buffalo keepers which inadvertently lead to the spread of brucellosis in Delhi

Nimita Kant[1], Parul Kulshreshtha[1*], Rashmi Singh[1], Anuradha Mal[2], Amita Dwivedi[1], Riya Ahuja[1], Rinkle Mehra[1], Mohit Tehlan[1], Paritosh Ahmed[1], Shilpa Kaushik[1], Shipra, Shashikant Kumar[1], Aas Mohammad[1], Shrikrishn Shukla[1], Damini Singh[3] and Rakesh Bhatnagar[3,4*]

Abstract

Background: India has the largest Buffalo population in the world, with every household in rural India owning buffaloes depending upon daily milk requirement – dairy farmers can own between 10 to 70 buffaloes. The health of Indian buffaloes is of economic importance since India is one of the largest buffalo meat exporters in the world, and Indian Buffalo semen is sold in the USA for breeding purposes. However, National Control Program on brucellosis is only active in South India and in Panjab (a North Indian state with high human brucellosis incidence). Our aim was to assess the knowledge and practices of the buffalo keepers of Delhi that make them susceptible to brucellosis.

Results: Amongst all the 11 districts of Delhi, there was 0% awareness about brucellosis and also about the S19 vaccine as the buffalo keepers had never heard of S19 vaccine which is available at minimal cost from Indian Veterinary Research Institute, Bareilly, India. Majority of the respondents drink raw milk, sleep in cattle sheds, do not isolate sick cattle, do not test buffaloes blood for any disease before purchasing them, apply intrauterine medication with bare hands to buffalo after abortion of foetus, never clean their cattle sheds with a disinfectant and believe that they can only acquire skin infections from cattle. All of these habits make them prone to brucellosis. While about 20 to 27% of respondents reported a history of abortions and retained placenta, disposed of the placenta with bare hands, and applied raw milk on cracked lips. It was surprising to note that majority of them never reared small ruminants like sheep and goat with buffaloes or *Bos* species as they were aware of the rapid spread of disease from small to big ruminants.

Conclusions: We found that buffalo keepers were ignorant of brucellosis, its causative agent, relevant vaccines and that they also involved in high-risk activities. As such, our findings highlight a need for buffalo keepers to be better educated via several awareness camps to minimize human exposure to *Brucella* in Delhi.

Keywords: Brucella, Delhi, Cattle keepers, Practices, Habits

Background

Brucellosis is a classified group III risk disease which has easy airborne transmission [6, 42]. *Brucella*, the causative agent of brucellosis is classified by Centers For Disease Control and Prevention or CDC as a category B pathogen that has the potential to be developed into a bioweapon. Brucellosis is endemic in India and affects dairy farming [19, 33, 40]. Bovine brucellosis has been discovered as the main reason for disease propagation in humans due to animal handling and consumption of bovine products [11]. Countries like Australia, Canada, Cyprus, Japan, Denmark, Finland, The Netherlands, New Zealand, Norway, Sweden and the United Kingdom have been able to eradicate human brucellosis because they have eradicated bovine brucellosis [7, 26]. Although some reliable reports are available from Western countries, brucellosis is always underreported in Asian countries. It has been identified that there is lack of data related to brucellosis from India, China and Sub-Saharan Africa [29].

* Correspondence: parulkuls26@gmail.com; rakeshbhatnagar@mail.jnu.ac.in
[1]Department of Zoology, Shivaji College, University of Delhi, 110027, New Delhi, India
[3]School of Biotechnology, Jawaharlal Nehru University, 110057, Delhi, India

The world organization for animal health (OIE) defines standards for surveillance, diagnosis, epidemiology, control, eradication efforts, and the reduction of risk for animal health [16]. It means that all these factors must amalgamate to establish a better healthcare system. India has been identified as one of the hotspots for emerging infectious diseases and brucellosis is one of the emerging diseases [38]. New Delhi, the National Capital Territory of India, boasts of one of the best health care system in India. But as yet the website of the Development department of Delhi Government [http://delhi.gov.in/wps/wcm/connect/lib_development/Development/Home/Citizen+Charter] does not give any information about brucellosis. The diagnostic innovations are of no use if people of the country have no knowledge about the disease. Therefore, spreading awareness is the most important aspect of a control program. Although the brucellosis control program is very aggressive in South India [14, 28], but it is very frail in North India except in Punjab which is a North Indian state with a high incidence of brucellosis [5]. During our study, we observed that the majority of the dairy farmers reared exclusively buffaloes in Delhi. Many among them reared *Bos* species in fewer numbers (to fulfil the household requirement) than buffaloes on the same farm. The reason for this practice was explained to be the thicker milk quality along with the higher milk volumes derived from the buffaloes. As the spread of infectious diseases can be regulated by amending the practices and taking precautions thus we did a survey to analyze the practices of the buffalo keepers that may be furnishing the spread of brucellosis from the infected buffaloes. At the same time, we informed the buffalo keepers of the ways to rectify their animal handling so as to improve animal health and to curtail the spread of brucellosis.

Methods

Informed consent

All the participants signed a consent form prior to responding to the survey. The questionnaires were signed by the cattle keepers after filling them.

Sampling

A purposive sampling of the buffalo keepers was conducted from August 2015 to December 2016 in order to analyze their practices which promote the spread of brucellosis in the National Capital Territory of New Delhi, the capital city of India. As brucellosis is regarded as an occupational hazard, therefore, this study was conducted by interviewing the buffalo keepers of different districts of Delhi. The buffalo keepers come in close contact with the livestock therefore they constitute the high risk population for brucellosis. The buffalo keepers here refer to the human population of Delhi who reared buffalo

exclusively for their household needs or for dairy farming. This human population also includes those who reared few *Bos* species on their cattle sheds to fulfil their household demands or for a few customers. This survey does not include data from the cattle keepers who exclusively reared domesticated cattle (*Bos* species) only. According to the census of 2012 there are 162,142 female and 20,445 male buffalos in rural and urban area of New Delhi. This includes the population of 11 districts of Delhi. The current census report is pending. We surveyed 1200 cattle sheds which included data of 5550 buffaloes (4828 females and 722 males) only. It is pin-pointed that the *Bos* species (usually 1 to 3 in number on each farm) on the surveyed cattle sheds were not included in this animal count. There is no report on the exact number of cattle sheds housing buffaloes per district in Delhi. Here, a 'cattle shed' is defined as the cattle establishment used to house buffaloes only or to house a few *Bos* species with buffaloes. These cattle establishments varied in the structure being close-house or open-house or some completely on the road-side or in the colony lanes between boundary walls of houses. The pictures of various cattle sheds are available as Additional files 1, 2, 3, 4 and 5. They do not resemble any buffalo farm advertised online by Indian companies as the common man cannot afford them. We tried to tap the maximum number of cattle sheds in each district but many people were not ready to interact. Therefore our study pertains to only those buffalo keepers who were ready to interact with the college students.

Questionnaire

A structured questionnaire was developed in English language and translated in the Hindi language (the native tongue of North Indians). The questionnaire contained both open-ended and close-ended questions. The face to face method of approach was employed to collect the data. The questionnaire is available as the supporting file. The questionnaire revolved around the issues affecting the spread of brucellosis like biosecurity, reproductive health of the livestock, maintenance of cattle sheds and knowledge about zoonoses. The issues addressed by the questionnaire and their relevance to brucellosis have been discussed in Table 1. Each buffalo keeper was asked the number of male and female buffalo housed in their cattle sheds, they were asked if they got their cattle vaccinated or not, if yes then they were asked to name the vaccines/ medicines they injected their cattle with. Each buffalo keeper was asked if their buffaloes were prone to miscarriages, if yes then in which month; they were asked if they separated sick animal from the healthy ones; they were also asked if they consumed unboiled/ unpasteurized milk; accessibility to veterinary doctors was also asked; buffalo keepers were asked how often do

Table 1 Questionnaire pertained to the matter of biosecurity, reproductive health of the livestock, maintenance of cattle sheds and knowledge of zoonoses

S.N.	Issue addressed	Relevance to Brucellosis
1.	Consumption of raw milk and Application of raw milk on cracked lips	Infected buffalo secrete large amounts of *Brucella* in their milk.
2.	Assisting animal birth, application of intrauterine medication post abortion, disposing aborted foetus and placenta with naked hands.	Uterine fluid, Placental membranes, aborted foetus of infected buffalo during parturition or abortion are rich sources of *Brucella*.
3.	History of abortion and retained placenta	The seroprevalence of brucellosis is found to be significantly higher in animals with a history of abortion and retained placenta.
4.	Knowledge about Brucellosis or any other zoonoses and S19 vaccine	Knowledge about a disease makes the high risk population cautious and thus prevents the spread. Vaccination of young animals is known to reduce burden of disease.
5.	Rearing small ruminants with large ruminants	Infectious diseases spillover from small ruminants to large ruminants and cause huge economic loss.
6.	Sleeping in cattle sheds	Close contact with buffalo is a risk factor identified for human Brucellosis.
7.	Blood testing before sale and purchase of cattle	Diagnosis of brucellosis may curb the sale of non-productive buffalo and curb the spread of disease.
8.	Separation of sick animals	Intermingling of sick buffalo or domesticated cattle like *Bos* sp. with healthy buffalo may facilitate the transmission of brucellosis to susceptible cattle.
9.	Use of disinfectant to clean the cattle shed	Disinfectants lyse the gram negative bacteria and thus remove infection from the environment of the cattle shed.

the Government Organizations come for blood tests. They were asked about the precautions they took while handling an aborted fetus. They were asked if the adult female buffaloes were milked; socio-economic data, medical histories. In order to receive an unbiased response, the disease of interest was not revealed to the respondents and the question regarding their knowledge about brucellosis was only asked at the end. On the completion of the questionnaire, the respondents were informed about brucellosis, safe livestock handling, S19 vaccinations and other measures to prevent the spread of brucellosis. Data validation was done during data collection in the field and also at the time of translation to English. Responses to the close-ended questions have been tabulated and responses to open-ended questions have been elaborated in result and discussion section. Data from questionnaires were entered in Microsoft Excel2010. "Yes" or "No" or "don't know" responses were recorded as per the response for each respondent. The 'countif' application was used to find the percentage of each response for each question.

Results
Income groups
As brucellosis is linked to the economic status of people, therefore, we enquired about the income of each respondent from districts tabulated in Table 2. Cattle sheds surveyed in New Delhi, Central Delhi, West Delhi and South East Delhi were fewer as these areas fall in the urban area with fewer cattle establishments. As is evident from Fig. 1, it can be seen that the 50% of the respondents belonged to the least income group of less than Rupees one Lac per

annum, while rest of the respondents fell into higher slabs. It is reiterated here that 1 Lac rupees are equivalent to 1500 US$. It can be seen that a meagre percentage of 2.41% of respondents belonged to the category of people who earned more than 7 Lac rupees per annum. About 5.79% did not know their income. Thus it can be concluded that most people belonged to the lower income group in our survey.

Knowledge about brucellosis
As is evident from Table 3, 0% of respondents knew about brucellosis. Thus they were not cognizant of *Brucella* infections. In our study, it was observed that 37% of cattle keepers got their cattle vaccinated but only 9.25% of respondents could name any vaccine. The vaccines named by these respondents included anthrax vaccine, Foot and Mouth Disease or FMD vaccine, Enterotoxaemia or ET and Black Quarter or BQ vaccine. None of the respondents could name S19 or RB51 vaccine. Exactly 13% of respondents expressed that they do not know if their cattle is vaccinated or not. While 15% of respondents agreed that they only sometimes get their cattle vaccinated. Therefore it could be concluded that awareness about brucellosis was completely absent amongst the buffalo keepers. Majority of respondents in our study, 98%, conveyed that they did not rear small ruminants with buffaloes because when a disease affects one small ruminant then it spreads to all the big and small ruminants in the cattle sheds resulting in a huge loss. Thus it can be said that the risk of *Brucella* infection spilling from small ruminants to large ruminants was not significant amongst the population surveyed. Thus the buffalo keepers are conscious of the danger of the rapid spread of infection

Table 2 The survey rate of the buffalo keepers per district of Delhi, India

S.N.	District	Headquarter	Subdivisions	Number of cattlesheds surveyed
1	New Delhi	Connaught Place	Chanakyapuri, Delhi Cantonment, Vasant Vihar	20
2	North Delhi	Narela	Model Town, Narela, Alipur	164
3	North West Delhi	Kanjhawala	Rohini, Kanjhawala, Sarawati Vihar	150
4	West Delhi	Rajouri Garden	Patel Nagar, Panjabi Bagh, Rajouri Garden	85
5	South West Delhi	Dwarka	Dwarka, Najafgarh, Kapashera	171
6	South Delhi	Saket	Saket, Hauz Khas, Mehrauli	128
7	South East Delhi	Defence Colony	Defense Colony, Kalkaji, Sarita Vihar	34
8	Central Delhi	Daryaganj	Kotwali, Civil lines, Karol Bagh	42
9	North East Delhi	Seelampur	Seelampur, Yamuna Vihar, Karawal Nagar	128
10	Shahdara	Shahdara	Shahdara, Seemapuri, Vivek Vihar	150
11	East Delhi	Preet Vihar	Preet Vihar, Gandhi Nagar, Mayur Vihar	128

from small to large ruminants. It was observed that many buffalo keepers of Delhi kept dogs on the cattle shed as an alarm system to alert the owner of cattle thieves. Most cattle sheds were open house thus there is a high probability of the contamination of the feed and water by infected stray dogs also. Dogs in the cattle shed form a parallel reservoir of *Brucella* because dogs shed *Brucella* in reproductive fluids and spread bovine brucellosis [31]. Despite its endemism it was confounding to observe ignorance towards brucellosis, thus during our study, we apprised the buffalo keepers about brucellosis and S19 vaccination in detail.

Availability of the veterinary services

In an open-ended question, all the respondents were asked to summarize the state of the veterinary services available to them. We were dismayed to know that the veterinary services in all the districts of Delhi were extremely poor. All respondents revealed that there have never been any awareness camps regarding brucellosis in their area. It is reiterated that the website of the

Development department of Delhi Government does not give any information about brucellosis. Respondents of our study from remote villages of South West Delhi divulged that the sweepers of government hospitals learn to deliver intravenous injections. Such sweepers act as veterinary doctors and visit the village in hope of mining money. Respondents also shared that these sweepers inject tetanus toxoid for every health problem and charge Rs.700 per buffalo. Thus, the respondents shared that in case of ill health they inject the buffaloes with antibiotics like terramycin by themselves without any veterinary intervention (Additional file 6). They also informed that they take the buffaloes to the hospital only during worst case scenario.

Practices

To assess if the practices of buffalo keepers increase the risk of *Brucella* infection, a number of questions were included in the questionnaire. It was found that 38% of respondents drank raw milk. These respondents agreed that they sometimes drank directly from the udders of cow or

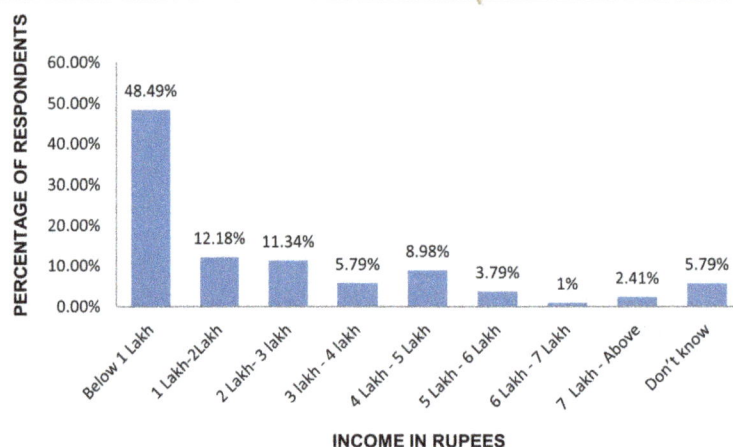

Fig. 1 Annual income of the respondents in Indian currency (Rupees)

Table 3 Knowledge and Practices of the buffalo keepers of Delhi, India

S.N.	Risk Factors	Yes %	No %	Sometimes %	Don't Know %
1	Drink raw milk	38	59	3	0
2	Milk the animal	60	35	5	0
3	Sleeping in Animal Sheds	79	11	10	0
4	Assisting Animal Birth	100	0	0	0
5	History of abortion in [3rd trimester] on farm	21	79	0	0
6	Disposed aborted fetus with naked hands	24	76	0	0
7	Incidence of retained placenta	20	70	10	0
8	Disposed placenta with naked hands	27	73	0	0
9	Applying raw milk on cracked lips	22	71	7	0
10	Vaccination of animals	37	35	15	13
11	Isolation of sick animals	32	68	0	0
12	Can you acquire disease from your cattle	37	38	25	0
13	Applied Intrauterine medication with naked hands after abortion	45	50	5	0
14	Use of disinfectant to clean cattle shed	13	85	2	0
15	Blood test before buying the animal	0	100	0	0
16	Do you rear goat and sheep with Buffalo and cow?	0	98	2	0
17	Have you heard of Brucellosis?	0	100	0	0

buffalo because they considered lactating animal as their religious mother. The practice of drinking raw milk driven by this belief puts these respondents at a risk of acquiring brucellosis. Respondents (65%) were also involved in milking the animal, 5% of respondents expressed that they sometimes milk the animal. As this activity involves touching the udders and coming in contact with raw milk thus this puts a large number of respondents at risk of acquiring brucellosis infection. Respondents from all the districts of Delhi revealed that their cattle sheds were prone to the theft of buffalo. Therefore, 79% of respondents always sleep in cattle sheds while 10% agreed that they only sometimes sleep in cattle sheds. *Brucella* can survive in soil for two to 6 months, therefore, sleeping in cattle sheds increases the probability of air-borne transmission or infection via an abrasion in the skin of buffalo keepers [23, 24]. All the respondents assisted the animal birth as it is celebrated as a special event in their family. Abortion in the 3rd trimester is a characteristic of brucellosis, and it was noted that 21% of surveyed cattle had abortions during this time. Thus it could be concluded that such buffaloes had the symptoms of brucellosis. Post-abortion the buffaloes are given tetanus toxoid intravenously only and no other medication was given. A low rate of abortion does not warrant a low prevalence rate as it is known that brucellosis may prevail as a silent disorder. Most buffalo keepers from Delhi confided that the buffaloes which suffered frequent abortions are sold off in local cattle fair. Thus such buffaloes stay in circulation as a major carrier of brucellosis. This practice will further deteriorate the state of affairs. We found that the aborted foetuses were

disposed of with naked hands by only 24% of respondents as most of them call doctor under such situation. Cattle keepers in rural areas (22% of respondents) of Delhi used raw milk to heal cracked lips while in urban areas most cattle keepers (71% of respondents) applied commercially manufactured creams on cracked lips while 7% of respondents agreed that they applied raw milk on cracked lips only sometimes. Applying raw milk to heal cracked lips is an age-old traditional therapy in India which can also cause *Brucella* infection in humans. The incidence of retained placenta in buffaloes was reported by only 20% of respondents as 70% denied any retention and 10% expressed that this happened only sometimes. Isolation of sick animals was not being followed by 68% of respondents, thus these cattle keepers put their healthy cattle at a risk of the infectious diseases harboured by sick animals. Many respondents (37%) expressed that they could acquire some disease from their cattle, 25% of respondents said that they may sometimes acquire infection from cattle. These respondents suggested that the skin infections could be acquired from the cattle. Many respondents (38%) thought that they cannot acquire any infection from cattle. They also shared that even the sickness of the buffalo does not hinder them from consuming its milk. Drinking milk of a sick buffalo is a huge health hazard which cannot be avoided until the buffalo keepers are not convinced about the potential of disease communication from buffalo to them. At the end of the survey, we informed the buffalo keepers about the routes of disease communication from buffalo to humans.

Only 45% of respondents agreed that they applied intrauterine medication with naked hands post -abortion, 5% of respondents said that they only sometimes practiced this. While 50% of respondents expressed that they do not apply any medication post abortion. Application of intrauterine medication with naked hands post-abortion again is a practice which may communicate brucellosis from buffalo to the buffalo keepers. Assisting parturition exposes the buffalo keepers to fetal membranes, aborted fetus and uterine fluid contaminated with *Brucella* species [3, 17]. Only 13% of respondents expressed that they used disinfectant to clean the cattle shed, a meagre 2% agreed that they utilized disinfectant for this purpose. Exactly 85% of respondent agreed that they never used disinfectants to clean the cattle shed rather they washed the sheds with water and throw dry sand over the sheds to clean. In such an environment the propagation of several infectious diseases becomes most probable.

Discussion

Brucellosis has not been listed amongst the neglected tropical disease in India and South East Asia [22], but it has been identified as a neglected tropical disease by WHO, the World Health Organization [43]. Brucellosis is known to cause huge economic losses in India ranging between a loss of US $ 6.8 per cattle and US$18.2 per buffalo [37]. Brucellosis has been listed amongst zoonosis that affects the health of poor and affects the trade of animal products [41]. In our study, most of the respondents hailed from a poor background, therefore, it was pertinent to assess the economic status of the respondents of our survey. India comprises a big geographical entity, consequently, the epidemiology of brucellosis varies from one region to another. Seroprevalence varies from 3.3–11.4% in Chennai only, while Isloor et al reported overall prevalence for Karnataka to be 1.9% in cattle and 1.8% in buffalo. The Indian Agricultural Research Institute reported a 13.5% of stable endemic equilibrium for brucellosis in India [15, 30, 36]. Rahman et al. recognized that Delhi has the highest seroprevalence but the exact data was not published [30]. The Project Directorate on Animal Disease Monitoring and Surveillance (PDMAS, India) under the Ministry of Agriculture launched "Vision 2030" in 2011 [30] to eradicate brucellosis from India by the year 2030. The obligatory prerequisite for the success of any control program is building awareness about the disease. According to our study, the buffalo keepers from Delhi were totally unaware of brucellosis. The same level of awareness was reported from Kenya in the year 2007 and recently from Tajikistan [18, 21]. As opposed to this, the cattle keepers and shepherds from Egypt declared that their animals have a history of *Brucella* infection.

They also confirmed that this infection was the main cause of abortions in their animals [10, 35]. Recent reports from Kenya show better awareness among the youngsters than the elderly. Better levels of awareness in recent years have been linked to higher seroprevalence of brucellosis in Kenya. Recent studies from other countries like Egypt, Tajikistan, and Kenya have reported that the livestock keepers were aware that brucellosis can spread from livestock to humans and that arthritis was a common symptom of the same [13, 21, 27]. In contrast to this, we found that cattle keepers from Delhi still think that only skin diseases can be acquired by handling cattle.

In western countries, there is a system of surveillance for brucellosis. Detailed data on brucellosis is compiled from time to time on demographics, the onset of symptoms, clinical signs, contact dates with the treating physicians, hospitalization, death, laboratory diagnosis, bacterial species, geographic origin and possible vehicle of infection. Standardized questionnaires containing these questions are sent to local health departments for every reported case of brucellosis. The same system needs to be developed in India as well. In terms of availability of information, India is 65 years behind the western countries [4, 12, 16, 32, 34]. Several countries identify their failures in controlling brucellosis [2] and discuss better ways along with newer possibilities to pin it down. This kind of model needs to be adopted by India as well. Other countries like the Gambia that report low prevalence for brucellosis should also be looked upon as a paradigm [9].

According to our study, the veterinary services in the rural areas of Delhi were not appropriate. Thus the buffalo keepers like to inject drugs by themselves in buffalo. Only when the problem escalates they take buffaloes to the remote veterinary hospitals. Studies from other countries like Egypt and Tajikistan also report reluctance on the part of the livestock keepers to contact veterinarians [11, 21]. It is also known that local dairies of India owned by the cattle keepers sell unpasteurized milk only. There have been reports of sale and purchase of unpasteurized dairy products in Iran, Egypt, Tajikistan, Uzbekistan, and Yemen also. This practice has been regarded as a prime risk factor for spread of human brucellosis [1, 8, 11, 39]. It is believed that even a sporadic abortion must be linked to brucellosis [3, 24] and the animal undergoing abortion must be culled. Many countries like India have reported the lack of official culling of infected sheep, goats, and buffalos as the main cause of high *Brucella* seropositivity [11]. Livestock keepers from other countries have reported that they feed the aborted foetuses to dogs or throw it in the water canal which adds to the spread of brucellosis [7]. This treacherous practice exposes the entire ecosystem to brucellosis. This habit was not reported by the buffalo keepers from Delhi. Studies from other parts of the world also report that gloves and masks are not being utilized while handling aborted foetuses and while assisting parturition [11, 20, 21]. All these practices make the spread of brucellosis convenient and rampant.

Most of the underdeveloped countries across the globe face the same situation but a developed country like the United States of America (USA) has identified brucellosis as a prioritized zoonoses. Furthermore, they have made a road map to combat not only the zoonoses but also the infectious diseases by gauging their own capabilities on the scale of surveillance and availability of diagnostics. Despite being a developed country, the USA has revealed an insufficiency of diagnostic capability [25]. Though the focus of our survey was brucellosis, we can conclude that the knowledge about most infectious diseases was insufficient, also the practices of these cattle keepers put them at high risk of acquiring infectious diseases. Thus, like developed countries, India must adopt a holistic approach to combat zoonoses and other infectious diseases as a whole. Our study may be regarded as only an elementary research due to several limitations. Our major limitation is the sampling bias. There is no report on the total number of cattle sheds in all the districts of Delhi. Therefore, it is impossible to pinpoint the percentage of the cattle keepers from Delhi who harbour the same opinion or perform the same practices documented in this study. Though, we tried to overcome this limitation by interviewing as many buffalo keepers as were ready to interact with us. Another limitation of this study has been the summation of responses of cattle keepers from household cattle sheds, small cattle sheds, and large cattle sheds together. As the situation and configuration of these cattle sheds differ, therefore, this may also be having a confounding effect on the inference.

Conclusion

In the NCT of Delhi, we found that the cattle keepers have never been surveyed for their opinions and practices regarding any infectious disease. On interviewing the cattle keepers we realized that they are sensitive towards the medical needs of their cattle but they do not have appropriate help. Our study indicates that the cattle keepers are oblivious to the practices which cause spillover infection of *brucella* from their cattle to them. It can be concluded from our survey that the cattle farmers are ignorant of brucellosis, its causative agent, the route of its transmission, its symptoms and vaccination. The emotional and religious belief of the cattle keepers further exposes them to the threat of brucellosis. The absence of culling, free cattle trade and absence of blood testing before buying the cattle makes the situation even gross. Thus we endeavoured to enlighten the cattle keepers with the harmful husbandry practices which endorse the spread of brucellosis. The

major problem faced by the control programs includes opinions of people and it is important to mould these opinions if brucellosis is to be combated successfully. In the current scenario, it is pertinent that the Indian Government must organize awareness camps before brucellosis becomes a bigger menace.

Additional files

Additional file 1: a) Open house cattle shed in North East Delhi b) and c) Open house cattle sheds of the South West Delhi district.

Additional file 2: A road-side establishment of North Delhi district.

Additional file 3: One side open cattle shed used for housing buffalo and *Bos* species together.

Additional file 4: Close house cattle shed in Shahdara district, arrow shows the opening of the housing area. This cattle shed was completely dark with light going through this opening.

Additional file 5: Buffalo residing in the lane between houses in the South Delhi district.

Additional file 6: One of the medicines shown by the respondents was Terramycin, Injectable solution.

Additional file 7: Questionnaire.

Abbreviations
BQ: Black Quarter; CDC: Centers For Disease Control and Prevention; ET: Enterotoxaemia; FMD: Foot and Mouth Disease; OIE: Organization for animal health; PDMAS: Project Directorate on Animal Disease Monitoring and Surveillance; USA: United States of America; WHO: World Health Organization

Acknowledgements
Due credit is extended to the Principal of Shivaji College, Dr.Shashi Nijhawan for her support.
Damini Singh, D.S., received fellowship from the ICMR-funded project no.BMS/FWIMM/2015-24280/AUG15/N.DELHI/11.

Funding
The Innovation Cell of University of Delhi, India, financially supported this work under its scheme of Innovation Projects (2015–16 cycle).

Authors' contributions
PK, NK and RB designed the work. PK, AM, DS, AD, RM, S., RA, AM, MT, PA, SS analyzed the data. AD, RM and RM helped in the preparation of figure and tables. PK and RS organized the references. PK wrote the manuscript and critically revised the content. All the authors conducted the survey. All the authors have read and approved the manuscript.

Consent for publication
Not applicable.

Competing interests
The authors declare that they have no competing interests.

Author details
[1]Department of Zoology, Shivaji College, University of Delhi, 110027, New Delhi, India. [2]Department of Botany, Shivaji College, University of Delhi, 110027, New Delhi, India. [3]School of Biotechnology, Jawaharlal Nehru University, 110057, Delhi, India. [4]Laboratory of Molecular Biology and Genetic Engineering, School of Biotechnology, Jawaharlal Nehru University, New Delhi 110067, India.

References

1. Al-Shamahy HA, Whitty CJM, Wright SG. Risk factors for human brucellosis in Yemen: case control study. Epidemiol Infect. 2000;125:309–13.

2. Azzam RA, El-Gamal AM, Elsheemy MT. Failure of control of Brucella melitensis infection in a dairy herd. Assiut Vet Med J. 2009;55(121):274–85.

3. Crawford RP, Huber JD, Adams BC. Epidemiology and surveillance. In: Nielsen K, Duncan JR, editors. Animal brucellosis. Boca Raton: CRC Press; 1990. p. 131–48.

4. Dahouk SA, Neubauer H, Hensel A, Schöneberg I, Nöckler K, Alpers K, Merzenich H, Stark K, Jansen A. Changing epidemiology of human brucellosis, Germany. Emerg Infect Dis. 2007;13:1962–2005.

5. Dhand NK, Gumber S, Singh BB, Aradhana BMS, Kumar H, Sharma DR, Singh J, Sandhu KS. A study on the epidemiology of brucellosis in Punjab (India) using Survey Toolbox. Rev Sci Tech. 2005;24(3):879–85.

6. Dean AS, Crump L, Greter H, Hattendorf J, Schelling E, Zinsstag J. Clinical manifestations of human brucellosis: a systematic review and metaanalysis. PLoS Negl Trop Dis. 2012;12:e1929.

7. Díaz AE. Epidemiology of brucellosis in domestic animals caused by *Brucella melitensis, Brucella suis* and *Brucella abortus*. Sci Tech Rev. 2013;32:53–60.

8. Earhart K, Vafakolov S, Yarmohamedova N, Michael A, Tjaden A, Soliman A. Risk factors of human brucellosis in Samarqand oblast, Uzbekistan. Int J Infect Dis. 2009;13:749–53.

9. Germeraad EA, Hogerwerf L, Faye-Joof T, Goossens B, van der Hoek W, Jeng M, Lamin M, Manneh IL, Nwakanma D, Roest HI, Secka A, Stegeman A, Wegmüller R, van der Sande MA, Secka O. Low Seroprevalence of Brucellosis in Humans and Small Ruminants in the Gambia. PLoS One. 2016; 11(11):e0166035.

10. Hegazy YM, Moawad A, Osman S, Ridler A, Guitian J. Ruminant brucellosis in the Kafr El sheikh governorate of the Nile Delta, Egypt: prevalence of a neglected zoonosis. PLoS Negl Trop Dis. 2011;5:e944.

11. Hegazy Y, Elmonir W, Hamid NHA, Elbauomy EM. Seroprevalence and "knowledge, attitudes and practices" (KAP) survey of endemic ovine brucellosis in Egypt. Acta Vet Scand. 2016;58:1.

12. Hernández-Mora G, Bonilla-Montoya R, Barrantes-Granados O, Esquivel-Suárez A, Montero-Caballero D, González-Barrientos R, Fallas-Monge Z, Palacios-Alfaro JD, Baldi M, Campos E, Chanto G, Barquero-Calvo E, Chacón-Díaz C, Chaves-Olarte E, Guzmán Verri C, Romero-Zúñiga JJ, Moreno E. Brucellosis in mammals of Costa Rica: an epidemiological survey. PLoS One. 2017;12(8):e0182644.

13. Holt H, Eltholth M, Hegazy Y, El-Tras W, Tayel A, Guitian J. *Brucella* spp. infection in large ruminants in an endemic area of Egypt: cross-sectional study investigating seroprevalence, risk factors and livestock owner's knowledge, attitudes and practices (KAPs). BMC Public Health. 2011;11:341.

14. Hundal JS, Sodhi SS, Gupta A, Singh J, Chahal US. Awareness, knowledge, and risks of zoonotic diseases among livestock farmers in Punjab. Vet World. 2016;9(2):186–91.

15. Isloor S, Renukaradhya GJ, Rajasekhar M. A serological survey of bovine brucellosis in India. Sci Tech Rev. 1998;17(3):781–5.

16. Jebara KB. Surveillance, detection and response: managing emerging diseases at national and international levels. Sci Tech Rev. 2004;23(2):709–15.

17. John K, Fitzpatrick J, French N, Kazwala R, Kambarage D, Mfinanga GS, Macmillan A, Cleveland S. Quantifying risk factors for human brucellosis in rural northern Tanzania. PLoS One. 2010;5:e9968.

18. Kang' Ethe EK, Ekuttan CE, Kimani VN, Kiragu MW. Investigations into the prevalence of bovine brucellosis and the risk factors that predispose humans to infection among urban dairy and non-dairy farming households in Dagoretti Division, Nairobi, Kenya. East Afr Med J. 2007;84:96–100.

19. Khurana SK, Srivastava SK, Prabhudas K. Seroprevalence of bovine brucellosis in Haryana by avidin-biotin serum ELISA and its comparison with RBPT and SAT. Indian J Anim Sci. 2012;82:448–50.

20. Kozukeev TB, Ajeilat S, Maes E, Favorov M. Risk factors for brucellosis-Leylek and Kadamay districts, Batken oblast, Kyrgystan, January-November. Morb Mortal Wkly Rep. 2006;28:31–4.

21. Lindahl E, Sattorov N, Boqvist S, Magnusson U. A study of knowledge, attitudes and practices relating to brucellosis among small scale dairy farmers in an urban and peri-urban area of Tajikistan. PLoS One. 2015;10:e0117318.

22. Lobo DA, Velayudhan R, Chatterjee P, Kohli H, Hotez PJ. The neglected tropical diseases of India and South Asia: review of heir prevalence, distribution, and control or elimination. PLoS Negl Trop Dis. 2011;5:e1222.

23. Mangalgi SS, Sajjan AG, Mohite ST, Gajul SJ. Brucellosis in Occupationally Exposed Groups. J Clin Diagn Res. 2016;10(4):DC24–7.

24. Makita K, Fevre EM, Waiswa C, Eisler MC, Thrusfield M, Welburn SC. Herd prevalence of bovine brucellosis and analysis of risk factors in cattle in urban and peri-urban areas of the Kampala economic zone, Uganda. BMC Vet Res. 2011;7:60.

25. Maxwell MJ, Freire de Carvalho MH, Hoet AE, Vigilato MA, Pompei JC, Cosivi O, Del Rio Vilas VJ. Building the road to a regional zoonoses strategy: a survey of zoonoses programmes in the Americas. PLoS One. 2017;12(3):e0174175.

26. Mohamed NS, Stephen MB, Nammalwar S. Brucellosis: a re-emerging zoonosis. Vet Microbiol. 2010;140:392–8.

27. Njuguna JN, Gicheru MM, Kamau LM, Mbatha PM. Incidence and knowledge of brucellosis in Kahuro district, Murang'a county, Kenya. Trop Anim Health Prod. 2017;49:1035–40.

28. Patil DP, Ajantha GS, Shubhada C, Jain PA, Kalabhavi A, Shetty PC, Hosamani M, Appannanavar S, Kulkarni RD. Trend of human brucellosis over a decade at tertiary care Centre in North Karnataka. Indian J Med Microbiol. 2016;34(4):427–32.

29. Pappas G, Papadimitriou P, Akritidis N, Christou L, Tsianos EV. The new global map of human brucellosis. Lancet Infect Dis. 2006;6(2):91–9.

30. Rahman H. DBT Network Project on Brucellosis: Indian Council of Agricultural Research, Project Monitoring Unit, Project Directorate on Animal Disease Monitoring and Surveillance, Annual Report; 2013. https://icar.org.in/files/Vision%202030_PDADMAS-11-01-2012.pdf.

31. Prior MG. Isolation of Brucella abortus from two dogs in contact with bovine brucellosis. Can J Comp Med. 1976;40:117–8.

32. Rana UV, Sehgal S, Bhardwaj M. A sero-epidemiological study of brucellosis among workers of veterinary hospital and slaughter houses of union territory of Delhi. Int J Zoonoses. 1985;12(1):74–9.

33. Renukaradhya GJ, Isloor S, Rajasekhar M. Epidemiology, zoonotic aspects, vaccination and control/eradication of brucellosis in India. Vet Microbiol. 2002;90:183–95.

34. Russo G, Pasquali P, Nenova R, Alexandrov T, Ralchev S, Vullo V, Rezza G, Kantardjiev T. Reemergence of human and animal brucellosis, Bulgaria. Emerg Infect Dis. 2009;15:314–6.

35. Hassan S, Meshref AR, Khoudair RM, Ashour HM. Multicenter study of brucellosis in Egypt. Emerg Infect Dis. 2008;14:1916–8.

36. Senthil NR, Narayanan SA. Seroprevalence study of bovine brucellosis in slaughter house. Int J Adv Vet Sci Technol. 2013;2(1):61–3.

37. Singh BB, Dhand NK, Gill JP. Economic losses occurring due to brucellosis in Indian livestock populations. Prev Vet Med. 2015;119(3–4):211–5.

38. Singh BB, Sharma R, Gill JPS, Aulakh RS, Banga HS. Climate change, zoonoses and India. Sci Tech Rev. 2011;30:779–88.

39. Sofian M, Aghakhani A, Velayati AA, Banifazl M, Elsamifar A, et al. Risk factors for human brucellosis in Iran: a case-control study. Int J Infect Dis. 2008;12: 157–61.

40. Trangadia B, Rana SK, Mukharjee F, Srinivasan VA. Prevalence of brucellosis and infectious bovine rhinotrachitis in organized dairy farms in India. Trop Anim Health Prod. 2010;42:203–7.

41. Thiermann A. Emerging diseases and implications for global trade. Sci Tech Rev. 2004;23(2):701–8.

42. World Health Organization (WHO), 2006. The control of neglected diseases. A route to poverty alleviation. Report of a joint WHO/DFID-AHP meeting with the participation of FAO and OIE. [http://www.who.int/zoonoses/Report_Sept06.pdf].

43. World Health Organization. The control of neglected zoonotic diseases. Geneva: report of a joint WHO/DFID-AHP; 2005. p. 54. Available: http://www.who.int/zoonoses/Report_Sept06.pdf. Accessed 28 May 2010

25

Development of within-herd immunity and long-term persistence of antibodies against Schmallenberg virus in naturally infected cattle

Kerstin Wernike[1]* (iD), Mark Holsteg[2], Kevin P. Szillat[1] and Martin Beer[1]

Abstract

Background: In 2011, the teratogenic, insect-transmitted Schmallenberg virus (SBV) emerged at the German/Dutch border region and subsequently spread rapidly throughout the European continent. In cattle, one of the major target species of SBV, first antibodies are detectable between one and three weeks after infection, but the duration of humoral immunity is unknown. To assess the course of immunity in individual animals and the development of the within-herd seroprevalence, cattle kept in a German farm with a herd size of about 300 lactating animals were annually blood sampled between December 2011 and December 2017 and tested for the presence of SBV-specific antibodies.

Results: During the monitored period, the within-herd seroprevalence declined from 74.92% in 2011 to 39.93% in 2015 and, thereafter, slightly increased to 49.53% in 2016 and 48.44% in 2017. From the animals that were tested in 2014 and 2015 for the first time (between 24 and 35 months of age) only 14.77% and 7.45%, respectively, scored positive. Thereafter, the seropositivity rate of this age group rose markedly to 58.04% in 2016 and 48.10% in 2017 indicating a circulation of SBV. Twenty-three individual animals were consistently sampled once per year between 2011 and 2017 after the respective insect vector season, 17 of them tested positive at the first sampling. Fourteen animals were still seropositive in December 2017, while three cattle (17.65%) became seronegative.

Conclusions: The regular re-emergence of SBV in Central Europe is a result of decreasing herd immunity caused by the replacement of animals by seronegative youngstock rather than of a drop of antibody levels in previously infected individual animals. The consequences of the overall decline in herd seroprevalence may be increasing virus circulation and more cases of fetal malformation caused by infection of naïve dams during gestation.

Keywords: Peribunyavirus, Immunity, Intra-herd prevalence, Antibody persistence, Serology, Epidemiology

Background

In 2011, an unidentified disease of cattle associated with fever, diarrhea and decreased milk production was reported in Germany and the Netherlands. The causative agent, a member of the Simbu serogroup within the family *Peribunyaviridae*, was eventually identified and named Schmallenberg virus (SBV) [1]. Clinical signs of an SBV infection are restricted to none or mild and transient symptoms in adult animals. However, an infection of naïve ruminants during a critical phase of pregnancy may lead to severe congenital abnormalities, abortion or stillbirth [2].

Like other Simbu serogroup viruses, SBV is primarily transmitted by *Culicoides* biting midges [3–5], while direct transmission between animals via the oral route is highly unlikely [6]. For the spread of potentially infected *Culicoides* midges over long distances, wind seems to play a relevant role [7], since several studies have linked wind movement to the spread of *Culicoides*-born viral diseases [8–10].

* Correspondence: kerstin.wernike@fli.de
[1]Institute of Diagnostic Virology, Friedrich-Loeffler-Institut, Greifswald - Insel Riems, Germany

The mammalian hosts comprise cattle, sheep, and goats as well as various wild and captive ruminants and some further ungulates and zoo animals [11–16].

In domestic ruminants, SBV-specific antibodies are induced during the first three weeks post infection [6, 17–19] and provide immunity against re-infection [6]. However, the duration of protection remains to be clarified, especially since a previously acquired herd immunity may play an important role in the cyclic re-emergence of the virus, which was observed in Central Europe during recent years [20–22]. Previous studies showed that SBV-specific antibodies are present in the majority of adult cattle for at least two to three years after a natural infection [23–25]. However, about 10% of animals became seronegative within three years [25]. This has raised doubts as to whether the acquired immunity persists for life.

In order to assess the course of immunity in individual animals and the development of the within-herd seroprevalence, cattle kept in a German farm located in the area initially affected most severely by SBV were regularly sampled over a period of six years and tested for the presence of SBV-specific antibodies.

Results

Development of within-herd seroprevalence and indications for SBV re-emergence

A private dairy cattle farm, located in the German federal state North Rhine-Westphalia, was monitored between 2011 and 2017 after the respective insect vector seasons, i.e. in the winter months December or January.

The animals included in the study were kept indoors year-round, but all animals younger than 24 months are kept outside. During the monitored period, the within-herd seroprevalence measured by a commercially available SBV antibody ELISA declined from 74.92% in 2011 and 83.08% in 2012 [25] to 39.93% in 2015 and, thereafter, slightly increased to 49.53% in 2016 and 48.44% in 2017. However, the seropositivity rate of animals that were tested for the first time in the respective year (between 24 and 35 months of age) rose drastically from 14.77% in 2014 and 7.45% in 2015 to 58.04% in 2016 and 48.10% in 2017 (Fig. 1).

A total of 23 animals were annually sampled between 2011 and 2017, six of them scored negative in the SBV-ELISA after the 2011 vector season (animal numbers 1–6, Fig. 2a), two of them seroconverted in 2012 (animals 5 and 6, Fig. 2a) and the remaining animals stayed seronegative until the end of the study. Of the other housed, seronegative animals (> 35 months of age), none seroconverted between the 2013 and 2014 and between the 2014 and 2015 sampling dates, but some of these cattle developed SBV-specific antibodies between December 2015 and December 2016 (Table 1). From 12 cattle that tested negative in 2013, five seroconverted in 2016, while from 26 initially in the year 2014 seronegative animals seven tested positive by ELISA in the year 2016, and of the additional 62 cattle seronegative in 2015, one seroconverted in 2016 (Table 1). No further seroconversions were detected between 2016 and 2017 (Table 1).

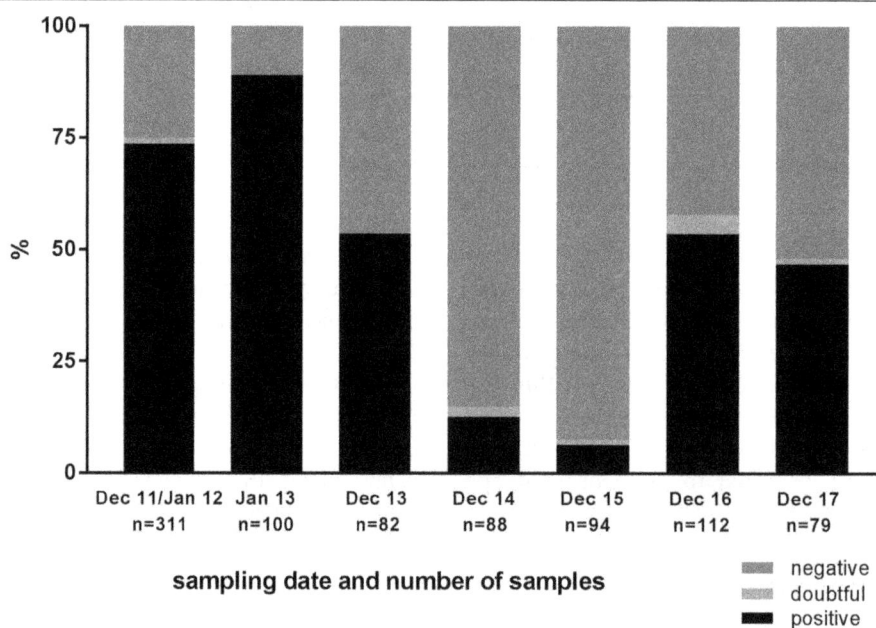

Fig. 1 Percentage of anti-SBV antibody positive, doubtful and negative animals among cattle that were sampled for the first time in the respective year. In the December 2011/January 2012 sampling, every animal older than 24 months was included, while thereafter only animals between 24 and 35 months of age were sampled for the first time

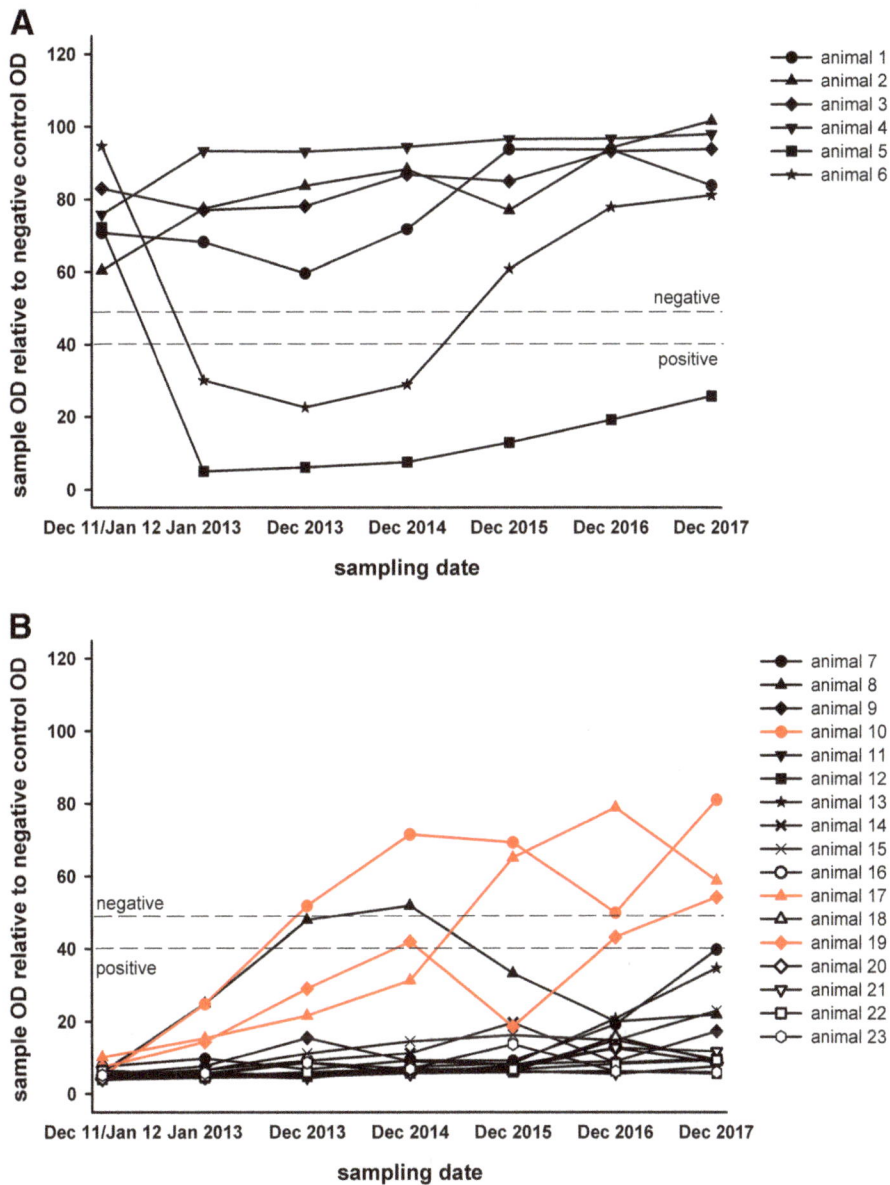

Fig. 2 SBV antibody ELISA results of all animals that were consistently sampled once per year between December 2011/January 2012 and December 2017. In figure panel A animals negative at the first sampling time point are shown, while in figure panel B animals that were seropositive at the first sampling are depicted. Animals that tested positive in December 2011/January 2012, but seronegative in December 2017 are marked in red

Long-term persistence of anti-SBV antibodies

Seventeen of the 23 animals that were annually sampled between 2011 and 2017 tested positive in the SBV antibody ELISA at the first sampling in December 2011/January 2012; 13 of them (76.47%) remained seropositive until December 2017, while three animals (17.65%) became seronegative (Fig. 2b). One animal (number 8 in Fig. 2b) tested positive in January 2013, scored ELISA-doubtful in December 2013, tested negative in December 2014, and scored positive again from 2015 onwards.

Discussion

After its emergence in 2011 in the Dutch/German border region, SBV spread very rapidly throughout the European continent [26]. In the following years, the virus further spread to previously unaffected regions [27–32], but also repeatedly re-appeared in the center of the initial epidemic [20–22].

In the present study, indications for the re-emergence of SBV in an individual cattle herd were found. A considerable increase of the seropositivity rate was observed in animals that were tested for the first time in 2016 or

Table 1 Number and status of animals tested for the presence of SBV-specific antibodies at every sampling time point

status at first sampling	number of samples (positive/doubtful/negative)						
	Dec 11/Jan 12	Jan 13	Dec 13	Dec 14	Dec 15	Dec 16	Dec 17
Positive	17/0/0	17/0/0	15/1/1	14/1/2	15/0/2	14/2/1	14/0/3
Negative	0/0/6	2/0/4	2/0/4	2/0/4	1/0/5	1/0/5	1/0/5
Positive		15/0/0	14/1/0	14/1/0	13/0/2	11/1/3	11/0/4
Negative		0/0/0	0/0/0	0/0/0	0/0/0	0/0/0	0/0/0
Positive			13/0/0	13/0/0	11/1/1	11/0/2	9/2/2
Negative			0/0/12	0/0/12	0/0/12	5/0/7	5/0/7
Positive				2/0/0	1/0/1	2/0/0	1/0/1
Negative				0/0/26	0/0/26	7/0/19	7/0/19
Positive					0/0/0	0/0/0	0/0/0
Negative					0/0/62	1/0/61	1/0/61
Positive						29/2/0	28/1/2
Negative						0/0/35	0/0/35
First sampled in 2017 (animals < 36 months of age)							37/1/41
> 36 months of age in 2017, but not consistently sampled in previous years							14/2/6

2017 compared to animals firstly tested in 2014 and 2015 indicating a re-circulation of SBV in that particular cattle herd leading to seroconversion of naïve young-stock. The age of the animals ranged from 24 to 35 months at the respective first sampling. Since all animals younger than 24 months are kept outside in the monitored herd and grazing increases the risk of an SBV infection for cattle compared to being housed in stables [33], the animals were most likely infected in their first two years of life, i.e. in the 2015, 2016 and/or 2017 vector seasons. But as only a small number of animals (7/94) sampled for the first time in December 2015 was SBV seropositive, large scale circulation of SBV in this year seems unlikely. Furthermore, none of the seronegative, older, housed animals seroconverted in 2015, but some of these cattle developed SBV-specific antibodies between December 2015 and December 2016. This observation and the fact that no further seroconversions were detected between 2016 and 2017 suggests virus circulation in the cattle herd in the 2016 vector season, which is in line with previously reported SBV-detections in North-Rhine Westphalia and further German federal states in that year [20]. Such patterns of cyclic re-circulation in a given area as currently seen in the case of SBV have also been described for other Simbu serogroup viruses. There are for example regular epidemics of Akabane virus (AKAV), Aino virus and Peaton virus in Japan [34–36] or of AKAV in Australia [37].

Regarding the long-term persistence of virus-specific antibodies, however, little information is available for these Simbu serogroup viruses, since the booster effect caused by re-infections of animals kept in endemic areas hampers attempts to measure the development of antibody levels in naturally infected, commercial cattle. For AKAV, it has been described that specific antibodies persist for at least two years [38]. The same holds true for SBV, and specific antibodies are detectable in the majority of cattle for at least two to three years [23–25].

In the present study, naturally infected cattle were monitored over a period of six years and only three out of 17 animals became seronegative in this time frame, while anti-SBV antibodies were still measurable after six years in the remaining 14 cattle. One of these animals tested negative at one sampling date (very close to the cut-off value) and again clearly positive at subsequent time points which might be caused by mixing-up animals during sampling or a false-negative test result or by a re-infection of that animal. A re-infection of all animals, however, is very unlikely, as the relatively low rate of seroconversions measured in 2016 in adult animals that are kept indoors compared to the higher rate seen in the youngstock, which is kept outside, further confirms that an SBV infection of housed animals is less likely than that of grazing cattle as was already previously reported [33]. Consequently, most of the seropositive housed animals, which were tested annually between 2011 and 2017, presumably were not re-infected and the measurable antibodies represent a specific humoral immunity acquired during their first infection in the 2011 or 2012 vector seasons. Hence, anti-SBV antibodies persist in the majority of cattle for at least six years, if not even lifelong.

Conclusions

Antibodies acquired following a natural SBV infection were detectable in more than 75% of the cattle monitored in the present study over a period of six years. Therefore, the regular re-emergence of SBV in given areas is more likely a result of decreasing herd immunity caused by replacement of animals by seronegative youngstock, which are susceptible once maternally-derived antibodies have declined at an age of 5 to 6 months [23, 24], than by a drop of antibody levels in previously infected animals. The result of the overall decline in the herd seroprevalence and the following renewed virus circulation may be again more frequent infections of naïve heifers during gestation and, as a consequence the induction of fetal malformations.

Methods

A private dairy cattle farm, located in the federal state North Rhine-Westphalia, was monitored between 2011 and 2017 after the respective insect vector seasons, i.e. in the winter months December and January. The cows included in the study were kept indoors year-round under routine production conditions, but all animals younger than 24 months are kept outside.

Serum samples of all cows older than 24 months were taken in December 2011 or January 2012, January 2013, December 2013, December 2014 [25], December 2015 (288 animals), December 2016 (317 animals), and December 2017 (320 animals).

From a total of 23 animals, routine diagnostic blood samples were available from every sampling date, the age of these animals ranged in December 2017 from 96 to 160 months.

Serum samples were taken by puncture of the vena coccygea and analyzed by a commercially available SBV antibody ELISA (ID Screen® Schmallenberg virus Competition, ID vet, France) according to the manufacturer's instructions. For the calculation of the within-herd seroprevalences, doubtful results were considered as positive.

Abbreviations

AKAV: Akabane virus; SBV: Schmallenberg virus

Acknowledgements

We thank Bianka Hillmann and Alrik-Markis Kunisch for excellent technical assistance and Dr. S. Hoppe (Agricultural Research and Training Center Haus Riswick) for providing the samples.

Funding

This work received no specific grant from any funding agency.

Authors' contributions

KW, MH and MB conceived and designed the experiments; KW and KPS performed the experiments; KW analyzed the data; MH contributed reagents/materials/analysis tools; KW wrote the paper. All authors read and approved the final manuscript.

Consent for publication

A written informed consent to use the animals in the study was obtained from the owner of the animals.

Competing interests

The authors declare that they have no competing interests.

Author details

[1]Institute of Diagnostic Virology, Friedrich-Loeffler-Institut, Greifswald - Insel Riems, Germany. [2]Chamber of Agriculture for North Rhine-Westphalia, Bovine Health Service, Haus Riswick, Kleve, Germany.

References

1. Hoffmann B, Scheuch M, Höper D, Jungblut R, Holsteg M, Schirrmeier H, Eschbaumer M, Goller KV, Wernike K, Fischer M, et al. Novel orthobunyavirus in cattle, Europe, 2011. Emerg Infect Dis. 2012;18(3):469–72.
2. Beer M, Conraths FJ, van der Poel WH. 'Schmallenberg virus' - a novel orthobunyavirus emerging in Europe. Epidemiol Infect. 2013;141(1):1–8.
3. Elbers AR, Meiswinkel R, van Weezep E. Sloet van Oldruitenborgh-Oosterbaan MM, Kooi EA: Schmallenberg Virus in Culicoides spp. Biting Midges, the Netherlands, 2011. Emerg Infect Dis. 2013;19(1):106–9.
4. Pagès N, Talavera S, Verdun M, Pujol N, Valle M, Bensaid A, Pujols J. Schmallenberg virus detection in Culicoides biting midges in Spain: first laboratory evidence for highly efficient infection of Culicoides of the Obsoletus complex and Culicoides imicola. Transbound Emerg Dis. 2017; 65(1):e1–6.
5. Rasmussen LD, Kirkeby C, Bodker R, Kristensen B, Rasmussen TB, Belsham GJ, Bøtner A. Rapid spread of Schmallenberg virus-infected biting midges (Culicoides spp.) across Denmark in 2012. Transbound Emerg Dis. 2014; 61(1):12–6.
6. Wernike K, Eschbaumer M, Schirrmeier H, Blohm U, Breithaupt A, Hoffmann B, Beer M. Oral exposure, reinfection and cellular immunity to Schmallenberg virus in cattle. Vet Microbiol. 2013;165(1–2):155–9.
7. Sellers RF. Weather, host and vector--their interplay in the spread of insect-borne animal virus diseases. J Hyg. 1980;85(1):65–102.
8. Sellers RF, Maarouf AR. Possible Introduction of Epizootic Hemorrhagic-Disease of Deer Virus (Serotype-2) and Bluetongue Virus (Serotype-11) into British-Columbia in 1987 and 1988 by Infected Culicoides Carried on the Wind. Can J Vet Res-Revue Canadienne De Recherche Veterinaire. 1991; 55(4):367–70.
9. Sellers RF, Pedgley DE. Possible windborne spread to western Turkey of bluetongue virus in 1977 and of Akabane virus in 1979. J Hyg. 1985;95(1): 149–58.
10. Braverman Y, Chechik F. Air streams and the introduction of animal diseases borne on Culicoides (Diptera, Ceratopogonidae) into Israel. Rev Sci Tech. 1996;15(3):1037–52.
11. Mouchantat S, Wernike K, Lutz W, Hoffmann B, Ulrich RG, Borner K, Wittstatt U, Beer M. A broad spectrum screening of Schmallenberg virus antibodies in wildlife animals in Germany. Vet Res. 2015;46(1):99.
12. Chiari M, Sozzi E, Zanoni M, Alborali LG, Lavazza A, Cordioli P. Serosurvey for Schmallenberg virus in alpine wild ungulates. Transbound Emerg Dis. 2014; 61(1):1–3.
13. Larska M, Krzysiak M, Smreczak M, Polak MP, Zmudzinski JF. First detection of Schmallenberg virus in elk (Alces alces) indicating infection of wildlife in Bialowieza National Park in Poland. Vet J. 2013;198(1):279–81.
14. Linden A, Desmecht D, Volpe R, Wirtgen M, Gregoire F, Pirson J, Paternostre J, Kleijnen D, Schirrmeier H, Beer M, et al. Epizootic spread of Schmallenberg virus among wild Cervids, Belgium, fall 2011. Emerg Infect Dis. 2012;18(12): 2006–8.
15. Laloy E, Braud C, Bréard E, Kaandorp J, Bourgeois A, Kohl M, Meyer G, Sailleau C, Viarouge C, Zientara S, et al. Schmallenberg virus in zoo ruminants, France and the Netherlands. Emerg Infect Dis. 2016;22(12):2201–3.
16. Molenaar FM, La Rocca SA, Khatri M, Lopez J, Steinbach F, Dastjerdi A. Exposure of Asian elephants and other exotic ungulates to Schmallenberg virus. PLoS One. 2015;10(8):e0135532.

17. Laloy E, Riou M, Barc C, Belbis G, Bréard E, Breton S, Cordonnier N, Crochet D, Delaunay R, Moreau J, et al. Schmallenberg virus: experimental infection in goats and bucks. BMC Vet Res. 2015;11(1):221.

18. Wernike K, Hoffmann B, Bréard E, Bøtner A, Ponsart C, Zientara S, Lohse L, Pozzi N, Viarouge C, Sarradin P, et al. Schmallenberg virus experimental infection of sheep. Vet Microbiol. 2013;166(3–4):461–6.

19. Poskin A, Martinelle L, Mostin L, Van Campe W, Dal Pozzo F, Saegerman C, Cay AB, De Regge N. Dose-dependent effect of experimental Schmallenberg virus infection in sheep. Vet J. 2014;201(3):419–22.

20. Wernike K, Beer M. Schmallenberg virus: a novel virus of veterinary importance. Adv Virus Res. 2017;99:39–60.

21. Wernike K, Hoffmann B, Conraths FJ, Beer M. Schmallenberg virus recurrence, Germany, 2014. Emerg Infect Dis. 2015;21(7):1202–4.

22. Delooz L, Saegerman C, Quinet C, Petitjean T, De Regge N, Cay B. Resurgence of Schmallenberg virus in Belgium after 3 years of epidemiological silence. Transbound Emerg Dis. 2017;64(5):1641–2.

23. Elbers AR, Stockhofe-Zurwieden N, van der Poel WH. Schmallenberg virus antibodies in adult cows and maternal antibodies in calves. Emerg Infect Dis. 2014;20(5):901–2.

24. Elbers AR, Stockhofe-Zurwieden N, van der Poel WH. Schmallenberg virus antibody persistence in adult cattle after natural infection and decay of maternal antibodies in calves. BMC Vet Res. 2014;10(1):103.

25. Wernike K, Holsteg M, Sasserath M, Beer M. Schmallenberg virus antibody development and decline in a naturally infected dairy cattle herd in Germany, 2011-2014. Vet Microbiol. 2015;181(3–4):294–7.

26. EFSA: "Schmallenberg" virus: analysis of the epidemiological data (November 2012). EFSA Supporting Publications 2012 EN-360 http://www. efsa.europa.eu/en/supporting/doc/360e.pdf, Accessed 15/07/2013 2012.

27. Balmer S, Vögtlin A, Thür B, Buchi M, Abril C, Houmard M, Danuser J, Schwermer H. Serosurveillance of Schmallenberg virus in Switzerland using bulk tank milk samples. Prev Vet Med. 2014;116(4):370–9.

28. Steinrigl A, Schiefer P, Schleicher C, Peinhopf W, Wodak E, Bago Z, Schmoll F. Rapid spread and association of Schmallenberg virus with ruminant abortions and foetal death in Austria in 2012/2013. Prev Vet Med. 2014; 116(4):350–9.

29. Wisloff H, Nordvik BS, Sviland S, Tonnessen R. The first documented clinical case of Schmallenberg virus in Norway: fetal malformations in a calf. The Veterinary record. 2014;174(5):120.

30. Fernandez-Aguilar X, Pujols J, Velarde R, Rosell R, Lopez-Olvera JR, Marco I, Pumarola M, Segales J, Lavin S, Cabezon O. Schmallenberg virus circulation in high mountain ecosystem, Spain. Emerg Infect Dis. 2014;20(6):1062–4.

31. Chaintoutis SC, Kiossis E, Giadinis ND, Brozos CN, Sailleau C, Viarouge C, Bréard E, Papanastassopoulou M, Zientara S, Papadopoulos O, et al. Evidence of Schmallenberg virus circulation in ruminants in Greece. Trop Anim Health Prod. 2014;46(1):251–5.

32. Bradshaw B, Mooney J, Ross PJ, Furphy C, O'Donovan J, Sanchez C, Gomez-Parada M, Toolan D. Schmallenberg virus cases identified in Ireland. The Veterinary record. 2012;171(21):540–1.

33. Veldhuis AM, Carp-van Dijken S, van Wuijckhuise L, Witteveen G, van Schaik G. Schmallenberg virus in Dutch dairy herds: potential risk factors for high within-herd seroprevalence and malformations in calves, and its impact on productivity. Vet Microbiol. 2014;168(2–4):281–93.

34. Kato T, Yanase T, Suzuki M, Katagiri Y, Ikemiyagi K, Takayoshi K, Shirafuji H, Ohashi S, Yoshida K, Yamakawa M, et al. Monitoring for bovine arboviruses in the most southwestern islands in Japan between 1994 and 2014. BMC Vet Res. 2016;12:125.

35. Hayama Y, Yanase T, Suzuki M, Unten K, Tomochi H, Kakehi M, Shono Y, Yamamoto T, Kobayashi S, Murai K, et al. Meteorological factors affecting seroconversion of Akabane disease in sentinel calves in the subtropical Okinawa Islands of Japan. Trop Anim Health Prod. 2018;50(1):209–15.

36. Kato T, Shirafuji H, Tanaka S, Sato M, Yamakawa M, Tsuda T, Yanase T. Bovine Arboviruses in Culicoides biting midges and sentinel cattle in southern Japan from 2003 to 2013. Transbound Emerg Dis. 2016;63(6):E160–72.

37. Geoghegan JL, Walker PJ, Duchemin JB, Jeanne I, Holmes EC. Seasonal drivers of the epidemiology of arthropod-borne viruses in Australia. Plos Neglect Trop D. 2014;8(11):e3325.

38. Inaba Y, Matumoto M. Chapter 43: Akabane virus. In: Dinter S, Morien B, editors. Virus infections of vertebrates, vol. III. Amsterdam: Virus infections of ruminants Elsevier Science Publishers BV; 1990. p. 467–80.

PERMISSIONS

The contributors of this book come from diverse backgrounds, making this book a truly international effort. This book will bring forth new frontiers with its revolutionizing research information and detailed analysis of the nascent developments around the world.

We would like to thank all the contributing authors for lending their expertise to make the book truly unique. They have played a crucial role in the development of this book. Without their invaluable contributions this book wouldn't have been possible. They have made vital efforts to compile up to date information on the varied aspects of this subject to make this book a valuable addition to the collection of many professionals and students.

This book was conceptualized with the vision of imparting up-to-date information and advanced data in this field. To ensure the same, a matchless editorial board was set up. Every individual on the board went through rigorous rounds of assessment to prove their worth. After which they invested a large part of their time researching and compiling the most relevant data for our readers.

The editorial board has been involved in producing this book since its inception. They have spent rigorous hours researching and exploring the diverse topics which have resulted in the successful publishing of this book. They have passed on their knowledge of decades through this book. To expedite this challenging task, the publisher supported the team at every step. A small team of assistant editors was also appointed to further simplify the editing procedure and attain best results for the readers.

Apart from the editorial board, the designing team has also invested a significant amount of their time in understanding the subject and creating the most relevant covers. They scrutinized every image to scout for the most suitable representation of the subject and create an appropriate cover for the book.

The publishing team has been an ardent support to the editorial, designing and production team. Their endless efforts to recruit the best for this project, has resulted in the accomplishment of this book. They are a veteran in the field of academics and their pool of knowledge is as vast as their experience in printing. Their expertise and guidance has proved useful at every step. Their uncompromising quality standards have made this book an exceptional effort. Their encouragement from time to time has been an inspiration for everyone.

The publisher and the editorial board hope that this book will prove to be a valuable piece of knowledge for researchers, students, practitioners and scholars across the globe.

LIST OF CONTRIBUTORS

Jessica Enright
Computing Science and Mathematics, University of Stirling, Stirling FK9 4LA, UK

Rowland R Kao
Institute of Biodiversity, Animal Health and Comparative Medicine, University of Glasgow, Jarrett Building, Glasgow G61 1QH, UK

Ueli Braun, Sonja Warislohner, Karl Nuss and Christian Gerspach
Department of Farm Animals, Vetsuisse-Faculty, University of Zurich, Winterthurerstrasse 260, CH-8057 Zurich, Switzerland

Paul Torgerson
Section of Epidemiology, Vetsuisse-Faculty, University of Zurich, Winterthurerstrasse 260, CH-8057 Zurich, Switzerland

Jens Raila and Florian J. Schweigert
Institute of Nutritional Science, University of Potsdam, Arthur-Scheunert-Allee 114-116, 14558 Nuthetal, Germany

Chiho Kawashima and Akio Myamoto
Obihiro University of Agriculture and Veterinary Medicine, Obihiro, Hokkaido 080-8555, Japan

Helga Sauerwein
Institute for Animal Science, Physiology and Hygiene, University of Bonn, Katzenburgweg 7-9, 53115 Bonn, Germany

Nadine Hülsmann and Christoph Knorr
Department of Animal Sciences, Biotechnology and Reproduction of Livestock, Georg-August-University Goettingen, Burckhardtweg 2, 37077 Goettingen, Germany

Hannah Trewby, Rowland R. Kao, Daniel T. Haydon and Roman Biek
Boyd Orr Centre for Population and Ecosystem Health, Institute of Biodiversity and Animal Health, University of Glasgow, Glasgow, UK

David M. Wright
School of Medicine, Dentistry and Biomedical Sciences, Queen's University Belfast, Belfast, UK

Robin A. Skuce
Veterinary Sciences Division, Agri-Food and Biosciences Institute, Stormont, Belfast, UK
School of Biological Sciences, Queen's University Belfast, Belfast, UK

Carl McCormick, Thomas R. Mallon and Eleanor L. Presho
Veterinary Sciences Division, Agri-Food and Biosciences Institute, Stormont, Belfast, UK

Maged R. El-Ashker
Department of Internal Medicine and Infectious Diseases, Faculty of Veterinary Medicine, Mansoura University, Mansoura 35516, Egypt

Mohamed F. Salama
Department of Biochemistry, Faculty of Veterinary Medicine, Mansoura University, Mansoura 35516, Egypt

Mohamed E. El-Boshy
Department of Clinical Pathology, Faculty of Veterinary Medicine, Mansoura University, Mansoura 35516, Egypt
Department of Laboratory Medicine, Faculty of Applied Medical Science, Umm Al-Qura University, Makkah 21955, Kingdom of Saudi Arabia

Eman A. Abo El-Fadle
Department of Animal Husbandry and Development of Animal Wealth, Faculty of Veterinary Medicine, Mansoura University, Mansoura 35516, Egypt

Leonardo Murgiano, Vidhya Jagannathan and Cord Drögemüller
Institute of Genetics, Vetsuisse Faculty, University of Bern, Bern, Switzerland

Christian Piffer
Servizio Veterinario dell'Azienda Sanitaria dell'Alto Adige, Bozen, Italy

Inmaculada Diez-Prieto
Laboratory on Urolithiasis Research, Department of Medicine, Surgery and Anatomy, Universidad de León, León, Spain

Marilena Bolcato and Arcangelo Gentile
Department of Veterinary Medical Sciences, University of Bologna, Ozzano dell'Emilia, Italy

Ueli Braun, Fredi Janett, Sarah Züblin and Michèle von Büren
Department of Farm Animals, Vetsuisse Faculty, University of Zurich, Zurich, Switzerland

Monika Hilbe
Institute of Veterinary Pathology, Vetsuisse Faculty, University of Zurich, Zurich, Switzerland

Reto Zanoni and Matthias Schweizer
Institute for Virology and Immunology and Department of Diseases and Pathobiology, Vetsuisse Faculty, University of Bern, Bern, Switzerland

Tariku Jibat Beyene
College of Veterinary Medicine and Agriculture, Addis Ababa University, Bishoftu/Debre Zeit, Ethiopia
Business Economics Group, Wageningen University, Hollandseweg 1, 6706 KN Wageningen, The Netherlands

Amanuel Eshetu, Amina Abdu, Etenesh Wondimu and Takele Beyene Tufa
College of Veterinary Medicine and Agriculture, Addis Ababa University, Bishoftu/Debre Zeit, Ethiopia

Ashenafi Feyisa Beyi
College of Veterinary Medicine and Agriculture, Addis Ababa University, Bishoftu/Debre Zeit, Ethiopia
Department of Animal Sciences, University of Florida, Gainesville, FL, USA

Sami Ibrahim
Cojengo Ltd, Glasgow, UK

Crawford W. Revie
Department of Health Management, Atlantic Veterinary College, University of Prince Edward Island, Charlottetown, PEI, Canada

David Gamarra, Andrés López-Oceja and Marian M. de Pancorbo
Biomics Research Group, University of the Basque Country (UPV/EHU), 01006 Vitoria-Gasteiz, Spain

Noelia Aldai and Luis Javier R. Barron
Lactiker Research Group; Lascaray Research Center, University of the Basque Country (UPV/EHU), 01006 Vitoria-Gasteiz, Spain

Aisaku Arakawa and Masaaki Taniguchi
Animal Genome Unit, Institute of Livestock and Grassland Science, National Agriculture and Food Research Organization (NARO), Tsukuba 305-0901, Japan

Carsten Kirkeby, Nils Toft and Tariq Halasa
National Veterinary Institute, Technical University of Denmark, Kemitorvet, bygning 204, 2800 Kgs. Lyngby, Denmark

Kaare Græsbøll
National Veterinary Institute, Technical University of Denmark, Kemitorvet, bygning 204, 2800 Kgs. Lyngby, Denmark
DTU Compute, Section for Dynamical Systems, Department of Applied Mathematics and Computer Science, Technical University of Denmark, Richard Petersens Plads, Bygning 324, 2800 Kgs. Lyngby, Denmark

Søren Saxmose Nielsen
Department of Large Animal Sciences, Section for Animal Welfare and DiseaseControl, University of Copenhagen, Grønnegaardsvej 8, 1870 Frb. C, København, Denmark

Mamadou Diallo, Younoussa Koné, Karim Tounkara and Mamadou Niang
Laboratoire Central Vétérinaire, Bamako, Mali

Bassirou Diarra, Moumine Sanogo, Antieme C. G. Togo, Anou M. Somboro, Mariam H. Diallo, Bréhima Traoré, Mamoudou Maiga, Yeya dit Sadio Sarro, Bocar Baya, Drissa Goita, Hamadoun Kassambara, Bindongo P. P. Dembélé, Sounkalo Dao, Souleymane Diallo and Anatole Tounkara
SEREFO, University of Sciences, Techniques and Technologies of Bamako (USTTB), Point-G, Bamako, Mali

Sophia Siddiqui
Division of Clinical Research, NIH, Bethesda, USA

Robert L. Murphy
Northwestern University, Chicago, IL, USA

Catherine M. McCann
Department of Infection Biology, Institute of Infection and Global Health, University of Liverpool, Liverpool, UK
Epidemiology Research Unit, Scotland's Rural College (SRUC), An Lòchran, Inverness Campus, Inverness IV2 5NA, UK

Helen E. Clough
Department of Infection Biology, Institute of Infection and Global Health, University of Liverpool, Liverpool, UK
Department of Public Health and Policy, Institute of Psychology, Health and Society, University of Liverpool, Liverpool, UK

Matthew Baylis
National Institute of Health Research Health Protection Research Unit in Emerging and Zoonotic Infections, University of Liverpool, Liverpool, UK
Department of Epidemiology and Population Health, Institute of Infection and Global Health, University of Liverpool, Liverpool, UK

Diana J. L. Williams
Department of Infection Biology, Institute of Infection and Global Health, University of Liverpool, Liverpool, UK

Ilona V. Gluecks
Vétérinaires sans Frontières Suisse, Nairobi, Kenya

Astrid Bethe
Institute of Microbiology and Epizootics, Centre for Infection Medicine, Free University Berlin, Berlin, Germany

Mario Younan
Vétérinaires sans Frontières Germany, Nairobi, Kenya

Christa Ewers
Institute of Hygiene and Infectious Diseases of Animals, Justus-Liebig University Giessen, Giessen, Germany

Hung-Hsun Yen, Elizabeth Washington, Wayne Kimpton, Evan Hallein, Joanne Allen, Silk Yu Lin and Stuart Barber
Faculty of Veterinary and Agricultural Sciences, The University of Melbourne, Parkville, VIC 3010, Australia

L. Bournez and P. Hendrikx
ANSES (French Agency for Food, Environmental and Occupational Health & Safety), Unité de coordination et d'appui à la surveillance, Direction des laboratoires, Maisons Alfort, France

L. Cavalerie
Ministère en charge de l'Agriculture, Direction générale de l'Alimentation, Bureau de la santé animale, Paris, France

C. Sailleau, E. Bréard and G. Zanella
ANSES, Laboratoire de santé animale, Université Paris-Est, Maisons-Alfort, France

R. Servan de Almeida and A. Pedarrieu
CIRAD, UMR ASTRE, Montpellier; Inra, UMR ASTRE, Montpellier, France

E. Garin
Coop de France, Paris, France

I. Tourette
GDS France, Paris, France

F. Dion
Races de France, Paris, France

D. Calavas
ANSES, Laboratoire de Lyon, Unité Epidémiologie, Laboratoire de Lyon, Lyon, France

Daniel Pakasi Nalapa and Clovice Kankya
Department of Biosecurity Ecosystem and Veterinary Public Health, College of Veterinary Medicine, Animal Resources and Biosecurity, Makerere University, Kampala, Uganda

Adrian Muwonge
Division of Genetics and Genomics, The Roslin institute, University of Edinburgh, Easter bush Campus, EH259RG Edinburgh, UK

Francisco Olea-Popelka
College of Veterinary Medicine and Biomedical Sciences, Department of Clinical Sciences & Mycobacteria Research Laboratories, Colorado State University, Fort Collins, CO, USA

F. Bonelli and G. Sarri
Dipartimento di Scienze Veterinarie, via Livornese snc, Università di Pisa, via Livornese, 56122, San Piero a Grado, Pisa, Italy

L. Turini, A. Serra and M. Mele
Centro di Ricerche Agro-ambientali "E. Avanzi" – Università di Pisa, via Vecchia di Marina, 6, 56122, San Piero a Grado, Pisa, Italy

A. Buccioni
Dipartimento di Scienze delle Produzioni Agroalimentari e Ambientali, Università di Firenze, via delle Cascine, 5, 50100 Florence, Italy

Casey L. Cazer
Department of Population Medicine and Diagnostic Sciences, College of Veterinary Medicine, Cornell University, Ithaca, NY, USA

Victoriya V. Volkova and Yrjö T. Gröhn
Department of Diagnostic Medicine/Pathobiology, College of Veterinary Medicine, Kansas State University, Manhattan, KS, USA

Nickala Best and Travis Beddoe
Department of Animal, Plant and Soil Science, Centre for AgriBioscience (AgriBio), La Trobe University, Bundoora, Melbourne, VIC, Australia

Lucas Zanandrez
Federal University of Minas Gerais, Belo Horizonte, Minas Gerais, Brazil

Jacek Gwozdz and Grant Rawlin
Department of Economic Development, Jobs, Transport and Resources Centre for AgriBioscience (AgriBio), Victorian Government, Bundoora, Melbourne, VIC, Australia

Eckard Klien and Nicky Buller
DAFWA Diagnostics and Laboratory Services, Biosecurity and Regulation, Department of Agriculture and Food, South Perth, Western Australia

Robert Sute
Agriculture Services and Biosecurity Operations, Department of Economic Development, Jobs, Transport and Resources, Attwood, Victoria, Australia

Kuastros Mekonnen Belaynehe, Seung Won Shin and Han Sang Yoo
Department of Infectious Diseases, College of Veterinary Medicine, Seoul National University, Seoul 08826, Republic of Korea

I. Nur Atikah, N. Abdullah, M. F. Jahromi and M. Ivan
Institute of Tropical Agriculture, Universiti Putra Malaysia, UPM, 43400 Serdang, Selangor, Malaysia

A. R. Alimon, H. Yaakub and A. A. Samsudin
Department of Animal Science, Faculty of Agriculture, Universiti Putra Malaysia, UPM, 43400 Serdang, Selangor, Malaysia

Tamirat Kaba
College of Veterinary Medicine, Haramaya University, Dire Dawa, Ethiopia

Berhanu Abera and Temesgen Kassa
Ethiopian Institutes of Agricultural Research, Holleta Agricultural Research Center, Holleta, Ethiopia

Caroline Marçal Gomes David, Celia Raquel Quirino, Wilder Hernando Ortiz Vega and Aylton Bartholazzi Junior
Universidade Estadual do Norte Fluminense, Centro de Ciências e Tecnologias Agropecuária Laboratório de Reprodução e Melhoramento Genético Animal, 2000, Alberto Lamego Ave, Campos dos Goytacazes, Rio de Janeiro 28016-811, Brazil

Aparecida de Fátima Madella-Oliveira
Instituto Federal de Educação, Ciência e Tecnologia do Espírito Santo, 482 Rod Br, Km 47, s/n - Rive, Alegre, Espirito Santo 29520-000, Brazil

Ricardo Lopes Dias Costa
Instituto de Zootecnia de Nova Odessa, 56, Heitor Penteado St., Nova Odessa, São Paulo 13460-000, Brazil

Nimita Kant, Parul Kulshreshtha, Rashmi Singh, Amita Dwivedi, Riya Ahuja, Rinkle Mehra, Mohit Tehlan, Paritosh Ahmed, Shilpa Kaushik, Shipra, Shashikant Kumar, Aas Mohammad and Shrikrishn Shukla
Department of Zoology, Shivaji College, University of Delhi, 110027, New Delhi, India

Anuradha Mal
Department of Botany, Shivaji College, University of Delhi, 110027, New Delhi, India

Damini Singh
School of Biotechnology, Jawaharlal Nehru University, 110057, Delhi, India

Rakesh Bhatnaga
School of Biotechnology, Jawaharlal Nehru University, 110057, Delhi, India
Laboratory of Molecular Biology and Genetic Engineering, School of Biotechnology, Jawaharlal Nehru University, New Delhi 110067, India

Kerstin Wernike, Kevin P. Szillat and Martin Beer
Institute of Diagnostic Virology, Friedrich-Loeffler-Institut, Greifswald – Insel Riems, Germany

Mark Holsteg
Chamber of Agriculture for North Rhine-Westphalia, Bovine Health Service, Haus Riswick, Kleve, Germany

Index

www.ingramcontent.com/pod-product-compliance
Lightning Source LLC
Chambersburg PA
CBHW061256190326
41458CB00011B/3684